Ref 670 How v.4
How products are made.

$115.00

How PRODUCTS Are MADE

How PRODUCTS Are MADE

An Illustrated Guide to Product Manufacturing

Volume 4

Jacqueline L. Longe, Editor

GALE

DETROIT • LONDON

STAFF

Jacqueline L. Longe, *Editor*

Maureen Richards, *Research Specialist*

Cynthia D. Baldwin, *Art Director*
Bernadette M. Gornie, *Page Designer*
Gary Leach, *Cover Designer*

Pamela A. Reed, *Photography Coordinator*

Electronic illustrations provided by Electronic Illustrators Group of Fountain Hills, Arizona

ISBN 07876-2443-8
ISSN 1072-5091

Printed in the United States of America
10 9 8 7 6 5 4 3 2

Contents

Introduction

About the Series

Welcome to *How Products Are Made: An Illustrated Guide to Product Manufacturing.* This series provides information on the manufacture of a variety of items, from everyday household products to heavy machinery to sophisticated electronic equipment. You will find step-by-step descriptions of processes, simple explanations of technical terms and concepts, and clear, easy-to-follow illustrations.

Each volume of *How Products Are Made* covers a broad range of manufacturing areas: food, clothing, electronics, transportation, machinery, instruments, sporting goods, and more. Some are intermediate goods sold to manufacturers of other products, while others are retail goods sold directly to consumers. You will find items made from a variety of materials, including products such as precious metals and minerals that are not "made" so much as they are extracted and refined.

Organization

Every volume in this series is comprised of many individual entries, each covering a single product. Although each entry focuses on the product's manufacturing process, it also provides a wealth of other information: who invented the product or how it has developed, how it works, what materials are used, how it is designed, quality control procedures, byproducts generated during its manufacture, future applications, and books and periodical articles containing more information.

To make it easier for you to find what you're looking for, the entries are broken up into standard sections. Among the sections you will find are the following:

- Background
- History
- Raw Materials
- Design
- The Manufacturing Process
- Quality Control
- Byproducts/Waste
- The Future
- Where To Learn More

Every entry is accompanied by illustrations. Uncomplicated and easy to understand, these illustrations generally follow the step-by-step description of the manufacturing process found in the text.

Bold faced items in the text refer to other entries in this volume.

A subject index of important terms, processes, materials, and people is found at the end of the book. Bold faced items in the index refer to main entries.

Main entries from previous volumes are also included in the index. They are listed along with their corresponding volume and page numbers.

About this Volume

This volume contains essays on 100 products, arranged alphabetically, and 15 special boxed sections. Written by Nancy EV Bryk, a curator at the Henry Ford Museum & Greenfield Village in Dearborn, Michigan, these boxed sections describe interesting historical developments related to a product. Photographs are also included.

Contributors/Advisor

The entries in this volume were written by a skilled team of technical writers and engineers, often in cooperation with manufacturers and industry associations. The advisor for this volume was David L. Wells, PhD, CMfgE, a long time member of the Society of Manufacturing Engineers (SME) and the Academic Dean at Focus: HOPE, a nonprofit civil and human rights organization dedicated to the technical training and education of the multicultural community of Detroit, Michigan.

Suggestions

Your questions, comments, and suggestions for future products are welcome. Please send all such correspondence to:

The Editor
How Products Are Made
Gale Research
27500 Drake Rd.
Farmington Hills, MI 48331-3535

Contributors

Nancy EV Bryk

Chris Cavette

Michael Cavette

Loretta Hall

Gillian S. Holmes

Kristin Palm

Perry Romanowski

Randy Schueller

Rose Secrest

David L. Wells

Angela Woodward

Acknowledgments

Armored Truck: Dunbar Armored; Lenco Industries. **Baseball Cap:** William Ault, Cooperstown Ball Cap Company, Cooperstown, OH. **Bowling Ball:** Del Warren, AMF; Vulcan Brunswick. **Bowling Pin:** Vulcan Brunswick; The American Bowling Congress (ABC). **Carousel:** Chuck Kaparich, Missoula, MT. **Corn Syrup:** Corn Refiners Association, Washington, DC. **Cushioning Laminate:** Sealed Air Corporation, Saddle Brook, NJ. **Fire Hose:** North American Fire Hose Corporation, Santa Maria, CA. **Fruitcake:** Collin Street Bakery, Corsicana, TX. **Globe:** Bryan Hollingsworth, George F. Cram Co., Indianapolis, IN; Colleen T. Sell, *Mercator's World* magazine. **Hammer:** Stanley Tools, New Britain, CT. **Hockey Stick:** Hockey Hall of Fame; Innovative Hockey; Itech Sports Products Inc. **Lace Curtain:** Linda Doucette and Mitch Buckwalter, Quaker Lace. **Lava Lamp:** Haggerty Enterprises, Chicago, IL. **Model Train:** Lionel Trains, Mt. Clemens, MI. **Pottery:** Mike Tkach, The Homer Laughlin China Company. **Rice Cake:** Tim O'Donnell, Lundberg Family Farms, Richvale, CA. **Shellac:** Gene Hoyas, Clear Finishes. **Toy Wagon:** Fred Michelau, Radio Flyer Inc., Chicago, IL. **Vinyl Floorcovering:** Congoleum Corporation, Trenton, NJ. **Water:** Alameda County Water District, Fremont, CA.

The historical photographs for the entries on Carousel, Cast Iron Stove, Dulcimer, Hot Dog, Insulated Bottle, Linen, Motorcycle, Oxygen, Photograph, Playing Cards, Polyester Fleece, Pottery, Raisins, Spandex, and Tin are from the collections of **Henry Ford Museum & Greenfield Village**, Dearborn, Michigan.

Electronic illustrations in this volume were created by **Electronic Illustrators Group (EIG)** of Fountain Hills, Arizona.

Acetylene

Background

Acetylene is a colorless, combustible gas with a distinctive odor. When acetylene is liquefied, compressed, heated, or mixed with air, it becomes highly explosive. As a result special precautions are required during its production and handling. The most common use of acetylene is as a raw material for the production of various organic chemicals including 1,4-butanediol, which is widely used in the preparation of polyurethane and polyester plastics. The second most common use is as the fuel component in oxy-acetylene welding and metal cutting. Some commercially useful acetylene compounds include acetylene black, which is used in certain dry-cell batteries, and acetylenic alcohols, which are used in the synthesis of vitamins.

Acetylene was discovered in 1836, when Edmund Davy was experimenting with potassium carbide. One of his chemical reactions produced a flammable gas, which is now known as acetylene. In 1859, Marcel Morren successfully generated acetylene when he used carbon electrodes to strike an electric arc in an atmosphere of hydrogen. The electric arc tore carbon atoms away from the electrodes and bonded them with hydrogen atoms to form acetylene molecules. He called this gas carbonized hydrogen.

By the late 1800s, a method had been developed for making acetylene by reacting calcium carbide with water. This generated a controlled flow of acetylene that could be combusted in air to produce a brilliant white light. Carbide lanterns were used by miners and carbide lamps were used for street illumination before the general availability of electric lights. In 1897, Georges Claude and A. Hess noted that acetylene gas could be safely stored by dissolving it in acetone. Nils Dalen used this new method in 1905 to develop long-burning, automated marine and railroad signal lights. In 1906, Dalen went on to develop an acetylene torch for welding and metal cutting.

In the 1920s, the German firm BASF developed a process for manufacturing acetylene from natural gas and petroleum-based hydrocarbons. The first plant went into operation in Germany in 1940. The technology came to the United States in the early 1950s and quickly became the primary method of producing acetylene.

Demand for acetylene grew as new processes were developed for converting it into useful plastics and chemicals. In the United States, demand peaked sometime between 1965 and 1970, then fell off sharply as new, lower-cost alternative conversion materials were discovered. Since the early 1980s, the demand for acetylene has grown slowly at a rate of about 2-4% per year.

In 1991, there were eight plants in the United States that produced acetylene. Together they produced a total of 352 million lb (160 million kg) of acetylene per year. Of this production, 66% was derived from natural gas and 15% from petroleum processing. Most acetylene from these two sources was used on or near the site where it was produced to make other organic chemicals. The remaining 19% came from calcium carbide. Some of the acetylene from this source was used to make organic chemicals, and the rest was used by regional industrial gas produc-

In 1991, there were eight plants in the United States that produced acetylene. Together they produced a total of 352 million lb (160 million kg) of acetylene per year. Of this production, 66% was derived from natural gas and 15% from petroleum processing.

ers to fill pressurized cylinders for local welding and metal cutting customers.

In Western Europe, natural gas and petroleum were the principal sources of acetylene in 1991, while calcium carbide was the principal source in Eastern Europe and Japan.

Raw Materials

Acetylene is a hydrocarbon consisting of two carbon atoms and two hydrogen atoms. Its chemical symbol is C_2H_2. For commercial purposes, acetylene can be made from several different raw materials depending on the process used.

The simplest process reacts calcium carbide with water to produce acetylene gas and a calcium carbonate slurry, called hydrated lime. The chemical reaction may be written as $CaC_2 + 2\ H_2O \rightarrow C_2H_2 + Ca(OH)_2$.

Other processes use natural gas, which is mostly methane, or a petroleum-based hydrocarbon such as crude oil, naphtha, or bunker C oil as raw materials. Coal can also be used. These processes use high temperature to convert the raw materials into a wide variety of gases, including hydrogen, carbon monoxide, carbon dioxide, acetylene, and others. The chemical reaction for converting methane into acetylene and hydrogen may be written $2\ CH_4 \rightarrow C_2H_2 + 3\ H_2$. The other gases are the products of combustion with oxygen. In order to separate the acetylene, it is dissolved in a solvent such as water, anhydrous ammonia, chilled methanol, or acetone, or several other solvents depending on the process.

The Manufacturing Process

There are two basic conversion processes used to make acetylene. One is a chemical reaction process, which occurs at normal temperatures. The other is a thermal cracking process, which occurs at extremely high temperatures.

Here are typical sequences of operations used to convert various raw materials into acetylene by each of the two basic processes.

Chemical reaction process

Acetylene may be generated by the chemical reaction between calcium carbide and water. This reaction produces a considerable amount of heat, which must be removed to prevent the acetylene gas from exploding. There are several variations of this process in which either calcium carbide is added to water or water is added to calcium carbide. Both of these variations are called wet processes because an excess amount of water is used to absorb the heat of the reaction. A third variation, called a dry process, uses only a limited amount of water, which then evaporates as it absorbs the heat. The first variation is most commonly used in the United States and is described below.

1 Most high-capacity acetylene generators use a rotating screw conveyor to feed calcium carbide granules into the reaction chamber, which has been filled to a certain level with water. The granules measure about 0.08 in x 0.25 in (2 mm x 6 mm), which provides the right amount of exposed surfaces to allow a complete reaction. The feed rate is determined by the desired rate of gas flow and is controlled by a pressure switch in the chamber. If too much gas is being produced at one time, the pressure switch opens and cuts back the feed rate.

2 To ensure a complete reaction, the solution of calcium carbide granules and water is constantly agitated by a set of rotating paddles inside the reaction chamber. This also prevents any granules from floating on the surface where they could overheat and ignite the acetylene

3 The acetylene gas bubbles to the surface and is drawn off under low pressure. As it leaves the reaction chamber, the gas is cooled by a spray of water. This water spray also adds water to the reaction chamber to keep the reaction going as new calcium carbide is added. After the gas is cooled, it passes through a flash arrester, which prevents any accidental ignition from equipment downstream of the chamber.

4 As the calcium carbide reacts with the water, it forms a slurry of calcium carbonate, which sinks to the bottom of the chamber. Periodically the reaction must be stopped to remove the built-up slurry. The

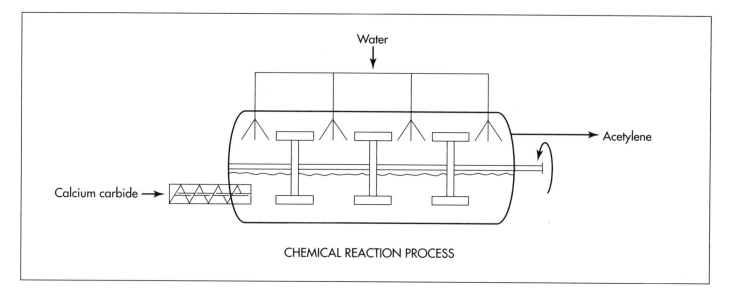

CHEMICAL REACTION PROCESS

slurry is drained from the chamber and pumped into a holding pond, where the calcium carbonate settles out and the water is drawn off. The thickened calcium carbonate is then dried and sold for use as an industrial waste water treatment agent, acid neutralizer, or soil conditioner for road construction.

Thermal cracking process

Acetylene may also be generated by raising the temperature of various hydrocarbons to the point where their atomic bonds break, or crack, in what is known as a thermal cracking process. After the hydrocarbon atoms break apart, they can be made to rebond to form different materials than the original raw materials. This process is widely used to convert oil or natural gas to a variety of chemicals.

There are several variations of this process depending on the raw materials used and the method for raising the temperature. Some cracking processes use an electric arc to heat the raw materials, while others use a combustion chamber that burns part of the hydrocarbons to provide a flame. Some acetylene is generated as a coproduct of the steam cracking process used to make ethylene. In the United States, the most common process uses a combustion chamber to heat and burn natural gas as described below.

1 Natural gas, which is mostly methane, is heated to about 1,200° F (650° C). Preheating the gas will cause it to self-ignite once it reaches the burner and requires less oxygen for combustion.

2 The heated gas passes through a narrow pipe, called a venturi, where oxygen is injected and mixed with the hot gas.

3 The mixture of hot gas and oxygen passes through a diffuser, which slows its velocity to the desired speed. This is critical. If the velocity is too high, the incoming gas will blow out the flame in the burner. If the velocity is too low, the flame can flash back and ignite the gas before it reaches the burner.

4 The gas mixture flows into the burner block, which contains more than 100 narrow channels. As the gas flows into each channel, it self-ignites and produces a flame which raises the gas temperature to about 2,730° F (1,500° C). A small amount of oxygen is added in the burner to stabilize the combustion.

5 The burning gas flows into the reaction space just beyond the burner where the high temperature cause about one-third of the methane to be converted into acetylene, while most of the rest of the methane is burned. The entire combustion process takes only a few milliseconds.

6 The flaming gas is quickly quenched with water sprays at the point where the conversion to acetylene is the greatest. The cooled gas contains a large amount of carbon monoxide and hydrogen, with lesser

Acetylene may be generated by the chemical reaction between calcium carbide and water. This reaction produces a considerable amount of heat, which must be removed to prevent the acetylene gas from exploding.

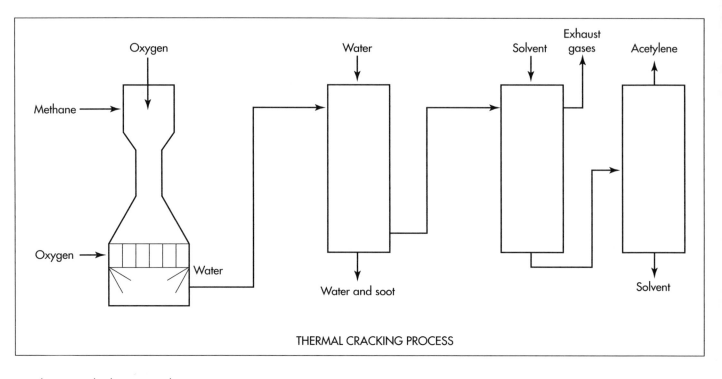

THERMAL CRACKING PROCESS

Acetylene may also be generated by raising the temperature of various hydrocarbons to the point where their atomic bonds break, or crack, in what is known as a thermal cracking process.

amounts of carbon soot, plus carbon dioxide, acetylene, methane, and other gases.

7 The gas passes through a water scrubber, which removes much of the carbon soot. The gas then passes through a second scrubber where it is sprayed with a solvent known as N-methylpyrrolidinone which absorbs the acetylene, but not the other gases.

8 The solvent is pumped into a separation tower where the acetylene is boiled out of the solvent and is drawn off at the top of the tower as a gas, while the solvent is drawn out of the bottom.

Storage and Handling

Because acetylene is highly explosive, it must be stored and handled with great care. When it is transported through pipelines, the pressure is kept very low and the length of the pipeline is very short. In most chemical production operations, the acetylene is transported only as far as an adjacent plant, or "over the fence" as they say in the chemical processing business.

When acetylene must be pressurized and stored for use in oxy-acetylene welding and metal cutting operations, special storage cylinders are used. The cylinders are filled with an absorbent material, like diatoma-

ceous earth, and a small amount of acetone. The acetylene is pumped into the cylinders at a pressure of about 300 psi (2,070 kPa), where it is dissolved in the acetone. Once dissolved, it loses its explosive capability, making it safe to transport. When the cylinder valve is opened, the pressure drop causes some of the acetylene to vaporize into gas again and flow through the connecting hose to the welding or cutting torch.

Quality Control

Grade B acetylene may have a maximum of 2% impurities and is generally used for oxy-acetylene welding and metal cutting. Acetylene produced by the chemical reaction process meets this standard. Grade A acetylene may have no more than 0.5% impurities and is generally used for chemical production processes. Acetylene produced by the thermal cracking process may meet this standard or may require further purification, depending on the specific process and raw materials.

The Future

The use of acetylene is expected to continue a gradual increase in the future as new applications are developed. One new application is the conversion of acetylene to ethylene for use in making a variety of polyethylene plastics. In the past, a small amount of acetylene had been generated and wasted as part of the steam cracking process used to make ethylene. A new catalyst developed by Phillips Petroleum allows most of this acetylene to be converted into ethylene for increased yields at a reduced overall cost.

Where to Learn More

Books

Brady, George S., Henry R. Clauser, and John A. Vaccari. *Materials Handbook*, 14th edition. McGraw-Hill, 1997.

Kroschwitz, Jacqueline I. and Mary Howe-Grant, ed. *Encyclopedia of Chemical Technology*, 4th edition. John Wiley and Sons, Inc., 1993.

Other

Acetylene Pamphlet G-1. Compressed Gas Association, 1990.

Compressed Gas Association. http://www.cganet.com.

—*Chris Cavette*

Antibacterial Soap

Today, annual sales of antimicrobial products in the United States have reached $600 million.

An antibacterial soap is a cleansing product designed to kill germs on the hands or body. These soaps are made in either liquid or bar form by blending detergent additives with ingredients, which have antimicrobial properties.

Background

Antibacterial soaps were originally marketed as deodorant soaps to control body odor caused by the action of bacteria on perspiration. These products, sold in bar form, gained popularity in the 1950s under such well-recognized brand names as Dial and Lifebouy. While many of these bar soaps are still available today, liquid antibacterial soaps used for disinfecting hands are becoming increasingly more popular. Major brands include Lever's Caress, Dove's Liquid and Proctor and Gamble's Oil of Olay liquid. In addition to these products intended for consumer use, other antibacterial cleansers are available for use by health care professionals. Such specialty products include surgical scrubs, wound disinfectants, and wound cleansers. In the United States, all soaps that make antibacterial claims are classified as over-the-counter (OTC) drugs because they are sold to kill germs. This designation means they can be purchased without a prescription as long as they adhere to guidelines set forth by the Food and Drug Administration (FDA). These guidelines are established in a document known as a monograph which specifies which active ingredients can be used, the claims that can be made, and so forth. As of 1998, the FDA has not issued a final monograph on antibacterial soaps, but over the last 25 years they have published a series of Tentative Final Monographs (TFMs).

The first proposed monograph, published in 1974, officially defined an antibacterial cleanser as a soap containing an active ingredient with invivo and invitro activity against skin organisms. It also suggested that antibacterial cleansers be grouped into the following seven categories: antimicrobial soaps, health-care personal hand washes, patient preoperative skin preparations, skin antiseptics, skin wound cleansers, skin wound protectants, and surgical scrub hand soaps. In 1978 the agency issued a TFM which began formal adoption of the definition and of the categories proposed in 1974. This tentative monograph allowed products to use terms like deodorancy and reduction of body odor. However, this version of the monograph was never finalized, and in 1991 the FDA issued another TFM with different rules. The 1991 TFM separated first aid antiseptics into a different category, which included skin antiseptics, wound cleansers, and wound protectants. Later monographs placed consumer and professional products into separate categories. Then, in a surprising move in 1994, the agency reversed the earlier tentative monographs that had recognized antibacterial cleansers specifically for consumers. The 1994 TFM does not directly allow antimicrobial soaps for home use. While it does not disallow these products, it does not set up separate rules for them. Therefore, consumer products must use the same active ingredients as professional health care products. More importantly, they are limited to the same types of claims that professional products can make. This is problematic because consumer products have different requirements than medical products. For example, deodorant soaps are intended to control body odor while professional surgical scrub products are not. On the other hand, professional products must be safe enough to use up to 50 times per day, whereas consumer products are typically not used more than two or three times per day. For these reasons and others, many soap manufacturers believe that consumer and professional products should be regulated separately. The FDA, however, tends to disagree. The regulatory status of antibacterial soaps is still tentative; the monograph has not been finalized and industry experts do not expect it to be ready until after the year 2000.

Today, annual sales of antimicrobial products in the United States have reached $600 million. However, the many antibacterial deodorant bar soaps for controlling body odor have largely been replaced by antiperspirants/deodorants. The growth in the antibacterial soap market has come primarily from the increased sales of liquid hand cleansers.

Design

An antibacterial soap is designed to safely kill germs and cleanse the skin. The formulator must therefore consider the types of organisms the product should be effective against and how much time is required for the product to work. The formulator must also consider factors related to cleansing such as foam quality, speed of foaming, rinsability, and skin feel, to name a few. In addition, the product's aesthetic qualities (how it looks and smells) must also be evaluated. The chemist formulating such products must address all of these factors. The chemist must design the formula, (a recipe which identifies the ingredients and the quantities used), the manufacturing procedure (which instructs how to make the product), and the product specifications (which describe the quality of the finished soap.)

Raw Materials

Water is the most abundant ingredient in antibacterial soaps because it is used as a carrier and a diluent for the other ingredients. Deionized or distilled water is used in cleansing products because the ions found in hard water can interfere with certain detergents. Formulas may contain 40-80% water.

Although the FDA has not yet officially ruled which active ingredients will be allowed by law, there are two ingredients commonly used in the industry at this time as antibacterial agents. One is 3,4,4'-trichlorocarbanilide (commonly called trichlocarban), which is used in bar soaps. The other, more common ingredient, is 2-hydroxy-2',4,4'-trichlorodiphenyl ether (commercially known as triclosan), which is used in liquids. These ingredients work by denaturing cell contents or otherwise interfering with metabolism of microbes. They are functional at levels as low as 0.5%. Both are effective against a broad range of microorganisms.

The same types of detergent ingredients used in common household and personal care cleansing products are used to make antibacterial soaps. Detergents and soaps are technically known as surfactants, which are materials that have the ability to solubilize dirt and oil. Surfactants are responsible for a product's ability to generate foam. The chemist must blend surfactants together to optimize foam and cleansing properties while minimizing negative effects like stripping skin of natural oils. Surfactants are loosely grouped into two categories: primary surfactants, which are responsible for foam and cleansing; and secondary surfactants, which work with the primaries to give the foam more creaminess, improve skin feel, and so on. Common primary surfactants include alkyl sulfates, alkyl ether sulfates, olefin sulfonates, and amphoterics. Blends of these materials can typically comprise 20-40% of the formula. Secondary surfactants may be materials such as amides, betaines, sultaines, and alkyl polyglucosides. These are typically blended to optimize foam and cleansing characteristics while maintaining cost guidelines. They are typically used in the range of 1-10%, depending on the requirements of the formula.

A variety of other ingredients are added to modify different aspects of the formula. These include thickeners, fragrances, colorants, pearlizing agents, preservatives, and featured ingredients.

Thickeners increase the viscosity of the product. Salt can be added to thicken systems containing anionic surfactants. In addition other materials may be added to con-

The first ingredient added to the tank is typically water because it is usually the most plentiful ingredient. The other ingredients are added to the tank as specified by the manufacturing procedure. Ingredients that are heat sensitive are added as the batch is cooled to room temperature.

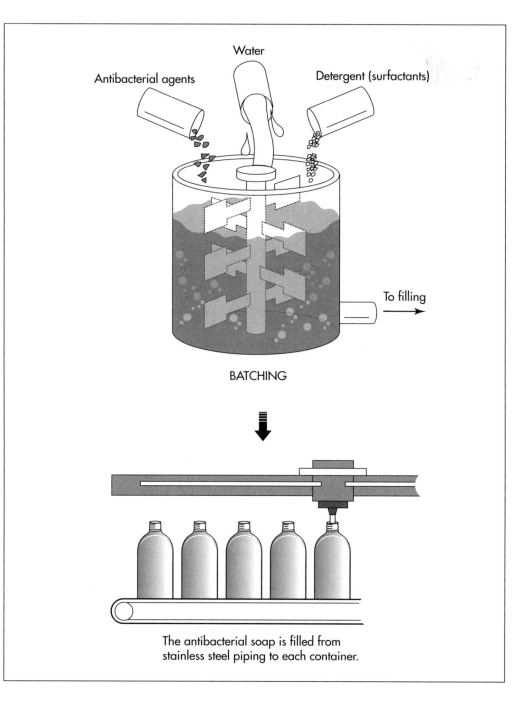

The antibacterial soap is filled from stainless steel piping to each container.

trol how the product flows. Gums, starches, and polymeric materials are used for this purpose at 0.1-1%.

Fragrances are aroma chemicals, which are added to mask the odor of the base and increase consumer appeal. These may be a variety of natural and synthetic materials blended together. In fact, a fragrance may consist of dozens of individual components. The compounded fragrance must be checked to make sure it is compatible with the detergent base. Fragrance is commonly used at levels ranging from 0.1-1%.

Colorants may also be included to improve the product's appearance. Some detergents have an inherent yellow color and dyes may be added to improve how the product looks. Colorants used in cosmetic products are controlled by the FDA and are designated as D&C (Drug and Cosmetic) or FD&C (Food, Drug and Cosmetic). These materials are used at very low levels (less than a hundredth of a percent.)

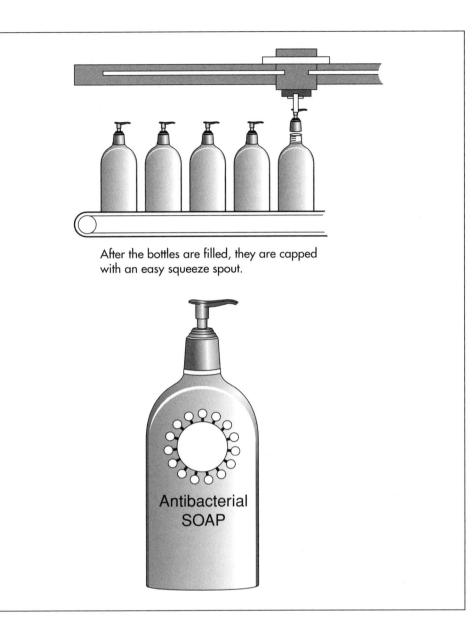

After the bottles are filled, they are capped with an easy squeeze spout.

Pearlizing agents are included to opacify the formula and give it a more pleasing appearance. These are typically fatty alcohol type materials such as glycol stearate, although titanium coated mica can also be used to give the product an attractive pearled appearance. These materials are used at 1% or less.

Preservatives are added to liquid soaps to prevent microbial growth. While the product contains other antibacterial agents, these are designed to kill skin organisms and may not be adequate to protect the product from other microbes such as molds and fungus. Therefore, additional preservatives may be added to the formula to provide broad spec-

trum protection. Preservatives are effective at 1% or less.

Various botanical extracts, proteins, natural oils, and other exotic materials may be added by the formulator to increase the product's appeal to the consumer.

The Manufacturing Process

Preparatory steps

1 Before manufacturing begins, the ingredients may be analyzed to ensure they meet quality standards. After approval, the ingredients may be pre-weighed and stored near the batching area to facilitate handling

during manufacture. The equipment must also be properly prepared before batching can begin. The tanks, plumbing lines, pumps, and transfer vessels must be cleaned and sanitized before manufacturing begins. This procedure helps to prevent microbial contamination of the batch and to make sure any traces of chemicals used in the previous batch are removed. Chemical contamination can ruin a batch because some ingredients (such as certain cationic and anionic surfactants) are incompatible and may react to form an insoluble precipitate which can clog filters, pumps, and transfer lines.

Batching

2 Liquid antibacterial soaps are typically manufactured in large stainless steel tanks. These vessels are available in a variety of sizes, but a typical batch could be as large as 3,000 gal (24,000 lb). Mixing and heating in these tanks must be carefully controlled during the batching process. Mixing is accomplished with a propeller or a sweep-style mixer, which provides good agitation without whipping an excessive amount of air into the product. Heating (or cooling) may be accomplished if the tank is equipped with a hollow jacket that can be filled with steam or chilled water. Sophisticated manufacturers employ automated computer systems to control mixing speed and temperature to ensure batch quality will be consistent.

The first ingredient to be added to the tank is typically water because it is usually the most plentiful ingredient. The other ingredients are added to the tank as specified by the manufacturing procedure; ingredients that are heat sensitive are added as the batch is cooled to room temperature. The batch is allowed to mix for a specified time at a controlled temperature to obtain an homogenous product. After batching has been completed, a sample is analyzed to ensure it meets specifications before the product is released for filling.

Filling/packaging

3 Once the product has been completed and is approved, it is ready to be filled into the appropriate packages. Filling can be initiated immediately after batching by pumping the product directly to a conveyor line outfitted with filling equipment. Alter-

nately, the product may be transferred to a temporary storage tank and filled later. The filling apparatus consists of a series of nozzles, which are metered to deliver a specific amount of product. Plastic bottles are fed into a hopper and directed under the filling nozzles via a conveyor system. The nozzles fill the appropriate amount of product into each bottle. After filling the bottles are sealed with screw-on caps or dispensing pumps. The bottles then move down the conveyor line where the lot number and an ink-jet printer codes expiration date on the side. The bottles are then packed into cardboard cartons for shipping.

Quality Control

Several steps are built into the manufacturing procedure of antibacterial soaps to guarantee high quality. Prior to commercialization the formula is stability tested to ensure the product's functional and aesthetic properties remain unchanged over time. OTC products are required to have a three-year shelf life. During the manufacturing process, the quality of the ingredients is chemically tested before batching begins. After batching is complete, the product is tested to make sure ingredients were added in the correct proportions. Some of these tests evaluate the product's physical properties such as viscosity and pH. Other tests are conducted to evaluate the product's antibacterial properties. Once such test, known as the glove juice test, is done by sampling the perspiration which collects inside a rubber glove worn by a volunteer who has used the test soap on one hand. This glove juice is then tested for microbial growth by applying it to a plastic plate that is specially coated to promote growth of skin organisms. If little or no growth is seen on the test glove, it can be assumed that the product is performing as it was intended.

Byproducts/Waste

A significant amount of waste can be generated if batches of antibacterial soaps are made incorrectly. Fortunately, many of these batches can be salvaged if they are within certain limits. For example, adjustments can be made to batches deficient in certain ingredients. Similarly, batches that are high in actives can be diluted to be within specifications. However, if batches are contaminated with extraneous materials they must be disposed of because the law does not allow the sale of adulterated products. Disposal must be done in accordance with appropriate local, state, and federal regulations because of the drug status of the product.

The Future

The future of antibacterial soaps depends on both chemical and regulatory factors. New chemicals for cleansing products are continually being developed. These new materials may offer improved foaming or cleansing properties, enhanced biodegradability, increased mildness, reduced cost, or other benefits. It is almost certain that such new ingredients will find use in future antibacterial soap formulations. Factors related to the regulatory status of antibacterial soaps are less certain. These factors will not be decided until the FDA finalizes the monograph, which is not expected to occur for the next few years. Once this happens, the products may have to be changed significantly to comply with the regulations.

Where to Learn More

Periodicals

"The Antimicrobial Debate." *Soap/Cosmetics/Chemical Specialties* (November 1995).

Jungerman, Eric. "Antimicrobial and Deodorant Soaps: Impact of Regulatory Developments." *Cosmetics & Toiletries*. Allured Publishing Corporation, May 1996.

Paulson, Daryl. "Developing Effective Topical Antimicrobials." *Soap/Cosmetics/Chemical Specialties* (December 1997).

Znaiden, Alex. "The War Against Germs." *Drug and Cosmetic Industry* 162. Advanstar Communications, January 1998.

—*Randy Schueller*

Antibiotic

Although the principles of antibiotic action were not developed until the twentieth century, the first known use of naturally-occurring antibiotics was by the Chinese over 2,500 years ago. Today, over 10,000 antibiotic substances have been reported. Currently, antibiotics represent a multibillion dollar industry that continues to grow each year.

Antibiotics are chemical substances that can inhibit the growth of, and even destroy, harmful microorganisms. They are derived from special microorganisms or other living systems, and are produced on an industrial scale using a fermentation process. Although the principles of antibiotic action were not discovered until the twentieth century, the first known use of antibiotics was by the Chinese over 2,500 years ago. Today, over 10,000 antibiotic substances have been reported. Currently, antibiotics represent a multibillion dollar industry that continues to grow each year.

Background

Antibiotics are used in many forms—each of which imposes somewhat different manufacturing requirements. For bacterial infections on the skin surface, eye, or ear, an antibiotic may be applied as an ointment or cream. If the infection is internal, the antibiotic can be swallowed or injected directly into the body. In these cases, the antibiotic is delivered throughout the body by absorption into the bloodstream.

Antibiotics differ chemically so it is understandable that they also differ in the types of infections they cure and the ways in which they cure them. Certain antibiotics destroy bacteria by affecting the structure of their cells. This can occur in one of two ways. First, the antibiotic can weaken the cell walls of the infectious bacteria, which causes them to burst. Second, antibiotics can cause the contents of the bacterial cells to leak out by damaging the cell membranes. Another way in which antibiotics function is by interfering with the bacteria's metabolism. Some antibiotics such as tetra-cycline and erythromycin interfere with protein synthesis. Antibiotics like rifampin inhibit nucleic acid biosynthesis. Still other antibiotics, such as sulfonamide or trimethoprim have a general blocking effect on cell metabolism.

The commercial development of an antibiotic is a long and costly proposal. It begins with basic research designed to identify organisms, which produce antibiotic compounds. During this phase, thousands of species are screened for any sign of antibacterial action. When one is found, the species is tested against a variety of known infectious bacteria. If the results are promising, the organism is grown on a large scale so the compound responsible for the antibiotic effect can be isolated. This is a complex procedure because thousands of antibiotic materials have already been discovered. Often, scientists find that their new antibiotics are not unique. If the material passes this phase, further testing can be done. This typically involves clinical testing to prove that the antibiotic works in animals and humans and is not harmful. If these tests are passed, the Food and Drug Administration (FDA) must then approve the antibiotic as a new drug. This whole process can take many years.

The large-scale production of an antibiotic depends on a fermentation process. During fermentation, large amounts of the antibiotic-producing organism are grown. During fermentation, the organisms produce the antibiotic material, which can then be isolated for use as a drug. For a new antibiotic to be economically feasible, manufacturers must be able to get a high yield of drug from the fermentation process, and be able to easily isolate it. Extensive research is usually required before a new antibiotic can be commercially scaled up.

History

While our scientific knowledge of antibiotics has only recently been developed, the practical application of antibiotics has existed for centuries. The first known use was by the Chinese about 2,500 years ago. During this time, they discovered that applying the moldy curd of soybeans to infections had certain therapeutic benefits. It was so effective that it became a standard treatment. Evidence suggests that other cultures used antibiotic-type substances as therapeutic agents. The Sudanese-Nubian civilization used a type of tetracycline antibiotic as early as 350 A.D. In Europe during the Middle Ages, crude plant extracts and cheese curds were also used to fight infection. Although these cultures used antibiotics, the general principles of antibiotic action were not understood until the twentieth century.

The development of modern antibiotics depended on a few key individuals who demonstrated to the world that materials derived from microorganisms could be used to cure infectious diseases. One of the first pioneers in this field was Louis Pasteur. In 1877, he and an associate discovered that the growth of disease-causing anthrax bacteria could be inhibited by a saprophytic bacteria. They showed that large amounts of anthrax bacilli could be given to animals with no adverse affects as long as the saprophytic bacilli were also given. Over the next few years, other observations supported the fact that some bacterially derived materials could prevent the growth of disease-causing bacteria.

In 1928, Alexander Fleming made one of the most important contributions to the field of antibiotics. In an experiment, he found that a strain of green *Penicillium* mold inhibited the growth of bacteria on an agar plate. This led to the development of the first modern era antibiotic, penicillin. A few years later in 1932, a paper was published which suggested a method for treating infected wounds using a penicillin preparation. Although these early samples of penicillin were functional, they were not reliable and further refinements were need-

ed. These improvements came in the early 1940s when Howard Florey and associates discovered a new strain of *Penicillium*, which produced high yields of penicillin. This allowed large-scale production of penicillin, which helped launch the modern antibiotics industry.

After the discovery of penicillin, other antibiotics were sought. In 1939, work began on the isolation of potential antibiotic products from the soil bacteria streptomyces. It was around this time that the term antibiotic was introduced. Selman Waxman and associates discovered streptomycin in 1944. Subsequent studies resulted in the discovery of a host of new, different antibiotics including actinomycin, streptothricin, and neomycin all produced by *Streptomyces*. Other antibiotics that have been discovered since include bacitracin, polymyxin, viomycin, chloramphenicol and tetracyclines. Since the 1970s, most new antibiotics have been synthetic modifications of naturally occurring antibiotics.

Raw Materials

The compounds that make the fermentation broth are the primary raw materials required for antibiotic production. This broth is an aqueous solution made up of all of the ingredients necessary for the proliferation of the microorganisms. Typically, it contains a carbon source like molasses, or soy meal, both of which are made up of lactose and glucose sugars. These materials are needed as a food source for the organisms. Nitrogen is another necessary compound in the metabolic cycles of the organisms. For this reason, an ammonia salt is typically used. Additionally, trace elements needed for the proper growth of the antibiotic-producing organisms are included. These are components such as phosphorus, sulfur, magnesium, zinc, iron, and copper introduced through water soluble salts. To prevent foaming during fermentation, anti-foaming agents such as lard oil, octadecanol, and silicones are used.

The Manufacturing Process

Although most antibiotics occur in nature, they are not normally available in the quantities necessary for large-scale production.

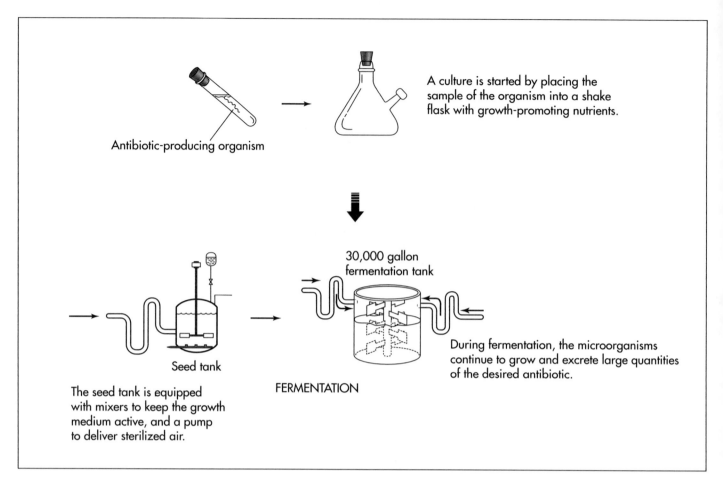

Antibiotic-producing organism

A culture is started by placing the sample of the organism into a shake flask with growth-promoting nutrients.

30,000 gallon fermentation tank

Seed tank

During fermentation, the microorganisms continue to grow and excrete large quantities of the desired antibiotic.

The seed tank is equipped with mixers to keep the growth medium active, and a pump to deliver sterilized air.

FERMENTATION

For this reason, a fermentation process was developed. It involves isolating a desired microorganism, fueling growth of the culture and refining and isolating the final antibiotic product. It is important that sterile conditions be maintained throughout the manufacturing process, because contamination by foreign microbes will ruin the fermentation.

Starting the culture

1 Before fermentation can begin, the desired antibiotic-producing organism must be isolated and its numbers must be increased by many times. To do this, a starter culture from a sample of previously isolated, cold-stored organisms is created in the lab. In order to grow the initial culture, a sample of the organism is transferred to an agar-containing plate. The initial culture is then put into shake flasks along with food and other nutrients necessary for growth. This creates a suspension, which can be transferred to seed tanks for further growth.

2 The seed tanks are steel tanks designed to provide an ideal environment for growing microorganisms. They are filled with the all the things the specific microorganism would need to survive and thrive, including warm water and carbohydrate foods like lactose or glucose sugars. Additionally, they contain other necessary carbon sources, such as acetic acid, alcohols, or hydrocarbons, and nitrogen sources like ammonia salts. Growth factors like vitamins, amino acids, and minor nutrients round out the composition of the seed tank contents. The seed tanks are equipped with mixers, which keep the growth medium moving, and a pump to deliver sterilized, filtered air. After about 24-28 hours, the material in the seed tanks is transferred to the primary fermentation tanks.

Fermentation

3 The fermentation tank is essentially a larger version of the steel, seed tank, which is able to hold about 30,000 gallons. It is filled with the same growth media

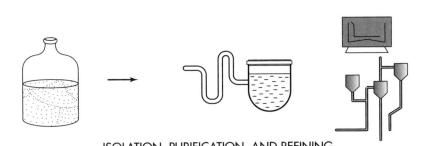

ISOLATION, PURIFICATION, AND REFINING

Once the antibiotic is isolated from the fermentation broth and purified using either the ion-exchange or solvent extraction method, a purified powder form of the antibiotic is produced.

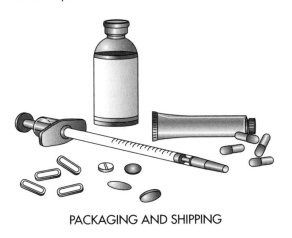

PACKAGING AND SHIPPING

found in the seed tank and also provides an environment inducive to growth. Here the microorganisms are allowed to grow and multiply. During this process, they excrete large quantities of the desired antibiotic. The tanks are cooled to keep the temperature between 73-81° F (23-27.2 ° C). It is constantly agitated, and a continuous stream of sterilized air is pumped into it. For this reason, anti-foaming agents are periodically added. Since pH control is vital for optimal growth, acids or bases are added to the tank as necessary.

Isolation and purification

4 After three to five days, the maximum amount of antibiotic will have been produced and the isolation process can begin. Depending on the specific antibiotic produced, the fermentation broth is processed by various purification methods. For example, for antibiotic compounds that are water soluble, an ion-exchange method may be used for purification. In this method, the compound is first separated from the waste organic materials in the broth and then sent through equipment, which separates the other water-soluble compounds from the desired one. To isolate an oil-soluble antibiotic such as penicillin, a solvent extraction method is used. In this method, the broth is treated with organic solvents such as butyl acetate or methyl isobutyl ketone, which can specifically dissolve the antibiotic. The dissolved antibiotic is then recovered using various organic chemical means. At the end of this step, the manufacturer is typically left with a purified powdered form of the antibiotic, which can be further refined into different product types.

Refining

5 Antibiotic products can take on many different forms. They can be sold in solutions for intravenous bags or syringes, in pill or gel capsule form, or they may be sold as powders, which are incorporated into topical ointments. Depending on the final form

of the antibiotic, various refining steps may be taken after the initial isolation. For intravenous bags, the crystalline antibiotic can be dissolved in a solution, put in the bag, which is then hermetically sealed. For gel capsules, the powdered antibiotic is physically filled into the bottom half of a capsule then the top half is mechanically put in place. When used in topical ointments, the antibiotic is mixed into the ointment.

6 From this point, the antibiotic product is transported to the final packaging stations. Here, the products are stacked and put in boxes. They are loaded up on trucks and transported to various distributors, hospitals, and pharmacies. The entire process of fermentation, recovery, and processing can take anywhere from five to eight days.

Quality Control

Quality control is of utmost importance in the production of antibiotics. Since it involves a fermentation process, steps must be taken to ensure that absolutely no contamination is introduced at any point during production. To this end, the medium and all of the processing equipment are thoroughly steam sterilized. During manufacturing, the quality of all the compounds is checked on a regular basis. Of particular importance are frequent checks of the condition of the microorganism culture during fermentation. These are accomplished using various chromatography techniques. Also, various physical and chemical properties of the finished product are checked such as pH, melting point, and moisture content.

In the United States, antibiotic production is highly regulated by the Food and Drug Administration (FDA). Depending on the application and type of antibiotic, more or less testing must be completed. For example, the FDA requires that for certain antibiotics each batch must be checked by them for effectiveness and purity. Only after they have certified the batch can it be sold for general consumption.

The Future

Since the development of a new drug is a costly proposition, pharmaceutical companies have done very little research in the last decade. However, an alarming development has spurred a revived interest in the development of new antibiotics. It turns out that some of the disease-causing bacteria have mutated and developed a resistance to many of the standard antibiotics. This could have grave consequences on the world's public health unless new antibiotics are discovered or improvements are made on the ones that are available. This challenging problem will be the focus of research for many years to come.

Where to Learn More

Books

Crueger, W. *Biotechnology: A Textbook of Industrial Microbiology.* Sunderland: Sinauer Associates, Inc., 1989.

Kirk Othmer Encyclopedia of Chemical Technology. New York: John Wiley & Sons, 1992.

Periodicals

Morell, Virginia. "Antibiotic Resistance: Road of No Return." *Science* 278 (October 24, 1997): 575-576.

Stinson, Stephen. "Drug Firms Restock Antibacterial Arsenal." *Chemical & Engineering News* (September 23, 1996): 75-100.

—*Perry Romanowski*

Armored Truck

An armored truck is a vehicle designed to securely transport currency and other valuables.

Background

Before the advent of armored vehicles, securely moving currency and valuables was achieved either by deceit or by force of arms. In the case of deceit, a courier in plain clothes would carry valuables disguised as a normal package or small piece of luggage. The courier traveled as a passenger on public means of transport, and although he was armed, his primary protection lay in appearing to be an average traveler. The main restriction of this method was size. In order to protect large shipments of valuables, man has long relied on a force of arms. Early caravans surrounded themselves with armed troops. Spanish galleons bristling with cannons carried treasure from the New World. In more modern times, stagecoaches carried locked but far from impregnable strong boxes of iron and wood, while a guard "rode shotgun" to ward off thieves. Railroad mail cars were outfitted with safes and were guarded by heavily armed government troops. The failure of this method was always twofold. First, although cargo was relatively safe while in its large, protected vessel, eventually it needed to be placed in smaller, more vulnerable vehicles to be carried to its final destination. The second problem was that no matter how many soldiers, swords, rifles, or cannons guarded a travelling precious cargo, a larger force of thieves with more swords, rifles, or cannons could be rallied to steal the cargo. As weapons became deadlier and more compact, this became more of a problem.

The first attempts at commercial armored trucks were inspired by the combat success of military armored cars in World War I. After the war, a marked increase in violent robberies of payroll clerks and messengers carrying deposits brought about the need for safer ways to transport cash. In 1920, a Chicago area delivery company called Brink's started converting school buses into security vehicles by attaching steel plates to the lower body panels and barring the windows. Each bus was followed by a Model-T automobile filled with armed guards. The first true commercial armored car was built that same year in Minneapolis, Minnesota, for a St. Paul police chief turned private detective Mike Sweeney. Sweeney designed the car and put it into service for his Sweeney Detective Bureau.

Early armored cars only wore steel plating on their body panels; they retained the wooden floors of the truck chassises on which they were built. This ended in 1927 when thieves buried explosives in the road and blew up a Brink's truck carrying $100,000 in payroll money. In the 1930s and 40s, manufacturers experimented with aluminum, which has a distinct weight advantage over steel. The metal was found to fatigue and crack after a short time and was discontinued. The 1970s worldwide fuel shortage spurred manufacturers to try lightweight plastic armor and smaller, more fuel-efficient chassises, but the results were similar to those with aluminum and the traditional steel regained its dominance in the industry.

Raw Materials

The material that makes up most of an armored car is also what makes it armored.

Surprisingly, the first part of an armored truck to need replacing is the chassis. The added weight of the armoring plus the weight of the cargo shortens the lifespan of the truck's suspension, braking, and mechanical systems.

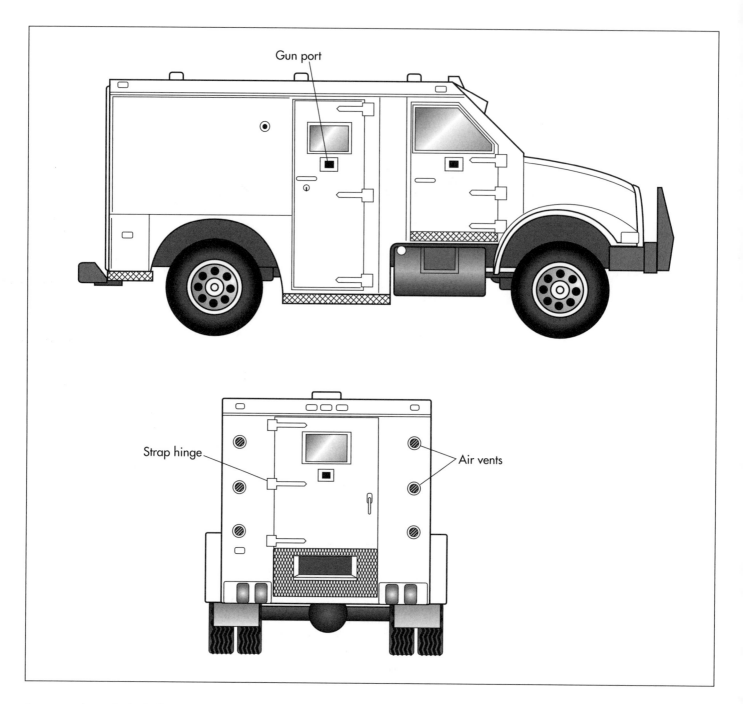

An armored car is basically a large, sealed metal box and is thereby very hot inside. The windows do not roll down for obvious reasons, so most trucks have four roof vents with a baffle to obstruct any direct lines of fire into the truck. The gun ports installed in each door employ a spring-loaded plate that must be slid open from the inside to prevent assailants from using them to fire into the vehicle.

The walls, floor, ceiling, and doors of an armored car are all made from steel. In recent years, both galvanized and stainless steel have been used to combat body rust and corrosion. The steel is hardened to increase its bullet resistance, either by heat treatment or by adding high levels of chromium (a very hard metal) and nickel (a very dense metal) during the forging process. In some applications, a ballistic fiberglass cloth known as woven roving is used to line the interior of the body. The windows of the truck are either made up of several layers of automotive glass or of layers of glass mixed with layers of bullet-resistant optical plastic.

Design

Four equally important goals must be considered in the making of an armored truck, several of which work against each other. The first consideration is ballistic resistance. Manufacturers use hardened steel in thicknesses varying from 0.125 in (0.317 cm) to greater than 0.25 in (0.635 cm), depending on the level of resistance required to build most of the body of an armored truck. Windows are made bullet resistant by using laminated glass in thicknesses between 1.50 in (3.81 cm) and 3 in (7.62 cm). Utilizing a mixture of glass and bullet-resistant optical plastic allows a much thinner and lighter window. Windshields are placed at a 45-degree angle to aid in deflection and to lower wind resistance. Tires are armored with a u-shaped hard plastic liner. If a tire deflates, it can run for several miles on the structure of this shell. Steel ram bumpers and front grill guards allow a driver to push through another vehicle that may be used as a roadblock. Increasing a truck's level of ballistic resistance means increasing the thickness of the steel and glass used and this works against the second design consideration—weight.

Gross Vehicle Weight (GVW) measures the maximum a motor vehicle can safely weigh, including its cargo and passengers. The average GVW for a "route truck," which makes everyday pickups and deliveries for banks and merchants, is 25,000 lb (11,350 kg). The average finished route truck weighs 12,000 lb (5,448 kg). This leaves 13,000 lb (5,902 kg) for cargo and guards. Reducing the finished weight of a truck allows for greater cargo weight. Decreasing the amount of armoring is not an attractive option, so trucks that need to haul heavier cargo, such as coins, must be built on larger, heavier chassises. A tandem-axle truck designed to haul large pallets of coin can have a GVW upwards of 55,000 lb (24,970 kg).

The third design consideration is security. Clearly, bullet resistance is of little use if the truck's operators and cargo are not securely separated from the outside world. With this in mind, security measures are aimed at installing and maintaining barriers. Most armored cars operate with two armed guards—a driver, who never leaves the vehicle, and a second guard known as a hopper, who rides in the cargo hold and carries valuables to and from the truck. The driver's compartment is separated from the cargo area by a steel bulkhead. The doors through which the hopper travels are fitted with slam locks, which automatically lock when the door swings shut. Once these doors are locked, the driver must reopen them electronically from the inside. Inside the cargo area, the hopper may place valuables inside a locked box, or he may use a drop safe, which has a one way chute and can only be opened once the truck has returned to its home facility. The gun ports installed in each door employ a spring-loaded plate that must be slid open from the inside to prevent assailants from using them to fire into the vehicle. The security requirements necessitate additional steps to accomplish the final design goal—crew comfort.

An armored car is basically a large, sealed metal box and is thereby very hot inside. The windows do not roll down for obvious reasons, so most trucks have four roof vents with a baffle to obstruct any direct lines of fire into the truck. Trucks are also fitted with dual air-conditioning and heating units, so the hopper and driver can independently adjust temperatures. The walls, ceiling, and floor of the cargo area are lined with lightweight foam-board insulation to further regulate inside temperatures.

The Manufacturing Process

Chassis

1 The production of an armored car begins with the delivery of a bare heavy duty truck chassis. The chassis arrives with complete drive train (engine and transmission) and suspension systems. Some manufacturers take delivery of a chassis with a full cab, which they cut off. Then they reuse many of the interior pieces. Others use a cowl chassis, which only arrives with fenders and a hood.

Body

2 The body of an armored truck is built much in the same way a house is framed. First, sections of square steel tubing are laid out vertically on a table known as a jig that represents the shape of an individual wall. Then lengths of steel channel called hat rails (because the cross section resembles a flat-brimmed hat) are laid horizontally at specific intervals across the vertical sections of tubing and tack-welded to hold them in place. Shortened sections of tubing and hat rail are used in certain areas to leave spaces that will become windows and doors. The process is repeated on the appropriate jig for each wall and for the roof, floor, and bulkhead.

3 Meanwhile, large sheets of hardened steel are being formed into outside body panels. The sheets are first cut to the correct size by enormous hydraulic shears. The cut panels are then rolled onto a table where the openings for windows and doors are cut with high temperature plasma torches. Some manufacturers employ welders to operate the torches, while some use computer-controlled robotic arms to handle the cutting. With this robotic system, plans are drawn on a computer; the computer then instructs the robotic arms to cut the exact shapes and dimensions to match the plans. The robotic arms slide vertically along an overhead track to accomplish vertical cuts, while rollers in the cutting table slide the steel across the path of the torch to handle horizontal cuts. The panels are then rolled onto various hydraulic presses where the necessary curves and angles are formed. Once the steel has been cut and formed into its appropriate shape, it is fitted against its corresponding frame and welded or riveted in place.

4 The floor is the first structure to be lifted onto the waiting chassis. First a sheet of hardwood is placed on the chassis' frame rails to insulate against vibration. Then the floor is placed on the hardwood and is attached to the chassis at several points with a number of c-shaped clamps. The walls are then each lifted onto the chassis and are tack welded or temporarily clamped where they join the floor and where their corners meet. Then the roof is placed on top of the walls. Once the entire structure has been checked for straightness and fit, all the joints and seams are thoroughly welded or riveted.

Outfitting the interior

5 Now that the body has been given its structure, the pieces that will make it a functioning armored truck are put in place. First, hinges are bolted to the door frames and the doors are hung and adjusted for straightness. Armored truck doors are hung on strap hinges, which extend horizontally across the face of the door to support the weight of the armoring. The hinges contain sealed grease fittings to allow the doors to swing smoothly.

6 Next, foam-board insulation is pressed into the spaces between the steel tubing of the body panels. Then an interior sheet of steel is welded or riveted to the tubing. Some manufacturers use a fiberglass ballistic cloth called woven roving in place of the interior steel lining. Layers of woven roving are infused with an epoxy and placed into a mold in the shape of the interior panels of the truck. The layers are pressed together in the mold and when dry, form a solid piece.

7 Once the interior has been lined, the bulkhead separating the cab from the cargo area is fastened in place. Then any shelves, bins, and safes are installed in the cargo area, and the vents are attached to the roof and the gun ports are fitted into the doors.

Finishing

8 The first step in creating a finished armored truck is to grind down any rough or irregular welds and to seal any seams with caulking. Then the interior is primed and painted, and the exterior is sprayed with numerous layers of sealant and primer before being painted to the customer's specifications.

9 Once the paint is dry, the electricians wire the truck (all wire in an armored truck is run through exposed conduit for ease of maintenance), and the heating and ventilation engineers install the rear air conditioning and heating unit. Next, the glass is installed; the locks are installed; the mirrors, bumpers, running boards, and grill guards are attached. The interior soft trim such as seats and belts, visors, and door handles are replaced last to avoid damage while other work proceeds. Finally, the finished truck is driven to a separate paint booth and the entire underside is sprayed with a corrosion resistant undercoating.

THE LIFE OF AN ARMORED TRUCK

Surprisingly, the first part of an armored truck to need replacing is the chassis. The added weight of the armoring plus the weight of the cargo shortens the lifespan of the truck's suspension, braking, and mechanical systems. The steel body, however, rarely wears out. For many years, fleet owners would remove the body from a worn chassis, refurbish it, and mount it on a new chassis, often as many as three times. Today, owners have found it more economical to sell older trucks in the burgeoning overseas markets. Trucks that are too old to be sold overseas are disassembled and the steel is sold as scrap.

Quality Control

Most manufacturers use the Underwriters Laboratories (UL) Standards for ballistic resistance as a point of reference for the armoring and windows on their trucks. UL tests various materials for their ability to withstand fire from a variety of weapons and rates the materials from Class 1 to Class 4. Class 1 offers the lowest resistance (a shot from a large caliber handgun) and Class 4 offers the highest. The trucks themselves are considered commercial vehicles and therefore must comply with standards set by the U.S. Department of Transportation. But most of the industry's quality control and design specifications are determined by what the insurance companies that underwrite armored carriers are willing to accept. The insurance underwriters determine acceptable armor levels, type and number of locks, and most stringently, operating procedures.

The Future

The armored truck has always relied on a show of force for its security. It is a massive, locked steel box filled with armed guards. Aside from stronger and lighter alloys of steel and laminates of glass used in its construction, the basic design will likely remain the same. The increased and varied placement of automated teller machines (ATMs) has created a demand for trucks based on smaller chassises to haul lighter but more numerous cargoes. Increased use of global positioning satellite (GPS) systems, which allow a dispatcher to track the exact position of each truck, will create greater efficiency in routing the growing number of trucks and may act as an additional deterrent to potential hijackers.

Where to Learn More

Other

"Independent Armored Car Operators Association, Inc. Home Page." http://www.iacoa.com/ (June 22, 1997).

"Streit Manufacturing—Custom Designed Armored Vehicles" http://www2.armored-vehicles.com/streit/.

—*Michael Cavette*

Artificial Snow

In 1975, a nucleating agent was discovered by Steve Lindow, a graduate student at the University of Wisconsin. While investigating a method to protect plants from frost damage, he found a protein that attracts water molecules and helps them form crystals. It was soon realized that this would be a useful material for making artificial snow.

Artificial snow is small particles of ice that are used to increase the amount of snow available for winter sports such as skiing or snow boarding. It is produced by a machine that uses a high-pressure pump to spray a mist of water into the cold air. The water droplets subsequently crystallize to form fake snow. The first commercially successful machines were developed in the 1950s and improvements in technology have steadily been introduced. With the increase in the popularity of winter sports, the artificial snow market is expected to show significant growth.

Background

The machines that produce artificial snow are designed to mimic the way that natural snow is made. In nature, snowflakes are formed when the temperature falls below 32° F (0° C). Atmospheric water then condenses on particles in the air and crystallizes. This action produces snowflakes that have a variety of sizes and shapes.

In a snow machine, water is first mixed with a nucleating material. It is then pressurized and forced through an atomizing nozzle. This breaks the water up into a mist, which is then injected with compressed air to break it up even further. As it exits the snow machine, the mist crystallizes on the nucleator and turns into tiny snow-like ice particles. Depending on the quality of the snow machine, the artificial snow can be as good as natural snow.

History

Although archeological evidence suggests that humans first skied about 4,000 years ago, interest in this activity as a sport did not begin until the middle of the nineteenth century. In 1883, the first international competition was held in Norway. The sport soon spread to the rest of Europe and America. As the popularity of skiing increased so did the need for a device that could provide snow when it was not naturally available. This need led to the development of the first artificial snow making machines.

One of the first machines was patented in the early 1900s. While it was functional, this machine was crude and unreliable. Steady improvements in design led to the development of a compressed air snow-making machine in the 1950s. This machine worked by using compressed air to force water through a nozzle. The nozzle would break the water up into smaller droplets, which would subsequently crystallize. The Pierce device, named after its inventor, was effective enough that most ski resorts used it. However, it did have its drawbacks, most notably, the nozzle tended to clog and it required a very high amount compressed air. This made it expensive to run. Additionally, the machine was quite noisy, and the snow that it produced tended to be wetter and icier.

During the 1970s, a variety of new innovations were introduced to the machines that improved the quality and method of producing artificial snow. One improvement was the addition of a rotating base and fan. The fan would blow the newly created snow farther away from the machine than compressed air alone and the rotating base allowed the direction of the snow to be changed. This made it possible to cover a much larger area with a single machine. Another improvement was the introduction of a ducted-fan machine. These machines were portable, making it possible to use them all over the ski run. They were superior to compressed air machines because they were significantly quieter and were less expensive to run.

In 1975, a nucleating agent was discovered by Steve Lindow, a graduate student at the University of Wisconsin. While investigating a method to protect plants from frost damage, he found a protein that attracts water molecules and helps them form crystals. It was soon realized that this would be a useful material for making artificial snow. The material was then trademarked and is now sold under the trade name Snomax.

As electronics improved, so did the controls for artificial snow-making machines. Computer controls were added, as were sensors that could automatically detect snow requirements. Higher powered fans were also added. Various other innovations led to machines that could produce better snow and more of it. Today, nearly all ski resorts employ some type of artificial snow-making system to improve skiing conditions and increase the length of the ski season.

Raw Materials

Water is the primary ingredient required to make artificial snow. Since ski areas are located on mountains however, finding an appropriate water supply is often a problem. If rivers or creeks are nearby they may be used. Otherwise, ponds or dams are created at the bottom of the mountain to produce a storage supply of water. The water is then pumped to the snow-making machines when needed.

In addition to water, compressed air and a nucleating material are also required to make snow. The compressed air is obtained using a pump. The nucleating agent is a biodegradable protein, which causes water molecules to form crystals at a higher temperature than normal. It is obtained from a nontoxic strain of a bacterium called *Pseudomonas syringae*. On average, this material can increase the amount of snow produced by a machine by 50%. It also helps produce lighter, drier flakes.

Design

The most important part of any snow-making system is the snow-making machine called a snow cannon or snow gun. A variety of designs are available, however most contain common elements including compressors, pumps, fans, and controls.

A central piece of the snow-making machine is the fan assembly. This part is responsible for converting the air/water mixture into tiny droplets and blowing it out onto the slope. It is similar to a typical portable house fan. It has a rotating propeller blade attached to a variable speed motor. Attached to the blades are curved vanes that direct the flow of air in a linear fashion. The fan is encased in an elongated steel duct that is open on both ends. As the blades of the fan move, air is drawn in from one side of the duct. This side is covered with a screen to prevent foreign objects from entering the assembly. The mechanisms controlling the main ingredients of the snow are located in the front, or discharge end, of the fan duct. This includes a water spray, compressed air pump, and a nucleating device. The nucleating device contains a reservoir filled with a nucleating agent. Water is pumped through this reservoir and the protein is incorporated.

During the snow-making process, the fan assembly is attached to a variety of pieces. To get water and air, hoses are hooked up to the fan assembly. These hoses are connected to a series of compressors and pumps that move air and water through pipes, up the mountain. To increase the coverage of the snow, the fan assembly is mounted on an oscillating stand, or yoke. Depending on the design, the placement of the yoke can be just off the ground or attached to a high tower. Levers may be connected to the yoke, which can adjust the angle at which snow exits the machine. A control box for the machine is typically located at the base of the yoke. This includes switches to operate things such as the water flow, fan rotation, and oscillation speed. The control box may be operated by a remote computer.

The Manufacturing Process

The production of artificial snow requires a series of devices that can move water and air up the mountain, combine them with a nu-

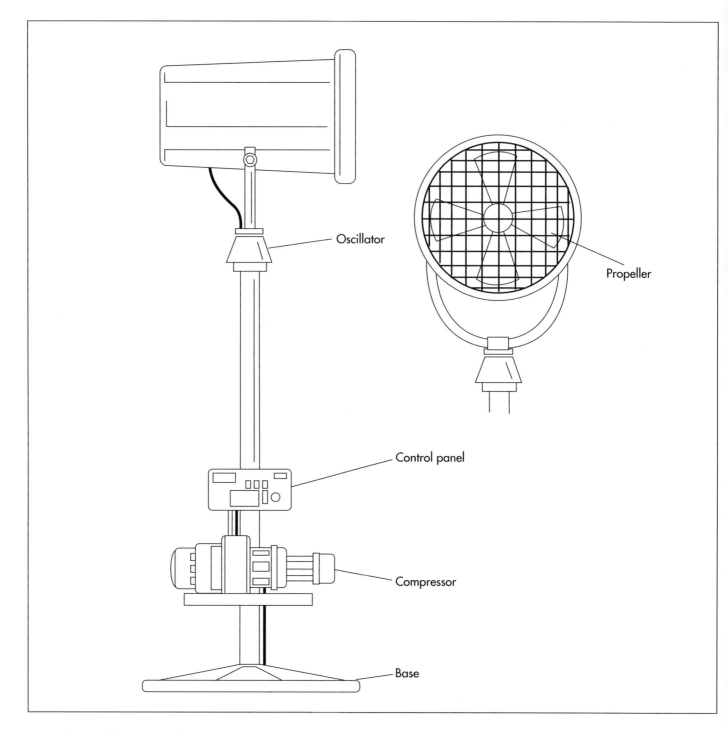

Oscillator

Propeller

Control panel

Compressor

Base

A central piece of the snow making machine is the fan assembly. This part is responsible for converting the air/water mixture into tiny droplets and blowing it out onto the slope. It is similar to a typical portable house fan.

cleating material, and spray them into the air as small droplets. Typically, the system is installed during the summer months and operated at night after the slopes have closed.

Installation of the system

1 Artificial snow making requires an entire system to be installed on the mountain slope. This system includes a series of water pipes, electric cables, pumps, and

compressors in addition to the snow making machines. First, plans showing the layout of the system are drawn. Then the water pipes and cables are laid in long trenches traversing the entire slope. The trenches must be dug significantly deep so water does not freeze during the winter months. At various points along the water line, valves and hoses are installed to bring water to the surface. Hay bales are placed around them for protection.

Mixing water with other components

2 Snow making is typically done at night and requires constant monitoring. It is typically only done when the outdoor temperature is 28° F (-2.2° C) or below. A number of snow machines are hooked up to the water lines all the way up the slope. When the machines are turned on, the snow making process begins. The water is first pumped up the mountain to the various machines. Depending on the type of machine, water may be mixed with the nucleating material prior to pumping or when it first enters the machine.

Creating the snow

3 The water is then mixed with compressed air and pumped through a high powered fan. The fan can spray the mixture nearly 60 ft (18.3 m) into the air. As it leaves the machine, the water crystallizes and forms snow. The snow is piled up is large mounds known as whales. At this point the snow may be analyzed and the machines are adjusted to produce the best quality snow.

4 When a pile of artificial snow is significantly high, the snow making machine is turned off. At optimal performance, a snow machine can produce enough snow to cover an acre in about 2 hours. The whale is then allowed to set, or cure, for two to three days. This lets excess water drain off and helps produce a softer snow.

Moving the snow

5 After the curing process, the snow pile is ready for grooming. Using a special plow, the snow is smoothed out onto the skiing surface. While it is being moved, it is sent through a tilling device. This fluffs up the snow, making it more skiable.

Quality Control

Producing artificial snow that is as good as or better than natural snow requires significant quality control measures. Prior to production, the nucleating material is checked to ensure that it meets the appropriate specifications. While the snow is being made, it is analyzed for crystal quality, appearance, and wetness. The air/water ratio may be adjusted to improve the quality of the snow. If the snow is of the highest quality, it will last longer, hold its shape better, and be easier to groom.

The Future

The shortcomings of the current artificial snow-making technology suggest possible improvements in the future. Currently, the noise generated by these machines is a problem. While attempts have been made to reduce the sound, future machines will be even quieter. Another limitation of the snow-making machines is their narrow temperature range of operation. New machines may be able to produce snow at temperatures over 28° F (-2.2° C). These machines may also produce higher quality snow in less time.

Where to Learn More

Periodicals

Brown, Rich. "Man Made Snow." *Scientific American* (January 1997): 119.

Hampson, Tim. "The Bacteria at the Heart of a Good Snowfall." *New Scientist* (January 1990): 38.

Other

Weaver, et. al. U.S. Patent #5,400,966, 1995.

VanderKelen, et al. U.S. Patent #5,167,367, 1992.

—*Perry Romanowski*

Asbestos

Asbestos usage in the United States fell from about 880,000 tons/yr (800,000 metric tons/yr) in 1973 to less than 44,000 tons/yr (40,000 metric tons/yr) in 1997. Worldwide usage of asbestos in 1997 was estimated at about 2.0 million tons/yr (1.8 million metric tons/yr).

Background

Asbestos is a general name that applies to several types of fibrous silicate minerals. Historically, asbestos is best known for its resistance to flame and its ability to be woven into cloth. Because of these properties, it was used to make fireproof stage curtains for theaters, as well as heat-resistant clothing for metal workers and firefighters. More modern applications of asbestos take advantage of its chemical resistance and the reinforcing properties of its fibers to produce asbestos-reinforced cement products including pipes, sheets, and shingles used in building construction. Asbestos is also used as insulation for rocket engines on the space shuttle and as a component in the electrolytic cells that make oxygen on submerged nuclear submarines. Much of the chlorine for bleach, cleansers, and disinfectants is produced using asbestos products.

The earliest known use of asbestos was in about 2500 B.C. in what is now Finland, where asbestos fibers were mixed with clay to form stronger ceramic utensils and pots. The first written reference to asbestos came from Greece in about 300 B.C. when Theophrastus, one of Aristotle's students, wrote a book entitled *On Stones*. In his book, he mentioned an unnamed mineral substance, which looked like rotten wood, yet was not consumed when doused with oil and ignited. The Greeks used it to make lamp wicks and other fireproof items. When the Roman naturalist and statesman Pliny the Elder wrote his comprehensive *Natural History* in about 60 A.D., he described this fireproof mineral and gave it the name asbestinon, meaning unquenchable, from which we get the English word asbestos.

Although the fireproof qualities of asbestos continued to fascinate the scientific community for hundreds of years, it wasn't until the 1800s that asbestos found many commercial uses. The first United States patent for an asbestos product was issued in 1828 for a lining material used in steam engines. In 1868 Henry Ward Johns of the United States patented a fireproof roofing material made of burlap and paper laminated together with a mixture of tar and asbestos fibers. It became an immediate success. Large-scale mining of asbestos deposits near Quebec, Canada, began in 1878 and spurred the development of other commercial uses. By 1900 asbestos was being used to make gaskets, fireproof safes, bearings, electrical wiring insulation, building materials, and even filters to strain fruit juices.

Technological developments in the early 1900s resulted in even more uses for asbestos. Many of the early plastic materials relied on asbestos fibers for reinforcement and heat resistance. Vinyl-asbestos tile became one of the most commonly used floor coverings and remained in use well into the 1960s. Automobile brake linings and clutch facings also used large amounts of asbestos, as did a multitude of building materials. After World War II, the use of asbestos in products continued to expand. Heart surgeons used asbestos thread to close incisions, Christmas trees were decorated with asbestos artificial snow, and a brand of toothpaste was marketed using asbestos fibers as an abrasive.

The widespread use of asbestos was not without a dark side, however. Health problems associated with exposure to airborne asbestos particles had been noted since the early 1900s, and resulted in the passage of the Asbestos Industry Regulations of 1931 in England. By the mid-1960s, health problems began to surface among shipyard workers who handled asbestos insulation during World War II. In the United States, the problem reached the crisis stage by the 1970s, forcing the Environmental Protection Agency (EPA) to place severe restrictions on the use of asbestos. Although the EPA lifted the ban for certain kinds of asbestos in 1991, the public's faith had been severely shaken, and most manufacturers had voluntarily removed asbestos from their products. As a result, asbestos usage in the United States fell from about 880,000 tons/yr (800,000 metric tons/yr) in 1973 to less than 44,000 tons/yr (40,000 metric tons/yr) in 1997.

In other countries, asbestos products are still widely used, especially in the construction industry. Worldwide usage of asbestos in 1997 was estimated at about 2.0 million tons/yr (1.8 million metric tons/yr). Most of this asbestos is used to make asbestos-reinforced concrete products, where the asbestos fibers are locked within the concrete.

Asbestos mining operations are found in 21 countries. The leading producers of asbestos are Russia (formerly the USSR), Canada, Brazil, Zimbabwe, China, and South Africa. Smaller deposits are found in the United States and several other countries.

Raw Materials

There are six types of asbestos: actinolite, amosite, anthophyllite, crocidolite, tremolite, and chrysolite. The first five types are known as amphiboles. They are characterized by having very strong and stiff fibers, which makes them a serious health hazard. Amphibolic asbestos fibers can penetrate body tissue, especially in the lungs, and eventually cause tumors to develop. The sixth type of asbestos, chrysotile, is known as a serpentine. Its fibers are much softer and more flexible than amphibolic asbestos, and they do less damage to body tissue. All six types of asbestos are composed of long chains of silicon and oxygen atoms, locked together with various metals, such as magnesium and iron, to form the whisker-like crystalline fibers that characterize this mineral.

Chrysotile is the most commonly used type of asbestos and accounted for about 98% of the worldwide asbestos production in 1988. It is usually white, and is sometimes known as white asbestos, although it can also be amber, gray, or greenish in color. Most chrysotile fibers are about 0.25-0.50 in (6.4-12.7 mm) long and are usually added to concrete mixes to provide reinforcement. Only about 8% of chrysotile fibers are long enough to be spun into fabric or rope.

Amosite, sometimes called brown asbestos, accounted for about 1% of worldwide production in 1988. It often has a light brown tinge, but is also found in dark colors, as well as white. Amosite has coarse fibers that are about 0.12-6.0 in (3.0-152.0 mm) long. The fibers are difficult to spin into fabric or rope and are mostly used as an insulating material, although that use is banned in many countries.

Crocidolite, sometimes called blue asbestos, accounted for the remaining 1% of worldwide production. It has a bluish tinge, and its fibers are about 0.12-3.0 in (3.0-76.0 mm) long. Crocidolite has very high tensile strength and excellent resistance to chemicals. One of its uses is as a reinforcement in plastics.

The other three types of asbestos—anthophyllite, actinolite, and tremolite—have no significant commercial applications and are rarely mined.

The Manufacturing Process

Asbestos deposits are found underground, and the ore is brought to the surface for processing using conventional mining practices. Chrysotile asbestos is usually found near the surface and can be accessed with an open-pit mine. Other asbestos deposits are found at varying depths and may require tunnels as deep as 900 ft (300 m) to gain access.

Asbestos fibers are formed by the gradual growth of mineral crystals in cracks, or veins, found in soft rock formations. The crystals grow across the vein, and the width of the vein determines the resulting asbestos fiber length. Because the minerals come from the surrounding rock, the chemical composition of the fibers is similar to the

ASBESTOS PROCESSING

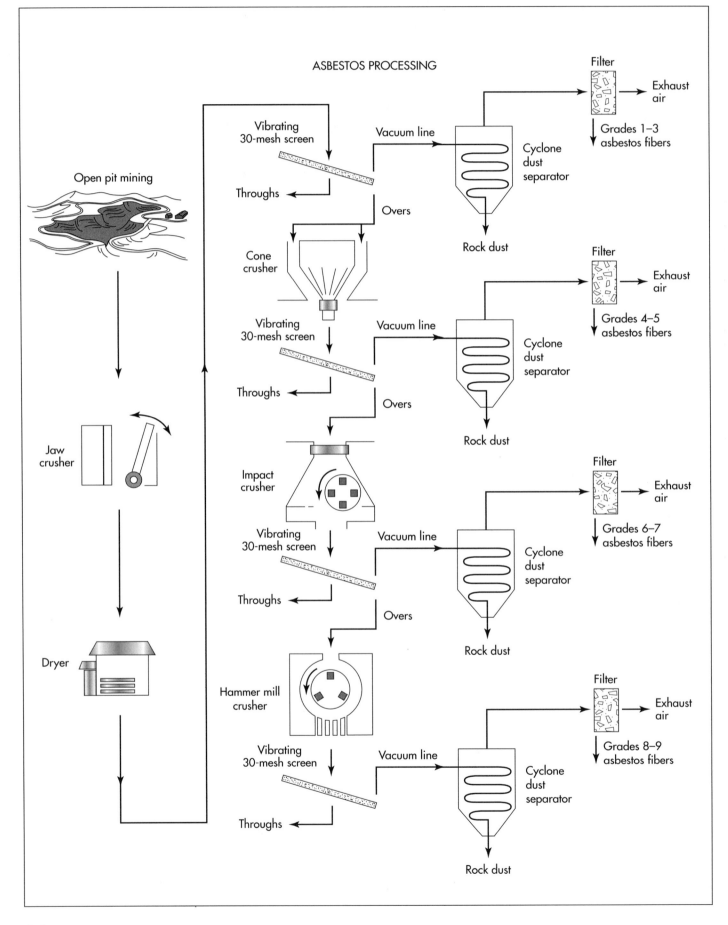

rock. As a result the asbestos must be separated from the rocky ore using physical methods, rather than the chemical methods sometimes used to process other ores.

Here are the steps used to process the chrysotile asbestos ore commonly found in Canada:

Mining

1 Chrysoltile asbestos deposits are usually located using a magnetic sensor called a magnometer. This method relies on the fact that the magnetic mineral magnetite is often found near asbestos formations. Core drillings are used to pinpoint the location of the deposits and to determine the size and purity of the asbestos.

Most chrysotile asbestos mining operations are conducted in an open-pit mine. A spiraling series of flat terraces, or benches, are cut into the sloping interior sides of the pit. These are used both as a work platform and as a roadway for hauling the ore up and out of the pit. The asbestos ore deposits are loosened from the surrounding rock by careful drilling and blasting with explosives. The resulting rocky debris is loaded into large rubber-tired haul trucks and brought out of the mine. Some operations use an excavation technique called block caving, in which a section of the ore deposit is undercut until it crumbles under its own weight and slides down a chute into the waiting haul trucks.

Separating

The ore contains only about 10% asbestos, which must be carefully separated from the rock to avoid fracturing the very thin fibers. The most common method of separation is called dry milling. In this method, the primary separation is done in a series of crushing and vacuum aspirating operations in which the asbestos fibers are literally sucked out of the ore. This is followed by a series of secondary separation operations to remove rock dust and other small debris.

2 The ore is fed into a jaw crusher, which squeezes the ore to break it up into pieces that are 0.75 in (20.0 mm) in diameter or less. The crushed ore is then dried to remove any moisture that may be present.

3 The ore falls on the surface of a vibrating 30-mesh screen, which has openings that are 0.002 in (0.06 mm) in diameter. As the screen vibrates, the loosened asbestos fibers rise to the top of the crushed ore and are vacuumed off. Because the crushed ore is much denser than the fibers, only the very smallest rock particles get vacuumed off with the asbestos.

4 The very fine silt and rock particles that fall through the vibrating screen are called throughs or tailings and are discarded. The crushed ore pieces that remain on the screen are called overs and are moved to the next stage of processing.

The crushed ore from the first screen is fed through a second crusher, which reduces the ore pieces to about 0.25 in (6.0 mm) in diameter or less. The ore then falls on another vibrating 30-mesh screen and repeats the process described in steps 3 and 4.

5 The process of crushing and vacuum aspiration of the asbestos fibers is repeated twice more. Each time the pieces of ore get smaller until the last asbestos fibers are captured and the remaining ore is so small that it falls through the screen and is discarded. This four-step process also separates the asbestos fibers by length. The longest fibers are broken free from the surrounding rock in the first crusher and are vacuumed off the first screen. Shorter length fibers are broken free and captured on each successive set of crushers and screens, until the shortest fibers are captured on the last screen.

6 The asbestos fibers and other material captured from each screen are carried suspended in a stream of air and run through four separate cyclone separators. The heavier debris and rock dust particles fall to the center of the whirling air stream and drop out the bottom of the separators.

7 The air then passes through four separate sets of filters, which capture the different length asbestos fibers for packaging.

Quality Control

Asbestos fibers are graded according to several factors. One of the most important factors is their length, since this determines the applications where they may be used and, therefore, their commercial value.

The most common grading system for chrysotile asbestos fibers is called the Quebec Standard dry classification method. This standard defines nine grades of fibers from Grade 1, which is the longest, to Grade 9, which is the shortest. At the upper end of the scale, Grades 1 through 3 are called long fibers and range from 0.74 in (19.0 mm) and longer down to 0.25 in (6.0 mm) in length. Grades 4 through 6 are called medium fibers, while Grades 7 through 9 are called short fibers. Grade 8 and 9 fibers are under 0.12 in (3.0 mm) long and are classified by their loose density rather than their length.

Other factors for establishing the quality of asbestos fibers include tests to determine the degree of fiber separation or openness, the reinforcing capacity of the fibers in concrete, and the dust and granule content. Specific applications may require other quality control standards and tests.

Health and Environmental Effects

It is now generally accepted that inhalation of asbestos fibers can be associated with three serious, and often fatal, diseases. Two of these, lung cancer and asbestosis, affect the lungs, while the third, mesothelioma, is a rare form of cancer that affects the lining of the thoracic and abdominal cavities.

It is also now generally accepted that different types of asbestos, particularly the amphiboles, pose a greater health hazard than chrysotile asbestos.

Finally, it is recognized that other factors, such as the length of the fibers and the duration and degree of exposure, can determine the health hazard posed by asbestos. In fact some studies have shown that some asbestos-induced lung cancers only occur when the exposure is above a certain level of concentration. Below that threshold, there is no statistical increase in lung cancer over that found in the general population.

Although not everyone agrees with these findings, overall concerns about the potential adverse health effects of inhaling asbestos fibers have led to stricter regulations on the amount of airborne asbestos allowable in the workplace. These regulations vary from one country to another, but they all mandate significantly lower levels than previously found. In the United States, the Occupational Health and Safety Administration (OSHA) set the maximum permissible exposure to fibers longer than 0.005 mm at 0.2 fibers/cubic centimeter during an eight-hour workday or 40-hour work week.

Airborne asbestos levels in the general environment outside the workplace are many times lower and are not considered a hazard.

The Future

Asbestos is still an important component in many products and processes, although its usage is expected to remain low in the United States. The stricter exposure regulations and improved manufacturing and handling procedures now in place are expected to eliminate health problems associated with asbestos.

Where to Learn More

Books

Brady, George S., Henry R. Clauser, and John A. Vaccari. *Materials Handbook,* 14th Edition. McGraw-Hill, 1997.

Hornbostel, Caleb. *Construction Materials,* 2nd Edition. John Wiley and Sons, Inc., 1991.

Kroschwitz, Jacqueline I. and Mary Howe-Grant, ed. *Encyclopedia of Chemical Technology,* 4th edition. John Wiley and Sons, Inc., 1993.

Periodicals

Alleman, James E., and Brooke T. Mossman. "Asbestos Revisited." *Scientific American* (July 1997): 70-75.

Other

http://www.asbestos-institute.ca.

http://www.epa.gov/ttnuatw1/hlthef/asbestos.html.

—Chris Cavette

Baby Formula

Background

Baby formula is a synthetic version of mothers' milk and belongs to a class of materials known as dairy substitutes. Dairy substitutes have been used since the early nineteenth century for products like oleomargarine and filled cheese. They are made by blending fats, proteins, and carbohydrates using the same technology and equipment used to manufacture real dairy products. Since the 1940s, advances in processing techniques such as homogenization, fluid blending, and continuous batching and filling have greatly improved the ways imitation dairy products, like formula, are made. The sales of infant formulas have also improved over the last several decades. Until the early 1990s, infant formula was sold only as a pharmaceutical product. Salespeople presented their brands to pediatricians who would then recommend the products to new mothers. In 1992 federal antitrust actions resulted in manufacturers shifting their marketing strategies toward more direct marketing techniques. Now, in addition to pharmaceutical sales, manufacturers rely heavily on direct mail campaigns and TV and print advertising to recruit new customers. In the United States alone, the infant formula industry is a $3 billion-a-year business with approximately another $1 billion in sales outside of the United States. There is some degree of controversy associated with marketing infant formula, however. There are concerns that formula is not as healthy for babies as breast milk and babies may actually become ill if the formula is improperly mixed or administered. Furthermore, once mothers have begun formula feeding on a regular basis it is difficult to return to breastfeeding. Leading authorities, including the World Health Organization (WHO), recommend that babies be completely breastfed for the first six months and that breast milk continue to be used as part of their diet until at least the beginning of the child's second year.

Design

It should be noted that the design of infant formulas is highly complex due to the nature of the biological requirements of the developing child. What follows is a generalized description of some of the key areas of infant formulations and is not meant to be an exhaustive review of the relevant nutritional chemistry. The key to successful formula design is to match as closely as possible the physical and nutritional properties of breast milk. Milk is a natural emulsion, which means it is a fine dispersion of tiny droplets of fats and oils suspended in water. Milk also contains important components including proteins, sugars, minerals, salts, and trace elements. Formula is made by blending similar materials in an attempt to match the characteristics of true milk. Formula design typically falls into one of three categories:

Milk based formulas (containing milk components such as casein or whey protein)

These formulas typically start with cow milk as a base since most infants have no problem ingesting cow's milk. This type of formula is fortified with extra nutritional elements.

Animal or vegetable fat based formulas (containing vegetable and/or milk components)

Some infants have a sensitivity, allergy, or potential allergy to formula based entirely

31

on cow's milk. Formula made with vegetable derived milk or a limited amount of cow's milk derived components may be more suitable for these children. Most vegetable derived formulas are soybean based. However, allergies to soybean milk also exist, so this approach does not guarantee the product will be trouble free. In general, using hydrolyzed proteins can minimize allergy concerns. They are less likely to cause allergic reactions.

Non-milk based (containing no milk components at all)

There are expensive, specialty formulas for infants who have a strong sensitivity to both cow's and soy milk, or other medical or digestive conditions that are formula related.

Formulas are available in three forms: powder, liquid concentrate, and ready-to-feed. Powder and liquid concentrate are less expensive but they require mixing/dilution prior to use. This may be a problem because they may be improperly mixed or mixed with water contaminated with bacteria. Ready-to-feed is the most expensive type but requires no mixing before use. This is an advantage because the mother can be sure the baby is getting the appropriate dose of nutrients and doesn't have to worry about contamination problems.

Raw Materials

Proteins

As described above, protein used in formulas can come from a variety of sources such animal milk or soybeans. Soy milk is made by taking soybeans, soaking them in baking soda, draining them, grinding the beans, then diluting the mixture with water and homogenizing it. The proteins, which come from soybeans, may be in the form of protein concentrates or protein isolates. The latter helps eliminate or reduce carbohydrates that can cause flatulence and abnormal stools. Other useful proteins can be derived from nuts, fish, and cottonseed oil but these have limited application in infant formulas.

Fats and carbohydrates

Fats and oils are an important dietary requirement for infants. Therefore formulations attempt to match the serum fatty acid profile of real breast milk. These fatty acids include eicosapentaenoic acid (EPA) which may be derived from fish oil and other sources. In actual breast milk there is a significant amount of fatty compounds known as triglycerides. For example, docosahexaenoic acid (DHA) is believed to be an important triglycerides. Triglycerides which are similar to (but not biochemically identical to) those found in breast milk can be derived from egg yolk phospholipids. Alternatively, fatty acid precursors (molecules which react to form dietary fatty acids) may be added to infant formula. These precursors (e.g., alpha and gamma linolenic acid) allow the infants' bodies to synthesize the necessary fatty acids. However, this method is not as efficient for delivering fatty acids as breast milk is.

Diluents

The diluent is the carrier or bulk of the liquid of the formula. For milk based formulations, skim milk may be used as the primary diluent. In milk free formulations, purified water is used.

Minerals

A number of essential minerals are added to infant formula. These include calcium, phosphate, sodium, potassium, chloride, magnesium, sulfur, copper, zinc, iodine, and iron. Iron is one of the most important components since all babies need a source of iron in their diet. Some parents are concerned that iron-fortified formulas cause intestinal problems in infants but this is a myth. In general parents can expect formula fed babies to experience more gastrointestinal problems than breastfed babies.

Vitamins

Vitamins are added to increase the nutritional value of formula. These include vitamins A, B_{12}, C, D, and E as well as thiamine, riboflavin, niacin, pyridoxine, pantothenate, and folacin.

Emulsifiers/stabilizers

A variety of materials are added to ensure the formula stays homogenous and that the oil and water soluble components do not

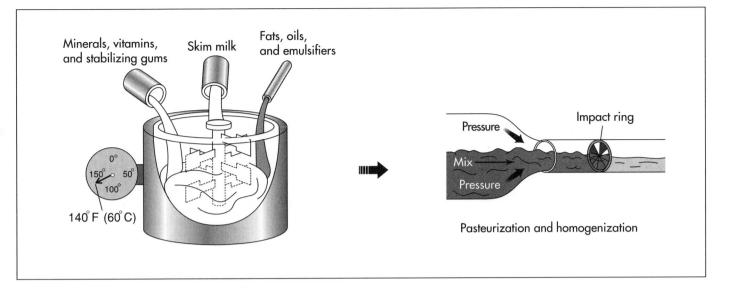

Minerals, vitamins, and stabilizing gums

Skim milk

Fats, oils, and emulsifiers

0°
50°
100°
150°

140° F (60° C)

Pressure

Impact ring

Mix

Pressure

Pasteurization and homogenization

separate. These include emulsifiers such as mono and di-glycerides as well as thickeners like natural starches and gums (e.g., such carrageenan.)

The Manufacturing Process

The method of manufacture depends on the type of formula being made. The following steps describe a general procedure for a ready-to-feed, milk-based formula.

Mixing ingredients

1 The primary ingredients are blended in large stainless steel tanks. The skim milk is added and adjusted to 140° F (60° C). Fats, oils and emulsifiers are added next. Additional heating and mixing may be required to yield the proper consistency. Minerals, vitamins, and stabilizing gums may be added at various points in the process depending on their sensitivity to heat. Once mixing is complete, the batch can be temporarily stored or transported via pipeline to pasteurization equipment.

Pasteurization

2 Pasteurization is a process that protects against spoilage by eliminating bacteria, yeasts, and molds. Pasteurization involves quickly heating and cooling the product under controlled conditions which microorganisms cannot survive. A temperature of 185-201.2° F (85-94° C), held for about 30 seconds, is necessary to adequately reduce

microorganisms and prepare the formula for filling. Several pasteurization methods are commercially available—one common method warms the formula by sending it through a tube adjacent to heat plate heat exchanger. Thus the formula is heated indirectly. Another method heats formula directly and then uses the heated liquid to preheat the rest of the incoming formula. The preheated formula is further heated with steam or hot water to the pasteurization temperature. After pasteurization is complete, the batch may be processed further by homogenization.

Homogenization

3 Homogenization is a process which increases emulsion uniformity and stability by reducing the size of the fat and oil particles in the formula. This process can be done with a variety of mixing equipment, which applies high shear to the product. This type of mixing breaks the fat and oil particles into very small droplets.

Standardization

4 The resulting composition is standardized to ensure key parameters, such as pH, fat concentration, and vitamin and mineral are correct. If any of these materials are at insufficient levels the batch can be reworked to achieve the appropriate levels. The batch is then ready to be packaged.

Once mixing is complete, the batch can be temporarily stored or transported via pipeline to pasteurization equipment. After pasteurization is complete, the batch may be processed further by homogenization.

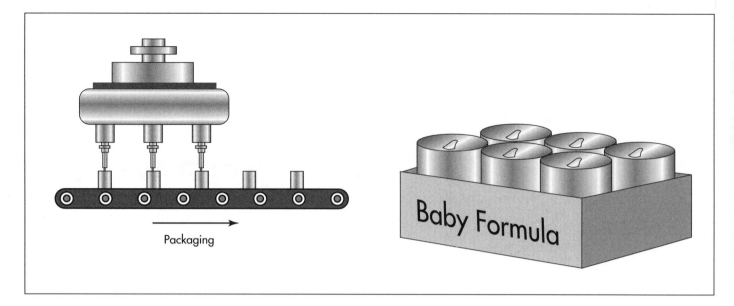

Packaging

Baby Formula

Conventional liquid filling equipment commonly used in the food and beverage industry are used to package ready-to-use baby formula.

Packaging

5 Packaging process depends on the manufacturer and type of equipment employed, but in general, the liquid formula is filled into metal cans which have lids crimped into place. These can be filled on conventional liquid filling equipment commonly used in the food and beverage industry.

Sterilization

6 The filled packages can be subsequently heated and cooled to destroy any additional microorganisms. The finished cans are then packed in cartons and stored for shipping.

Quality Control

Quality of infant formula is ensured at three levels, which have some degree of overlap. First, in the United States, there are governmental standards, which establish the nutritional quality of infant formulas and other dairy substitutes. Specific details of these standards can be found in the Code of Federal Regulations; more information is available from the Food and Drug Administration (FDA) which regulates infant formula as a special diet food. The FDA publishes a monograph detailing everything from the mandated nutrient list to label copy and artwork used on packaging. Second, the dairy industry sets its own industry-wide quality control standards. The industry is self-

policing and has its own regulatory organization, the International Dairy Federation, which sets industry standards for manufacturing and quality control. Third, individual companies set their own standards for quality control. For example Martek, one producer of triglycerides used in formula, has microbiologists and engineers monitor 30 different checkpoints of triglyceride production, 24 hours a day.

The Future

Future developments in infant formula manufacturing techniques will be driven, in part, by business and marketing concerns. This dependence on the marketing climate may be a benefit to the industry because there is tremendous opportunity for expansion. It is estimated that the total worldwide market for infant formula could be as high as $80 billion. Therefore, the current estimated world sales of formula of $4 billion represents only 5% of the total potential sales. Should the market grow even close to $80 billion, it is likely to spur manufacturers to find better ways to simulate breast milk. One such future improvement is being developed by scientists who have recently identified an important fatty acid in breast milk that is not found in infant formulas. This particular fatty acid appears to be important for the development of cell membranes in eye, brain, and nerve tissue. Addition of this material could be a significant advance in formula technology. Formula manufacturers can continue to make their products better by incorporating breakthrough research findings such as this one. However, even though there is great potential for growth, there is no guarantee it will be realized. The industry is experiencing criticism from groups which claim that formula is unnecessary and, in fact, may be harmful to infants. Should this trend negatively impact formula sales, manufacturers may be less likely to make significant investments in product and process development.

Where to Learn More

Books

Carbohydrates in Infant Nutrition. Springer-Verlag New York, Inc., 1997.

Infant Nutrition. Mosby-Year Book, Inc., 1993.

Periodicals

Black, Rebecca F., Jill P. Blair, Vicki N. Jones, and Robert H. DuRant, "Infant Feeding Decisions among Pregnant Women from a WIC Population in Georgia." *Journal of the American Dietetic Association* 90 (February 1990): 255-260.

—*Randy Schueller*

Baby Stroller

A study conducted by Henry Ford Hospital in Detroit, Michigan, showed that jogging with a pram or stroller increased runners' heart rates by 3-5% over those jogging without strollers.

Background

Babies have needed to be carried for as long as parents have needed to go places, and different cultures have devised ingenious methods to ease the burden of bearing the weight of a small child on long, or even short, walks. Many Native Americans used a cradleboard, a highly decorated board covered in cloth in which the baby could be secured, typically by laces running across the cloth. The cradleboard was then strapped to the carrier's back. Members of other cultures carried, or sometimes still carry, children wrapped in shawls, slings, or strips of cloth tied around the carrier's shoulders, neck, or hip. The adult Inuits of northern Canada often carry children in the large, furry hoods of their caribou-skin parkas. The Papuans of Papua New Guinea carry their children in nets woven from string made from the inner bark of trees. The net hangs from the carrier's head and the baby nestles against the carrier's chest. This allows nursing mothers to feed their children as they carry them.

An even less strenuous method of carrying small children came about in England in the 1700s, when the first baby carriage was introduced. The first baby carriages, designed to be pulled by dogs or Shetland ponies, were large and bulky. Naturally, the design has been streamlined over the past three centuries. Today, many varieties of baby carriages, or strollers, are available, from models that allow the child to lie down to those that keep the child upright. The invention has become more versatile as well, with collapsible versions, models that convert into car seats, and small trailer-like models, called prams, that can be pulled behind bicycles, as well as pushed. Prams are popular with bicyclists and joggers and can carry more than one child, as well.

A study conducted by Henry Ford Hospital in Detroit, Michigan, showed that jogging with a pram or stroller increased runners' heart rates by 3-5% over those jogging without strollers. The stroller also helped runners increase oxygen consumption by 2%. Women burned between 4-21 extra calories per 30-minute jog and men burned between 4-41 calories, depending on the make of the stroller.

History

English architect William Kent introduced the first baby carriage in 1733. Designed for the child of the third Duke of Devonshire, the carriage was shaped like a shell and was intended to be pulled by a dog or a Shetland pony. Over the years, the idea caught on for children not born of royalty, although a human-powered baby carriage was not developed in the United States until Charles Burton came up with the idea in 1848. Pedestrians, in America at least, did not take kindly to Burton's invention at first as inexperienced operators tended to run into them. Burton left such troubles behind, however, and moved to London where he continued to build baby carriages. Burton acquired a royalty-studded clientele overseas, designing baby carriages for Queen Victoria, Queen Isabel of Spain, and the Pasha of Egypt.

Improving upon Kent's and Burton's designs, Parisian E. Baumann devised the first collapsible baby carriage in 1906 after recognizing the difficulties in storing earlier models in cramped lodgings. Baumann dubbed his invention "The Dream."

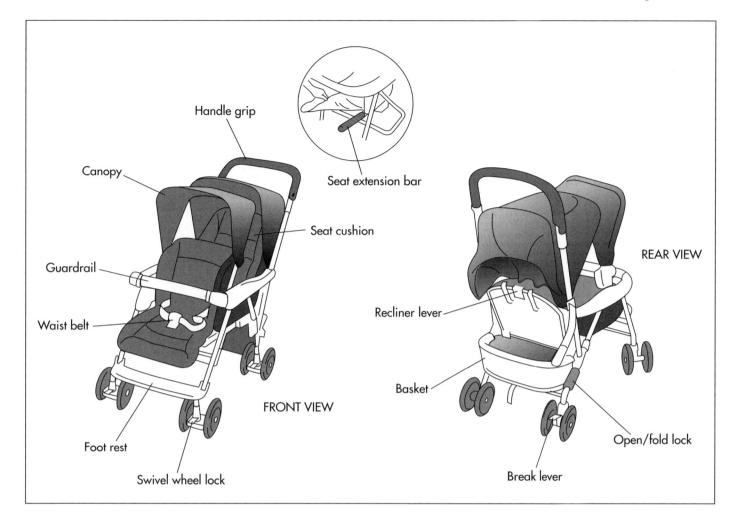

Handle grip

Canopy

Seat extension bar

Seat cushion

REAR VIEW

Guardrail

Recliner lever

Waist belt

Basket

Open/fold lock

FRONT VIEW

Foot rest

Swivel wheel lock

Break lever

The first pram appeared in Landau, Germany, just after World War I. The four-wheeled carriage featured two facing seats placed parallel to the axles and also included a hood that could be replaced by a sunshade in hot weather. Other accessories that became available included a mosquito net, an adjustable umbrella stand, and a spare wheel.

Today's prams are highly modernized and meet the needs of an active society. Most are equipped with small bicycle-type wheels and are aerodynamic. A three-wheeled version shaped like a needle-nosed race car is popular with joggers, while a small carriage-type unit can be attached to the bicycle, allowing cyclists to spend time with the kids and exercise at the same time.

Raw Materials

The primary materials used in manufacturing a baby stroller are aluminum or steel for

the frame, cloth for the seat and/or hood and rubber and plastic for handles and wheels.

The Manufacturing Process

Baby strollers come in many varieties and are typically manufactured on an assembly line. An upright, collapsible stroller would typically be made according to the following manufacturing process:

The frame

1 Steel or aluminum tubing is bent, using a press, to form the components of the frame.

2 The metal frame components are then dipped into an antioxidizing solution that prevents rusting and also helps the paint adhere to the frame.

Baby strollers come in many varieties and are typically manufactured on an assembly line. The primary materials used in manufacturing a baby stroller are aluminum or steel for the frame, cloth for the seat and/or hood and rubber and plastic for handles and wheels.

3 The frame components are spray-painted using a process known as powder-coating (this is the same process by which automobiles are painted). A powder-coating apparatus works like a large spray-painter, dispersing paint through a pressurized system evenly across the metal frame.

4 The frame components are pre-heated then coated with a powder resin finish.

5 The components are heated again to bake the finish onto the frame.

6 The various components of the frame are assembled using rivets, screws and/ or bolts.

The seat and hood

7 The seat and hood are cut from a large swath of material with an overhead cutting apparatus known as a cutting die. The die can cut 10-20 pieces at one time.

8 The seat and hood are stitched together on a large, mechanical sewing machine.

9 Trim is stitched onto the seat and hood, again using a large, mechanical sewing machine.

Final production

10 The seat and hood are attached to the assembled frame.

11 The wheels are attached to the assembled frame.

12 The final products are placed in polybags (plastic bags) and boxed for shipping.

Quality Control

Because the baby stroller must be safe for a small child to ride in, it is subject to strict quality control regulations. All baby strollers produced in the United States must meet American Society for Testing and Materials standard number 833-97. The components of the stroller are inspected by the supplier and certified as meeting the ASTM standards before being shipped to the manufacturer.

The Future

Americans are as active as ever and innovations in the stroller have contributed to such activity. Joggers, bicyclists, and casual walkers can all bring the kids along in different forms of prams and it is anticipated that future models will continue to be developed to meet the needs of the public, just as they have for the past three centuries.

Where to Learn More

Books

Bernhard, Emery and Durga. *A Ride on Mother's Back: A Day of Baby Carrying around the World.* San Diego: Harcourt Brace & Company, 1996.

Travers, Bridget, ed. *World of Invention.* Detroit: Gale Research Inc., 1994.

Periodicals

Creager, Ellen. "Rolling and Strolling the Distance." *Detroit Free Press*, May 12, 1998, pp. 13-14F.

—*Kristin Palm*

Bagel

Background

The bagel is a dense ring of bread, often rather bland, raised with yeast and containing almost no fat. In fact, the average bagel is about 4 oz (113.4 g) and 200 calories and contains no cholesterol (unless it is an egg bagel) and no fat (unless it is a specialty bagel such as cheese). The bagel's peculiar crustiness and density results from regulating the amount the yeast is allowed to rise so the bagel does not become too bready (not a desirable trait in a bagel). Whether handmade at home or with the aid of machinery in a bagel bakery, bagel dough is always boiled in water then baked until it is golden brown.

The popularity of the bagel is staggering. The appetite for bagels has increased 37% since 1994, and it is estimated that in the near future sales may increase as much as 7% over the previous year's figures to reach $840 million by the year 2000. Bagels are purchased by 46% of all consumers—and most purchase frozen bagels from their local supermarket. However, the fresh bagel market is expanding and the bagel bakery is visible in most communities. Once the product of small specialty bakeries in ethnic communities, the bagel is now seen on the menus of donut and cake bakeries and baked by restaurants all over the country.

Bagels are made in three different places. These include the large commercial bakery that bakes bagels then freezes them for transport across the region or country in plastic bags, the local bagel bakery that bakes fresh bagels for immediate consumption (from dough made there or made elsewhere), and at home. The fresh bagel bakery's traditional flavors-salt, egg, poppy seed, onion, plain, and rye-are now sold alongside new flavors like chocolate chip, spinach and cheese, cinnamon raisin, dried tomato and herb, and maple walnut. The cream cheese (the schmear in Yiddish), which often imparted the bagel with some pizzazz, now comes in many new varieties, including jalapeno and vegetable.

History

The history of the bagel is not clear. Bagel folklore tells us that the roll was devised as a tribute to Jan Sobieski, a Polish general, who saved Vienna from the invading Turks in 1683. As the triumphant hero rode through town, the grateful townspeople clung to his stirrups—called *breugels*. The king had a baker fashion bread in the shape of Sobieski's stirrups as a tribute. Eventually the stirrup-shaped breugel became round and was known as a bagel. Other stories indicate that the name comes from *beigen*, the German word for to bend, and could be a descendant of the pretzel. Still others believe the round hole was perfect for Russian and Polish bakers to skewer them on a long pole and walk the streets hawking their fresh bread.

Eastern European immigrants brought their skills as bagel bakers to the New World—by 1915 a bagel bakers union #338 had formed in New York City. Some of these bagel bakers and their apprentices began baking bagels in parts of the country—particularly the East Coast—when they moved out of the city. Harry Lender, a Polish immigrant, saw interest in the bagel and he and his son Murray baked bagels in quantity and packaged them for sale to supermarkets. In 1960 Dan

The appetite for bagels has increased 37% since 1994, and it is estimated that in the near future sales may increase as much as 7% over the previous year's figures to reach $840 million by the year 2000.

39

Thompson invented the first machine for making bagels. Until that time, all bagels were hand rolled. By 1962 the Lenders were baking and freezing their bagels and distributing their goods nationally. Throughout the 1960s and 1970s, bagels made a slow trek across the country via bagel entrepreneurs.

Now bagel bakery chains ranging from New York state to Colorado have sprung up to accommodate the needs of bagel connoisseurs. There are cookbooks devoted to making homemade bagels, including recipes for making bagels in bread makers.

Raw Materials

Ingredients for bagels vary tremendously according to who makes the bagel, whether it is made at home or in a commercial bakery, and the flavor of the bagel. Generally, all bagels must contain at least the following: water, salt, flour, and yeast. Water is needed to both soften the dry yeast and add moisture to the batter. Salt must be present to slightly inhibit the action of the yeast-without salt, yeast can rise too much. The flour the bagel baker uses matters little-various recipes call for bread flour, regular flour, bromated flour, whole wheat flour, and rye flour. Some call for a pinch of sugar to assist the yeast in rising.

Of course, the flavor of the bagel determines the remainder of the ingredients. This can vary from maple syrup, to jalapenos, to walnuts. The flavors are only as limited as one's imagination.

Design

The design and marketing of commercial bagel bakeries is extensive. Many bagel bakeries bring in competitors' bagels for blind survey by the general populous. These guests are served a variety of bagels and asked as series of questions regarding important characteristics of bagels including texture, chewiness (density), flavor, value, and fat and nutritional content. Answers to these questions help the bagel bakery determine the direction of product development. These bakeries cannot produce an infinite number of flavors within their facilities. Thus, these taste surveys help the bakeries determine the bagel flavors they will offer to

the public. Customer surveys and continual blind tastings insure that the companies can offer the consumer what he or she is looking for in a bagel.

The Manufacturing Process

The bagel franchises prepare bagel dough and bake them in a variety of ways. Essentially, the dough must be created with the raw ingredients, the yeast must rise, the bagels likely stored for some period of time before baking (as it is unlikely a new batch is made each time bagels are baked), and the then the bagels boiled and baked.

Some bagel bakery chains make the dough in regional commissaries in very large quantities-they mix the ingredients, form the bagels, activate the yeast, then cool it for storage until it is ready to be transported to small bakeries which produce the fresh, hot bagels. Thus, all but the baking of the bagels occurs at the regional commissaries. Here we'll look at this method of fresh bagel baking in which bagels are mixed and formed in one place and then sent to the store for baking.

Mixing the ingredients at the regional commissary

1 Many, but not all, bagel bakeries use fresh ingredients and fresh dough as opposed to frozen dough to make bagels. All ingredients, flour, salt, yeast, water, and various flavorings, are mixed together in a batch. The definition of a batch is determined by the amount of flour included. At one national bakery chain, a batch is defined as using 200 lb (90.8 kg) of flour, which makes 316 lb (143.5 kg) of dough.

2 Once the ingredients are mixed in a batch it must be closely monitored for temperature—too hot and the dough will rise too high too quickly or even be killed off because of the heat, too cool and it won't rise sufficiently. Since yeast is a living organism, it also has a limited life span as the yeast dough must be used within about 48 hours after mixing for the best bagel product.

The mixing can occur with a machine purchased for mixing bagel dough, such as a

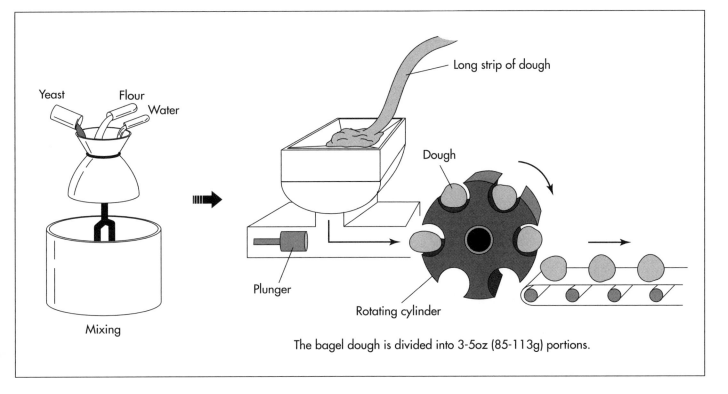

Yeast
Flour
Water
Long strip of dough
Dough
Plunger
Rotating cylinder
Mixing

The bagel dough is divided into 3-5oz (85-113g) portions.

spiral mixer. Mixing takes about 10-12 minutes and is carefully timed.

Dividing the dough

3 When the mixing is finished, the dough is taken from the mixer, put on a table and laid in strips. The large strips are fed into a divider, which perfectly portions out the amount of dough required to make an individual bagel, either 3-5 oz (85-113 g) of dough. The bagel is just a clump of dough at this point.

Forming the bagel shape

4 Next, the dough rolls under a pressure plate, which rolls the dough into a cigar-like form. This cigar shape drops onto the former, a round forming tube, which rolls it around and meshes the ends together to form the bagel shape. The forming machine shapes one bagel per every second.

5 A person picks the bagels off of the belt (coming at a fast and furious pace) and puts them onto a cornmeal-coated board. Those bagels deemed irregular or too small are rejected and remade. When the wooden board, which accommodates 25 bagels, is full, the bagels undergo a proofing process in order to get the yeast to rise.

Proofing the yeast and stopping the proofing

6 The boards are put into a proofer and subjected to heat (so the yeast will activate) and humidity (so the bagels won't dry out) for about 20 minutes. The bagels are then taken out of the proofer and subjected to a quick chill to about 20-25° F (-6.7- -3.9° C) for about 15 minutes in order to stop the yeast from activating too much and rising too much. Note that all bagel bakeries do not subject their bagels to this chilling process. Without this chill, the product is bready rather than chewy.

Ready for transport to the stores

7 The product is taken from the chilling process and placed in a holding cooler adjusted to just under 40° F (4.4° C) where it waits to be transported to smaller branches in the bakery system. There it is baked fresh on premises. The bagels are not permitted to rise in the holding cooler as they cannot rise under 40°. The boards (with 25 bagels each) are tagged as to their age, shrink - wrapped and when ready for shipment are placed in a transport rack (there are 30 boards on a transport rack) and put on re-frigerated trucks destined for a bakery.

Once the ingredients are mixed into a dough, the dough is separated into predetermined portions.

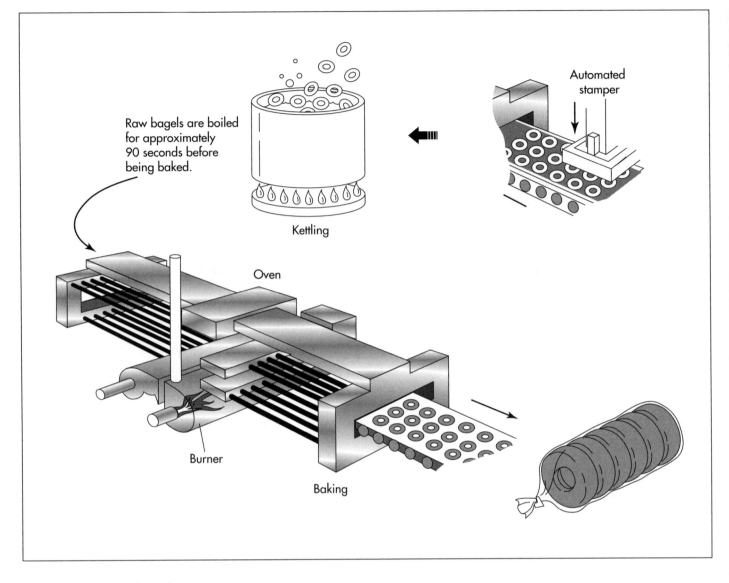

Raw bagels are boiled for approximately 90 seconds before being baked.

Kettling

Automated stamper

Oven

Burner

Baking

An automatic stamper forms the raw bagel. After kettling, the bagels are baked.

Distribution to the store

8 Once the transport racks come into the store, they are ready to be baked. Their tag, which indicates flavor and age, tells the bagel bakers how much time they have before the life span of the yeast is over and that the dough is not usable.

Kettling

9 Each bakery has at least one huge kettle filled with boiling water and malt for reactivating the bagels. Water in these kettles must be at a rolling boil. The bagels (usually one board or 25 bagels at a time) are dropped into the kettle. In this hot kettle, the dormant yeast is reactivated. After about 90 seconds, the bagel comes up to the surface of the water in a "float." Kettling with malt in the water helps put on a hard crust and retards drying.

Baking

10 These floating bagels are scooped out and laid out onto sticks, which are burlap-covered aluminum. The toppings (poppy seeds, sesame seeds, salt, etc.) are sprinkled on the top of the bagel, which is face up on the sticks. The sticks are then flipped out onto the shelves of the bagel oven; thus, the bottom of the bagel (the side without topping) is face up in the oven. The bagels take about 20 minutes to dry out and cook. The bagel cook generally eyes the bagels and decides when they are finished. Huge wooden paddles called peels then lift the baked bagels off the shelves into the wire bins for purchase.

Quality Control

Perhaps most important for quality control is that all ingredients are up to the minimum standards required by the franchise or bakery. Good quality flour and yeast are of the utmost importance. Second, temperatures for water, for the proofer, the cooler, and even the temperature of the flour before mixing must be precisely monitored or yeast will not activate properly. Third, the life span of the yeast must dictate handling priorities. As one baker put it, bagels just mixed and proofed are like "teenagers" with robust yeast waiting to rise; however, bagels that were proofed nearly 48 hours prior are like "90-year old grandpas"—they have little "zing" in them and may not make the best bagels. Thus, it is imperative to know the age of the raw product as indicated on the tags attached to the boards. Lastly, the bagels are only as good as the experienced bagel baker who must pull inferior or malformed bagels off moving belts or who monitors baking regardless of what the timer reads.

Where to Learn More

Books

Bagel, Marilyn and Tom. *The Bagel Bible*. Old Saybrook, CT: Globe Pequot Press, 1992.

Mellach, Dona Z. *The Best Bagels are Made at Home*. San Leandro, CA: Bristol Publishing Enterprises, 1995.

—Nancy EV Bryk

Baseball Cap

In 1993, ball boys at Wimbledon were first permitted to wear baseball caps on the court—quite a departure from Wimbledon's bare-headed tradition. These tennis enthusiasts knew that the caps are not just for decoration or camaraderie—they keep glare from the eyes and keep the heat off the head.

Background

A baseball cap is a soft hat that consists of a soft fabric crown sewn of several sections of fabric and a visor that protects the eyes from the sun. Some special order caps are made to fit the wearer in specific sizes, but most of the mass-produced models have a plastic extender in the center back that can be make bigger or smaller according to the wearer's needs. A fabric-covered button sits on the top of the crown where all the fabric sections converge. Team logos may be embroidered on the center front of the crown, or the letters associated with the team may be applied on center front.

Baseball team uniforms, whether amateur or professional, always include baseball caps. They help identify players on the field and are essential in order to keep the sun from players' eyes. There are currently two different styles of caps available today. One which has a rather tall, boxy crown and is often horizontally striped (the "old style" cap still popular with those who love the traditions of the game) or the conventional soft crown of six or eight triangular-shaped sections of fabric. Interestingly, most baseball cap manufacturers still make baseball caps in wool, as they have for over a century, as well as more easily washed fabrics such as cotton or cotton-synthetic combinations.

History

An ancestor of the game we know as baseball was familiar to some in the 1700s. Called "base ball," both the British and the Americans knew of a game with sticks and balls. Soldiers during the American Revolution played a game similar to baseball. A game called rounders, which was not too dissimilar from the game we know as baseball, was popular in Great Britain and the New World in the early 1800s, however the runner was considered "out" if they were physically hit with the ball. As early as the 1840s, amateur teams were formulated in the East Coast; during the Civil War, when New York and New Jersey soldiers shared their understanding of the game, a "national style" of baseball was established. Baseball, as we know it today, was played just after the American Civil War.

From the beginning of the development of amateur teams after the Civil War, the baseball uniform was considered an extremely important part of the game. Players and managers understood that the uniforms imparted the team members with a sense of pride, legitimized this new game of baseball much as police uniforms help sanction the work of policemen, and imbued the team members with a sense of camaraderie. Photographs of teams from the later nineteenth century shows each member of the team proudly wearing matching uniforms including the baseball cap.

Catalogs of sports equipment from the 1880s show baseball uniforms that include the shirt, the trousers, and the cap. Three different styles of caps are featured in these catalogs. A boxy cap made from horizontal strips of fabric was referred to as the college cap. A conventional triangular-sectioned cap with a front visor was called the Boston cap, and a soft crown cap, resembling a rain hat, with a brim running all the way around the hat was called the base ball hat.

The sectional cap we know as the baseball cap resembles other soft sports caps popular

44

in the 1800s and may be a variation of the yachting cap or a generic outdoor sport cap. These soft outdoor sports caps shielded the eyes, were relatively easily washed (were easier to wash than stiffened and blocked wool felt derbies), were popular throughout the early twentieth century, and remain relatively unchanged today. However, today the baseball cap can be adjustable or made-to-order.

Raw Materials

Baseball caps are still often made of solid-color wool. Some caps not used expressly for baseball, but are of the baseball cap style and have company or other logos on them are of cotton or cotton-synthetic combinations. The visors of these caps always include some kind of stiffening. Sometimes this stiffening is buckram, other times it is a plastic insert (the material depends on the company). Some caps are stiffened in center front to accommodate the embroidered logo and this may be accomplished by sewing various kinds of stiffeners in the two front panels. Sweatbands may be cotton or even thin leather depending on the company or product.

Design

Because the design of baseball caps has varied little, the design often changes only in the color of fabric used for the hat (based on the needs of the team) and the logo on the center front of the hat. Some companies have tinkered with the basic baseball cap design in order to get a better-fitting cap on the head (one company reduced the sections of the soft crown from eight to six sections for a better fit). Still others experiment with the insert into the visor in order to render the visor flexible but sturdy. Elaborately embroidered logos, some of which include over 30,000 stitches in order to produce the logo, are carefully designed with the aid of a computer in order to perfectly and consistently sew that logo onto the front of the caps.

Interestingly, a few companies actually reproduce old-style baseball caps that were popular over 100 years ago. These companies cannot take apart the old caps in order to draft a prototype pattern (sometimes of plain cardboard) but must work from whole pieces that are visually examined and measured in order to make these patterns.

The Manufacturing Process

1 The fabric layers—wool, cotton, or a cotton-synthetic mix (depending on the company and the style of hat)—are cut. Many dozens of layers of fabric are carefully laid on top of each other, and then they are cut at once using a fabric-cutting saw.

2 Some cap designs require a mesh inner layer behind the two front panels. This mesh acts as a stiffener in these front panels in order for them to stand up to the stitching requirements of the embroidered logo. The mesh is put against the back of the panels before the panels are sewn to one another.

3 The sections of the soft crown, generally of long triangular shaped gores, are sent to the eyelet department where a machine pierces each panel creating a small hole and binding the hole completely with thread. The number of eyelets in each gore or section of crown varies according to the manufacturer, however there is at least one eyelet in each section but may be as many as forty or fifty in total. These eyelets serve as vents so the perspiration and heat that builds up under the cap can be released through the eyelets.

4 The two front panels that are to receive embroidery are then sent to the embroidery department. Here a computer-driven embroidery machine embroiders a logo or an entire word onto one or two of the front panels. One company reveals that these logos are very complex and precise—some logos require 8,000 stitches, others require 31,000 stitches. Still other companies apply fabric letters or other applied insignia to the front panel.

5 The panels of the crown, generally either six or eight panels, are then sent to the sewing department where they are stitched together. With stiffening and logo, these wool panels may be rather heavy to sew through and require human skill to sew the panels together and an industrial grade sewing machine. When all six or eight panels are sewn together, the soft crown is completed but the entire hat is not. This hat

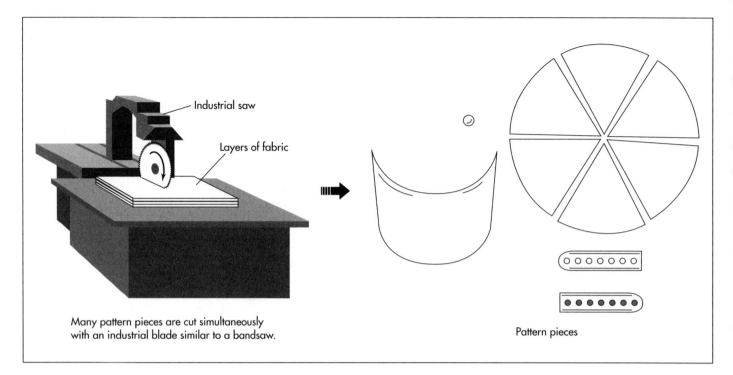

Many pattern pieces are cut simultaneously with an industrial blade similar to a bandsaw.

Industrial saw

Layers of fabric

Pattern pieces

A baseball cap is a soft hat that consists of a soft fabric crown sewn of several sections of fabric, a visor that protects the eyes from the sun, and a plastic extender in the center back that can adjust the size of the cap.

without a visor is referred to in the industry as a "beanie."

6 Then, these beanies are sent to the binders or the binding department at which the raw seams of the soft crown are covered or hidden with a binding tape, generally cotton, that is applied over the raw edges of the crown. This tape gives the hat a finished look (no raw edges are seen when one examines the inside of the cap) and ensures that the seams won't unravel due to hard wear, perspiration, or washing.

7 A galvanized steel button is self-covered (covered over in the same color as the rest of the cap) and it is then applied at the dead center of the baseball cap on top of the beanie crown at the place in which all the sections of the cap converge.

8 Visors are die-cut according the desired size and then sewn onto the cap next. Some companies make the visor of two pieces with a stiffener such as plastic in the center; other might put a thin stiffener inside and stitch the visor a few times for strength and to prevent the stiffener from moving around and bunching up in one spot. The proper color of visor and cap are matched up and sewn together. (Some ball players roll or curl their visors, "telescoping" them, so that when they're searching for the fly ball

or grounder that rolled visor keeps the glare out from the player's eyes. Thus, some companies experiment with inserting various materials into the visor to get the best telescope effect.)

9 Those companies that offer the "one size fits all" hat will sew the adjustable plastic band in the back of the cap. Other companies that make hats to order according to head size will only sew the front sections of the cap together and will vary the size or breadth of the back cap sections according to the size required by the customer.

10 Finally, a sweatband of some sort is sewn onto the inside of the cap. This is done on industrial grade sewing machines as well and may include placing buckram (a thick, stiff mesh fabric) behind a sweatband such as inexpensive cotton or even soft leather. A label may be added at this time as well. A label may be sewn in if needed.

11 Some companies add different steps to ensure a superior product. For example, one company blocks and steams the finished, sewn product so that its shape is "set." They claim it renders a great, long-lasting, attractive fit. Some companies add a step in which they stitch along different edges so that fabric doesn't roll back on itself and appear unsightly.

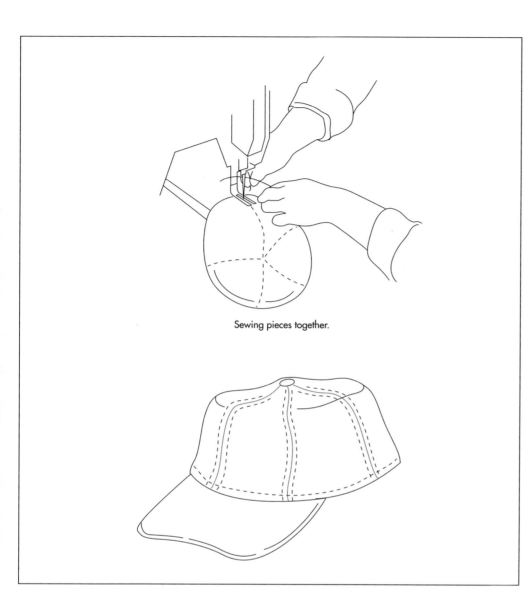

Sewing pieces together.

12 The cap is now ready to be stored or shipped as needed.

Quality Control

A remarkable number of these steps are accomplished by hand using a sewing machine. Thus, the sewing machine operative can and does notice any quality inconsistencies in the product. Of course, all raw materials are held to the standards the company requires (including color fastness-important if sportsmen are playing in the rain or are perspiring in the caps).

Where to Learn More

Books

A Century of Baseball. Macmillan Publishing Co., 1976.

Goldstein, Warren. *Playing for Keeps: A History of Early Baseball*. Ithaca, NY: Cornell University Press, 1991.

—*Nancy EV Bryk*

Basket

The craft of basketry gave rise to pottery making because baskets were used as molds for some of the earliest pots. Consequently, the history of pottery and basketry, as unearthed and decoded by archaeologists, is irrevocably interwoven.

Background

The basket is one of humankind's oldest art forms, and it is certainly an ethnic and cultural icon filled with myth and motif, religion and symbolism, and decoration as well as usefulness. Basketry, in fact, encompasses a wide range of objects from nearly rigid, box-like carriers to mesh sacks. Baskets range in size from "burden baskets" that are as much as 3 ft (91.44 cm)in diameter to tiny collectibles 0.25 in (0.64 cm) in diameter.

Some baskets are manufactured by machines, however part of the tradition is that baskets are defined as receptacles that are woven by hand of vegetable fibers. Although baskets may have distinct bottoms and tops, they are essentially continuous surfaces. They are woven in that their fibers are twisted together, but, unlike the weaving of textiles, tension is not placed on lengthwise threads (the warp) because the fibers are less flexible than threads.

Baskets are part of the heritage of nearly every native people, and types of construction differ as radically as other customs and crafts. Uses for baskets may be the most uniting feature. Dry food is gathered, stored, and served in baskets; liquids are also retained in baskets that have been waterproofed. Basket-making techniques are used for clothing, hats, and mats. Openwork baskets are made to function as filters (for tea in Japan) and as sieves and strainers. Their variety and clever construction also makes baskets desirable as decorations in primitive cultures as well as modern homes.

History

Baskets are the children of the gods and the basis of our earth, according to the ancient Mesopotamians. They believe that the world began when a wicker raft was placed on the oceans and soil was spread on the raft to make the land masses. Ancient Egyptian bakers used baskets to hold baked loaves of bread. The single, most famous basket may well have been the basket made of bulrushes and mud in which the baby Moses was floated to safety. All ancient civilizations produced baskets; the Romans cultivated willow for their baskets, and the Japanese and Chinese also counted basketry among their many handicrafts with ancient origins.

The craft of basketry gave rise to pottery making because baskets were used as molds for some of the earliest pots. Consequently, the history of pottery and basketry, as unearthed and decoded by archaeologists, is irrevocably interwoven. Where the vegetable fibers have not survived, many pots that show the patterns of the baskets used to mold them have been found.

The Native Americans may well have left the greatest legacy to the world of baskets. The Indians of Arizona and New Mexico made basket-molded pottery from 5000 to 1000 B.C. as part of the earliest basket heritage. Their baskets (many of which have survived in gravesites) are heralded as a pure art form and one that was created not only by a primitive people but also by women. Basketry extended into the making of many other materials the Indians used daily including fishing nets, animal and fish snares, cooking utensils that were so finely woven that they were waterproof, ceremonial costumes and baskets, and even plaques. In the Northwest, the Tlingit and Chilkat made twined baskets from the most delicate of fibers. In the Southwest, the Hopi, Apache, and other Pueblo tribes made coiled baskets with bold decorations and geometric patterns of both dyed and natural fibers.

In the late 1800s, the basketry of Native Americans became popular as decorative objects with the disadvantage that there were fewer Indian craftspeople remaining to meet the demand. In 1898, after the Spanish American War, the Philippines, which also had a strong basket-making tradition, were governed by the United States. Rural dwellers grew their own basket-making materials and manufactured baskets for sale in the cities. The mutual need for baskets in the United States and the strengthening of the economy of the Philippines caused schools with classes in basket weaving to be established. The only books on the subject were about the baskets made by Native Americans, so the schools taught traditional Indian basketry to the Filipinos. Eventually, native Filipino weavers became the teachers as well, and both broad ranges of styles found a new homeland for manufacture and a ready market in the United States. The Philippine Islands remain a major basket-making center today. Basket weaving has never been found suitable to mechanization, but standardization of hand methods and concentrated production centers and facilities produce uniform, high-quality products.

Raw Materials

Raw materials include a wide range of plant fibers including roots, cane, twigs, and grasses; reeds, raffia, and basket willows may be the best known. Concentrated cloth dyes are also used in some types of manufacture, and vegetable dyes are sometimes made by hobbyists to reproduce unique colorations imitating historic baskets. Wood is also used for some designs, particularly when the type of basket needs a solid bottom and for some types of handles. Other than raw materials, the basket maker needs tools like saws, awls, planes, knives, and beaters for hammering or bending pieces of willow. A tub is required for soaking fibers. If coiled baskets are to be made, sewing tools like blunt tapestry needles and thread are required. The manufacturer also needs patterns or designs. For the hobbyist, many of these items can be purchased in basket-making kits.

Historically, most Native American baskets have been made with willow (which is, in fact, the most popular basket-making material worldwide), twigs, and native grasses. Raffia and rattan have been substituted for these, with raffia taking the place of the grasses and rattan substituting for the more rigid fibers. Raffia is the fiber of the raffia palm, which is native to Southeast Asia. It produces durable, clean strands and can easily be dyed. Rattan is also a tropical palm; its leaves and stems are used in basket making, and it is often called reed or wicker. Rattan does not accept dye as well, and its fibers are hard to work. Usually, it is soaked and woven while the fibers are still damp.

Design

Every basket has a character that is largely determined by the kind of fiber used to make it. Design, therefore, may depend on the available fibers, or, conversely, to produce a particular design, appropriate fibers need to be purchased or found. Fibers are round, flexible, or flat. Round rods are usually woven among other round rods. Similarly, flat strips can be woven together or twisted around stiff rods. Grasses, crushed stems, or other flexible fibers are wrapped around each other to form a coil then the coil is stitched to itself in a rising spiral to form the basket sides. The designer, therefore, has determined what fibers are available and plans the basket accordingly.

Designs can be based on existing baskets, photos of historic types, a particular purpose or use for the basket, or a size and shape required for practical uses or desired for decorative ones. Another aspect of design is any pattern or coloration that may be worked into the shape of the basket. Again, materials, their natural colors, and their susceptibility to dyeing need to be considered.

The Manufacturing Process

Many baskets are made in very standard shapes and sizes, some unique to various

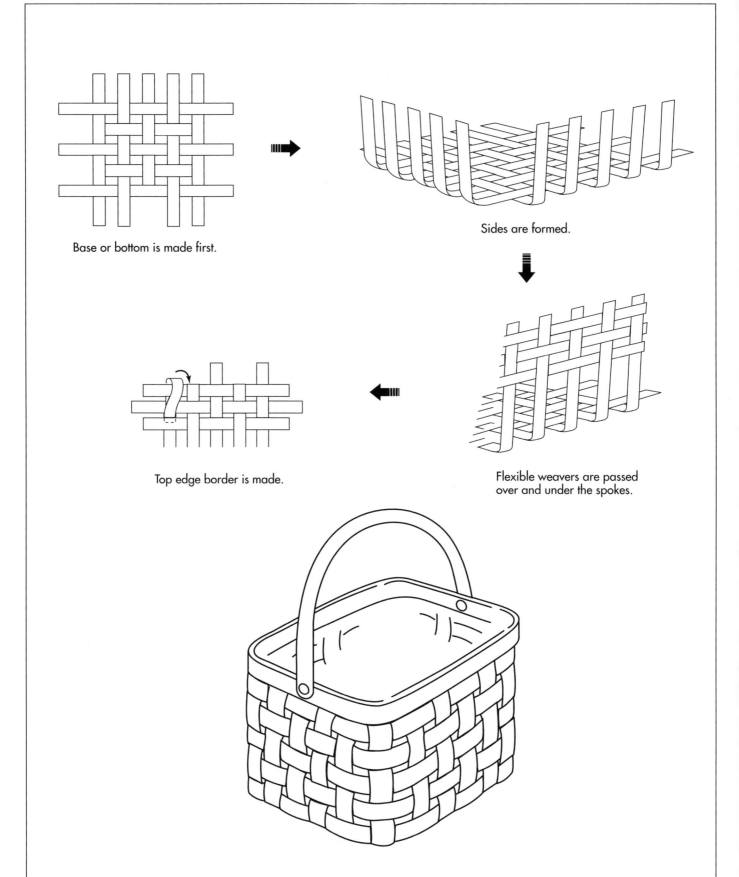

Base or bottom is made first.

Sides are formed.

Flexible weavers are passed over and under the spokes.

Top edge border is made.

parts of the world, and they look so much alike that they could have come from machines. They are indeed mass-produced objects but made by hand.

1 The process begins by choosing a design or standard pattern including shape and size. Materials are also gathered or purchased, and the necessary tools for working those materials are assembled. If the fibers are such that they need to be soaked, then soaking is done in advance of basket making, depending on the nature of the fiber. Fibers are also dyed in advance of weaving or coiling.

2 If the design calls for a wood base, the base is shaped, and holes are bored in the wood to accommodate the spokes forming the sides of the basket.

3 A basket is built from the ground up. Its base or bottom is made first. For a round basket with a flat bottom (as an example of any of hundreds of types of baskets that may be manufactured), the base is made by laying out a series of spokes that are stiff and work like rods to support more flexible woven material. Other rods called weavers are woven in and out among the spokes; the weavers are lighter, thinner, and more flexible, so that they can be woven and so they won't be strong enough to distort the spokes.

4 The sides of this kind of basket can be formed in either of two ways. Initially, the spokes for the base can be cut to be long enough to form the sides as well. When the base is finished, the spokes are soaked to soften them, squeezed with pliers at the perimeter of the base, and then bent up to form the sides.

5 The sides are also formed by cutting side spokes and weaving them down through the base perimeter fibers and then up again so they form side spokes. Side spokes are essential if the base spokes are large. The sides are then woven with flexible weavers that are passed over and under the side spokes. Again, these weavers need to be smaller than the material forming the spokes so the spokes are not distorted. The side spokes are longer than the finished basket is tall; the remaining ends of the spokes are used to finish the top edge of the basket with a border. The spoke ends need to be soaked before the border can be made so the spokes can more easily be woven in and out of each other and the ends turned down into the basket sides.

6 The handle of the basket is chosen of the best available reed to be strong, durable, attractive, and relatively smooth to the touch so it can be held. The ends of the handle reeds are soaked in water and threaded down into the sides of the basket. The overlap has to be long enough to prevent the handle from pulling out of the sides when the basket is filled and used.

7 If the basket has a lid, the lid is made in the same manner as the base, but the rods and weavers should be of the same sizes as those in the sides of the basket to match the appearance of the basket.

Quality Control

The individual basket weaver may set the standards for making a particular basket. In some cases, basket styles are somewhat rough or primitive and may allow for quality variations; for other styles, a high level of detail or conformity is required, and irregularities in materials or workmanship will be readily apparent. Where baskets are mass-produced, the quality is protected by working from a standard pattern or design, selecting uniform materials, and cutting or preparing the materials in quantities and to a quality standard. A supervisor may oversee a number of basket weavers and reject imperfect baskets; however, as in the case of most handicrafts, basket weavers take pride in their profession and demonstrate their skills in each product. Even mass-produced baskets are prized for their uniqueness, so some variations are to be expected and treasured.

Byproducts/Waste

Byproducts do not usually result from basket manufacture, although a basket maker may produce several different styles to make economical use of materials. Fibers are often imperfect, and there are many trimmings that comprise the waste from basket weaving. Some fibers can be finely ground and composted.

The Future

As packing and transporting devices, baskets have been replaced with cardboard cartons, synthetics, woods like plywood, and lightweight metal alloys. Despite the extreme decline in practical uses, the appreciation of handcrafted items has continued to grow. Baskets are widely used as decorations in the home. Baskets are also treasured as collectibles with areas of specialization including historic baskets, baskets of various forms, or the baskets of a particular culture. Among those that are particularly collectible are the baskets made by the Shakers, a religious community that immigrated to the United States and made baskets until about 1925. Shaker sewing baskets and baskets made of split ash and shaped to carry pies and cakes are highly prized.

Overall, the demand for baskets seems to remain constant. Companies that produce baskets find their products are in demand, but there is a shortage of worker trainees. Individual basket makers can take a wide variety of classes to learn designs and methods of meeting the specialized demand for traditional, detailed baskets. Collectors and decorators should not, however, view baskets as inexpensive. Cultivation of basket willows and other plants used for basketmaking is considerably more limited as the availability of agricultural land diminishes, and skilled weavers all over the world have recognized the value of their labor and their products.

Where to Learn More

Books

Couch, Osma Palmer. *Basket Pioneering.* New York: Orange Judd Publishing Company, Inc.,1940.

Rossbach, Ed. *The New Basketry.* New York: Van Nostrand Reinhold Company, 1976.

Wright, Dorothy. *The Complete Guide to Basket Weaving.* New York: Drake Publishers Inc., 1972.

Other

Clarson Enterprise, Inc. http://www.philexport.org/clarson/clarson.htm.

In a Hand Basket. http://www.inahandbasket.com/.

Marion Steinbach Indian Basket Museum. http://www.tahoecountry.com/nlths/baskets.htm.

The Weaving Network. http://www.weavenet.com.

—*Gillian S. Holmes*

Bath Towel

Background

Bath towels are woven pieces of fabric either cotton or cotton-polyester that are used to absorb moisture on the body after bathing. Bath towels are often sold in a set with face towels and wash cloths and are always the largest of the three towels. Bath towels are generally woven with a loop or pile that is soft and absorbent and is thus used to wick the water away from the body. Special looms called dobby looms are used to make this cotton pile.

Bath towels are generally of a single color but may be decorated with machine-sewn embroidery, woven in fancy jacquard patterns (pre-determined computer program driven designs) or even printed in stripes. Since towels are exposed to much water and are washed on hot-water wash settings more frequently than other textiles, printed towels may not retain their pattern very long. Most towels have a two selvage edges or finished woven edges along the sides and are hemmed (cut and sewn down) at the top and bottom. Some toweling manufacturers produce the yarn used for the toweling, weave the towels, dye them, cut and sew hems, and ready them for distribution. Others purchase the yarn already spun from other wholesalers and only weave the toweling.

History

Until the early nineteenth century, when the textile industry mechanized, bath toweling could be relatively expensive to purchase or time-consuming to create. There is some question how important these sanitary linens were for the average person—after all, bathing was not nearly as universally popular 200 years ago as it is today! Most nineteenth century toweling that survives is, indeed, toweling probably used behind or on top of the washstand, the piece of furniture that held the wash basin and pitcher with water in the days before indoor plumbing. Much of this toweling was hand-woven, plain-woven natural linen. Fancy ladies' magazines and mail order catalogs feature fancier jacquard-woven colored linen patterns (particularly red and white) but these were more likely to be hand and face cloths. It wasn't until the 1890s that the more soft and absorbent terry cloth replaced the plain linen toweling.

As the cotton industry mechanized in this country, toweling material could be purchased by the yard as well as in finished goods. By the 1890s, an American housewife could go to the general store or order through the mail either woven, sewn, and hemmed Turkish toweling (terry cloth) or could purchase terry cloth by the yard, cut it to the appropriate bath towel size her family liked, and hem it herself. A variety of toweling was available—diaper weaves, huckabacks, "crash" toweling— primarily in cotton as linen was not commercially woven in this country in great quantity by the 1890s. Weaving factories began mass production of terry cloth towels by the end of the nineteenth century and have been producing them in similar fashion ever since.

Raw Materials

Raw materials include cotton or cotton and polyester, depending on the composition of the towel in production. Some towel factories purchase the primary raw material, cotton, in 500 lb (227 kg) bales and spin them

A typical towel-weaving machine has 350 shuttle insertions in a single minute—nearly six shuttles fired across each second. In one small towel-making factory, 250 dozen bath towels can be made in one loom in a single week—and there are 50 looms in the factory.

with synthetics in order to get the type of yarn they need for production. However, some factories purchase the yarn from a supplier. These yarn spools of cotton-polyester blend yarn is purchased in huge quantities in 7.5 lb (3.4 kg) spools of yarn. A single spool of yarn unravels to 66,000 yd (60,324 m) of thread.

Yarn must be coated or sized in order for it to be woven more easily. One such industry coating contains PVA starch, urea, and wax. Bleaches are generally used to whiten a towel before dyeing it (if it is to be dyed). Again, these bleaches vary depending on the manufacturer, but may include as many as 10 ingredients (some of them proprietary) including hydrogen peroxide, a caustic defoamer, or if the towel is to remain white, an optical brightener to make the white look brighter. Synthetic or chemical dyes, of complex composition, which make towels both colorfast and bright, may also be used.

Design

Most towels are not specially designed in complex patterns. The vast majority is simple terry towels woven on dobby looms with loop piles, sewn edges at top and bottom. Sizes vary as do colors depending on the order. Increasingly, white or stock towels are sent to wholesalers or others to decorate with computer-driven embroidery or decorate with applique fabric or decoration. This occurs in a different location and is often done by another company.

The Manufacturing Process

Spinning

1 As mentioned above, some factories spin their own yarn for bath towels. If this is done at the factory, the manufacturer receives huge 500 lb (227 kg) bales of either high or "middling grade" (of medium quality) cotton for conversion into yarn (quality depends on the manufacturer and quality of the towel in production). These bales are broken open by an automated Uniflock machine that nips a bit off the top of each bale, opens it up and then lays it down. The Uniflock opening machine blends the cotton fibers together by repeatedly beating it so

impurities fall out or are filtered out (these bales contain many impurities within the raw cotton). The more pure fibers are blown through tubes to a mixing unit where the cotton is blended together before they are spun. Higher quality towels use cotton with fibers that are blended together three times before spinning. In some factories, the cotton is blended with polyester during this blending process.

2 The mixed fibers are then blown through tubes to carding machines where revolving cylinders with wire teeth are used to straighten the fibers and continue to remove impurities before spinning. The cotton fibers, while not yet yarn, are shaping up into parallel fibers in preparation for spinning.

3 These parallel fibers are then condensed into a sliver—a twisted rope of cotton fibers. These slivers are sent into another machine in which they are blended again and sent between other rollers for straightening. The ultimate goal is long, straight, parallel fibers because they produce stronger yarns. (Stronger yarns require less twisting which also produces strong yarns but makes them less soft and absorbent.) The fibers are wound on a large roll and sent on a cart and fed into the combing machine.

4 Fibers are combed here, further straightening the fibers with a finer set of wire teeth than used on the carding machine. Combing removes the shorter fibers, which are coarser and woollier, leaving the finer, longer, silkier cotton fibers for spinning into yarn. Once combed, the fibers are formed into a twisted rope sliver again.

5 The slivers travel to roving machines where the fibers are further twisted and straightened and formed into rovings. The roving frame also slightly twists the fibers. The result is a long roving of cotton, which is then wound onto bobbins in the final step before spinning.

6 Now the roving is ready for spinning. The bobbin is spun on a ring-spinning machine, which mechanically draws out or pulls the cotton roving out into a single strand. The fibers essentially catch one another to form one continuous thread and twists the thread slightly as it is pulled or

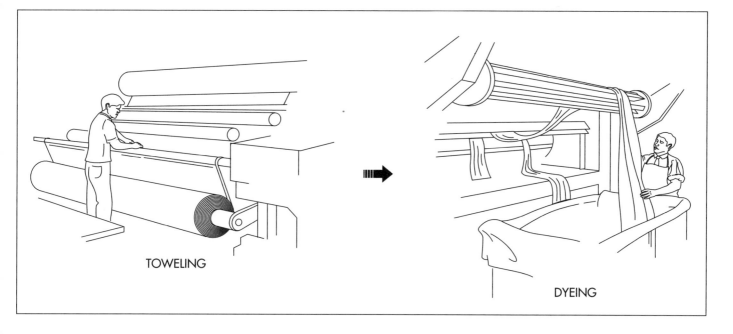

TOWELING

DYEING

spun. Once the yarn is spun, it is automatically wound on large wheels that resemble rounds of cheese when full of thread.

Warping

7 Warp is longitudinal threads in a piece of woven material that are tightly stretched or warped on a beam. Latitudinal threads called weft or filler are passed under and over the warp to form the fabric. The large spools of just-spun cotton are ready to be warped or wound on a beam that will be inserted into the loom for weaving. If the yarn is purchased, the 7.5 lb (3.4 kg) spools are readied for warping. A warping beam is then warped in which threads are anchored and wrapped to a large beam in hundreds of parallel rows. Different towel widths require different numbers of warp threads.

8 These huge beams, full of wrapped warp threads, are placed into a rack that holds up to 12 beams and sized in preparation for weaving. The threads must be sized or stiffened to make the piece easier to weave. PVA starch, urea, and wax are rolled onto and pressed into the yarn. The threads are then run over drying cans—Teflon-coated cans with steam heat emanating from within. This helps to dry the warp threads quickly. (1,000 warp ends are pulled over nine cans to dry.) These beams, with coated threads, are now sent to the looms.

Weaving

9 The beams are picked up by a pallet jack or hydraulic lift truck and transported to looms. These looms vary in width but may be as narrow as 85 in (216 cm) or as wide as 153 in (389 cm). (Not surprisingly, the wider the loom, the slower the weaving as it takes longer for weft threads to cross the warp.) The beams are lifted onto the looms mechanically with a warp jack, which can bear the weight and size of the beam.

10 Towels are woven on dobby looms, meaning each loom has two sets or warp and thus two warp beams—one warp is called the ground warp and forms the body of the towel and the other is called the pile warp and it produces the terry pile or loop. Each set of warp threads is carefully fed through a set of metal eyes and is attached to a harness. (Harnesses are separate, parallel frames that can change in their vertical relationships to one another.) These harnesses mechanically raise and lower these warp threads so that the weft or filler can be passed between them. The intersection of the warp and weft is woven fabric.

The filler yarn is programmed so that it is loosely laid into the woven fabric. When this loose filler is beaten or pressed into the fabric, the slack is pushed up becoming a little loop.

Once the toweling is made, it is wound on an off-loom take-up reel. It is then transported to bleaching as huge rolls of fabric and put into a water bath with bleaching chemicals such as hydrogen peroxide, caustic defoamers, and other proprietary ingredients. All toweling must be dyed pure white before it is dyed any color.

After being dyed, the towel is hemmed and cut into standardized sizes.

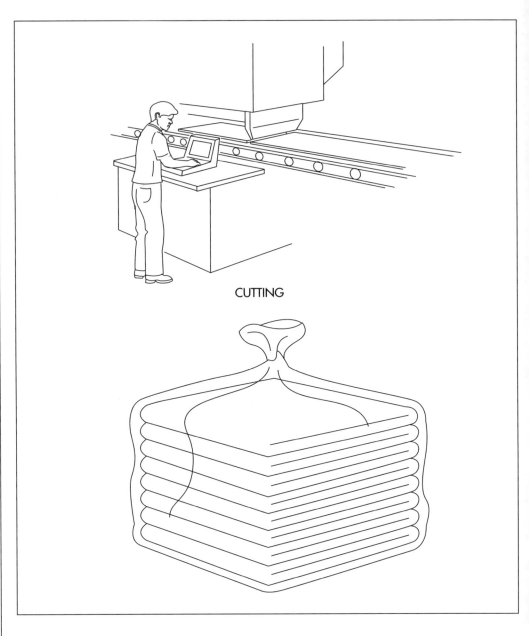

CUTTING

Shuttles, which carry the filler threads, are truly shot across these large looms at top-speeds—these towel-making looms may have 18 shuttles fired across the warp from a firing cylinder. One shuttle follows right behind the next. As soon as the one shuttle shoots across the warp threads, the shuttle drops down and is transported back to firing cylinder and is shot across again. A typical towel-weaving machine has 350 shuttle insertions in a single minute—nearly six shuttles fired across each second. Thus, towels are woven very quickly on these large mechanized dobby looms. In one small towel-making factory, 250 dozen bath towels can be made in one loom in a single week—and there are 50 looms in the factory.

Bleaching

11 Once the toweling is made (it is one long terry cloth roll and has no beginning or end), it is wound on an off-loom take-up reel. It is then transported to bleaching as huge rolls of fabric and put into a water bath with bleaching chemicals such as hydrogen peroxide, caustic defoamers, and other proprietary ingredients. All toweling must be dyed pure white before it is dyed any color. The wet toweling laden with chemicals is then subjected to tremendously high temperatures. The heat makes the chemicals react, bleaching the towel. The roll is then washed at least once and as many as three times in a large wash-

er to get all chemicals out of the toweling. The toweling is dried, and if it is to remain white toweling, it is ready to be cut at the top and bottom, lock-stitched sewn, and have a label attached (all of this is done with one machine).

Dyeing

12 If it is to be dyed, the large, dried uncut rolls are taken to large vats of chemical dyes, which have proven over time to provide colorfast toweling after extensive residential laundering. After being immersed in the vat, the toweling is removed and pressed between two heavy rollers which forces the dye down into the toweling. A thorough steaming sets the color. The toweling is again steam-dried, fluffed in the drying process, and then the dyed towels are ready for cutting, hemming, and labeling.

Cutting, folding, and packaging

13 Final visual inspection of the cut and hemmed towels occurs and they are handfolded and conveyed to packaging, where automatic packaging equipment forms a bag around the towels and UPC labels are attached to the bags. These packaged towels are sent to the stock room, awaiting transport out of the plant.

Quality Control

Towels are rigorously checked for quality control throughout the production process. If yarn is purchased, it is randomly checked for weight and must be the standard established by the company (lighter yarn spools indicate the yarn is thinner than desired and may not make as sturdy toweling). Bleach and dye vats are periodically checked for appropriate chemical constitution.

During the weaving process, some companies pass the cloth over a lighted inspection table. Here the weavers and quality inspectors monitor the towel for weaving imperfections. Slightly unevenly woven towels may be straightened out and touched up. But those that cannot may be labeled "seconds" or imperfect or completely rejected by the company. As in all aspects of the process, visual checks are a key to quality control—all involved in the process understand minimum standards and monitor the product at all times.

Byproducts/Waste

Potentially harmful byproducts are often mixed in the water that is used to bleach, wash, and dye the towel fabric. Particularly, the bleaching process includes ingredients (peroxides and other caustics) that cannot be discharged untreated into any water supply. Many toweling factories run their own water treatment plants to insure that the water the plant discharges meets minimum standards for pH, temperature, etc.

Where to Learn More

Books

Montgomery Ward & Co. *Spring and Summer 1895 Catalogue and Buyer's Guide.* NY: Dover Publications, Inc. 1969.

Tate, Blair. *The Warp: A Weaving Reference.* Ashville, NC: Lark Books, 1991.

Other

Fieldcrest Cannon. "The Making of Royal Velvet Towels." Unpublished script for a video on towel production. Kannapolis, NC, 1998.

—Nancy EV Bryk

Beef Jerky

Beef jerky is part of the rapidly growing meat snack market which experienced yearly sales of over $240 million in 1996.

Background

Beef jerky is a type of snack food that is made by marinating beef in a curing solution and drying it. Meat treated in this way has a long shelf life and a unique flavor. Invention of this meat processing technology is attributed to Native Americans who smoke-dried meat to preserve it. Beef jerky is part of the rapidly growing meat snack market which experienced yearly sales of over $240 million in 1996. This growth has been attributed to the healthy aspects of beef jerky such as its high protein content and low fat level.

Beef jerky is a specific type of a more general kind of meat snack called jerky. Jerky is any type of meat, which has been cured with a salt solution and has had the moisture reduced to less than 50% of the total. It is typically brown colored and has a rough texture. Compared to unprocessed meat, it is tougher and has a more powerful flavor. This is primarily due to the flavor concentrating effect of the moisture removal process. Beef jerky is promoted as a nutritious, low calorie product, which is low in cholesterol and fat and high in protein and energy.

To increase the consumer acceptance of the product, there are many different flavors of beef jerky that are produced. One of the most common is the pepper flavored jerky. Other types include teriyaki jerky, hot styles, and barbecue styles. Hickory smoked and maple spice flavors are also produced. In addition to these flavors, other meats are used in jerky making. Currently, the most popular jerky meat after beef is turkey.

History

Beef jerky is thought to have originated in South America during the 1800s. The Quechua tribe, who were ancestors of the ancient Inca empire, produced a meat similar to beef jerky called *ch'arki*, or *charqui*. It was made by adding salt to strips of muscle tissue from game animals such as deer, buffalo, and elk, and allowing them to dry in the sun or over fires for extended periods of time. This method of preparation enabled the people to preserve meats during times when it was readily available and eat it when food was scarce. When the Spanish encountered this method of meat preservation, they adopted it and made it available to the rest of the world. It became a staple foodstuff for American cowboys and pioneers. Early explorers built smoke huts and hung cuts of meat over a fire to smoke cure the meat. True jerky was made when the meat was first flavored and then cured. Over the years, people discovered that the meat could be made more palatable by the addition of various spices.

Raw Materials

The meat and a curing solution are all the components needed for making meat jerky. The primary starting material for making jerky is the meat. For the best tasting jerky, the beef used is range fed, 100% premium, visually lean, United States Department of Agriculture (USDA) inspected, flank steak. There are some manufacturers who produce a slightly different textured beef jerky using ground beef. In most cases, it requires over five pounds of meat to produce one pound of jerky. While most meat jerky is made using beef, various other types of meat are also used. Turkey has become popular because it is perceived as a healthier alternative to beef. It produces a jerky that is not as tough as beef jerky. Pork is another meat that is made into jerky. This jerky has a slightly different taste.

The curing solution is used to provide a better taste and a longer shelf life for the beef jerky. It is also responsible for the final color of the product. The cure solution also has an antimicrobial effect, which prevents the growth of harmful bacteria. A typical curing solution is composed of water and salt plus sodium nitrite. The salt has a dehydrating effect on the meat. The sodium nitrite helps to retard the development of rancidity and stabilizes the color. Sodium ascorbate may also be added to increase the pink color of the meat.

The curing solution is mixed with a brine, which is an aqueous solution of seasonings, spices, salt, sugars, and phosphates. Some common flavorants used include soy sauce, lemon juice, pepper, monosodium glutinate (MSG), or garlic powder. Worcestershire sauce is also used, as is teriyaki sauce. The sugars that are used to provide a level of sweetness include sucrose, dextrose, brown sugar, and dark corn syrup. Flavored salts such as hickory salt or onion salt are also included in the brine. Sodium phosphates may also be used. Some manufacturers use a material known as liquid smoke, which is made by dissolving smoke in water. This material gives the meat a smoke cooked taste without requiring smoke during cooking. Since jerky is very tough, a tenderizing agents such as polyphosphates or papin enzyme can be included. The use of these materials is limited however, because it increases the amount of time required for drying.

The Manufacturing Process

While many different techniques can be used to make jerky, each have the primary steps of preparing the meat, soaking it in a curing solution, and cooking it until it is dried.

Initial meat preparation

1 Jerky can be prepared from either whole muscle or ground beef. In both cases, the meat is first treated to remove the bones and connective tissue, then the fat is removed. There are generally three methods for defatting the meat. In one instance, the meat is put into a large centrifuge. This device creates a spinning motion, which causes the liquid fat particles to separate from the meat. In another method, the meat is pressed to squeeze out all of the fat. Fat can also be removed from the meat by filtering it.

In addition to deboning and defatting, the meat is subjected to other processes, which help remove foreign bodies and other unwanted materials. The meat is put on a conveyor and workers inspect it as it passes by. It may then be put on a metal screen and shaken to remove unwanted material. Other processes may include water separation and the removal of any metals via magnets. Some plants even use x-ray examination to ensure the purity of the meat prior to use.

Preparing the curing solution

2 While the meat is being processed, the curing solution may be prepared. This is typically done in a large tank equipped with mixing blades. Water is filled into the tank and the salt, seasonings, and other materials are mixed in. The solution is heated as required until it is ready to be used. Since some of the materials are not water soluble, it is often necessary to mix before use.

Meat processing and curing

3 At this point, the meat can either be frozen and cut up into chunks by an automatic cutting machine or it can be ground up using a bowl chopping machine. Using the frozen meat method, the meat is allowed to partially thaw, which causes a release of the natural juices. The meat can then be dipped in the curing solution. It must be left in for an appropriately long time to allow total penetration of the liquid, however, not so long as to risk meat contamination. Another method of curing the meat is by injecting it with the curing solution by means of a multi-needle device. Meat treated as such is then sent to a large, stainless steel tumbling device that contains additional curing solution. This helps tenderize the meat and ensures the total penetration of the solution. If ground meat is used, the curing solution can be mixed directly into the meat to produce a workable paste. While ground meat is easi-

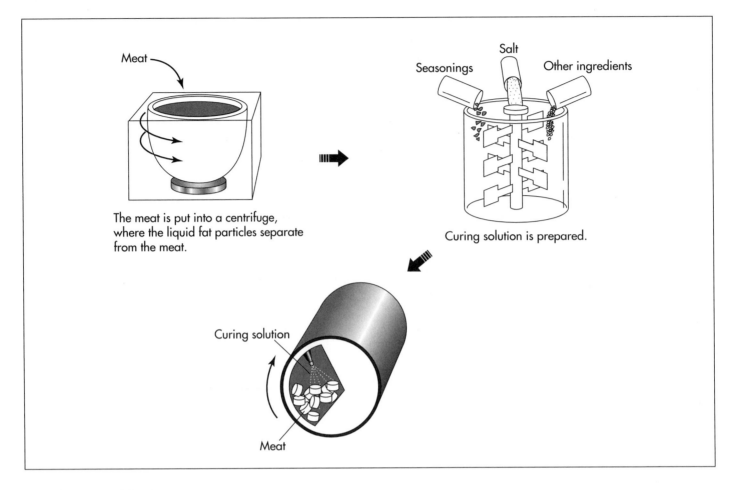

Meat

The meat is put into a centrifuge, where the liquid fat particles separate from the meat.

Seasonings   Salt   Other ingredients

Curing solution is prepared.

Curing solution

Meat

Once the meat is defatted, it is mixed with the curing solution.

er to work with, it produces a jerky with somewhat undesirable characteristics.

4 After seasoning, the meat is molded into blocks and cooled to a temperature of 18-28° F (-8 - -2.2° C). When it is adequately frozen, it is sliced into strips. The strips are preferably cut in line with the fiber of the meat. This makes the final product have a more natural looking texture. These strips are then spread out onto wire mesh trays and sent to drying ovens to be cooked. Here the meat strips are heated to 160° F (71.1° C) and gradually cooled to about 90° F (32.2° C). Depending on the method by which the meat was initially prepared, cooking can take as long as 12 hours. During cooking, the moisture in the meat is reduced to 20-40%.

Packaging

5 Many different types of packaging are used for beef jerky. To preserve freshness, most jerky is packaged in a vacuum-sealed bag. One manufacturer uses a triple barrier bag in which the meat is put in, the package is evacuated, filled with nitrogen, and sealed. This method removes all oxygen from the system thereby preventing spoilage due to oxidation. Recently, manufacturers have introduced a resealable type package. These bags are initially vacuum sealed, but also have a zip lock so they can be closed after opening. After the individual meat pieces are packed, they are placed in boxes, put on pallets, and shipped via truck to retail stores.

Quality Control

In any food processing facility, quality control is extremely important. Governmental regulations require that certain minimum standards be met for any raw material that will be used. Meats in particular are heavily regulated because the use of poor quality meat represents a significant health risk. Most companies will use only high quality meat to assure their jerky is of similar quality. All the initial raw materials are checked prior to processing for such things as such as pH, percentage moisture, odor, taste, and ap-

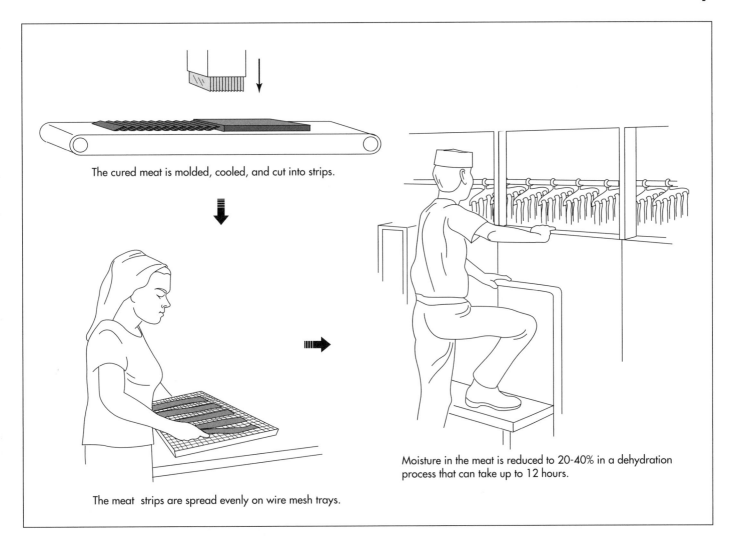

The cured meat is molded, cooled, and cut into strips.

The meat strips are spread evenly on wire mesh trays.

Moisture in the meat is reduced to 20-40% in a dehydration process that can take up to 12 hours.

pearance. Additionally, sanitation procedures for the production equipment are also subject to regulation. Since the beef jerky will be ingested, steps must be taken to ensure that it will have an appealing taste and be free from contamination. For this reason, tests similar to the ones run on the initial raw materials are performed on the final product.

The Future

It is anticipated that future developments in meat jerky processing will be found in a few key areas. An important area of product development is focused on developing new flavors of meat jerky. This involves creating different recipes and using various kinds of meats. To further boost the image of meat jerky as a healthy snack, manufacturers will attempt to find ways of reducing the salt content of the final product. In the manufacturing area, a more continuous

process is being developed. These techniques should result in more consistent product produced in a much shorter time period. Also, environmental concerns should lead to the development of waste-minimizing technologies.

Where to Learn More

Books

Bell, Mary. *Just Jerky: The Complete Guide to Making It*. Dry Store Publishing. 1996.

LeMaguer, M. and T. Jelen, editors. *Food Engineering and Process Applications*. London: Elsevier Applied Science Publishers, 1986.

Prowse, Brad. *Jerky Making: For the Home, Trail, and Campfire*. Naturegraph Publishing, 1997.

—*Perry Romanowski*

The cured meat is cut, spread out on trays, and dried into a jerky.

Birth Control Pill

Oral contraceptives, or birth control pills, have been used by more than 60 million women worldwide, and are considered by many to be the most socially significant medical advance of the twentieth century.

Background

Oral contraceptives, or birth control pills, have been used by more than 60 million women worldwide, and are considered by many to be the most socially significant medical advance of the twentieth century. The birth control pill is a tablet taken daily by a woman to prevent pregnancy. The birth control pill does this by inhibiting the development of the egg in the woman's ovary during her monthly menstrual cycle. During a woman's menstrual cycle, a low estrogen level normally triggers the pituitary gland to send out a hormone that initiates development of an egg. The birth control pill releases enough synthetic estrogen to keep that hormone from being released during the monthly cycle. The birth control pill also contains a second synthetic hormone, progestin, which increases the thickness of cervical mucus and impedes development of the uterine lining to further prevent pregnancy. Studies have shown that the birth control pill is 99% effective in preventing pregnancy. The results of studies on the safety of the birth control vary. Some studies show that its use increases the risk of certain types of cancer, while others show that risk to be minimal. There are also claims that the birth control pill increases risk of stroke and heart attacks.

History

The Planned Parenthood Federation of America commissioned Dr. Gregory Pincus and Dr. John Rock to develop a simple and reliable form of contraception in 1950. Over the next several years, the doctors worked on formulating a birth control pill at the Worcester Foundation for Experimental Bi-ology in Massachusetts. They tested their invention on 6,000 women in Puerto Rico and Haiti. The invention was then marketed in the United States in 1960 as Enovid-10.

Many attribute the changing social landscape in the United States during the 1960s to the widespread acceptance and use of the birth control pill. As sexual relations outside of marriage and for reasons other than childbearing became more socially acceptable and women seeking careers sought family planning methods, the environment was ripe for introduction of this discreet, easy-to-use form of contraception.

Despite its popularity, soon after the birth control pill was introduced, the public began to raise concerns about side effects and safety. As early as 1961, reports had begun to circulate that the birth control pill increased a woman's risk of suffering a stroke or a heart attack by causing blood clotting. In 1965, the federal Food and Drug Administration (FDA) provided a scientist at Johns Hopkins School of Hygiene and Public Health to study the side effects of the birth control pill. The agency also established an Advisory Committee on Obstetrics and Gynecology to study the relationship between oral contraceptives and blood clotting, as well as whether the birth control pill increased risk of breast, cervical, or endometrial cancer. The committee, the first-ever advisory committee established by the FDA, reported in 1966 that it had found no evidence to render the birth control pill unsafe for human use.

Unsatisfied, the FDA called for a larger study of the effects of the birth control pill on blood clotting. The agency also deter-

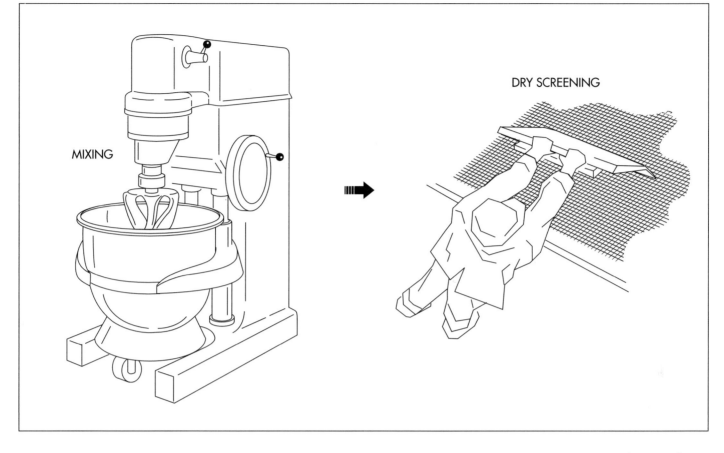

MIXING

DRY SCREENING

mined, however, that the birth control pill had not been in use long enough for a study of its relationship to cancer to be observed. At the same time, the World Health Organization (WHO) also determined that the effects of the birth control pill on blood clotting warranted study. By 1968, a British study revealed an increase in blood clots among women taking oral contraceptives. The FDA required that packages of birth control pills contain warning labels. In 1969, the agency concluded that the amount of estrogen affected the level of blood clotting and that birth control pills containing lower dosages of estrogen were as effective as their high-estrogen counterparts. The agency began advising doctors to prescribe the lowest estrogen dosage possible to their patients.

An oral contraceptive containing only progestin was introduced in the early 1970s. Dubbed the mini-pill, this form of oral contraceptive prevented pregnancy solely by causing changes in the uterus and cervix. An egg was produced, but the changes caused by the mini-pill made it difficult for the egg to unite with sperm from the male. While the mini-pill eliminates the risks posed by

estrogen, it has been found to be less effective in preventing pregnancy than pills containing estrogen. Throughout the 1970s, pills containing consistently lower doses of estrogen were introduced on the market.

In 1982 a biphasic birth control pill was introduced, followed by a triphasic pill in 1984. These low-dose pills contained varying ratios of progestin to estrogen. In 1988 all three drug companies still manufacturing high-dose birth control pills withdrew their high-dose products from the market, at the FDA's request. By 1990, the amount of estrogen in birth control pills had been reduced by at least two-thirds. Studies show that the risk of blood clotting in women taking the birth control pill has decreased accordingly. Further studies have shown that high-dose birth control pills actually reduced a woman's risk of ovarian and endometrial cancers, benign cysts of the ovaries and breasts, and pelvic inflammatory disease. The risk of breast or cervical cancer is still disputed.

The pill is still unsafe for certain groups of women, including those who smoke; are obese; have a history of health problems

Using a process known as the wet granulation method, the active ingredients are mixed together with a dilutant and a disintegrant in a large mixer. Once mixed, the powder mass is forced through a mesh screen.

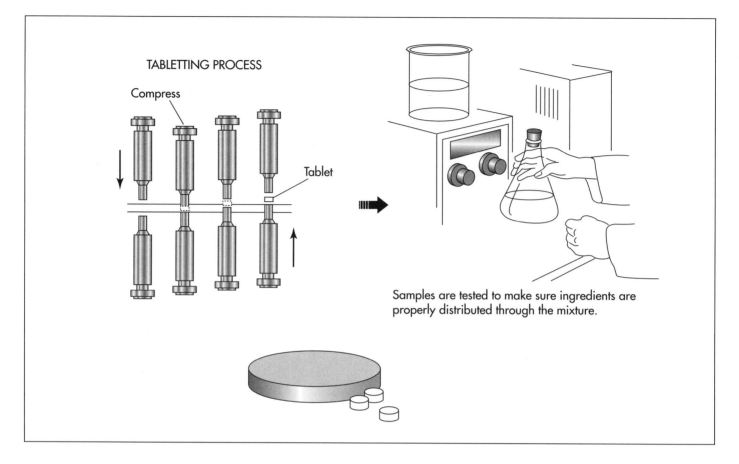

TABLETTING PROCESS

Compress

Tablet

Samples are tested to make sure ingredients are
properly distributed through the mixture.

Once tested, the mixture is molded
into tablet form and packaged.

such as diabetes, high blood pressure, or
high cholesterol; or have a history of blood
clots heart attack, stroke, liver disease, breast
cancer or cancer of the reproductive organs.

In addition to preventing pregnancy, the
birth control pill can also relieve symptoms
associated with premenstrual syndrome.
There are at least 30 varieties of birth control pills on the market today.

Raw Materials

The main ingredients of the birth control pill
are powders containing synthetic versions of
the hormones estrogen and progestin.

The Manufacturing Process

1 Using a process known as the wet granulation method, the active ingredients—
the powders containing synthetic versions of
estrogen and progestin—are mixed together
with a dilutant and a disintegrant (products
that dilute the powders and cause them to
dissolve in liquid) in a large mixer resem-

bling the mixmaster found in many
kitchens. For larger batches, a device known
as a twin-shell blender may be used.

2 Solutions carrying a binding agent (the
material that will cause the contents of
the tablet to cohere) are stirred into the powder mass, which is wetted until it takes on
the consistency of brown sugar.

3 The powder mass (known as wet granulation) is forced through a mesh screen.

4 The moist material is then placed on
shallow trays covered with large sheets
of paper and placed in drying cabinets.

5 A lubricant, in the form of a fine powder, is screened onto the dried material
(known as dry granulation).

6 The lubricant and the dry granulation are
then mixed in a blender, using a tum-
bling-type action.

7 Tablets are formed from the mixture,
typically using a method known as direct compression. Direct compression uses

steel punches and dies in large machines, which press tablets directly from the powdered mixture. The physical composition of the powdered mixture is not altered in any way. The punch and die system is often computerized.

8 The tablets are inspected to ensure compliance with federal regulations and packaged for shipment to pharmacies.

Quality Control

Like medications, birth control pills are subject to strict regulations set forth by the FDA. Production of birth control pills occurs in a highly sterile environment and samples are taken throughout the production process to ensure each batch of pills meets federal regulations. Factors examined include weight, coloration, and other cosmetic concerns. Many computerized tablet machines can provide weight information. The tablet punch and die systems are regularly inspected as well. In addition, the environment in which the tablets are produced is heavily controlled to avoid the influx of contaminants.

The Future

A relatively recent innovation in the field of birth control is the introduction of Norplant, a contraceptive that works on the same time-release concept as the pill but is inserted under the skin of the upper arm and releases the proper dosage into the body's system each day. Another innovation that is new in the United States, although it has been used in Europe, is RU486, a type of birth control that can be taken after intercourse to prevent pregnancy.

Where to Learn More

Books

Gennaro, Alfonso R., ed. Remington: *The Science and Practice of Pharmacy.* Eaton, PA: Mack Publishing Co., 1995.

Harris, Carla D. "The Birth Control Pill Revisited." In *NAACOG's Clinical Issues*, 1992: pp. 246-50.

Ketzung, Bertram. *Basic and Clinical Pharmacology*. Stamford, CT: Appleton and Lange, 1998.

—*Kristin Palm*

Bowling Ball

Sixty-five million people fling heavy balls down bowling lanes in the United States each year at speeds up to 20 miles an hour.

Background

Sixty-five million people fling heavy balls down bowling lanes in the United States each year at speeds up to 20 miles an hour. Other than the finger holes and eye-catching colors, the balls look simple—deceptively so. At prices ranging from less than $50 to around $300, the balls are much more than solid spheres.

Bowling balls are designed to perform best on various types of surfaces (lanes are not as simple as they look, either) and to compliment the style and strength of an individual bowler. Wooden bowling lanes are treated with mineral oil daily to protect them from the action of the balls. Typically, the first two-thirds of the lane is oiled rather heavily (the exact degree varies by establishment), while the final third is oiled lightly. As a result, a properly thrown ball will slide straight down the lane until it encounters the less-oiled surface, and then curve toward the pins as it gains better traction. Matching the rotational characteristics of the ball to the release style and strength of the individual bowler gives the best results.

History

Both lawn bowling (in which balls are rolled at a target ball) and pin bowling have been played for thousands of years. The excavated grave of an Egyptian child buried 5,200 years ago yielded a set of stone pins apparently used for a form of bowling.

Lawn bowling was quite popular in Europe during the Middle Ages. In 1366, King Edward III outlawed the game so his troops would pay more attention to their archery practice. Similarly, ninepin bowling (with the pins arranged in a diamond pattern) was outlawed in Connecticut and New York during the early 1800s because it was associated with heavy gambling. This led to the addition of a tenth pin (arranged in the now-common triangular pattern) to circumvent the law.

Lawn-bowling balls are either weighted or shaped asymmetrically so that they will curve when rolled. Balls used in pin bowling must be exactly round in shape, but they contain hidden weights that affect their balance and rotation. They also differ from lawn-bowling balls by having finger holes; they may have two (for the thumb and middle finger) or, more popularly, three (for the thumb and middle and ring fingers). When a bowler purchases a ball, the holes are drilled to fit his or her hand.

Structural Evolution

Historically, most bowling balls were made of *Lignum vitae*, a very hard wood. In 1905, the first rubber bowling ball (the Evertrue) was produced, followed nine years later by Brunswick Corporation's rubber Mineralite ball. Hard rubber balls dominated the market until the 1970s, when polyester balls were developed. In the 1980s, urethane bowling balls were introduced. Around 1990, dramatic changes were made in the design of the ball cores (dense blocks within the ball that modify the ball's balance). Shortly thereafter, reactive urethane was introduced as a new coverstock (the ball's surface layer) option.

Also referred to as resin, the new reactive urethane coverstock was used in combination with innovative core designs, drastically changing the sport. During the first full winter season in which reactive balls were used, the number of perfect games (12 successive throws in which all 10 pins are knocked down) increased by nearly 20%—the American Bowling Congress reported 14,889 in 1991-92 and 17,654 the following year.

Some examples of core shapes are lightbulb, spherical, and elliptical. Combination cores are made by enclosing a core of one shape and density within a second core of another shape and density. The main core may be supplemented by adding a collar or weight block to the core or by embedding small counterweights separately in the interior of the ball.

Since about 1993, bowling ball manufacturers have been using computerized design software to generate frequent improvements in core design. Designs have become so sophisticated that even for one model of ball, a different core design may be used for different ball weights (e.g., one for 12- and 13-pound balls, another for 14-pound balls, and a third for 15- and 16-pound balls). A developmental chemist working for a major manufacturer was quoted in a 1996 *Design News* article as saying, "Not long ago, a company could introduce a good ball and keep the market for two years. Now products come out so quickly that you need to have new designs ready all the time."

Raw Materials

Manufacturers currently use three types of plastics as coverstock material. Polyester, the least expensive, produces the smallest amount of hook on the back third of the lane because it is relatively unaffected by varying amounts of oil on the lane surface. In the middle of the price range, urethane balls offer more hooking action than polyester balls but are more durable and require less maintenance than reactive urethane balls. At the top of the price scale, reactive urethane (resin) balls provide the greatest hooking ability and deliver more power to the pins on impact. Various ball manufacturers have formed alliances with chemical suppliers to formulate proprietary materials by blending various resins with urethane.

Cores are made by adding a heavy substance such as bismuth graphite or barium either to

resin, making a very dense type of plastic, or to a ceramic material. An article written by ball manufacturer Track Inc. asserts that fired ceramic cores result in harder-hitting balls because no energy is absorbed by the ceramic part of the core. It also explains that fired ceramic cores cannot be altered during finger-hole drilling, whereas cores made of millable ceramic alloys can be. Millable ceramic alloys are made by blending ceramic powder with a binding agent. These types of ceramic cores are softer and less adhesive than fired ceramics, and that they do absorb energy upon impact with the pins.

In some balls, 2-4 oz (56.7-113.4 g) of iron oxide is used as a weight block to shift the ball's center of gravity toward one side of the core. Zirconium is used by one manufacturer for counterweights.

The Manufacturing Process

Between the early 1800s and the early 1990s, most bowling balls were made of three-piece construction. A small amount of dense material was poured into spherical core mold to create a pancake-like core. Then the remainder of the core mold was filled with a less-dense core material. Finally, the core was centered in a mold and a layer about 1 in (2.54 cm) thick of coverstock was poured around it. Since being pioneered by manufacturer Faball Inc. in the early 1990s, a two-piece construction method has become more popular.

Making the core

1 For the particular model of ball being manufactured, a mold is formed to the core shape developed during the computerized design process. The appropriate material is poured into the core mold and allowed to harden. The solid core is removed from its mold.

2 A second step may be necessary to finish the core. For example, some ceramic cores are fired in a kiln. A compound core may be formed by inserting the first core into a second mold and pouring material of a different density around all or part of it.

Forming the shell

3 The finished core is placed inside a spherical mold called the coverstock

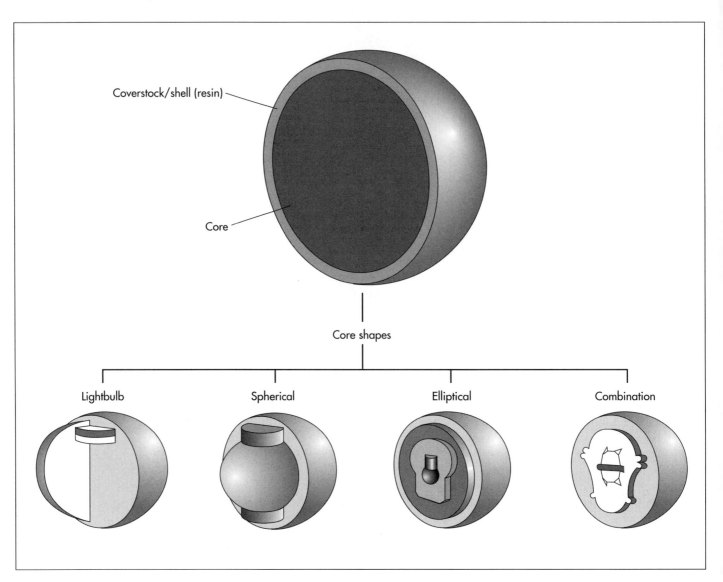

Some examples of bowling bowl core shapes are lightbulb, spherical, and elliptical. Combination cores are made by enclosing a core of one shape and density within a second core of another shape and density. The main core may be supplemented by adding a collar or weight block to the core or by embedding small counterweights separately in the interior of the ball.

mold. The core is attached to a pin that projects inward from the shell of the mold. The pin holds the core in the correct position. If the pin points toward the center of the mold, the core is said to be pin in; if it is tilted away from the center, the core is pin out.

4 The coverstock material is poured into the mold, encasing the core, and is allowed to harden. The thickness of the coverstock may be as little as 1 in (2.54 cm) or as much as 2 in (5.08 cm), depending on the design of the particular ball.

Filling the gaps

5 When the ball is removed from the coverstock mold, there is a hole where the core-holding pin had been. A plastic dowel is inserted into the hole and cemented in

place. The pin is a different color than the coverstock. After the ball has been purchased, the pin will be used as a guide for positioning the finger holes to take advantage of the core design.

6 Fill material is added to the logo imprint that was molded into the ball. This may be the same color as the pin, or it may be a different color. The logo is located at the top of the ball, that is, above its center of gravity.

Finishing

7 The ball is finished to the proper size specification by turning it on a lathe and shaving off enough coverstock to achieve the right shape or it may be done on a centerless grinder that scours the ball into the desired size and roundness.

8 Finally, the surface of the ball is finished to the desired texture. It is sanded to either a matte finish or to an appropriate degree of polish, indicated by the roughness of the sanding material (generally ranging from 240-600 grit).

9 The ball is boxed and shipped to the company's distributor.

Quality Control

When the American Bowling Congress (ABC) was founded in 1894, one of its primary missions was to standardize the sport by developing equipment specifications. The current rules require a ball to have a diameter between 8.500-8.595 in (about 21.6-21.8 cm), and to have a weight of 16 lb (about 7.3 kg) or less. No minimum weight is specified, and some balls weigh as little as 6 lb (about 2.7 kg). In order to earn the ABC/WIBC (Women's International Bowling Congress) seal of approval, sample balls of each model must be sent to the ABC for testing and verification of meeting the official standards.

In response to the dramatic changes in ball designs that began in the early 1990s, the ABC issued additional regulations in 1994. For example, the new rules establish limits on the ball's radius of gyration, which is the distance between the ball's rotational axis and its center of gravity. The rule limits this value to 2.430-2.800 in (6.2-7.1 cm). Other ABC specifications govern such technical characteristics as the ball's coefficient of restitution (a measure of the energy transferred from the ball to the pins), surface hardness, and hooking potential.

The Future

The dramatic innovations in bowling ball design and materials since the early 1990s have been credited with leveling the playing field for bowlers of all sizes and strengths. Writing in *Popular Mechanics* magazine, John G. Falcioni noted that some bowlers unhappily refer to the new-generation balls as cheaters. He summed up the impact of ball refinements by writing, "The sport has become so sophisticated that knowledge of engineering and physics is likely to prove more helpful in throwing strikes than doing curls with a dumbbell."

Where to Learn More

Periodicals

"CAD Helps Bowlers Improve Their Form." *Design News* (August 26, 1996): 29.

Falcioni, John G. "Strike Force." *Popular Mechanics* (March 1994): 60-63+.

Other

"Ball Tech....." Bowling This Month magazine. http://users.aol.com/phorvick/balltech/htm (15 Feb. 1998).

"Bowling Equipment Specifications." Bowling Page-Information. http://www.icubed.com/users/allereb/equip.html (15 Feb. 1998).

"Frequently Asked Questions." 19 Aug. 1995. Bowling Page. http://www.icubed.com/users/allereb/faq.html (15 Feb. 1998).

"Reactive, Urethane & Polyester. Match the Ball to the Conditions." The Complete Bowling Index. http://www.bowlingindex.com/products/balls/balls.htm (15 Feb. 1998).

"Why Use a Ball with Ceramics?" The Complete Bowling Index. http://www.bowlingindex.com/products/balls/ceramic.htm (15 Feb. 1998).

—Loretta Hall

Bowling Pin

A bowling pin is a complex assembly of several maple wood pieces; each made of still smaller pieces. Each sub-assembly is glued and clamped into a pressurized mold to build it into the next larger piece.

Background

The first recorded reference to a game in which a ball is thrown at pins dates from 300 A.D. The reference comes from Germany where the game was part of a religious ceremony practiced by monks. During the ceremony, the pins signified the bowler's sins, which were to be struck down with a ball. Once discovered, bowling spread in various forms throughout Europe. In 1366, King Edward III of England is said to have outlawed any form of bowling (apparently it was distracting his troops).

In the early 1800s, the modern game of bowling with nine pins first appeared in the United States. There was not any standardization of the game and regional rules and specifications prevailed until 1895. At that time, restaurateur Joe Thum assembled representatives of various bowling clubs in New York City and created the American Bowling Congress (ABC).

The pins used in this early competition and for another half century were made from a single block of hard rock maple. These solid pins were durable enough and were simple to manufacture. The only significant drawback was that, with the varying density of wood, the weight of an individual pin was difficult to standardize. Then in 1946, American Machine and Foundry purchased the patent for an automatic pin setter and began producing a machine that would both popularize the game and change the way bowling pins were made. The new machine was not an immediate success. Bowling alleys had plenty of pin boys to reset the lanes, and the original machinery was relatively undeveloped. However, patrons enjoyed the speed of the new pin setters, and the ma-

chines could work long hours without rest. As the automatic pin setters became more widespread, a new problem arose. The pin setters were much harder on the pins than human hands had been, so the pins began splintering and cracking much more rapidly. In turn, the splinters from the pins were fouling the machinery of the pin setters.

In 1954, Vulcan Manufacturing produced the first pin to address the problem, the Vulcanate. Vulcan sawed a standard solid pin in half before its final shaping and glued a flat piece of maple between the halves. With this design, Vulcan could drill wood out of the center piece as needed to reduce the pin's final weight. The design also presented more edge grain to the outside of the pin, making the pins more resistant than the solid wood ones. That same year Henry Moore introduced a plastic-coated pin. As with the pin setter, the new coating was not immediately embraced by the bowling industry. The liquid plastic used in Moore's pin added significant weight and chipped easily. However, the concept behind the pin design was sound and soon manufacturers began searching for a lightweight and durable coating. Even the lightest of coatings added weight to the pins, which decreased scoring, so manufacturers moved away from solid maple and began to follow Vulcan's example of gluing smaller pieces together into the shape of a pin. By 1961, the solid pin had disappeared altogether.

Raw Materials

Core

As mandated by the American Bowling Congress and the Women's International

70

Bowling Congress (ABC/WIBC), all bowling pins are made "of new (unused), sound, hard maple." Other materials may be used if they are given ABC/WIBC approval. The maple used is harvested above the 45th parallel, which runs through the northernmost states of the United States, above the areas where high mineral concentrations lessen the wood's density. Other core materials have been tried. In the late 1960s, manufacturers attempted to further the lamination concept by using particle lumber. Particle lumber is made by shredding wood, mixing it into a bonding agent, and pressing it into shape. The problem was that a dense mixture made the pins durable but low scoring, while light pins had the opposite problem. A polypropylene-fiberglass foam core known as a dykehouse core was also attempted. The foam was molded in two pieces, top and bottom, and the halves were fitted over a wooden dowel. Durability tests were impressive, but the project never developed into production.

Coating

When pins were made of solid blocks of maple, they were simply coated with a layer of white lacquer and finished with a layer of clear lacquer. The advent of the automatic pin setter spurred a need for more durable coatings, and the first successful one was ethylcellulose. Ethylcellulose coating was a seven-layer process. Pins were first dipped into a coat designed to seal the pores of the wood and then into one designed to provide better adhesion for the following layers. Then five layers of ethycellulose were applied. After the first layer, a nylon mesh sock was stretched over the pin where it acted like iron reinforcing bars in concrete or like straw in adobe brick, providing a structure for the ensuing layers.

The two coatings in use today are nylon and DuPont's Surlyn. Nylon has the advantage of being more forgiving in the molding process and having a lower cost than Surlyn. But cores must be dipped in latex to prevent the nylon from cracking, and nylon does not provide the same rigidity as Surlyn.

Design

The three considerations in designing a bowling pin are durability, scoring, and sound. The durability of pins has been vastly improved with the use of multi-piece cores and more resilient coatings. Multipiece cores are stronger and more durable than durable single blocks because of the way in which wood grows. A tree grows larger in layers or rings, each ring following the contours of the one under it. If the previous ring had a kink or curve in a certain spot, the next ring will also have that kink or curve, as will all the following rings. When you use a single block of wood, all of the layers in that wood will flex and eventually break in the same spot because they are all shaped the same, but if you take several small pieces and glue them into a block, each piece has its natural bends and weak spots in a different place, so each reinforces the other. An additional benefit of this method for bowling pins stems from the way in which the rings hold to one another. When a tree is in its natural state, its cross section is a series of concentric rings. Once milled into lumber, the cross section only contains portions of the rings, so it more closely approximates stacked sheets of cardboard. If you were to push against the side of a stack of cardboard, each sheet would be fairly rigid but each would slide out of the stack fairly easily, and you could lift off the top sheets with no trouble at all. If you were to press on the top of the stack, you could not easily split it in half but it would compress with little effort. This is exactly the situation with bowling pins. The old solid pins would dent on the face of their grain layers, and the large, flat layers would peel and split at the edge. Gluing together several pieces of wood gives the compression resistant advantage of presenting all edge grain to the outside of the pin, but because no one layer runs all the way through the pin, they are less prone to splitting.

Scoring and sound are both shaped by bowler's preferences and are primarily controlled by the core material. A bowling pin's whole purpose is to fall over. If it never falls over, the game becomes far less interesting. Although an obvious solution to the problem of durability would be to make a pin of a material stronger than wood, such as steel, most other materials are too heavy, so they do not fall over as readily as bowlers have come to expect. What has discouraged the use of many other materials, though, is

A bowling pin is a complex assembly of several maple wood pieces; each made of still smaller pieces. Each sub-assembly is glued and clamped into a pressurized mold to build it into the next larger piece.

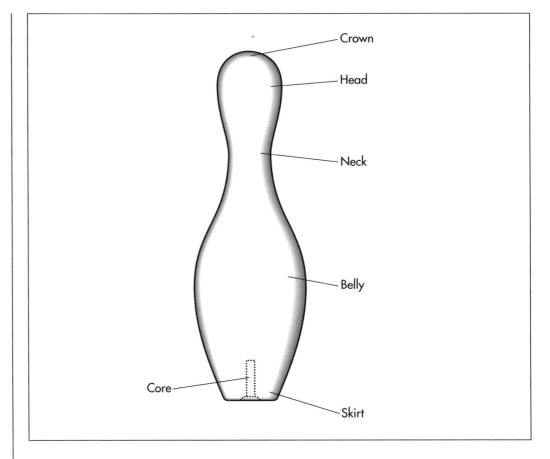

sound. The distinctive sound of pins crashing is part of the aura of the game, and that sound comes from wood. Presently, no other material has been able to match it.

The Manufacturing Process

A bowling pin is a complex assembly of several maple wood pieces; each made of still smaller pieces. Each sub-assembly is glued and clamped into a pressurized mold to build it into the next larger piece. Lumber is dried in large kilns for 8-10 days before being shipped to the plant. When it reaches the plant, entire planks are cut at once into strips and then into blocks of specific size by multi-bladed automatic saws.

Core

1 The core of a pin begins with the post. The post is made up of three sections of maple, each 2.8125 in (7.14 cm) wide by 0.875 in (2.22 cm) thick by 15.5 in (39.3 cm) tall to form a slightly rectangular 15.5 in (39.3 cm) tall block. At the same time,

eight strips of 16 in (41 cm) long 0.875 in. (2.22 cm) by 1.25 in (3.17 cm) thick maple are glued to form a 7 in (18 cm) wide by 16 in (41 cm) long board called the panel. The panel is run through an electric planer to make its top and bottom surfaces flat and even, and then it is cut with a table saw lengthwise and in half into two 3 in (8 cm) wide pieces and two 4 in (10.2 cm) wide pieces called cheeks. The cheeks will be glued to the post to form the large bottom end of the pin known as the belly.

2 First, the 3 in (8 cm) cheeks are glued 1 in (2.54 cm) from the bottom on the longer side of the post and the assembly is clamped into a jig to ensure alignment of all the pieces.

3 Once this assembly has dried, it is run through the planer to make its edges flush and to make the post a perfect 2.625 in (6.67 cm) square. Then the piece is run through a machine that weighs it, calculates its density, and determines the amount of wood that needs to be removed to bring the final pin within specifications.

4 The machine drills into the post to adjust its weight then repeats the process until the pin falls within a 4 oz (113.4 g) weight range. The 4 in (10.2 cm) cheeks are glued over the holes, overlapping the sides of the 3 in (8 cm) cheeks and the new assembly is again clamped into a jig to properly align all the pieces. At this step, the core is called the billet and has the basic shape of a bowling pin except that it is square. A hole is drilled into the center of the bottom of the billet to be used later to center the pin during the coating process.

5 The billet is then placed on a lathe and a single blade matching the profile of an entire pin cuts it into its final shape.

Coating

6 The primary method used today for coating bowling pins is injection molding. The process involves placing an object into a two-piece metal mold, with each piece carrying an impression of half the object to be molded. In the case of bowling pins, the halves are identical. If nylon is used as the coating, the pins are first dipped in latex to prevent the nylon from cracking.

7 The turned core is placed in the bottom half of the mold. A retractable pin slides into the center hole that was drilled into the bottom of the billet. This pin and several raised pieces in the mold called hold downs will hold the core in the center of the mold while the coating is applied. The mold is then closed and clamped shut.

8 The coating is injected under high pressure through numerous small openings called gates. This high pressure not only ensures an even coating but it evenly compresses the wood, effectively making it denser and more resistant to denting.

9 Once the predetermined amount of coating has been injected, water is run through tubes in the mold to cool it. The cooled, coated pin is removed from the mold and a worker trims the stalks of coating left by the gates.

10 The surface is then sanded to remove any remaining excess left by the gates and to smooth the slight impressions made by the hold-downs. The pin maker's emblem, the ABC/WIBC approval stamp, and any decorative markings are silk-screened onto the surface.

11 Then, a final protective gloss coat is sprayed onto the pin. The bottom of the pin is trimmed flat and a hard plastic ring is glued into a notch at the bottom of the pin called the skirt. The ring will both protect the skirt and will provide the 5/32 in (0.4 cm) radius the ABC/WIBC requires along the outside edge of the pin's base.

Byproducts/Waste

Much of the wood used in the manufacture of bowling pins is itself a byproduct. The flooring industry mills thousands of board feet of maple every year. Because consumers prefer light-colored wood, the darker parts of the lumber are trimmed. Since a bowling pin is coated, color does not matter. Mills have begun to trim these pieces to the size used in pins, so what had been scrap is now a useful product.

Quality Control

The ABC/WIBC sets strict standards for pin height, weight, moisture content, coating thickness, center of gravity, and myriad other details. Each new pin design must meet these standards and must pass specific durability field tests. The manufacturer inspects each block of wood before it enters the assembly process. Pieces with knots, cracks, mineral deposits, or irregular grain are rejected. Once in use, a pin will last six months or so before it needs to be patched or recoated. Afterwards, it will last another six months before breaking. Most breaks happen in the thin area just below the top called the neck, where the whipping action of being knocked over flexes the wood by as much as 0.25 in (0.63 cm).

The Future

Maple has been the core material of choice for bowling pins for well over 200 years. Despite experiments with such diverse materials aluminum, plastic, and magnesium, nothing has been found to provide the durability, sound, and scoring range of maple.

Where to Learn More

Books

Borden, Fred and Ackerman, John. *Bowling*. Madison, WI : Brown & Benchmark, 1997.

Stallings, Howard. *The Big Book of Bowling*. Salt Lake City: Gibbs Smith, 1995.

Steele, H. Thomas. *Bowl-O-Rama : the visual arts of bowling*. New York: Abbeville Press, 1986.

Williams, Donald. *The Bowling Trivia Book*. Novi, MI: Williams Bowling Group, 1995.

Other

International Bowling Museum and Hall of Fame. http://www.bowlingmuseum.com/.

—*Michael Cavette*

Braille Publication

Background

Braille is a tactile writing system used by the blind that was invented by Louis Braille in France in 1824. It gradually spread beyond France, and it is now in widespread use across the globe.

The blind read Braille by feeling letters with their fingertips. Letters in Braille are formed by raised dots arranged in specific places in a six-position matrix. The matrix consists of two vertical lines of three points each. Various combinations of raised dots in the matrix stand for each letter in the Roman alphabet. For example, the letter A is indicated by one raised dot in the upper left of the matrix; the letter B by two dots, the upper left and the one beneath it; the letter T by four raised dots, the middle and lower left and the middle and upper right. In standard English Braille, some common words such as and, of, and the are also represented by a single Braille character, as are some dipthongs and vowel-consonant combinations. For compactness, Braille also makes use of many phonetic or syllabic abbreviations, such as "ing," as in shorthand. There are 189 of these abbreviations in standard English Braille. Braille can also indicate punctuation and accent marks. The first 10 letters of the alphabet double as numbers in Braille. There is also a modified Braille code, the Nemeth Code, that incorporates signs and symbols used in scientific notation, used by blind scientists and mathematicians.

Before the invention of Braille, the blind were generally not taught to read and write, and many European cultures considered blind people to be mentally deficient. A precursor to Braille was invented by Valentin Haüy, a Frenchman who founded a school for the blind in Paris in the 1770s. Haüy printed books in large, embossed type, so that his students could feel the outlines of the letters. The principal drawback to this system was the size of the letters. Because the letters are so large, one sentence might take up an entire page. A cumbersome volume written in Haüy letters might consist of no more than a few paragraphs of actual text.

Louis Braille, born in 1809, was blinded in an accident at the age of three. An exceptionally bright child, he was sent to the Royal Institute for Blind Youth in Paris at the age of 10. He expected to learn to read there, but was disappointed to find that the Institute's library consisted of only a few of the Haüy books. The young Braille became acquainted with a different system of writing using raised dots, invented by a military signalman named Barbier. Barbier had developed a code made of groups of raised dots and dashes punched in cardboard, for use by an army sending messages at night, when a light to read by might be dangerous. He showed his "night writing" system to the head of the Royal Institute for the Blind, hoping it could be used by the students there. Barbier's system used the dots and dashes to denote sounds, instead of letters, and it took considerable time and patience to read or write even a simple message. Louis Braille, introduced to night writing at the age of 13, struggled to modify Barbier's system, coming up with the simpler six-position matrix for letters, and eventually incorporating signs for accent marks and punctuation. Braille unveiled his writing system when he was only 15 years old, and it instantly revolutionized how the blind

John Gardner, a physicist blinded late in life, has created an improved Braille system for writing equations. Called Dots Plus, it uses conventional Braille for letters and numbers, but renders mathematical symbols just as they appear to sighted people, only magnified and raised. This renders mathematical formulas more compact, and so they are easier to read and re-read.

could learn. Using a tablet and stylus, blind students could quickly and easily write in Braille. Whole books could easily be transcribed, for the blind to read. Braille was officially adopted in France in 1854, and in the English-speaking world in 1932.

Louis Braille also invented a form of typewriter, which he called a raphigrapher. This embossed large Roman letters on paper, and both the blind and the sighted could read the results. Early Braille writing machines were similar, but embossed Braille letters. The first was the Hall Braille writer, invented in 1892 by Frank Hall, a superintendent of the Illinois School for the Blind. Braille writing machines currently in use employ only six keys, one for each position in the Braille letter matrix. Skilled operators who have gone through a certification process use these Braille writers to produce manuscripts for Braille publication.

Raw Materials

Raw materials for a Braille publication are not significantly different from those used in other publications. Standard size paper for Braille books is 11 x 11.5 in (28 x 29.21 cm), and the weight is heavier than for other books. Some Braille is printed on more specialized paper, such as swell paper, a heat-sensitive paper that rises where printed upon. Zinc is an important raw material for Braille books, because the maste of a Braille text is punched on a zinc plate. Because Braille books are large format, they are often bound in plastic ring binders rather than in the hard or soft-cover format of conventional books.

The Manufacturing Process

Translation

1 Unless a manuscript has been written originally in Braille, it needs to be translated. This is generally done in one of three ways. A typist using a special six-key typing machine may re-type the manuscript in Braille. Braille transcribers are specially trained and certified. Professional Braille typists must take a two-year training course administered by the National Library Service for the Blind, and pass a accreditation examination. So though this method is gen-

erally more laborious than the newer, computer-based alternatives, Braille typists bring their insight and experience to the work, and assure a high level of quality. The two alternate methods utilize computer software to make the translation. Text in English may be fed into a computer program through a scanner, which reads the text electronically and stores it in Braille form. In this way, previously published material such as a novel by Dickens or an article from *The New York Times* can be converted into Braille. In the case of a new publication such as a journal released simultaneously in Braille and in conventional format, the text may be already stored on a computer disk. Special software developed for this purpose converts the text on the disk into Braille. Braille conversion technology is becoming increasingly sophisticated as well as speedy. Entire books can be converted in seconds.

Proofreading

2 If the manuscript has been typed by a Braille transcriber, it is now ready for proofreading. If the manuscript has been converted using computer software, it must be printed out. Braille printers similar to other computer printers produce the manuscript. Then the manuscript is proofread so that any errors can be corrected or changes made before the manuscript is published. At this stage, a blinder reader and a sighted reader work side by side, comparing the original text with the Braille version.

Making the master

3 After the manuscript is completely proofed and corrected, a master copy of it is made for the printing press. The master is cast on a zinc plate. A special machine, separate from the actual printing press, is used to stamp the Braille impressions in the metal. Each page of the manuscript has its own zinc master. The zinc plate is bowed in the middle. It fits onto a rotating barrel on the printing press.

Printing

4 After the zinc plates are fitted onto the press, a worker running the press feeds paper into the machine. The press is not significantly different from a conventional printing press, except that the letters are em-

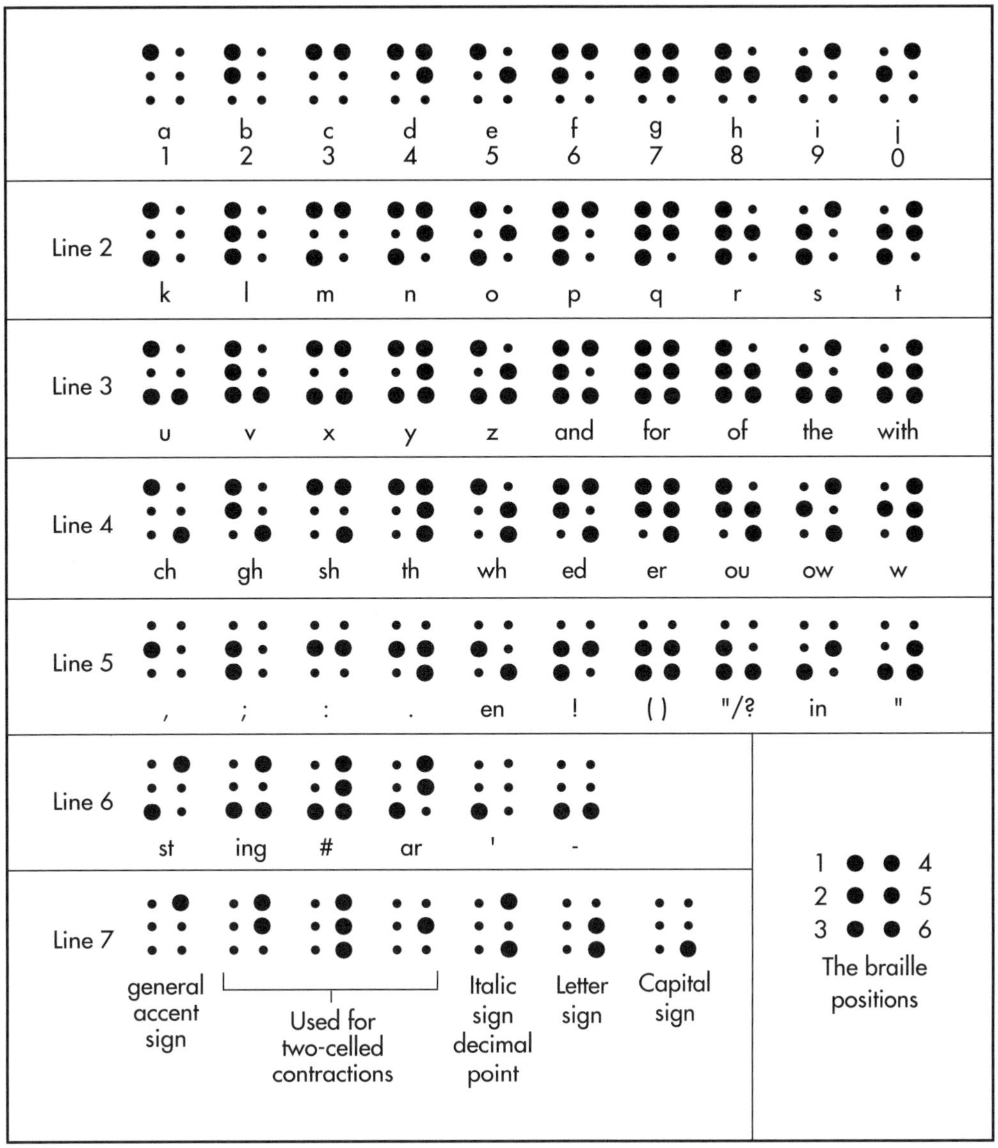

Braille Characters

a / 1	b / 2	c / 3	d / 4	e / 5	f / 6	g / 7	h / 8	i / 9	j / 0

Line 2: k l m n o p q r s t

Line 3: u v x y z and for of the with

Line 4: ch gh sh th wh ed er ou ow w

Line 5: , ; : . en ! ( ) "/? in "

Line 6: st ing # ar ' -

Line 7: general accent sign | Used for two-celled contractions | Italic sign decimal point | Letter sign | Capital sign

The braille positions
1 ● ● 4
2 ● ● 5
3 ● ● 6

bossed. There is no ink. The paper is pressed against the zinc master as the barrel rotates, and the impressions of the raised dots are transferred to the paper. Then the sheet of paper is ejected.

Collating and assembling

5 The pages of a Braille publication must be collated by hand. Though this is extremely labor intensive, mechanical collators are not adequate for Braille books. Because a mechanical collator would hold and handle stacks of pages, it tends to mash the dots, thus destroying the text. Instead the pages are carefully placed in order by hand. Then the book can be finished in a number of ways. Some books are bound in a three-ring binder, and for these, the pages need to be punched. Other publications are saddle-stitched and bound in a conventional hardback book format. Finished books can then be boxed and shipped to customers or to a warehouse for distribution.

Quality Control

One aspect of quality control in Braille publication is the training of Braille typists.

Letters in Braille are formed by raised dots arranged in specific places in a six-position matrix. The matrix consists of two vertical lines of three points each. Various combinations of raised dots in the matrix stand for each letter in the Roman alphabet.

Though some Braille typists are volunteers, often producing books for a visually impaired family member, professional transcribers go through a rigorous training course. Then they must pass a national examination. Another aspect of quality control for Braille text is adherence to common editing standards. Whereas in conventional print text, there are any number of ways text can be embellished, with bold print, italics, margin size, use of headers and sidebars, different fonts, etc., in Braille there are only a few possibilities. For example, a blank line in Braille text is used to separate distinct blocks of text, and is never simply decorative. Text size and width of indentations is fairly standardized. Non-standard text layout may be confusing to Braille readers, or just make the reading process more difficult. Taking care of layout problems may be done at the proofreading stage. New computer programs are also under development that can reliably convert widely differing print manuscripts into standardized Braille formats.

The Future

There are many new developments in Braille publication technology. Software for converting text to Braille is still undergoing improvement and refinement. As software gets better, more of it is available to individuals. Blind users need not rely only on specialized printing houses to provide Braille materials if they have a personal computer, conversion software, and a Braille printer. While many printers and programs are still too expensive for many users to own their own, some manufacturers now specialize in low-cost equipment. Some Braille printers meant for individual use utilize a narrower paper than conventional Braille paper, because this is more economical. Common in Europe, and increasingly so in the United States, is so-called paperless Braille. A handheld unit attached to a computer can raise a line or so at a time of Braille text on a board, using small pins. Another Braille technology just introduced in Korea prints Braille using a clear, glue-like substance on glossy paper.

Braille printing and conversion technology is still advancing, and Braille itself is under development. John Gardner, a physicist blinded late in life, has created an improved Braille system for writing equations. Called Dots Plus, it uses conventional Braille for letters and numbers, but renders mathematical symbols just as they appear to sighted people, only magnified and raised. This renders mathematical formulas more compact, and so they are easier to read and re-read. Gardner, in collaboration with a blind mathematician, is also developing a different Braille language for math that uses an eight-dot matrix instead of six.

Voice output technology, which allows computers to speak text, is also improving rapidly, and is very useful for the blind. But even as great leaps are taken in voice input, Braille maintains its importance. A listener using voice output technology has to rewind or go back, to have text repeated. Some text, such as complex mathematical formulas, are difficult to represent in speech. Even as voice technology gets more sophisticated, Braille is not likely to give way to other approaches.

Where to Learn More

Books

Bryant, Jennifer Fisher. *Louis Braille, Teacher of the Blind.* Chelsea House Publishers, 1994.

Periodicals

Kumagai, Jean. "Inventions Born of Necessity Offer New Tools for the Blind to Study and Do Science." *Physics Today* (March 1995): 82-84.

Lazzaro, Joseph J. "Unix Helps the Disabled." *Byte* (April 1997): 51-52.

"Printer Helps the Blind Help Themselves." *Design News* (April 11, 1994): 44.

—*Angela Woodward*

Carbon Fiber

Background

A carbon fiber is a long, thin strand of material about 0.0002-0.0004 in (0.005-0.010 mm) in diameter and composed mostly of carbon atoms. The carbon atoms are bonded together in microscopic crystals that are more or less aligned parallel to the long axis of the fiber. The crystal alignment makes the fiber incredibly strong for its size. Several thousand carbon fibers are twisted together to form a yarn, which may be used by itself or woven into a fabric. The yarn or fabric is combined with epoxy and wound or molded into shape to form various composite materials. Carbon fiber-reinforced composite materials are used to make aircraft and spacecraft parts, racing car bodies, golf club shafts, bicycle frames, fishing rods, automobile springs, sailboat masts, and many other components where light weight and high strength are needed.

Carbon fibers were developed in the 1950s as a reinforcement for high-temperature molded plastic components on missiles. The first fibers were manufactured by heating strands of rayon until they carbonized. This process proved to be inefficient, as the resulting fibers contained only about 20% carbon and had low strength and stiffness properties. In the early 1960s, a process was developed using polyacrylonitrile as a raw material. This produced a carbon fiber that contained about 55% carbon and had much better properties. The polyacrylonitrile conversion process quickly became the primary method for producing carbon fibers.

During the 1970s, experimental work to find alternative raw materials led to the introduction of carbon fibers made from a petroleum pitch derived from oil processing. These fibers contained about 85% carbon and had excellent flexural strength. Unfortunately, they had only limited compression strength and were not widely accepted.

Today, carbon fibers are an important part of many products, and new applications are being developed every year. The United States, Japan, and Western Europe are the leading producers of carbon fibers.

Classification of Carbon Fibers

Carbon fibers are classified by the tensile modulus of the fiber. Tensile modulus is a measure of how much pulling force a certain diameter fiber can exert without breaking. The English unit of measurement is pounds of force per square inch of cross-sectional area, or psi. Carbon fibers classified as "low modulus" have a tensile modulus below 34.8 million psi (240 million kPa). Other classifications, in ascending order of tensile modulus, include "standard modulus," "intermediate modulus," "high modulus," and "ultrahigh modulus." Ultrahigh modulus carbon fibers have a tensile modulus of 72.5-145.0 million psi (500 million-1.0 billion kPa). As a comparison, steel has a tensile modulus of about 29 million psi (200 million kPa). Thus, the strongest carbon fiber is about five times stronger than steel.

The term graphite fiber refers to certain ultrahigh modulus fibers made from petroleum pitch. These fibers have an internal structure that closely approximates the three-dimensional crystal alignment that is characteristic of a pure form of carbon known as graphite.

Carbon fiber-reinforced composite materials are used to make aircraft and spacecraft parts, racing car bodies, golf club shafts, bicycle frames, fishing rods, automobile springs, sailboat masts, and many other components where light weight and high strength are needed.

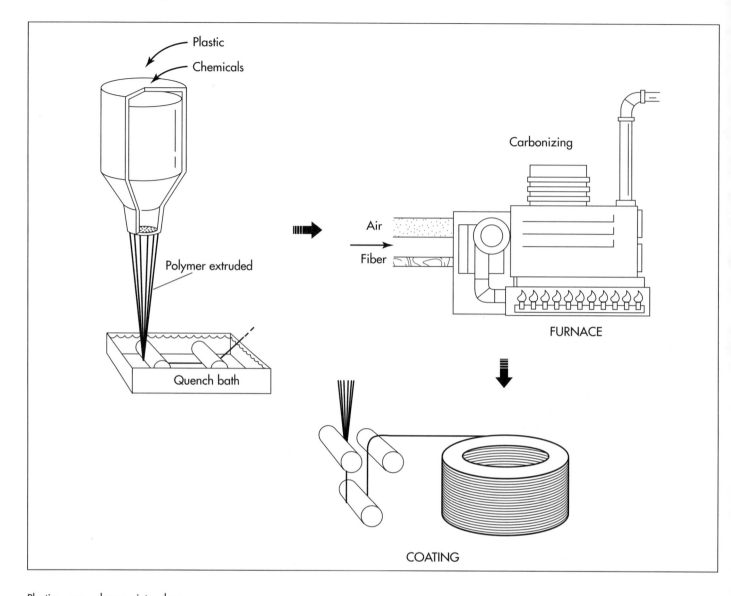

Plastics are drawn into long strands or fibers and then heated to a very high temperature without allowing it to come in contact with oxygen. Without oxygen, the fiber cannot burn. Instead, the high temperature causes the atoms in the fiber to vibrate violently until most of the non-carbon atoms are expelled.

Raw Materials

The raw material used to make carbon fiber is called the precursor. About 90% of the carbon fibers produced are made from polyacrylonitrile. The remaining 10% are made from rayon or petroleum pitch. All of these materials are organic polymers, characterized by long strings of molecules bound together by carbon atoms. The exact composition of each precursor varies from one company to another and is generally considered a trade secret.

During the manufacturing process, a variety of gases and liquids are used. Some of these materials are designed to react with the fiber to achieve a specific effect. Other materials are designed not to react or to prevent certain reactions with the fiber. As with the precursors, the exact compositions of many of these process materials are considered trade secrets.

The Manufacturing Process

The process for making carbon fibers is part chemical and part mechanical. The precursor is drawn into long strands or fibers and then heated to a very high temperature without allowing it to come in contact with oxygen. Without oxygen, the fiber cannot burn. Instead, the high temperature causes the atoms in the fiber to vibrate violently until most of the non-carbon atoms are expelled. This process is called carbonization and leaves a fiber composed of long, tightly

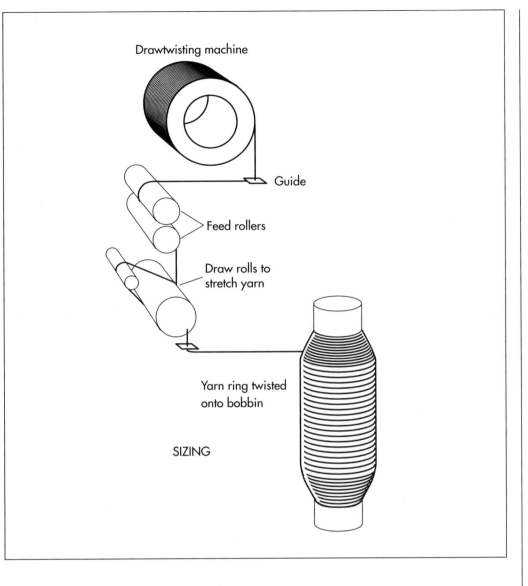

Drawtwisting machine

Guide

Feed rollers

Draw rolls to
stretch yarn

Yarn ring twisted
onto bobbin

SIZING

The fibers are coated to protect them from damage during winding or weaving. The coated fibers are wound onto cylinders called bobbins.

inter-locked chains of carbon atoms with only a few non-carbon atoms remaining.

Here is a typical sequence of operations used to form carbon fibers from polyacrylonitrile.

Spinning

1 Acrylonitrile plastic powder is mixed with another plastic, like methyl acrylate or methyl methacrylate, and is reacted with a catalyst in a conventional suspension or solution polymerization process to form a polyacrylonitrile plastic.

2 The plastic is then spun into fibers using one of several different methods. In some methods, the plastic is mixed with certain chemicals and pumped through tiny jets into a chemical bath or quench chamber where the plastic coagulates and solidifies into fibers. This is similar to the process used to form polyacrylic textile fibers. In other methods, the plastic mixture is heated and pumped through tiny jets into a chamber where the solvents evaporate, leaving a solid fiber. The spinning step is important because the internal atomic structure of the fiber is formed during this process.

3 The fibers are then washed and stretched to the desired fiber diameter. The stretching helps align the molecules within the fiber and provides the basis for the formation of the tightly bonded carbon crystals after carbonization.

Stabilizing

4 Before the fibers are carbonized, they need to be chemically altered to convert

their linear atomic bonding to a more thermally stable ladder bonding. This is accomplished by heating the fibers in air to about 390-590° F (200-300° C) for 30-120 minutes. This causes the fibers to pick up oxygen molecules from the air and rearrange their atomic bonding pattern. The stabilizing chemical reactions are complex and involve several steps, some of which occur simultaneously. They also generate their own heat, which must be controlled to avoid overheating the fibers. Commercially, the stabilization process uses a variety of equipment and techniques. In some processes, the fibers are drawn through a series of heated chambers. In others, the fibers pass over hot rollers and through beds of loose materials held in suspension by a flow of hot air. Some processes use heated air mixed with certain gases that chemically accelerate the stabilization.

Carbonizing

5 Once the fibers are stabilized, they are heated to a temperature of about 1,830-5,500° F (1,000-3,000° C) for several minutes in a furnace filled with a gas mixture that does not contain oxygen. The lack of oxygen prevents the fibers from burning in the very high temperatures. The gas pressure inside the furnace is kept higher than the outside air pressure and the points where the fibers enter and exit the furnace are sealed to keep oxygen from entering. As the fibers are heated, they begin to lose their non-carbon atoms, plus a few carbon atoms, in the form of various gases including water vapor, ammonia, carbon monoxide, carbon dioxide, hydrogen, nitrogen, and others. As the non-carbon atoms are expelled, the remaining carbon atoms form tightly bonded carbon crystals that are aligned more or less parallel to the long axis of the fiber. In some processes, two furnaces operating at two different temperatures are used to better control the rate of heating during carbonization.

Treating the surface

6 After carbonizing, the fibers have a surface that does not bond well with the epoxies and other materials used in composite materials. To give the fibers better bonding properties, their surface is slightly oxidized. The addition of oxygen atoms to the surface provides better chemical bonding properties and also etches and roughens the surface for better mechanical bonding properties. Oxidation can be achieved by immersing the fibers in various gases such as air, carbon dioxide, or ozone; or in various liquids such as sodium hypochlorite or nitric acid. The fibers can also be coated electrolytically by making the fibers the positive terminal in a bath filled with various electrically conductive materials. The surface treatment process must be carefully controlled to avoid forming tiny surface defects, such as pits, which could cause fiber failure.

Sizing

7 After the surface treatment, the fibers are coated to protect them from damage during winding or weaving. This process is called sizing. Coating materials are chosen to be compatible with the adhesive used to form composite materials. Typical coating materials include epoxy, polyester, nylon, urethane, and others.

8 The coated fibers are wound onto cylinders called bobbins. The bobbins are loaded into a spinning machine and the fibers are twisted into yarns of various sizes.

Quality Control

The very small size of carbon fibers does not allow visual inspection as a quality control method. Instead, producing consistent precursor fibers and closely controlling the manufacturing process used to turn them into carbon fibers controls the quality. Process variables such as time, temperature, gas flow, and chemical composition are closely monitored during each stage of the production.

The carbon fibers, as well as the finished composite materials, are also subject to rigorous testing. Common fiber tests include density, strength, amount of sizing, and others. In 1990, the Suppliers of Advanced Composite Materials Association established standards for carbon fiber testing methods, which are now used throughout the industry.

Health and Safety Concerns

There are three areas of concern in the production and handling of carbon fibers: dust inhalation, skin irritation, and the effect of fibers on electrical equipment.

During processing, pieces of carbon fibers can break off and circulate in the air in the form of a fine dust. Industrial health studies have shown that, unlike some asbestos fibers, carbon fibers are too large to be a health hazard when inhaled. They can be an irritant, however, and people working in the area should wear protective masks.

The carbon fibers can also cause skin irritation, especially on the back of hands and wrists. Protective clothing or the use of barrier skin creams is recommended for people in an area where carbon fiber dust is present. The sizing materials used to coat the fibers often contain chemicals that can cause severe skin reactions, which also requires protection.

In addition to being strong, carbon fibers are also good conductors of electricity. As a result, carbon fiber dust can cause arcing and shorts in electrical equipment. If electrical equipment cannot be relocated from the area where carbon dust is present, the equipment is sealed in a cabinet or other enclosure.

The Future

The latest development in carbon fiber technology is tiny carbon tubes called nanotubes. These hollow tubes, some as small as 0.00004 in (0.001 mm) in diameter, have unique mechanical and electrical properties that may be useful in making new high-strength fibers, submicroscopic test tubes, or possibly new semiconductor materials for integrated circuits.

Where to Learn More

Books

Brady, George S., Henry R. Clauser, and John A. Vaccari. *Materials Handbook.* McGraw-Hill, 1997.

Kroschwitz, Jacqueline I. and Mary Howe-Grant, ed. *Encyclopedia of Chemical Technology.* John Wiley and Sons, Inc., 1993.

Periodicals

Ebbesen, T.W. "Carbon Nanotubes." *Physics Today* (June 1996): 26-32.

Other

American Carbon Society website. http://www.ems.psu.edu/carbon.

Carbon Composites website. http://www.carb.com.

—Chris Cavette

Carbon Monoxide Detector

According to the U.S. Consumer Product Safety Commission, more than 2,500 people will die and 100,000 will be seriously injured by carbon monoxide over the next 10 years.

Background

A carbon monoxide detector is an electronic device that senses the presence of carbon monoxide (CO) in a building and sounds an alarm to warn the occupants to escape. Carbon monoxide is an odorless, poisonous gas, which can be generated by gas furnaces and water heaters, ranges, space heaters, or wood stoves if they are malfunctioning or not vented properly. Cars, portable generators, and gas-powered gardening equipment also generate carbon monoxide and can cause problems if they are operated in enclosed areas or attached garages. Once inhaled, carbon monoxide inhibits the blood's ability to carry oxygen by replacing oxygen in the red blood cells, preventing the oxygen supply from reaching the organs in the body. This oxygen deprivation can cause varying amounts of damage depending on the level of exposure. Low level exposure can cause flu-like symptoms including shortness of breath, mild headaches, fatigue, and nausea. Higher level exposure may cause dizziness, mental confusion, severe headaches, nausea, and fainting. Prolonged high level exposure can cause death. According to the U.S. Consumer Product Safety Commission, more than 2,500 people will die and 100,000 will be seriously injured by carbon monoxide over the next 10 years.

Technology used to detect carbon monoxide was originally developed for industrial applications. For example, the chemical industry uses a number of electronic gas sensors for analytical applications. Early industrial sensors involved a dual chambered sensor, which oxidized carbon monoxide and compared the heat of oxidation from the test chamber to a reference chamber. This type of oxidation requires a special platinum oxide catalyst and a heat source to burn the carbon monoxide. These systems were unacceptable for home use due to their complexity of operation, expense, and lack of sensitivity. However, in the last decade or so, home carbon monoxide detectors have become possible through improvements in advanced gas sensing technology. Other key factors have also contributed to the increased popularity of CO detectors. One is the rise in the use of other home safety appliances, such as smoke alarms. Another is the increased awareness of the dangers of carbon monoxide. Today, relatively inexpensive CO detectors can be purchased for as little as $30-$80. In fact, many cities are now requiring that at least one smoke detector be installed in every home, apartment, and hotel.

Design

The most important design factor for a CO detector is the type of sensor it employs. Home detectors may be designed with several different types of sensors. The simplest type is known as a detection card. These are fiberboard cards printed with a dot that chemically changes color when exposed to carbon monoxide. This type of detector does not sound an alarm and requires regular checks to determine if it has been exposed to carbon monoxide. While they are inexpensive ($4-$18), they do not offer sufficient protection to be used as a primary detector. The bio-mimetic gel sensor is a more sophisticated technology, which is designed to mimic the body's response to carbon monoxide by continually absorbing gas. However, because this type of sensor constantly absorbs carbon monoxide, it cannot reset itself to zero properly and is therefore more prone to false alarms. Furthermore, the bio-mimetic gel sensor can take up to 48 hours to reset after exposure during which time the occupants of the home are unprotected. Metal oxide sensors are more accurate and are the common type of sensor employed in home models. This type of sensor uses solid state tin-dioxide circuits, which clear quickly and continually monitor the air for presence of carbon monoxide. Detectors built with this technology can display CO concentration as a digital readout. When a specific CO level is reached, the detector sounds an alarm. However, these detectors have limited self-diagnostic capability to determine the efficiency or working condition of the sensor. Furthermore, they may be sensitive to gases other than carbon monoxide that are found in the home such as hair spray propellants. Finally, the accuracy of this type of sensor can drift by up to 40% after six months of use. Another type of sensor employed by certain manufacturers is the Instant Detection and Response (IDR) electrochemical sensing technology, which is claime to be the most effective detection method. IDR technology is used as an industry standard for professional sensing equipment and will instantly detect the presence of carbon monoxide. Detectors built with this technology will not react to other gases and are accurate to within plus or minus 3%.

Another important design factor is the type of power source for the detector. Both battery powered and AC powered detectors are available. Battery operated detectors are easy to install, easy to move, and continue to operate during power outages when emergency heating systems might be in use. However, batteries must be replaced at least every two years. On the other hand, AC powered plug-in detectors do not require battery replacement. These electrically powered units are able to clear a false reading within minutes. Plug-in detectors with battery back up are also available at a slightly higher cost. In addition to battery and plug in models, some models are available which can be hardwired in place. This style allows multiple detectors to be wired together so that they all sound an alarm when carbon monoxide is detected by any of one of the detectors.

Components

Carbon monoxide detectors are assembled from the following components: a sensor, which is capable of measuring the concentration of the gas and sending a signal when the carbon monoxide concentration reaches predetermined levels; a microprocessor, which is able to receive electrical signals from the sensor and able to send signals to the alarm horn and the control panel; a visual display (usually a Liquid Crystal Display (LCD) panel), which communicates CO level and other operating information; an alarm circuit capable of generating a sound loud enough to wake people sleeping in areas adjacent to the detector; a power connection (either an AC plug, battery connection, or both); a circuit board, which serves as a base for the electronic components; and a plastic housing which holds all the components together.

The Manufacturing Process

The production of a carbon monoxide detector involves three major steps. The first step is the fabrication of the individual electronic components and attachment of these components onto the circuit board. The second is the fabrication of the plastic housing. The third step involves the assembly of all the components, testing to confirm performance, and packaging for shipment.

Component construction

1 The circuit board is assembled from schematic plans by plating (or adding) copper onto the surface of the substrate in the desired pattern. The various diodes and circuit components are inserted into holes in the board and soldered into place. The major detector components such as the sensor chamber, alarm horn, and LCD display panel are usually manufactured separately by companies specializing in electronic componentry. These are often purchased by the smoke detector manufacturer.

Plastic housing fabrication

2 The plastic housing is made by an injection molding processes in which plastic resin and other additives are mixed together, melted, and injected into a two piece mold

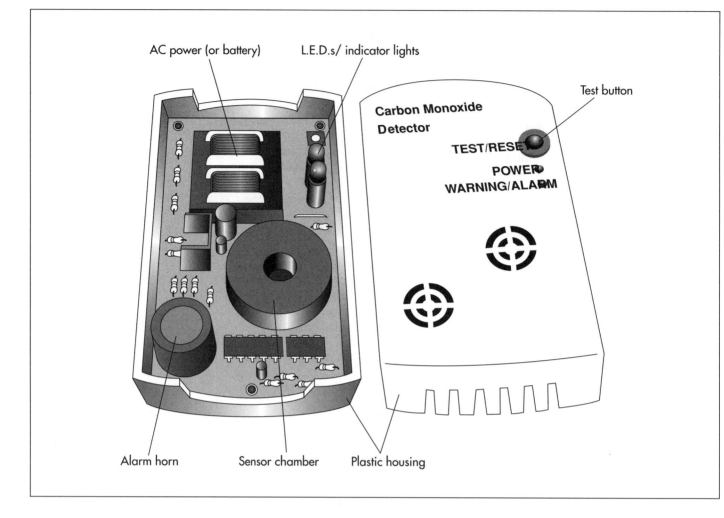

AC power (or battery)   L.E.D.s/ indicator lights

Test button

Carbon Monoxide Detector

TEST/RESET

POWER
WARNING/ALARM

Alarm horn   Sensor chamber   Plastic housing

The production of a carbon monoxide detector involves three major steps. The first step is the fabrication of the individual electronic components and attachment of these components onto the circuit board. The second is the fabrication of the plastic housing. The third step involves the assembly of all the components, testing to confirm performance, and packaging for shipment.

under pressure. After the plastic has cooled, the mold is opened and the plastic housing is ejected. The pieces may require manual brushing to smooth small imperfections or burrs in the plastic.

Final assembly and packaging

3 The circuit board is installed in the housing and appropriate connections are made. A test button is installed, a mounting bracket is added, and a cover is put in place to complete the assembly. The appropriate identification and warning labels are applied with pressure sensitive adhesive. A representative number of units are performance tested prior to packaging. Finally, the units are boxed and shipped to distributors.

Quality Control

The key quality control feature of CO detector manufacture is the calibration of the sensor. The higher quality CO detectors are

really gas monitors, which continually assess local concentration of CO compared to an internal standard. This calibration process allows the sensors to discriminate between a normal background level of CO and a dangerously high concentration. Under normal conditions, an acceptable background level may be as high as 25-35 parts per million (ppm). Harmful exposure can result if the concentration reaches the 75-100 ppm range. The Underwriter Laboratory standards for CO detectors requires them to sound an alarm within 90 minutes of exposure to 100 ppm CO; within 35 minutes when exposed to 200 ppm; and within 15 minutes when exposed to 400 ppm. Early CO detectors had to be manually calibrated by placing the instrument in an environment of known CO concentration and measuring the results. However, this process was expensive and time consuming and was therefore only used for expensive industrial equipment. With the increasing popularity of home units, more efficient calibration methods were required. High quality modern detectors are equipped with internal calibration features, which can perform low-level gas emission diagnostic tests on a regular basis to confirm the accuracy and operating status of the sensor. If the detector discovers any problem with the sensor, it emits a special sound pattern to alert the occupants the sensor is defective. In addition, each detector is fitted with a test button to allow the alarm circuit to be manually evaluated.

Underwriters Laboratory (UL) has issued quality standards which have been adopted by the CO detector industry. As of October 1, 1995, a detector should bear the number "UL 2034" if it complies with current safety and quality standards.

The Future

The future of carbon monoxide detectors is constantly evolving as improvements are made in gas sensing electronics. The IDR technology previously described is one example of this cutting edge technology. Future detectors will also incorporate similarly advanced features. Increased controllability offered by computer-controlled interfaces will make future devices more user friendly. These offer consumer benefit in the combined safety devices. For example, future generations of computer-controlled detectors may be linked with the household appliances, which are most likely to generate carbon monoxide, such as gas furnaces or hot water heaters. When the unit senses unacceptably high CO levels, it will send a signal to the appliance to terminate the combustion process and shut off the gas flow so no further carbon monoxide is released. As new models become available with improved sensitivity and other value-added features, CO detectors will be made even more user-friendly and will be even more useful as lifesaving devices.

Where to Learn More

Periodicals

"Air Conditioning." *Heating and Refrigeration News* (March 6, 1995).

Other

US Patent #5132231 (Carbon Monoxide Detector).

US Patent #5659125 (Automatic Calibration Method for Carbon Monoxide Monitors).

—Randy Schueller

Carousel

Today, only two or three carousel makers practice their craft in the United States although there are many hobbyists who carve their own horses and refurbish antiques.

Background

The precursors to the carousel may be as much as 1,500 years old when baskets lashed to a center pole were used to spin riders around in a circle in ancient Byzantium. During the twelfth century in Turkey and Arabia, men and their horses played a game in which delicate balls filled with perfumed water were tossed between riders. Losers would sport a definite aroma, and winners were presumably the better horsemen. The game was called *carosello*, or little war in Italian. At the French court in about 1500, this game blossomed into an elaborate pageant with spectacularly outfitted horses and riders. Horsemen added the challenge of trying to lance a small ring while galloping at full tilt. If the rider snagged the ring, it pulled away from a tree or posts with a stream of ribbons behind it. Contestants could practice this game by mounting wooden "horses" that were legless and resembled vaults used in gymnastics and that were mounted to a circular platform. As the platform rotated, the riders would try to spear the brass ring.

Craftsmen observed this play among the nobility and began building platforms with wooden horses mounted on them for commoners and their children to ride. These carousels were quite small because the power source for turning the carousel was a mule, man, or horse. In 1866, Frederick Savage, an English engineer, combined steam power with his carousels and drew crowds to the European fairs he toured with his machine. Steam-driven carousels reached the United States in about 1880. Savage was also responsible for developing the system of overhead gears and cranks that allow the suspended horses to move up and down as the carousel turns and simulate an actual ride on horseback. As carousels became more popular, they acquired a number of names including *karussell* (Germany), *carrousel* and *manèges de chevaux* (France), gallopers and roundabouts (England), and merry-go-rounds, whirligigs, spinning or flying jennies, dip-twisters, and flying horses (United States). Today, preservationists tend to prefer the name carousel over these others for its historic context.

The jewels of the carousel have always been the horses. Thanks to the stream of immigrants from Europe, the United States had a thriving carousel industry by the 1870s. Expert carvers, such as Gustav Dentzel from Germany, had practiced cabinetry and carousel crafting in their homelands and quickly established businesses in America. Carousel factories like The American Merry-Go-Round & Novelty Company were full-time manufacturers, but other makers including Charles Looff and Charles Dare in New York City, Dentzel in Philadelphia, and Allan Herschell in upper New York state transformed their furniture businesses and machine shops into at least part-time carousel production. Wood workers and carvers prided themselves on fashioning beautiful crested horses with flashing eyes, flying manes, realistic poses (for both standers and jumpers), and ornate ornamentation from flowers to heraldic crests, French fleurs-de-lys, jeweled saddles and tassels, and patriotic symbols like eagles and profiles of presidents. Of the carousel figures made in the United States, 80% were horses and 20% were made up of a menagerie. The Herschell-Spillman Company produced kangaroos, pigs, giraffes, sea monsters, frogs, and dogs and cats.

The carousel's zenith in America was from about 1900 to the Depression. During this period, jobs were plentiful, motor transport was available, and amusements for the family were sought. Craftsmen were also still in demand, but as technology advanced, it also invaded the carousel business. Factories began to build cast aluminum horses (and animals cast in fiberglass and plastic soon followed), and the carvers had to find other trades. Repair work was available as the wooden horses aged, but often amusement park operators resorted to patchwork maintenance instead.

In the early 1970s, the National Carousel Association was formed. Antique horses began to sell on the auction blocks of Sothebys and Christies at phenomenal prices, and collectors sought to acquire originals by carvers like Salvatore Cernigliaro or Marcus Charles Illions. For those with smaller pocketbooks, bisque porcelain figures and small-scale carousel horses also became collectible. Today, only two or three carousel makers practice their craft in the United States although there are many hobbyists who carve their own horses and refurbish antiques.

The Basics of Carousel Operation

The carousel revolves around a stationary center pole made of metal or wood. An electric motor drives a small pulley that is controlled by a clutch for smooth starts. This pulley turns a drive belt and a larger pulley that turns a small-diameter, horizontal shaft. The end of the shaft is a pinion gear that turns a platform gear. The platform gear supports a vertical shaft that turns another pinion gear and final drive gear attached to the support beams of the carousel, called sweeps, which extend outward from the center pole like the ribs of an umbrella and support the platform, horses, and riders. The sweeps hold cranking rods that are turned by small gears at the inner ends that are driven by a stationary gear on the center pole. Horse hangers are suspended from the cranks, and as they turn, the horses move up and down about 30 times per minute. A typical carousel platform with horses and riders may weigh 10 tons and be driven by a 10-horsepower electric motor. After the motor's revolutions are reduced by the series of gears, the riders on the outer row

Carousel griffin made by the Hershell-Spillman Co. of North Tonawanda, New York. (From the collections of Henry Ford Museum & Greenfield Village, Dearborn, Michigan.)

Carousels, with colorful figures attached to a revolving horizontal mechanism, have amused the masses since the end of the 1700s. By 1800, carousels were advertised as amusements as well as an activity that got the blood circulating. After the Civil War, a number of merry-go-round manufacturers started up businesses and popularized the carousel.

The griffin depicted here is the product of Hershell-Spillman Co. of North Tonawanda, NY, a well-known carousel manufacturer nearly a century ago. It is part of a 1913 merry-go-round now operating six months of the year in Greenfield Village in Dearborn, Michigan. With its one-man band mechanism playing charming turn-of-the-century tunes, this carousel remains a great favorite. It was a "park-style model," manufactured for permanent placement in a park. (Some, like one still operating in Story City, Iowa, were designed to travel and were not for permanent installation.) Children can ride horses, lions, tigers, zebras, and even leaping frogs—all exquisitely carved and painted. Today, carousel figures are treasured for their colorful beauty, and carousel figure collectors pay thousands of dollars for a single animal.

Nancy EV Bryk

of mounts will gallop along at about 5-11 miles per hour.

Raw Materials

The two primary materials for a carousel are metal and wood. The metal mechanism includes the electric/hydraulic motor, gears, bearings, and crankshafts. Horse hangers and platform suspension rods are metal with brass sleeves, and the center pole is steel. The wood parts of the carousel include the

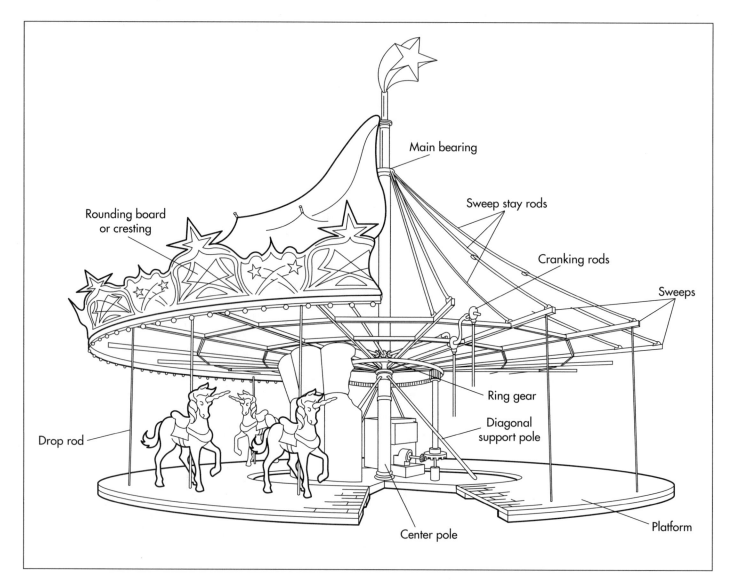

A typical carousel platform with horses and riders may weigh 10 tons and be driven by a 10-horse-power electric motor.

horses, which are carved from basswood, the oak platform, sweeps, rounding boards, panels, and mirror frames. The platform and various panels and gingerbread work were made of wood or plaster in the old days, and today they may be made of these same materials or may be cast in plastic or fiberglass. The tent-like top is made of canvas. Music is supplied by a band organ that is also electrically or mechanically powered and plays much like those of a player piano. Specialty manufacturers provide the organs.

Design

Design of a carousel begins in the middle at the center pole. A bearing at the top of the pole bears the entire weight of the carousel. The sweeps (arms or umbrella-like ribs) of the carousel are suspended from the top bearing, and two rods extending down from each sweep support the platform. About half-way down the center pole is a center bearing or hub that keep the works from shifting from side to side. The motor, of course, spins the whole umbrella structure around. From the midpoint, a series of diagonals keep the center pole aligned with a cross-brace that rests on the ground. A center pole that is 15 in (38.1 cm) in diameter will support about 50 horses and riders.

The Manufacturing Process

The basic process of manufacturing a carousel has not changed despite the fact that few are built today. No new carousels

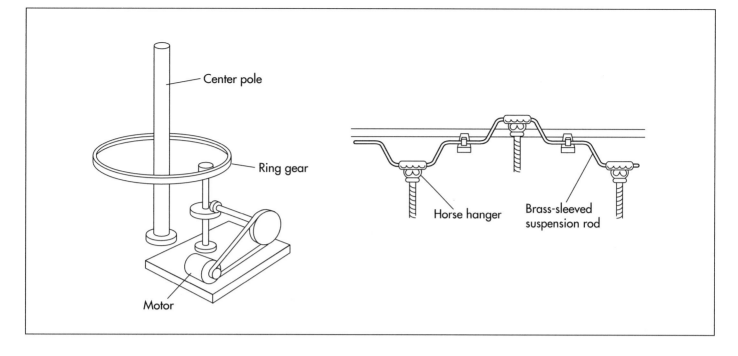

Center pole

Ring gear

Motor

Horse hanger

Brass-sleeved suspension rod

with wooden animals were made in the period from the 1930s to 1994, and only about 10 new wood-horse machines have been made in the United States in the last 60 years. All new wooden carousels are, therefore, custom-made, although refurbishing old machines is also a part of this industry.

1 When a carousel builder receives an order, he works with the customer to determine the size of the machine based on cost and maintenance considerations. Most carousels today may be built from scratch or from refurbished mechanisms. In either case, today's maker must be a metal fabricator with a shop to concoct the metal pieces. Traveling carousels can also be manufactured with little variation from those fixed in place except that the steel center pole is hinged so it will fold in half to be transported in a van. All the other pieces can be broken down by two men in about 3 hours and carried in a truck including a platform made in sections and the horses that have the metal hangers removed from them.

2 The carousel builder purchases a band organ for the "new" carousel from a specialty manufacturer. Today, the Stinson Organ Company of Ohio is the only manufacturer to still make custom and product-line band organs; five or six manufacturers including Wurlitzer made band organs during the heyday of the American carousel,

and many other organs were imported. Band organs are played by air pushed by a bellows through wooden pipes, stops, and valves. Because the wood parts are highly subject to temperature and humidity changes, the organs are constantly out of tune and require considerable maintenance. They make music by forcing air through perforated paper rolls, much like a player piano, and the rolls cycle continuously from one to another, thanks to a device called a tracker frame developed by Wurlitzer. Although there is no sound like a real carousel band organ, maintenance costs often force buyers of "new" carousels to use tapes or compact discs for their music.

3 The romance of the carousel rides with its animals, however, and the auction prices of up to $60,000 per horse for antique ponies have brought a new awareness to the importance of equipping a new carousel with the genuine article. Amateur carvers and woodworkers have also been attracted to the carving of carousel critters; and body blocks that include the body, head and neck, tail, and legs can be purchased in various sizes for carving.

4 Assuming the carousel builder is carving a wooden horse from scratch, he chooses a size and weight suitable to the overall design of the carousel and selects an appropriate artistic design. This may be based on

The carousel revolves around a stationary center pole made of metal or wood. An electric motor drives a small pulley that is controlled by a clutch for smooth starts. Horse hangers are suspended from the cranks, and as they turn, the horses move up and down about 30 times per minute.

a theme for the carousel or on the customer's favorite historic model. The outside of the carousel animal is called the "romance" side and is seen by onlookers. This display side is usually more ornately decorated than the inside. Many of the original carousel designers made full-scale sketches of their horses so that details were properly conceptualized and scaled and so that several carvers could work on parts of the same horse. Today, one-eighth scale models are sometimes made for the designer's models and the customer's approval.

5 Full-scale paper patterns are glued to pieces of basswood to cut the body, legs, and other parts. Basswood is used because it is hard and close-grained, and the grain must run the length of the part for strength. A jigsaw or coping saw is used to cut the parts, and the parts are glued together to form the carving block. In the old days, the "glue-up" was done by a skilled craftsman who was an expert at the types of glue, amounts, and pressures required to prepare the carving block.

6 Each carver has his own preferences for how to proceed with carving. Many start by using the paper patterns still glued to the animal to rough-cut the shapes, a process called boasting. Detailed carving follows, and this is usually done without reference to patterns but with a sense of the wood grain and the artistic creation that has been trapped in the wood. The completed carving block is sanded, and sometimes other small details are glued on. If the animal is for display, rather than for a working carousel, a base or stand suitable to the size and configuration of the beast is also made. A footrest is also made for each animal (unless it has stirrups or other substitutes), and these are carved and painted to match the animal. Preparation of a carousel animal up to the painting stage typically takes about 35 hours per animal.

7 The carved horse is stained, primed and painted, and varnished to suit the design of the carousel. Platforms are painted to complement the carving, but overhangs are sometimes brightly colored to highlight the details of both the overhangs and the animals. Traditionally, the animals are painted bright colors, and the paints are chosen for durability and safety as well as appearance.

Removal of old paint may be one of the most time-consuming tasks in refurbishing old animals; some have as many as 30 coats of paint that have filled the finely carved details. Dappling, addition of gold or silver leaf, placement of horsehair tails, and burnishing of metallic leaf provide other touches of realism and elegance. Rhinestones and other jewels are often added to the romance side of the finished horse.

The Future

According to carousel maker Chuck Kaparich, carousels and carousel animals are experiencing a resurgence. Thanks to the drawing power of the colorful carousel display and music, historic town centers and shopping malls are commissioning new carousels or refurbishing their old ones to attract customers to these areas. Kaparich expects the romance of the carousel to always remain with the American public, but, realistically, he acknowledges the limited demand and the likelihood that the present resurgence may only have a lifespan of 10 to 20 years before carousels are again temporarily forgotten. It is hoped that new generations of carousel aficionados will recall the current boom and add the magic of their mounts, music, and motion to a bank of undying memories.

Where to Learn More

Books

Fraley, Nina. *The American Carousel.* Benicia, CA: Redbug Publishing, 1979.

Fraley, Tobin. *The Great American Carousel: A Century of Master Craftsmanship.* San Francisco: Chronicle Books, 1994.

Fried, Frederick. A *Pictorial History of the Carousel.* Vestal, NY: Vestal Press, Ltd., 1964.

Marlow, H. LeRoy. *Carving Carousel Animals From 1/8 Scale to Full Size.* New York: Sterling Publishing Co., Inc., 1989.

Other

Carousel News & Trader. http://www.carousel.net/trader/.

Chuck Kaparich, Carousel Man of Missoula, Montana. http://www.carousel.net/kaparich.

Classic Carousel Collectibles. http://www.finest1.com/carousel/.

—*Gillian S. Holmes*

Cast Iron Stove

The pollution produced by woodburning stoves led the Environmental Protection Agency (EPA) to issue regulations in 1988, which required all newly manufactured woodburning stoves to meet standards for emissions. Stove manufacturers developed improved technology to produce cast iron stoves which were highly efficient and which produced very little pollution.

Background

A cast iron stove is a device, built from a material consisting of iron mixed with carbon, in which a solid fuel such as wood or coal is burned to produce heat for warmth or cooking. The stove usually consists of a grate, which holds the fuel, a hollow interior in which the fuel burns, flues through which hot air flows, and baffles to slow down the flow of hot air, allowing the stove to produce more heat.

Human beings have burned wood and other natural fuels to provide warmth and to cook food since prehistoric times. At first, open fires were used. A major drawback with this simple method was the fact that much of the heat of the fire was wasted as it escaped in the form of hot, rising air. Prehistoric people soon learned to build the fire against a flat rock standing up in such a way as to reflect back heat. For cooking, the fire could be used to heat a pit dug into the ground or a hearth made of thin, flat rocks.

The ancient Romans developed a heating system known as a hypocaust, consisting of a series of flues beneath a tiled floor, which carried hot air from a fire to all parts of a room. A hypocaust could also be used to heat a cauldron of water for cooking or bathing. Similar systems of heating are still used in China and Korea.

Despite the early development of this sophisticated device, until the Middle Ages most Europeans relied on open fires on a central hearth beneath a hole in the roof to let out smoke. Fireplaces with chimneys began to appear in castles in Northern Europe around 1000 A.D. For hundreds of years, fireplaces were limited to large hous-es owned by the wealthy. In England, as late as 1600 A.D., fireplaces were still fairly uncommon.

The wealthiest homeowners had fireplaces with chimneys made of stone or brick, while the less prosperous had fireplaces made of mud and wattle. Wattle, a material consisting of vertical wooden rods or poles interwoven with horizontal sticks or reeds, was dangerous because it was flammable. Despite this hazard, mud and wattle chimneys were common in the United States as late as 1800.

Fireplaces were not much more efficient than open fires because much of the heat went out the chimney with the hot, rising smoke. Small fireplaces with gently burning fires were more efficient than large fireplaces with quickly burning fires. A device known as a curfew, consisting of a sheet of brass or tin, could be used to limit the flow of hot air, resulting in a steady, slowly burning fire. Curfews were used to keep a fire burning throughout the night without being tended.

The first stove to appear in historical records was built of brick and tile in 1490 in Alsace, a part of Europe on the border between France and Germany. In Scandinavia, stoves were built with tall iron flues and iron baffles. In Russia, stoves as large as 8 ft (2.4 m) tall containing as many as six thick-walled masonry flues were placed at the intersection of walls to heat four rooms at once.

Cast iron was first produced in China in the sixth century B.C. and in Europe during the twelfth century, but it was not used to any great extent until the seventeenth century.

The first cast iron stove was manufactured in Lynn, Massachusetts, in 1642. Early cast iron stoves consisted of flat rectangular plates bolted or pinned together to form a box. The plates were made by pouring molten cast iron into molds made of sand, a method still used today.

In 1744, Benjamin Franklin invented a more efficient cast iron stove known as a Pennsylvania fireplace. This device controlled the flow of hot air so that the smoke from the fire burned more completely, resulting in the release of more heat. This design was extremely successful and is still used today. A more complex device invented by Franklin in 1786 was not as successful, but its design anticipated modern cast iron stoves, which burn almost all the smoke from the fire.

Cast iron stoves remained mostly unchanged in basic design for about 200 years. In the 1970s, large increases in the price of heating oil led to an increase in the use of woodburning stoves. The pollution produced by these stoves led the Environmental Protection Agency (EPA) to issue regulations in 1988, which required all newly manufactured woodburning stoves to meet standards for emissions. Stove manufacturers developed improved technology to produce cast iron stoves which were highly efficient and which produced very little pollution.

Raw Materials

Cast iron is a substance consisting of iron and between 2-4% carbon. Various small amounts of silicon, manganese, sulfur, and phosphorus are also present. For special applications, various amounts of nickel, chromium, and molybdenum may be included to produce cast iron which is resistant to heat, wear, and corrosion.

A modern cast iron stove may contain as little as one-third cast iron. The rest of the stove consists mostly of steel. Steel is a substance consisting of iron and, in most cases, between 0.01-1.2% carbon. Some special forms of steel may contain as little as 0.003% carbon or as much as 2% carbon. Steel may also contain various amounts of manganese, silicon, aluminum, nickel, chromium, cobalt, molybdenum, vanadium, tungsten, titanium, niobium, zirconium, ni-

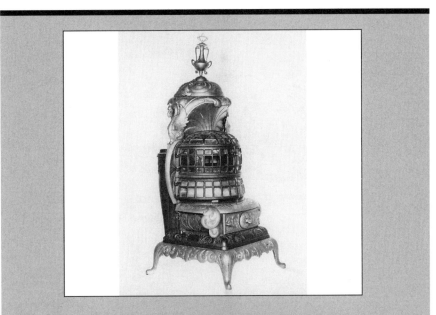

A cast iron stove produced by the Michigan Stove Co. in 1882. (From the collections of Henry Ford Museum & Greenfield Village, Dearborn, Michigan.)

Many of us lick our lips at the thought of waking up to a farm-style breakfast made on a cast-iron kitchen stove. After the 1850s, stove manufacturers produced large models on which a farmwife might cook bacon, eggs, hash browns, and corn beef hash on top, with cinnamon rolls rising in the stove. The position of the burners on these stoves dictated the temperature of the burners, so the cook knew the best burner for "simmering the coffee." Since there was no thermostat on the stove, the cook learned to regulate the temperature based on the look and feel of the fire (wood or coal, depending on the model).

Cast iron stoves were also used to keep rooms warm. Fireplaces are notoriously inefficient ways to heat rooms without central heat. By 1860, many families boarded up the fireplace and put a parlor stove like this one on the hearth and vented the stove through the chimney. This "art garland" model was so called because of its decorative cast iron scrollwork, nickel, and fashionable shape. Produced by the Michigan Stove Co. in 1882, it includes isinglass, or sheets of thinly-sliced mica, which serves as glass in the iron grills in front so that one could see the glow of the flames but not feel the full intensity of the heat.

Nancy EV Bryk

trogen, sulfur, copper, boron, lead, tellurium, and selenium.

Other materials which may be used in cast iron stoves include: ceramics (various materials made from nonmetallic substances subjected to high temperatures), firebrick (a type of brick made from heat-resistant clay),

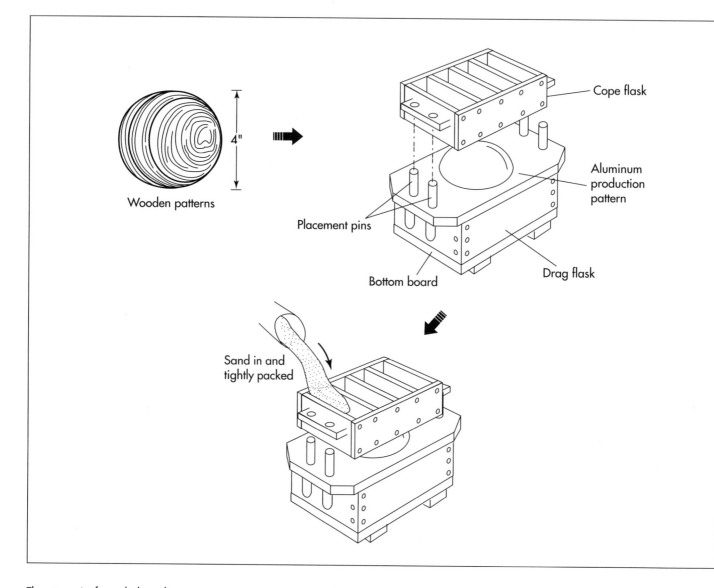

Wooden patterns

4"

Cope flask

Aluminum production pattern

Placement pins

Bottom board

Drag flask

Sand in and tightly packed

The stove is formed through a process called sand casting. Once sand is hardened into a mold in the shape of a stove, molten cast iron is poured into the cavity and allowed to cool.

and platinum or palladium (metallic elements used in catalytic converters which cause smoke to burn at a lower temperature, resulting in the release of less pollution).

The Manufacturing Process

Making cast iron

1 Iron ore is dug out of the earth in surface mines. The ore is obtained in lumps ranging in size from more than 40 in (1 m) in diameter to less than 0.04 in (1 mm) in diameter. To produce cast iron or steel, the lumps must be 0.3-1 in (7-25 mm) in diameter. Lumps of ore, which are too large, are crushed and pass through sieves, which separate the resulting material by size. Lumps

that are too small, known as fines, are melted together into larger lumps, a process known as sintering.

2 The lumps of iron ore are mixed with coke, a carbon-rich substance produced by heating coal to a high temperature in the absence of air. A conveyor belt moves the mixture, known as charge, to the top of a blast furnace. A blast furnace is a tall, vertical, steel shaft lined with firebrick and graphite. Air is heated to a temperature of 1,650-2,460° F (900-1,350° C) and blown into the blast furnace. As the charge descends the coke burns in the hot air to produce carbon monoxide and heat. The carbon monoxide reacts with the iron oxides in the iron ore to produce free iron and carbon dioxide. The final result of this process is molten

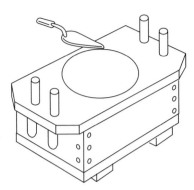

The result is imperfect castings in sand that are carefully perfected.

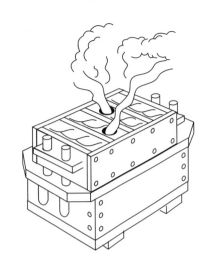

The cope flask and the drag flask are resealed. A cavity is formed where the aluminum pattern has been removed. Molten alloy is injected or poured into the cavity, creating a casting.

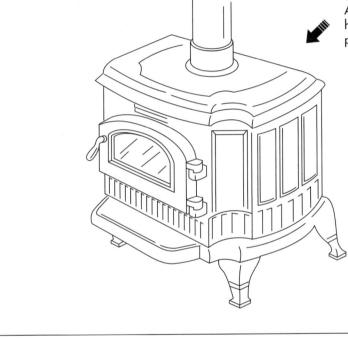

pig iron, which consists of at least 90% iron, 3-5% carbon, and various impurities.

3 The molten pig iron is poured into large molds and allowed to cool into a solid. It is then mixed with scrap metal, which has been selected to give the mixture the desired combination of raw materials. This mixture is moved by conveyor belt to the top of a cupola, which resembles a small blast furnace. The pig iron and scrap metal fall on a bed of hot coke through which hot air is blown. This process removes the impurities and a small amount of the carbon, resulting in molten cast iron.

Shaping cast iron

4 Cast iron, as its name suggests, is usually shaped by pouring the molten metal into a mold, a process known as casting. The most common method is known as sand-casting. A pattern in the shape of the desired final product is shaped form wood, metal, or plastic. It is then firmly packed into sand held together with various substances known as bonding agents. The sand is hardened by heat or by chemical bonding with various substances known as bonding agents. The sand is hardened by heat or by chemical bonding produced by including sodium silicate in the original sand mixture.

After hardening, the pattern is removed, leaving a cavity in the sand. Molten cast iron is poured into the cavity and allowed to cool, resulting in solid cast iron in the desired shape.

Assembling the stove

5 Cast iron components and steel components are shipped from the iron and steel company to the stove manufacturer and inspected. Before assembly, the cast iron components must be polished. A surface grinder is used to remove about one-sixteenth of an inch (1.6 mm) of the cast iron, resulting in a very smooth, glossy surface. A typical surface grinder is a sheel about 14 in (35.6 cm) wide consisting of a hard, abrasive material known as grinding rock. It spins at about 1,800 revolutions per minute as it grinds away the surface of the cast iron.

6 The polished cast iron components are assembled with steel bolts. The bolts are started by hand, then tightened by machine to ensure that the cast iron components are firmly bonded together with no leaks. Various other components, such as firebrick linings or catalytic converters, are assembled into the stove at the same time.

7 The completed stove is inspected again for any cracking, which may have taken place during the tightening of the bolts. It is then treated with oil to prevent rusting and packed with expanded polystyrene foam (a very light but strong plastic foam) into cardboard boxes to be shipped to the consumer.

Quality Control

During the manufacturing of cast iron, the most important factor in producing metal with the desired characteristics is controlling the amount of elements other than iron and carbon present in the final product. In particular, the amount of silicon present produces two very different forms of cast iron.

White cast iron (named for the bright surfaces seen when the metal is fractured) contains more than 2% silicon. It is not as hard as white cast iron but it is easier to cast and shape with machines. Gray cast iron is the material used to make cast iron stoves.

During the assembly of the stove the cast iron components are inspected for pits, cracks, and rust. After bolting the components together the stove is inspected to ensure that the stove is airtight. This prevents smoke from escaping from the body of the stove rather than moving through the flues. When the stove is oiled to prevent rusting, it is inspected to see if any oil seeps through the metal, indicating that a crack is present.

The Future

EPA regulations introduced in the 1980s required manufacturers of woodburning stoves to reduce the amount of emissions produced. Manufacturers have complied with these regulations in three ways. Some have installed catalytic converters, in which a ceramic honeycomb coated with platinum or palladium causes smoke to burn more completely. Some have created so-called "high tech" stoves, which improve existing technology so that fuel burns more efficiently.

The most radical change is the development of pellet-burning stoves. Instead of ordinary wood, these stoves burn small pellets formed from wood chips, sawdust, bark, and other wood scrap, which is dried, pulverized, and compressed. Because they are much drier than ordinary wood, these pellets burn extremely cleanly. Pellet-burning stoves are also easier to use than traditional woodburning stoves. The owner only needs to purchase the pellets and load them into the stove's automatic hopper, which controls the rate at which fuel is added to the fire. This recent innovation in the technology of woodburning stoves ensures that the seemingly old-fashioned cast iron stove will continue to be used well into the twenty-first century.

Where to Learn More

Books

Adkins, Jan. *The Art and Ingenuity of the Woodstove*. Everest House, 1978.

Sanders, Clyde A. and Dudley C. Gould. *History Cast in Metal: The Founders of North America*. Cast Metals Institute, 1976.

Periodicals

Turbak, G. "A New Generation of Wood Stoves." *Mother Earth News* (December/January 1992): 66-70.

Vivian, J. "A Clean Burn." *Country Journal* (November/December 1991): 60-64.

—Rose Secrest

Change Machine

Background

A change machine is a device used to exchange one form of money for another, typically paper currency for coins. Sensors in the machine detect the type of bill that is fed into it and relay this information to a microprocessor. The processor then sends commands to coin hoppers to dispense the appropriate coins. Similar devices for accepting coins and bills were originally developed for use in coin-operated amusement and gaming devices, such as slot machines. These devices evolved into machines that could quickly dispense proper change without the need for a cashier. These early machines were similar to devices used today in that they were required to accept bills and dispense proper coin equivalents. However, they differed in the sophistication of their components. For example, they used a simple pay out device such as a long tube full of coins with a solenoid located at the bottom. The solenoid, which consists of a coil of wire wrapped around a metal plunger, acted like a shutter. When the machine logged in a bill, it sent a pulse of electricity through the solenoid coil, which caused it to become magnetized. Once magnetized, the plunger moved, allowing a coin to fall into the collection chamber below. Another early version used a rotating disk below a coin-filled hopper instead of a solenoid. The disk would turn one notch in response to a signal from the machine's electronic "brain." This action would allow the appropriate coins to fall into the collection area. These original devices were simple but advances in electronics over the last 30 years have resulted in significant improvements. Today, change machines are very sophisticated, with microcomputers controlling the way the change is dispensed. Modern change machines are capable of quickly dispensing large amounts of change while carefully discriminating acceptable bills from counterfeit or tampered ones.

Design

While the demand for money-exchanging equipment is relatively high, these machines are durable and last a long time. In fact, the average lifetime of a change machine is about 20 years. Therefore, the market is limited to a small group of companies in the United States and abroad. These companies design their change machines based on modular components to best suit their customers' needs. The specific components used in each machine can be varied to deliver the desired features. When the manufacturer receives an order for a new machine, they must determine the customer's requirements and select the appropriate components to include. While specific requirements may vary from machine to machine, the basic elements are similar. For example, for a machine intended to change U.S. dollars into coins, the following design factors must be taken into consideration.

Bill/coin input

One key consideration is the type of bill and/or coin the machine will be required to accept. This will impact the type of sensor employed in the machine. Most change machines are designed to accept one and five dollar bills, although their sensors can be programmed to accept any denomination. Some machines are designed to change a larger coin (quarters for dimes and nickels). In this case, the machine must be outfitted with a coin discriminator as well.

Pay out

Pay out is the term used to describe the coins dispensed by the machine. Coins are stored inside the machine in hoppers separated by denomination. Therefore, the number of different coins that the machine is required to pay out will determine the number of hoppers in the machine. Most machines require three hoppers (for quarters, dimes, and nickels). However, some models incorporate as many as eight hoppers. The additional capacity may be useful for dealing with certain foreign currencies. In addition to the number of hoppers, other considerations related to pay out include the speed with which the coins are dispensed and the coin storage capacity. For applications which require fast pay out, change machines are available which can dispense up to 500 coins per minute. For other applications, it is more important to be able to go longer periods of time without refilling the coin hoppers. The customer must decide which features are most desirable.

Location

The location of the machine is also important. Different sizes and shapes are available; the larger units are designed as free standing floor models; smaller units can be mounted on the wall. These wall-mounted units take up little space and can be conveniently installed flush with the surface. However, they have less coin holding capacity and fewer advanced features. The floor models are larger and heavier with better security measures, more advanced features, higher coin output, and larger capacity.

Security

Given the large amounts of money stored in these machines security is a major design consideration. The machines must be built solidly to deter mechanical break in and they must be able to detect and reject counterfeit bills or bills that have been tampered.

Components

Housing

The change machine is housed in a metal cabinet constructed of steel plate, the dimensions of which vary depending on customer requirements and cost. The sidewalls of the housing may be constructed from 7.9 in (2mm) thick steel plate with 11.8 in (3 mm) plate used for the folded steel doors. These doors may be equipped with a double edge to restrict forced entry. All cabinets are equipped with some form of locking mechanism and an alarm is usually built into the cabinet. Some units are fitted to base stands that include a locking cupboard for storage. The cabinets are available in a range of stock colors or they can be custom painted to provide the customer with a specific finish. These cabinets can even be configured so that they are suitable for outdoor operation. One manufacturer even offers a model fitted with a heater and ventilator to keep bills free from moisture when used outdoors. The cabinet also houses the power source for the bill changer. These machines are alternating current (AC) powered and may be plugged into a normal outlet, or they may be hard wired in place.

Input sensors

For the machine to operate properly, it must be capable of recognizing the bills or coins from which it is to make change. To accomplish this, machines are equipped with input sensors, which determine the identity of the item being entered. For paper currency, a cluster of light-emitting diodes (LEDs) is used to shine light onto the bill as it is inserted. An optical sensor analyzes the light that is reflected off the bill's surface to determine its denomination. Sophisticated electronics and software allow the sensors to differentiate badly worn or torn bills from counterfeit ones. These systems are easily programmable to allow the user to teach the machine to recognize new bills. Coins, on the other hand, require a more involved identification technique because sensors can not read the metal surface the way they can paper currency. Instead, coin identification is done by measuring electromagnetic characteristics of the metal coin, as well as its physical properties such as weight and/or diameter. When a coin enters the input slot, it rolls down a chute until it reaches two hook-shaped balance arms. These arms are designed so that a coin of the correct weight will depress the right arm enough to move a counter-weight on the left. The coin can then move past the correct arm and travel

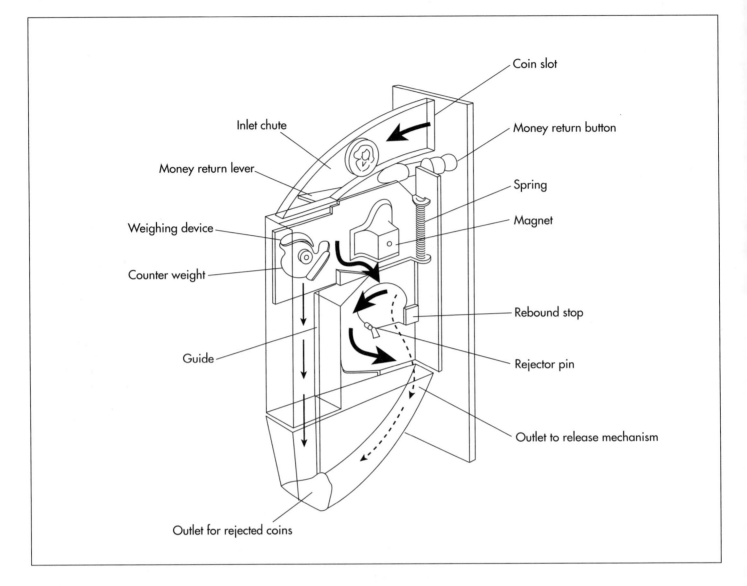

A cross section of a change machine mechanism.

further down the chute. The coin must also have the correct diameter to continue. Coins, which are too narrow, slip down an exit channel and are rejected. Coins of the right size and weight will continue to move down the chute until they pass a powerful magnetic field. If the coin is a fake and is too rich in iron, a magnet will grab it and pass it out the reject slot. If the coin is genuine, it will continue down the slot at a certain speed, which will allow it to jump over a rejector pin. A release mechanism signals the machine to proceed.

Bill transport and storage

The bill-input slot is equipped with small electrically powered rollers to pull the bill forward. The rollers are activated as soon as a bill is placed at the slot entrance. If the sensors reject the bill, the rollers reverse direction and spit it out. If the bill is accepted as valid, the rollers pull it inside the machine and feed it into a stacking chamber. The stacking chamber is a secure area that holds all the cash in a neat pile until an operator physically empties. Because of their weight, coins are easier to transport. When a coin is placed in the machine, it essentially falls down a shaft, which leads it to the coin discriminator. Once the coin has been accepted, it is allowed to fall into a storage bin.

Coin storage

Coins are stored inside the machine in metal hoppers; the total coin capacity varies depending on the model, but higher end machines can hold up to 12,000 coins. Access to these hoppers is through a door on top of the machine. This design maximizes the machine's coin capacity and permits coins to be added without opening any of the main panels.

Coin dispenser/pay out device

The pay out of coins is controlled by the microprocessor. The processor issues a signal to the pay out device in the form of electronic pulses. These pulses activate a rotating conveyor belt that travels in a loop from the bottom to the top of the hopper. The belt has small flights on it that act as scoops for the coins. As the belt travels around the loop, it passes through the pile of coins at the bottom of the hopper. The appropriate coins are caught on these protruding flights and are moved to the top of the hopper. At the top of the hopper the belt reaches the end of the track and turns back down. This causes the coins to fall off the belt and drop into a dispensing chamber. This type of pay out device can dispense up to 500 coins per minute.

Display/control features

All change machine functions are controlled by a microprocessor, which receives information from the input sensors and relays instructions to the pay out device. The computer also controls display information such as user instructions and error messages. Some simple machines are equipped with only an indicator light, which is extinguished when any hopper is nearly empty. More advanced models have illuminated 40-character Liquid Crystal Display (LCD) panels which provide the user with additional information including, but not limited to, step by step instructions. Some models even allow users to select the combination of coins to be dispensed and provide usage instructions in multiple languages. In addition, these advanced models have sophisticated onboard computer controls featuring extensive auditing facilities. Optionally, an audit printer can be connected to the machine to download and print a complete diagnostic profile or a summary of all audit information. Other advanced control features include built in burglar alarms and customized lighting arrays.

The Manufacturing Process

Computer-aided design (CAD)

1 The first step in the manufacturing process is to design the machine based on the customer's requirements. Since every customer's requirements are slightly different, each machine is custom built. The design factors listed above are discussed with the customer in order to optimize performance. Once the design is finalized, the necessary components are ordered from the appropriate vendors. A CAD program is used to determine the layout of the components and the size and shape of the cabinet.

Cabinet fabrication

2 The final design information is sent to a CNC metal-cutting machine, which cuts the steel plate into the appropriate sizes. These pieces are primed and painted as necessary. During this stage, they may be screwed or welded in place.

Component assembly

3 The various electronic and mechanical parts are installed in the cabinet on a partially automated assembly line. Each piece is welded, screwed, bolted, or clipped into place. This assembly process is partly done by hand by three to six people. At maximum capacity, approximately 100 machines can be made per month.

Quality control testing

4 After assembly is complete, each machine is plugged in and tested. The software is loaded and the machine is tested to ensure all the components are working properly. At that time if the customer has any special requirements, the additional programming can be checked. The machine is then crated and prepared for shipping.

Byproducts/Waste

The manufacturing process for change machines generates little if any waste material. Some scrap steel is created during fabrication of the cabinets, but this is minimized by the computer design and efficient automated cutting equipment. There are few wasted components because they are checked at the electronics vendor before shipping to the change machine manufacturer.

The Future

New developments in technology are changing the future of change machines. Improvements in sensor and bill-moving technology have allowed many vending machines to directly accept paper money. This advance may impact the change machine industry by requiring fewer freestanding machines to generate change. Another major factor impacting the future of this equipment is the advent of electronic-based money. Change machines of the future will require advanced computer linkage to track debits and credit electronically. A third factor affecting the future of change machines is the possibility that change machines of the future may dispense much more than just change. Already, some automatic teller machines (ATMs) are able to deal out movie tickets. Finally, change machine technology is being used by a new generation of automatic coin-sorting machines. These devices accept large quantities of loose change, count the coins at a rate of 600 per minute, and issue credits for cash.

Where to Learn More

Other

Coinage Limited, 5 Millfield Trading Estate Chard, Somerset TA20 2BB United Kingdom. +44 (0)1460 61771. http:\\www.coinage.co.uk.

—Randy Schueller

Charcoal Briquette

Background

Charcoal is a desirable fuel because it produces a hot, long-lasting, virtually smokeless fire. Combined with other materials and formed into uniform chunks called briquettes, it is popularly used for outdoor cooking in the United States. According to the barbecue Industry Association, Americans bought 883,748 tons of charcoal briquettes in 1997.

Basic charcoal is produced by burning a carbon-rich material such as wood in a low-oxygen atmosphere. This process drives off the moisture and volatile gases that were present in the original fuel. The resulting charred material not only burns longer and more steadily than whole wood, but it is much lighter (one-fifth to one-third of its original weight).

History

Charcoal has been manufactured since prehistoric times. Around 5,300 years ago, a hapless traveler perished in the Tyrolean Alps. Recently, when his body was recovered from a glacier, scientists found that he had been carrying a small box containing bits of charred wood wrapped in maple leaves. The man had no fire-starting tools such as flint with him, so it appears that he may have carried smoldering charcoal instead.

As much as 6,000 years ago, charcoal was the preferred fuel for smelting copper. After the invention of the blast furnace around 1400 A.D., charcoal was used extensively throughout Europe for iron smelting. By the eighteenth century, forest depletion led to a preference for coke (a coal-based form of charcoal) as an alternative fuel.

Plentiful forests in the eastern United States made charcoal a popular fuel, particularly for blacksmithing. It was also used in the western United States through the late 1800s for extracting silver from ore, for railroad fueling, and for residential and commercial heating.

Charcoal's transition from a heating and industrial fuel to a recreational cooking material took place around 1920 when Henry Ford invented the charcoal briquette. Not only did Ford succeed in making profitable use of the sawdust and scrap wood generated in his automobile factory, but his sideline business also encouraged recreational use of cars for picnic outings. Barbecue grills and Ford Charcoal were sold at the company's automobile dealerships, some of which devoted half of their space to the cooking supplies business.

Historically, charcoal was produced by piling wood in a cone-shaped mound and covering it with dirt, turf, or ashes, leaving air intake holes around the bottom of the pile and a chimney port at the top. The wood was set afire and allowed to burn slowly; then the air holes were covered so the pile would cool slowly. In more modern times, the single-use charcoal pit was replaced by a stone, brick, or concrete kiln that would hold 25-75 cords of wood (1 cord = 4 ft x 4 ft x 8 ft). A large batch might burn for three to four weeks and take seven to 10 days to cool.

This method of charcoal production generates a significant amount of smoke. In fact, changes in the color of the smoke signal transitions to different stages of the process. Initially, its whitish hue indicates the presence of steam, as water vapors are driven out of the wood. As other wood components such as

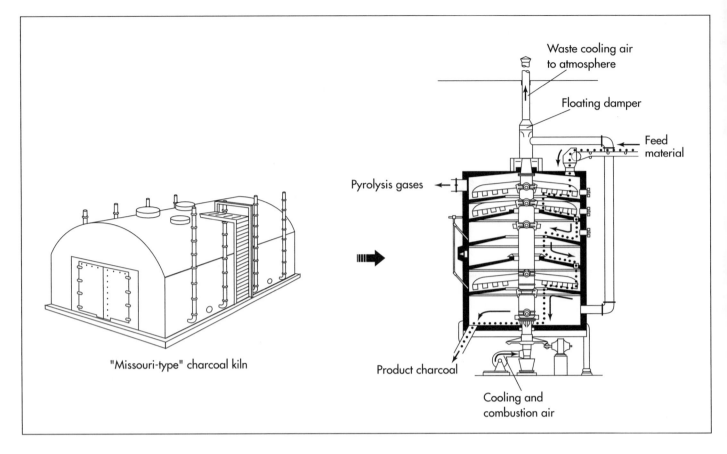

Waste cooling air
to atmosphere

Floating damper

Feed
material

Pyrolysis gases ←

Product charcoal

Cooling and
combustion air

"Missouri-type" charcoal kiln

Basic charcoal is produced by burning a carbon-rich material such as wood in a low-oxygen atmosphere. This process drives off the moisture and volatile gases that were present in the original fuel. The resulting charred material not only burns longer and more steadily than whole wood, but it is much lighter (one-fifth to one-third of its original weight).

resins and sugars burn, the smoke becomes yellowish. Finally the smoke changes to a wispy blue, indicating that charring is complete; this is the appropriate time to smother the fire and let the kiln's contents cool.

An alternative method of producing charcoal was developed in the early 1900s by Orin Stafford, who then helped Henry Ford establish his briquette business. Called the retort method, this involves passing wood through a series of hearths or ovens. It is a continuous process wherein wood constantly enters one end of a furnace and charred material leaves the other; in contrast, the traditional kiln process burns wood in discrete batches. Virtually no visible smoke is emitted from a retort, because the constant level of output can effectively be treated with emission control devices such as afterburners.

Raw Materials

Charcoal briquettes are made of two primary ingredients (comprising about 90% of the final product) and several minor ones. One of the primary ingredients, known as char, is basically the traditional charcoal, as de-

scribed above. It is responsible for the briquette's ability to light easily and to produce the desired wood-smoke flavor. The most desirable raw material for this component is hardwoods such as beech, birch, hard maple, hickory, and oak. Some manufacturers also use softwoods like pine, or other organic materials like fruit pits and nut shells.

The other primary ingredient, used to produce a high-temperature, long-lasting fire, is coal. Various types of coal may be used, ranging from sub-bituminous lignite to anthracite.

Minor ingredients include a binding agent (typically starch made from corn, milo, or wheat), an accelerant (such as nitrate), and an ash-whitening agent (such as lime) to let the backyard barbecuer know when the briquettes are ready to cook over.

The Manufacturing Process

The first step in the manufacturing process is to char the wood. Some manufacturers use the kiln (batch) method, while others use the retort (continuous) method.

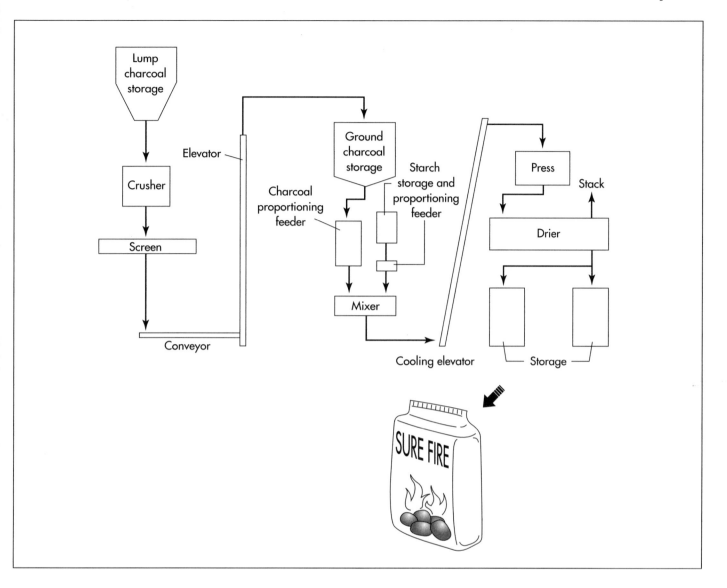

Lump charcoal storage

Elevator

Crusher

Screen

Conveyor

Ground charcoal storage

Charcoal proportioning feeder

Starch storage and proportioning feeder

Mixer

Cooling elevator

Press

Stack

Drier

Storage

SURE FIRE

A schematic diagram illustrating the manufacturing processes necessary to create charcoal briquettes.

Charring the wood

1 (Batch process) It takes a day or two to load a typical-size concrete kiln with about 50 cords of wood. When the fire is started, air intake ports and exhaust vents are fully open to draw in enough oxygen to produce a hot fire. During the week-long burning period, ports and vents are adjusted to maintain a temperature between about 840-950° F (450-510° C). At the end of the desired burning period, air intake ports are closed; exhaust vents are sealed an hour or two later, after smoking has stopped, to avoid pressure build-up within the kiln. Following a two-week cooling period, the kiln is emptied, and the carbonized wood(char) is pulverized.

2 (Continuous process) Wood is sized (broken into pieces of the proper dimen-

sion) in a hammer mill. A particle size of about 0.1 in (3 mm) is common, although the exact size depends on the type of wood being used (e.g., bark, dry sawdust, wet wood). The wood then passes through a large drum dryer that reduces its moisture content by about half (to approximately 25%). Next, it is fed into the top of the multiple-hearth furnace (retort).

Externally, the retort looks like a steel silo, 40-50 ft (12.2-15.2 m) tall and 20-30 ft (6.1-9.14 m) in diameter. Inside, it contains a stack of hearths(three to six, depending on the desired production capacity). The top chamber is the lowest-temperature hearth, on the order of 525° F (275° C), while the bottom chamber burns at about 1,200° F (650° C). External heat, from oil- or gas-fired burners, is needed only at the beginning and ending stages of the furnace; at the intermediate levels, the evolving wood gases burn and supply enough heat to maintain desired temperature levels.

Within each chamber, the wood is stirred by rabble arms extending out from a center shaft that runs vertically through the entire retort. This slow stirring process (1-2 rpm) ensures uniform combustion and moves the material through the retort. On alternate levels, the rabble arms push the burning wood either toward a hole around the central shaft or toward openings around the outer edge of the floor so the material can fall to the next lower level. As the smoldering char exits the final chamber, it is quenched with a cold-water spray. It may then be used immediately, or it may be stored in a silo until it is needed.

A typical retort can produce approximately 5,500 lb (2.5 metric tons) of char per hour.

Carbonizing the coal

3 Lower grades of coal may also be carbonized for use in charcoal. Crushed coal is first dried and then heated to about 1,100° F (590° C) to drive off the volatile components. After being air-cooled, it is stored until needed.

Briquetting

4 Charcoal, and minor ingredients such as the starch binder are fed in the proper proportions into a paddle mixer, where they are thoroughly blended. At this point, the material has about a 35% moisture content, giving it a consistency somewhat like damp topsoil.

5 The blended material is dropped into a press consisting of two opposing rollers containing briquette-sized indentations. Because of the moisture content, the binding agent, the temperature(about 105° F or 40° C), and the pressure from the rollers, the briquettes hold their shape as they drop out the bottom of the press.

6 The briquettes drop onto a conveyor, which carries them through a single-pass dryer that heats them to about 275° F (135° C) for three to four hours, reducing their moisture content to around 5%. Bri-

quettes can be produced at a rate of 2,200-20,000 lb (1-9 metric tons) per hour. The briquettes are either bagged immediately or stored in silos to await the next scheduled packaging run.

Bagging

7 If "instant-light" briquettes are being produced, a hydrocarbon solvent is atomized and sprayed on the briquettes prior to bagging.

8 Charcoal briquettes are packaged in a variety of bag sizes, ranging from 4-24 lb. Some small, convenience packages are made so that the consumer can simply light fire to the entire bag without first removing the briquettes.

Byproducts/Waste

During the late nineteenth and early twentieth centuries, recovery of acetic acid and methanol as byproducts of the wood-charring process became so important that the charcoal itself essentially became the byproduct. After the development of more-efficient and less-costly techniques for synthesizing acetic acid and methanol, charcoal production declined significantly until it was revitalized by the development of briquettes for recreational cooking.

The batch process for charring wood produces significant amounts of particulate-laden smoke. Fitting the exhaust vents with afterburners can reduce the emissions by as much as 85%, but because of the relatively high cost of the treatment, it is not commonly used.

Not only does the more constant level of operation of retorts make it easier to control their emissions with afterburners, but it allows for productive use of combustible off-gases. For example, these gases can be used to fuel wood dryers and briquette dryers, or to produce steam and electricity.

Charcoal briquette production is environmentally friendly in another way: the largest briquette manufacturer in the United States uses only waste products for its wood supply. Woodshavings, sawdust, and bark from pallet manufacturers, flooring manufacturers, and lumber mills are converted from piles of waste into useful briquettes.

The Future

Charcoal and briquette production methods have changed little in the past several decades. The most significant innovation in recent years has been the development of "instant-light" briquettes. A new version being introduced in 1998 will be ready to cook on in about 10 minutes.

Where to Learn More

Books

Emrich, Walter. *Handbook of Charcoal Making: The Traditional and Industrial Methods*. Hingham, MA: Kluwer Academic Publishers,1985.

Moscowitz, C. M. *Source Assessment: Charcoal Manufacturing: State of the Art.* Cincinnati, Ohio: Environmental Protection Agency, Office of Research and Development, Industrial Environmental Research Laboratory, 1978.

Periodicals

Scharabok, Ken. "Amaze Your Friends and Neighbors: Make Your Own Charcoal!" *Countryside & Small Stock Journal* (May 1997): 27-28.

Zeier, Charles D. "Historic Charcoal Production Near Eureka, Nevada: An Archaeological Perspective." *Historical Archaeology* 21(1987): 81-101.

—*Loretta Hall*

Children's Clothing

Designs for today's children's clothes are motivated by factors such as ease of laundering, designer-name labels (sometimes worn on the outside of clothes), safety including fire-retardant sleepwear, popularity of TV and sports characters, and adjustability.

Background

Children's clothes are a relatively recent invention. From the rudimentary beginnings of clothing all the way to the nineteenth century, children wore miniature versions of adult costumes. There were a few minor exceptions. Children's clothes often had leading strings sewn in them so the child could be tethered out of harm's way, but the costume itself was still a small copy. Early in the 1800s, fashions for both adults and children became lighter in weight and freer of restrictions. Several popular styles were developed just for children, notably the sailor's suit and the hussar's or Eton jacket. Babies probably had more clothing challenges than any other group. They wore many layers, some of them wool (due to the parents' general fear of colds), two caps were worn with a set for daywear and another for nighttime, and babies even wore corsets or stomach bands. Infants wore long clothes until they were eight months old when the need to crawl and walk made shorter garments more practical. Boys as well as girls wore skirts well past babyhood.

In the United States, a wide range of social changes impacted children's clothes. The sewing machine (both home and factory varieties) eased the burden of sewing clothes for the family, and, by the Civil War, paper patterns were readily available for children's clothes. The Civil War itself changed children's garments because standard sizes of uniforms were made for soldiers. Soon, all clothing was sized, and styles for children began to differ because of size ranges. Transportation methods diversified and necessitated new clothing styles. Sports clothes were developed with train travel in mind, and tailored suits for girls and box coats for boys were made specifically for travel. The bicycle and the baby carriage were both popular by mid-century, and clothes for children to wear while riding were created. Dolls and paper dolls became popular toys, and, as the shape of the "baby doll" changed from a diminutive grown-up to a more baby-like shape, costumes for both real babies and their toy counterparts were modified to suit baby's needs. Bonnets as absolutes for American girls also dropped out of fashion when the Civil War eliminated cotton production.

By the 1870s, styles had again become very restrictive. Young girls wore laced and boned corsets to shape their waists from an early age even though doctors had become aware of the damage caused to growing bones, circulation, and breathing. Children also wore mourning clothes complete with veils. Girls being sent to boarding school had all the accoutrements of their mothers including fans, stockings, pantaloons, bustles, and feathered hats. Starched clothing appeared neater and cleaner, so shirts and dresses were stiff to the point of discomfort. Changes were beginning to occur rapidly in clothing manufacture, however, as the long centuries of handmade clothes gave way to factory-made garments. Those factory-made garments were also available to anyone with a mail-order catalog.

By 1900, fashion began to be a true cultural mirror that reflected war, depression, revolution, the emancipation of women, the evaporation of class distinctions, and the growth of cities and decline of agriculture. Bicycle-riding clothes were extended to girls who wore Turkish trousers. Overalls were initially advertised as bike-riding attire and soon became work clothes for adults; for children, they were called Brownie suits and they revolutionized playwear. The pullover sweater was also created about this time, and open-necked pullovers, turtle-necks, sweaters, and cardigans soon followed for all ages.

The changing status of women altered the clothing worn by their children. Women had begun to work outside the home and had less time to make their children's clothing. Children's activities were more liberated, so a wider variety of clothing for play and school was needed. Concern for children's comfort grew as well, so soft and loose nightclothes were just as important as tough-wearing jeans. These clothes had to be easy to clean and durable because mother's time was limited.

Surprising developments influenced clothing. Rubber allowed the development of elastic waistbands so young boys could wear trousers instead of skirts. Underwear became more secure and less restrictive, and mother and child saved time when the child could easily pull on clothes rather than waiting for mother and the button hook. Tennis shoes with rubber soles became classic casual wear for this century. Synthetics resulted in wrinkle-resistant clothes, weatherproof garments for outdoors, and soft-spun underclothes. Furthermore, factory-made clothes became less expensive, so a wider wardrobe was opened to children as well as adults. Designs for today's children's clothes are motivated by factors such as ease of laundering, designer-name labels (sometimes worn on the outside of clothes), safety including fire-retardant sleepwear, popularity of TV and sports characters, and adjustability.

Raw Materials

Manufacture of children's clothes calls for a wide variety of textiles, including specially treated material, thread, zippers and other fasteners, and decorations. Most manufacturers do not make their own textiles although they may have established relationships with textile manufacturers who produce fabrics treated with fire-retardant, ecologically-friendly fabrics and chemically safe dyes, soft or brushed materials for babies' clothes and sleepwear, and other specialty items. Patterns are also required and may be designed and printed in house or provided by suppliers. Body shapes for children are standardized depending on age and weight, so manufacturers rely on other design aspects to catch the attention of the purchaser. In raw materials, these may include lace, pre-made collars and cuffs, pockets, belts, ribbons, frills, trimmings, closings (buttons, zippers, snaps, etc.), bows, brooches, artificial flowers, knee and elbow patches, and an endless array of other add-ons.

Design

Designs are developed around some fundamental concepts. The comfort, safety, and appearance of the child comprise the primary focus; but design approaches to these may vary with the age of the child. Ease of care is also an important consideration. because the most inexpensive and eye-catching outfit won't sell. Durability is important for playclothes and outerwear, whereas uniqueness and decoration help sell party dresses. Seasonal factors influence design, not only in terms of winter versus summer but trends throughout the fashion year. Because many clothes enjoy a second life as hand-me-downs, features like extra-long hems and cuffs are considered in design. Nothing in design is simple whether the clothing item is plain or fancy; details like oversewn seams are selected in design even for cottons and playclothes.

The Manufacturing Process

Clothing manufacture is completed in several basic processes including cutting, sewing, assembling, decorating, and finishing the garments.

1 Fabric for a garment is stacked one length on top of another in reaches or lays that may be over 100 ft (30.5 m) long and hundreds of plies (fabric pieces) thick. These stacks of fabric are constructed by spreading machines. A pattern, called a marker or cutting lay, is fastened to the fabric with adhesive stripping or staples; the pattern includes all the pieces of the garment carefully arranged so a minimal amount of fabric is

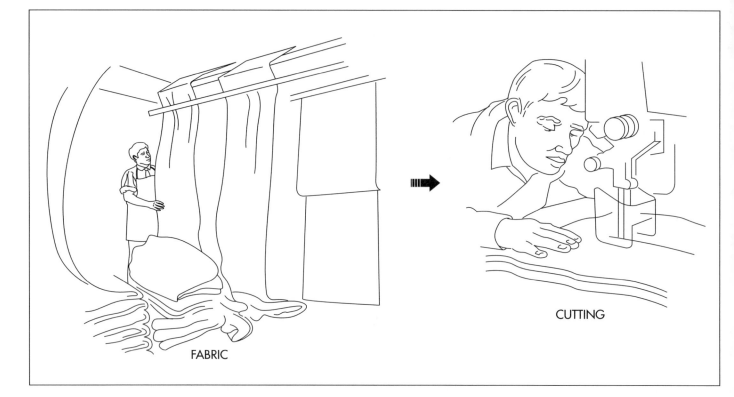

FABRIC

CUTTING

The fabric is cut by an industrial saw into pattern pieces. Many pieces are cut at once.

wasted. Any one of several types of machines may be used to cut out the garment components. The machine is selected depending on the type of fabric and other considerations. These machines include band cutters that work much like band saws, cutters with rotary blades, machines with reciprocal blades that saw up and down, die clickers that are a form of die or punch press, and computerized machines that use either blades or laser beams to cut.

2 The cut pieces are taken to sewing stations where operators typically perform only one operation on one piece or set of pieces being sewn. The industrial sewing machines vary in the type of stitch they make and the configuration of the frame. Both are factors in the manner in which the machine sews and, therefore, in the part of the garment that can be sewn at that station. Operator A may make only straight seams, Operator B may make sleeve insets, Operator C sews the waist seams, and Operator D only makes buttonholes. Some machines work in sequence and feed their finished step directly to the next machine, while others (called "gang machines") have multiple machines performing the same operation under the oversight of a single operator.

3 The final sewing step is assembly when all the pieces put together in segments (like sleeves or pant legs) are assembled into the final product. Either finishing or decorating is performed next depending on the configuration of the garment. Finishing may include a variety of steps that manufacturers call "molding"; molding changes the finished surface of the garment by applying pressure, heat, moisture, or some combination. Pressing is a basic molding process, as are pleating and creasing. Creasing may be done before other finishing if the finishing involves stitching a cuff; creasing is also done before a decoration, like a pocket, is stitched in place. Other decorations that may be damaged by molding processes, like silk flowers or raised embroidery, may be added after the product has been finished. Completed garments are collected by size and type and are bagged or otherwise packed.

Quality Control

For many manufacturers, more care goes into the production of children's clothes than into adult lines. Fabrics and decorations must be chosen for safety as well as appearance, and details like overstitched seams add to comfort for young wearers.

The garments are sewn by workers, using industrial sewing machines.

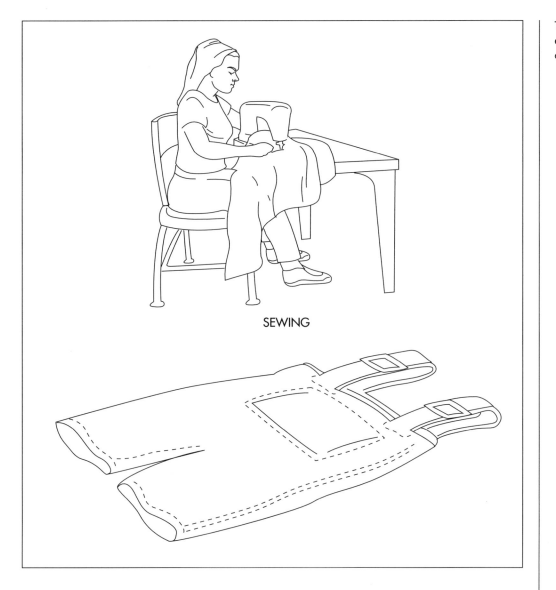

SEWING

Quality is checked throughout the production process; although sewing machine operators may be paid by the piece, their work is inspected at quality stations and rejected for flaws. Machinery is also inspected regularly, and most have shutoffs built into their operation if they run out of thread or perform poorly. Finishing steps are also routinely checked, and, before clothing items are bagged or boxed for sale, a team of inspectors scrutinizes the clothing for loose threads, flaws, and general appearance.

The Future

In the several centuries in which children's clothes have been specially made for them, children have advanced from being fashion followers to trend-setters. Their tastes in colors, fabrics, designers, accessories, and all aspects of clothing are carefully observed by fashion mavens and copied by adults. Trends now go global in a matter of minutes, thanks to the Internet, and children are attracted to the adventure of changing fashion as well as being as appearance- and comfort-conscious as their parents. Improvements in technology are certainly in the future of all clothing manufacture, but, to spot the latest fashion, keep an eye on your favorite six-year-old.

Where to Learn More

Books

Worrell, Estelle Ansley. *Children's Costume in America 1607-1910*. New York: Charles Scribner's Sons, 1980.

Yarwood, Doreen. *European Costume: 4000 Years of Fashion.* New York: Bonanza Books, 1975.

Other

CAF Textile and Ready Made Clothing Company. http://www.caf-tr.com.

The Children's Place. http://www.tcp.com.

Ecobaby. http://www.ecobaby.com.

Rainbow Rags. http://www.promotion.com/rainbow/.

Roosters. http://www.roosters.com/.

Spencers Wee Bear Baby Wear. http://www.spencers.com/.

Vaco. http://www.vaco.be/.

—*Gillian S. Holmes*

Chopsticks

Chopsticks are a pair of sticks, usually wooden, used for eating Asian food. They originated in China sometime during the Shang dynasty (1766-1122 B.C.). As Chinese culture spread, chopsticks were introduced to other countries, and quickly became common across Asia. The English term chopsticks apparently is derived from the Pidgin English spoken in British Chinese colonies. A Chinese term, *kuai-tzu*, or quick ones became chop (Pidgin for quick) sticks.

Background

Much lore surrounds chopsticks, especially in Japan. Their use is said to promote a child's intellectual development, and at home each member of the family has his or her own pair of chopsticks which are suited to his or her hand size. Many taboos govern the use of chopsticks. For instance, the two sticks must not be grasped in one fist or laid across a bowl. It is also forbidden to stab food with chopsticks, to lick the tips, or to beat on a plate or bowl with them to get someone's attention. The shape, size, and material of chopsticks indicate specialized uses. Chopsticks for personal use may be quite ornate and beautiful, hand carved, inlaid, and coated with lacquer in traditional patterns. Plain, long wooden chopsticks with blunt tips are used for cooking. For eating out, Asian restaurants provide disposable single-use chopsticks made of light wood. There are even special long chopsticks used only for cleaning out cat litter boxes in Japan. The sticks worn in the hair of Japanese Samurai warriors in pre-modern times were apparently used for grasping the severed head of a vanquished enemy.

Raw Materials

The most prevalent material used to make chopsticks is aspen wood. Aspen is used to make the disposable chopsticks used in restaurants. About 20-billion pair are used yearly, mostly in Japan. Many other materials are used to make chopsticks designed for more than one use. Metal chopsticks are common in some areas, and elaborate chopsticks may be carved of precious materials such as ivory or jade. Most chopsticks are made of some variety of wood, and coated with oil, paint, or lacquer. Some varieties of chopstick wood have superstitions related to them. Chestnut chopsticks are said to bring wealth, black persimmon chopsticks, long life. Other typical woods used for chopsticks are pine, cedar, cherry, sandalwood, and paulownia. A traditional Japanese material is a sandwich of thin boards of maple, pine, and cedar called shuboku wood. In general, the wood used needs to be relatively hard and impervious to water. The color and grain of the wood is also important for fine quality chopsticks.

The Manufacturing Process

This is the process for fine quality, hand-crafted chopsticks.

Milling the wood

1 The chopstick maker begins with wood that has been only roughly cut into a board or block. It may have previously been cured, dried, or aged. This depends on the variety of wood, its hardness and imperviousness to water. The maker may study the piece of wood to find the best pattern of

About 20-billion pairs of chopsticks are used yearly, mostly in Japan.

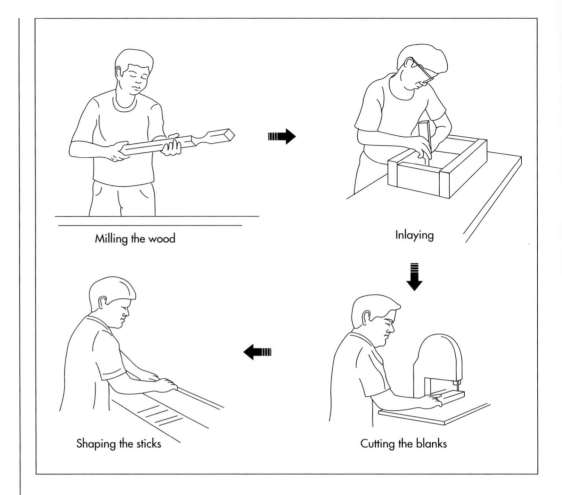

Milling the wood

Inlaying

Shaping the sticks

Cutting the blanks

grain in it. Then the maker mills the wood on a bandsaw, cutting it into a smooth rectangular block just longer than the finished sticks and with the width of several pairs.

Inlaying

2 If the wood is to be inlaid with a contrasting wood, this happens next. The inlay provides an interesting pattern to the finished stick. The maker cuts a trench in the milled block. Then the maker glues a thin sheet of the contrasting wood in the trench. The maker clamps the inlay to the block and lets it dry.

Cutting the blanks

3 After the inlaid wood is completely dry and secure, the chopstick maker cuts blanks. The blanks are wood pieces that are roughly the length of the finished chopstick, and about 0.25 in (0.64 cm) wide. Using a table saw, the maker cuts the milled block of wood into a number of blanks, one for each

finished chopstick. They come out as long, thin rectangles.

Shaping the sticks

4 The rectangular blanks must next be shaped. This can be done with hand tools or mechanically. The shape of the chopstick varies according to what the maker or customer desires. Some chopsticks are square at the end, tapering to a cylindrical point, and others are cylindrical overall. The degree of taper is also a matter of discretion. The maker shapes the sticks using a handheld scraper or holds them against the belt of a sanding machine.

Sanding

5 After shaping, the chopsticks need to be made smooth, so that they don't splinter in the mouth of the user. Handmade chopsticks are sanded very carefully, using several grades of sandpaper. Now they are ready for finishing with oil, paint, or lacquer.

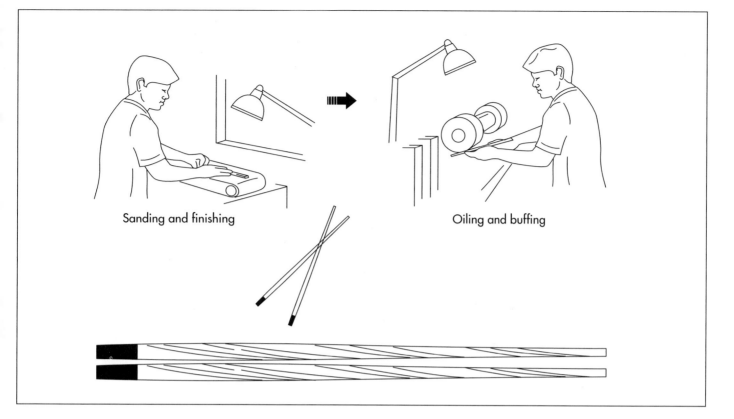

Sanding and finishing

Oiling and buffing

Finishing

6 Some chopsticks will be dipped in lacquer at this point. Lacquer is a natural substance made from the sap of a variety of sumac. It is nontoxic, and has long been used as a decorative wood finish in Japan and China. There are several traditional styles of chopstick lacquering in Japan, with distinctive patterns belonging to different traditional schools. The chopsticks are dipped in the liquid lacquer and hung to dry until the finish is clear and hard. The chopsticks may also simply be dipped in nontoxic paint. Fine wooden chopsticks are often finished with oil, which gives them a soft polish and brings out the beauty of the wood grain. The oil may be applied with a soft cloth, and then the maker buffs the chopsticks either by hand or by holding them against the wheel of a mechanical buffer.

Mass-produced chopsticks

Mass-produced chopsticks, especially the disposable kind, are made rapidly in a fully automated process. Aspen wood is harvested, and the finest grade wood selected. This wood is fed into a mill, which cuts it into blocks. This process typically happens at the site where the wood is grown. Then the aspen blocks are exported to the country where they will be used. The blanks are cut, sanded, and finished at a chopstick factory, which may churn out millions of pairs a year. Disposable chopsticks are typically "halfsplit." That is, the two halves of the chopstick pair are only half separated, and they are only snapped apart when ready to be used. So the blank in this case is actually for the pair of chopsticks, not the individual sticks.

Quality Control

The quality of the wood is very important to how well a chopstick will wear. Fine makers inspect the wood carefully before beginning, and are able to observe it throughout the manufacturing process. The maker picks the wood for a pleasing color and grain, and strives to bring out these characteristics in the shaping and finishing.

Byproducts/Waste

The disposable chopstick industry has been accused of exceedingly wasteful foresting practices. Because only very fine-grained wood is suitable for chopsticks, only some trees, or only parts of some trees, can be forested. In some cases, the forest is clear-cut, though only one quarter of the wood is then fed into the chopstick mill. The remaining lumber is left to rot or burn. The bulk of disposable chopsticks are sold in Japan, where using someone else's chopsticks is considered disagreeable. Restaurants almost always provide their customers with one-use chopsticks, but because of environmental concerns, some Japanese consumers are foregoing disposable chopsticks. Some corporations are providing their workers with reusable plastic chopsticks in company lunchrooms. Another replacement product growing in popularity is disposable chopsticks that are made only from wood obtained from forest thinning. This is supposed to represent wood that would otherwise be wasted, so the product is environmentally sound. Consumer boycotts and voiced concerns have already made disposable chopsticks a prominent environmental issue. Faced with growing opposition to their wasteful practices, chopstick manufacturers may be forced to come up with alternative.

Where to Learn More

Books

Amaury, Saint-Gilles. *Mingei: Japan's Enduring Folk Arts.* Boston: C.E. Tuttle, 1989.

Periodicals

"Chopped Chopsticks." *The Economist* (August 4, 1990): 56.

Karliner, Joshua. "God's Little Chopsticks." *Mother Jones* (September 1994): 16.

—*Angela Woodward*

Cider

Cider is a natural, liquid beverage that is obtained from the pressing of a finely ground fruit such as apples. Under the proper conditions, it undergoes a natural fermentation process, which yields an alcoholic juice. Cider has been made for thousands of years, however it has only recently seen a significant rise in popularity.

Background

Cider is the sweet juice of apples that can be consumed as a beverage or used as a raw material in vinegar making. It is typically a clear, golden drink, which can range in color from a pale yellow to a dark amber rose. It has a fruity flavor and a varying degree of taste from very sweet to tart. Sweet cider is the non-alcoholic versions of cider and it can be made into apple juice by pasteurizing it and adding preservatives to stop the natural fermentation process. Hard cider is the product that results when the juice is allowed to undergo fermentation. This cider contains alcohol. Additionally, it is often effervescent due to the activity of the natural yeasts present.

People have known how to make cider for thousands of years. Archaeological evidence shows that ancient European and Asian cultures used apples to make a crude version of cider as early as 6500 B.C. The art of cider making improved over the years as people developed a better understanding of the factors that impact cider flavor. During the sixth century, a profession of skillful brewers was established in Europe. These people made beer-like beverages and also cider.

By the sixteenth century, Normandy became one of the largest cider-making areas in the world. Experimentation with different types of apples ensued, which resulted in better tasting ciders. England and colonial America also produced cider during this time and it became an important part of each culture. The ciders of this time period were inconsistent however, as small farmers each had their own methods of manufacture. The technology of cider production made significant improvements over time as people developed a better understanding of each step in the cider making process. Today, it is a highly controllable operation, which results in a dependable, good-tasting product.

Raw Materials

Apples are the primary raw material used in cider making. Suitable apples vary in size with diameters from about eight inches wide to less than two inches. Nearly all of the characteristics of the final cider product depend on the quality of the apples from which it is made. To produce the best cider, these apples must be juicy, sweet, well ripened and have adequate levels of natural acids and tannins. The skin of the apples contains many of the compounds that contribute to the taste of the cider so apples are not peeled before being used for cider manufacturing. The seeds are not removed either however, in typical milling machines, they are not broken open, and do not significantly contribute to taste. It should be noted that pears and sweet cherries are also occasionally used to make cider.

A full-bodied cider requires the use of several different types of apples to give it a balanced flavor. This is because certain varieties of apples have flavor characteristics that work well together. There are four dif-

Archaeological evidence shows that ancient European and Asian cultures used apples to make a crude version of cider as early as 6500 B.C.

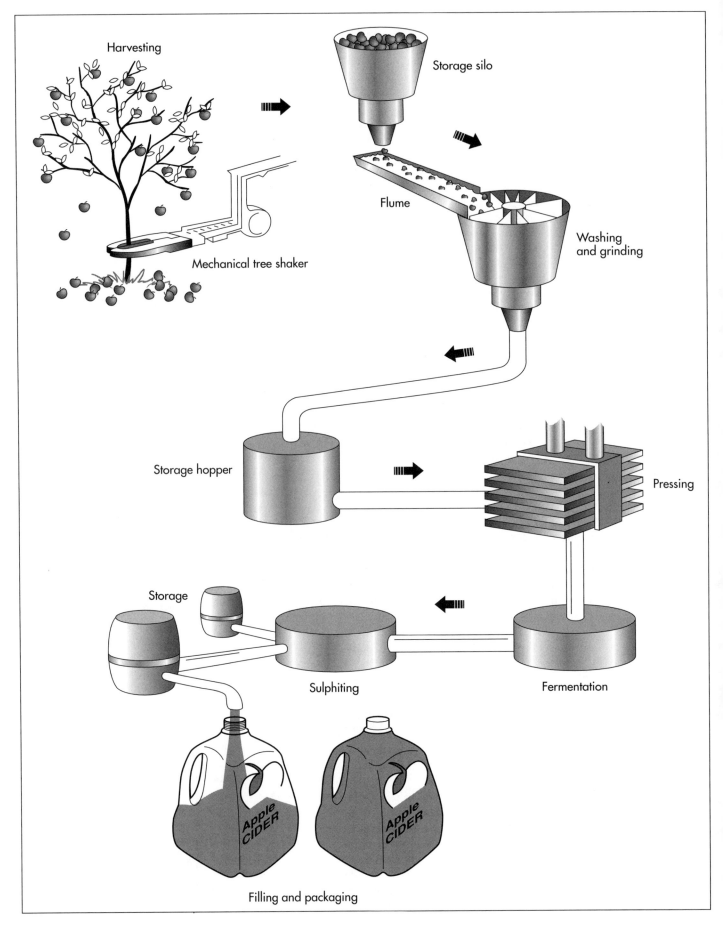

Harvesting

Mechanical tree shaker

Storage silo

Flume

Washing and grinding

Storage hopper

Pressing

Storage

Sulphiting

Fermentation

Apple CIDER

Apple CIDER

Filling and packaging

ferent types of apple juices including aromatic, astringent, acid-tart, and neutral tasting. Generally, sweet and tart apples are blended together to create a balanced cider. A typical blend might include 50% neutral base, 20% tart, 20% aromatic, and 10% astringent. In this cider, the flavor is a balance between tartness and sweetness. Beyond apple blending, some cider producers may also improve flavor by adding tannic, malic, and other natural acids. Tannins add a slight bitter taste and astringency to cider. Malic, citric, and tartaric acid give a zesty tingle. They also help to inhibit microbial contamination.

Producing a gallon of cider requires 11-14 lb (5-6.4 kg)of apples depending on the juiciness of the fruit. Fresh cider will remain in its full-bodied state for several weeks if it is refrigerated. After this time natural fermentation process begins. If a non-alcoholic cider is desired, the juice may be pasteurized or preserved by the addition of potassium sorbate. This material effectively kills undesirable organisms. For some cider manufacturers, the alcoholic cider is preferred. Alcoholic cider is made by either letting the inherent fermentation process continue without the addition of any other ingredients, or by adding a variety of ingredients, which give more controllable results.

Fermentation of apple cider is the process by which yeast converts the apple sugars into ethyl alcohol and carbon dioxide. It occurs in two steps. First, yeast converts the sugar to alcohol and then lactic acid bacteria convert the natural malic acid into carbon dioxide. This hard cider contains 2-3% solids and 2-8% alcohol. Fermentation aids include components such as sulfur dioxide, yeast, sugar, and natural acids. Sulfur dioxide is typically added to the freshly pressed juice before fermentation is allowed to begin. It has the effect of killing most of the bacteria and yeasts present in the freshly squeezed juice, or must. Enough of the desirable yeast survives the sulfur dioxide treatment and these organisms will go on to ferment the sweet juice.

Natural yeasts are present in apples, but sometimes cider manufacturers add their own yeast to ensure that a consistent fermentation will be achieved. Some of these strains have been around for generations and they are repeatedly used to produce a distinctive tasting cider. To help yeasts grow and speed up fermentation, yeast nutrients such as ammonium sulfate and thiamine may also be added. For similar reasons extra sugar, honey or other sweeteners may also be added to the unfermented juice. This will improve fermentation and increase the alcohol content of the final product.

The Manufacturing Process

The cider making process typically involves three stages including crushing the fruit, pressing out the juice, and allowing it to ferment. To begin however, the fruit must be harvested, sorted, and washed.

Harvesting

1 In the United States, apples are typically harvested in the fall. At this time, field workers pick the apples by hand and transfer them to large storage bins, which can hold about half a ton of fruit. When these bins are filled, they are transported by tractor to the processing plant. At the plant they are stored outside for about a week which allows them to soften. This makes the apples easier to process and increases the amount of sugar in the juice.

Washing

2 After the apples have mellowed, they must be washed to remove leaves, twigs, insects, spray residues, and harmful bacteria. To this end, they are automatically poured out from the bins onto a scrubber. This machine rinses and scrubs each apple, removing most chemical residues from the skin. From there, they are moved along a conveyor to a hopper filled with water. A worker is on hand to see that an even stream of apples flows into the bath and makes sure that the apples are thoroughly washed.

From the hopper, the apples are put on a conveyor and moved to another worker. Jets of water aid in moving the apples. During manufacture, only whole apples are used because they have not been exposed to the flavor-damaging effects of oxidation. This means that each apple is inspected and any rotten or moldy fruits are removed. Since

Opposite page:
Once the apples are harvested and washed, the fruit is crushed and pressed. The remaining juice is fermented, creating cider.

cider taste can be negatively effected by many different factors, cleanliness is essential during manufacturing.

Grinding

3 Next, the apples are put in a large mill and ground to a fine pulp with the consistency of applesauce. This is done to ensure that the maximum amount of juice can be extracted from the apples. The finer the pulp, the greater the yield of juice. Fine grinding has the added benefit of reducing damage caused by oxidation. The pulp is put into appropriately labeled 55 gal (208 l) steel drums with plastic liners. Some of these drums continue on through the cider making process while the rest are sent to a freezer to be used later. The frozen pulp ensures that cider can be produced throughout the year when apple supplies are low.

Pressing

4 To remove the juice from the pulp, or pomace, it is pressed. Depending on the desired cider flavor, the pomace from various types of apple pulp are used. Typically, anywhere from three to six different types are blended together in a large tank. This blend is then taken by the press operator and stacked for pressing. Wooden racks and forms are used for stacking the pomace. Each form is lined with a nylon cloth. Nylon is used because it is easy to clean and sturdy enough to withstand many pressings. To start, several barrels of pomace are poured onto the cloth. The corners are then folded up and the form is removed. As a result, a square-shaped layer of pomace called a cheese is formed. A rack is placed on top of the cheese and another form is put in place. The process is repeated until 10-12 cheeses are in a stack. The whole stack is put in a large stainless steel tray that has been designed to hold the cider as it is pressed from the pomace. A worker puts the stack under the cider press, called a wring, and turns it on. This delivers as much as 30,000 lb (13,620 kg) of pressure from a hydraulic pump.

Cooling and Filling

5 The cider is expelled from the pomace and pumped through plastic tubes to a cooling tank. As the cider is transferred to the cooling tanks, it is passed through a screen mesh to remove any pulp pieces from the liquid. It is then chilled and stored at 33° F (0.6° C). This helps to inhibit the contamination by undesirable microorganisms. If this cider is of the unfermented variety, meant to be unfermented, it is sent to a mixing tank and pasteurized. Preservatives such as potassium sorbate are added and the juice is sent off to the filling lines.

Fermentation

6 Before fermentation is allowed to proceed, the various fermentation-assisting chemicals are added. Depending on the manufacturer, the cider may be allowed to ferment in a large, sealed bulk tank, or in the individual bottles. If it is fermented in the bottles, the product will be sold with a bottom layer of sediment. The sediment is the remains of the fermentation yeast. In bulk fermentation, the cider is siphoned off after the yeast has died. This allows for a sediment-free product. Complete fermentation may take one month or more.

Filling and packaging

7 When the cider is ready for filling, it is filtered again and pumped into the appropriate packaging. In this filling process, the empty, sterile bottles move along a conveyor and are passed under a filling machine. The machine pumps cider into the bottles to the desired volume. The caps are then put on the bottles and then labeled. The jugs are put in boxes, then pallets, and stored at just above freezing until the next day when they are delivered to stores.

Quality Control

There are standard quality control measures, which are performed at various points in the manufacturing process. At the beginning, the apples are checked by line inspectors. This ensures that rotten fruit, twigs, and leaves do not make it into the grinding mill. The pomace may also be inspected before being pressed. This is particularly important when using pomace that has been frozen for many months. For fermented cider, the level of sugar is determined. Since the amount of sugar is directly proportional to the amount of alcohol, this allows the manufacturer to correctly label the product for alcoholic content. Acid testing equipment is also used at this stage to ensure the juice has not been contaminated with acetic acid producing bacteria. After the final packaging, the alcohol level of the cider is determined. The taste, appearance, and other physical and chemical characteristics are verified by trained quality control tasters.

The Future

Current data suggest that cider production will show significant growth in the near future. This will be a result of the expected continuing movement toward more natural products. Refinements in the manufacturing process should also be expected. This would include more efficient methods for harvesting and sorting apples and improved presses, which will squeeze even more juice out of the pomace. Manufacturers will also develop more useful yeast cultures, which will produce better tasting cider with increased alcohol content.

Where to Learn More

Books

Macrae, R. et al., editors. *Encyclopedia of Food Science, Food Technology and Nutrition*. San Diego: Academic Press, 1993.

Proulx, Annie and Lew Nichols. *Sweet & Hard Cider*. Charlotte, SC: Garden Way, Inc. 1980.

Valentas, Kenneth. *Food Processing Operations and Scale-up*. New York: Marcel Dekker, 1991.

Periodicals

Curtis, Lauren. "Pop Art: Designing Soft Drinks." *Food Product Design* (January 1998): 41 - 66.

—Perry Romanowski

Clothing Pattern

One single clothing pattern printing facility can print 100,000 complete patterns (meaning all the tissue pieces) in a single day; it produces 23-million patterns in one year.

Background

Clothing patterns are used to sew stylish garments that fit well. Individual pattern pieces are used to cut fabric pieces, which are then assembled and sewn to create a wearable garment. Today, clothing patterns are usually mass-produced of thin tissue packaged in envelopes, and are sold according to standard body sizes (size 4, 6, 8, 10, etc.) Garment illustrations and pertinent information such as purchase of closure and notions are printed on the outside of the envelopes. General instructions are included in the package, and individual pattern pieces contain specific information pertaining to seam allowance and alignment of the fabric according to the grain or warp of the material. Sewing instructions are keyed to numbered or lettered pattern pieces so they are easy to understand. Patterns are distributed through fabric stores (they are shown in catalogs there) or by mail.

The actual printing of the paper pattern pieces is not time-consuming, nor expensive. Rather, the design of the pattern is the most time-consuming and costly part of production. Essentially, a designer's sketch must be translated into a standard-size pattern that must be stylish and easy to construct. A successful pattern enables a sewer to produce an article of clothing for a fraction of the cost it would take to purchase a garment ready-made in a store.

History

For centuries, obtaining fashionable clothing that also fit properly was difficult to do. The wealthy hired tailors or professional dressmakers to sew custom-fit fashions. However, those of lesser means muddled through with old clothes, makeshift fashions that were ill-fitting, or lived with re-made hand-me-downs. The ready-to-wear industry was not in full swing and therefore did not produce affordable women's dress until about 1880 (some men's garments were available earlier in the century).

However, by the early nineteenth century, some women's magazines included pattern pieces for garments such as corsets in order to assist women in obtaining fashionable dress. Since the pieces were simply illustrated on a small magazine page and just a few inches in size, they were not easy to use. By the 1850s, Sarah Josepha Hale's famous women's magazine *Godey's Lady's Book* offered full-size patterns, but they were one size only—the reader would have to size it according to individual measurements.

About the time of the Civil War, tailor Ebenezer Butterick developed the mass-produced tissue-paper pattern sized according to a system of proportional grading. These first patterns were cut and folded by members of the Butterick family. The Buttericks established a company in New York City and began mass-producing ladies' dress patterns by 1866. It is reputed that Butterick alone sold six million clothing pattern by 1871. James McCall, another pattern entrepreneur, produced women's clothing patterns shortly thereafter as well. At last American women could obtain a well-fitting, rather stylish garment by using a mass-produced clothing pattern. Amazingly after 120 years, both McCall and Butterick remain giants in the pattern industry.

Innovations in the pattern industry since the late nineteenth century include superior marketing through women's magazines, opening branch offices throughout this country as well as Europe to keep abreast of styles, improvements in instruction sheets, the development of different product style lines, and the addition of designer lines based on the pattern of a couture creation.

Raw Materials

The paper pattern, envelope, and instructions are made of paper of varying grades. The most important component, the tissue paper pattern, is made from the lightest and thinnest paper commercially available (it is not made at the pattern companies). It is called 7.5 lb (3.4 kg) basis paper, meaning that a ream of it (500 sheets) only weighs 7.5 lb (3.4 kg).

Design

The design of the mass-produced paper pattern includes many steps. Furthermore, the creation of an easy to use, fashionable, of good fit pattern is the result of collaboration of many departments and many talents.

At the outset of the design process of any garment, the pattern company's product development department must evaluate three key elements: the typical customer profile (lifestyle, skills, taste, etc), the current fashion trends, and last season's sales figures. These all factor in to making a profitable pattern—the goal of the company.

Pattern companies vary in the number of new pattern collections launched each year; many launch four new collections a year. The in-house designers are inspired by observing people and their physical movements, learning about their needs, and understanding trends in their customers' lifestyle. Designers attend fashion shows, read magazines, newspapers, and trade journals to keep abreast of fads and fashions.

Many designs are created for a proposed collection. Preliminary sketches are discussed by marketers, dress designers, dressmakers, etc. Sales histories on previous styles and patterns are examined and compared. Some patterns may remain in a line for more than a season based on sales alone.

If a design goes through the review and appears to be a viable candidate for a pattern, it is assigned to a line, which earmarks it for a particular customer profile. The final selections are assigned a style number and returned to the design department.

Next, the illustrators create the first sketches of the creation. These sketches are known as *croquis*, which is the French word for beginning. The croquis contains all critical information for each pattern and will form the basis of the worksheet to construct the item.

In order to make the actual pattern, members of all technical departments (design merchandising, product standards, patternmaking, dressmaking) hold a construction meeting to decide details of a style and determine construction. Decisions are made on the number of pattern pieces, the style number based on degree of difficulty, suitable fabrics, sizes the patterns will be graded to, and how it will be constructed.

A folder is begun for each design so that crucial information is contained within and passed to appropriate departments. The folder with the notes from the construction meeting is given to the patternmaking department.

The Manufacturing Process

Preliminary pattern

1 Culling information from the construction meetings, the patternmaker creates the first pattern. The paper pattern is drafted onto muslin (a plain fabric) and drapes up a sample garment. The drape is pinned in place and basted (hand stitched) to keep it in place. The drape is thoroughly reviewed by both the patternmaker and designer. Adjustments are made where needed.

2 When the drape is approved, the pattern draft is turned over to computer-aided design (CAD/CAM). The technician digitizes the basic pattern pieces. Then all the separate pattern pieces are blocked, which means they are created with all the information and additions (seam allowances, fold lines, dart lines, etc.) needed to make them usable pattern pieces. It is important to note that each is initially made up in a standard

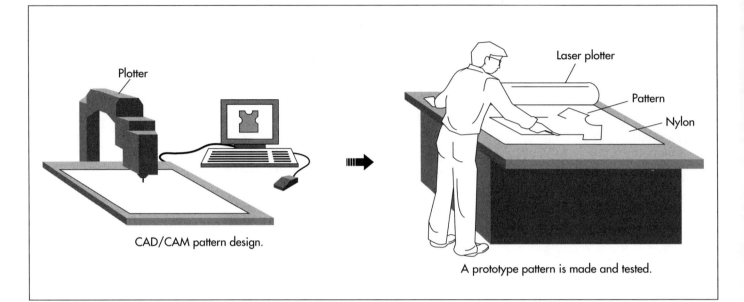

Plotter

CAD/CAM pattern design.

Laser plotter

Pattern

Nylon

A prototype pattern is made and tested.

size 10. After blocking, the pieces are plotted using a laser plotter.

3 The completed size 10 pattern is sent to the dressmaking department, where it is tested using several different fabrics. Techniques of the home sewer and domestic-use sewing machine are simulated to insure that the design will work using various fabrics and that it is not too complicated to construct.

4 After passing the home-sewing test, the pattern is then graded to the various sizes using a computer program. Thus, the complicated task of grading patterns that used to be manually performed by the patternmakers is now computerized.

5 The measuring department determines fabric yardage and notions needed. Computer software helps the technicians create the optimum fabric layout to suggest so fabric can be used efficiently. Once all information for step-by-step instructions is known, they are written up for the consumer in easy-to-understand language.

Printing the pattern

6 A computer template (or plot) is used to plot out the pattern. Pattern pieces are laid out in such a way that little tissue will be wasted in the printing process. The computerized plot and the instruction sheets are physically sent to the printer. The pattern

envelope, however, is sent to the pattern printer electronically.

7 The plot is unrolled on a pre-sensitized aluminum plate that varies in size according to the size of the tissue sheet to be printed. Plates are as small as 30 in x 90 in (76.2 cm x 229 cm) or as large as 50 in x 90 in (127 cm x 229 cm). A vacuum frame adheres the plot to the aluminum plate, lights expose the plate, and the plate is etched where the lines on the plot are printed. Thus, the plate is essentially burned with the image of the pattern pieces.

8 The plate is then printed using off-set lithography. The image is inked, transferred to a felt blanket, and is then transferred from the felt blanket to paper. This saves wear on the metal plate.

9 Pattern tissues are printed in units of 1,300 sheets. These units are kept together using clamps and are transported together. Some units may be cut down into smaller tissue pieces with a sharp saw. All tissue pieces must be folded to fit into the envelopes and may be either folded by hand or by machine. Instruction sheets are also printed using off-set lithography.

10 Envelopes, however, are sent electronically from the design offices. A film of the envelope design is created off the computer information and is used to expose the aluminum plate. The four-color enve-

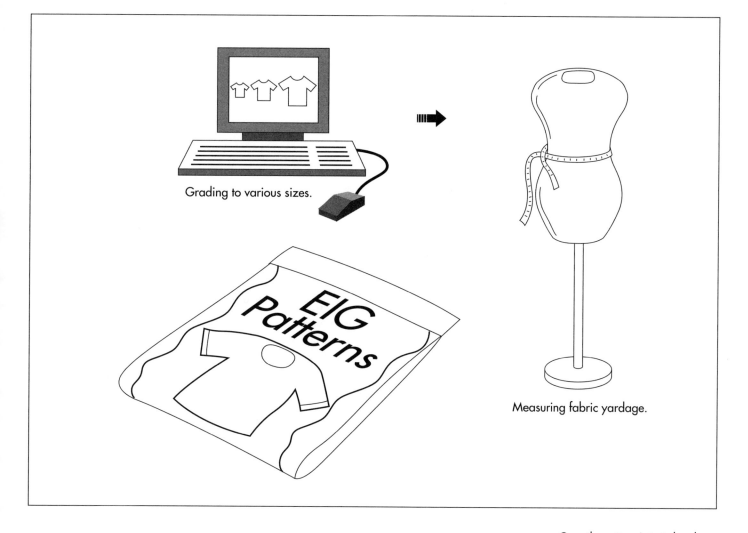

Grading to various sizes.

Measuring fabric yardage.

lope is then printed with off-set lithography. Once printed, the envelopes are folded, glued, and readied to receive the folded tissue patterns.

One single clothing pattern printing facility can print 100,000 complete patterns (meaning all the tissue pieces) in a single day; it produces 23 million patterns in one year.

Quality Control

The pattern companies rely heavily on their consumer service departments to address questions, concerns, and problems with patterns and pattern instructions. Service representatives have a thorough knowledge of sewing and of all company patterns. All customer problems, comments, or concerns are reviewed, and feedback on patterns and instructions are continually re-analyzed in order to improve the functionality of the pattern.

Where to Learn More

Books

Bryk, Nancy Emelyn Villa. *American Dress Pattern Catalogs 1873-1909.* NY: Dover Publications, Inc., 1988.

Kidwell, Claudia Brush. *Cutting a Fashionable Fit.* Washington, D.C.: Smithsonian Institution Press, 1979.

Periodicals

Gould, Donna. "The Making of a Pattern. " *Vogue Patterns* (January/February 1996 - September/October 1996).

—*Nancy EV Bryk*

Once the pattern is tested and approved, a computerized grading program creates the various size patterns and the desired amount of fabric needed is measured.

Coffin

Over the past 50 years, the coffin industry in the United States has become increasingly centralized. A few manufacturers with large, automated plants now dominate the market.

In reaction to this centralization, many small casket makers have recently tried to reach the public directly, selling coffins either through showrooms, by mail, or over the Internet.

Background

Coffins, or funeral caskets, are containers in which the dead are buried. Burial practices differ markedly across cultures and through history, but many peoples have used wooden, stone, or metal boxes for burial. Beautifully decorated stone boxes called sarcophagi were used in ancient Egypt. Stone coffins were also used in Europe in the Christian era, and later lead or iron coffins became common. Only wealthier people could afford elaborate coffins, and in Western cultures since the Middle Ages, poorer people were buried in simple wooden boxes. The very poor had no coffins at all, and might be laid in the grave wrapped in a blanket.

The making of a wooden coffin is not significantly different from any other type of carpentry or cabinetry . In some parts of the world, skilled carpenters specialize in elaborate coffins. Italy has a vanishing tradition of hand-built burial caskets, and master craftsmen in Ghana continue to create coffins in fanciful shapes such as birds, cars, and ears of corn. In the United States, coffins were traditionally built only as needed, by the local carpenter. The carpenter "undertook" to take care of the deceased, hence the origin of the term undertaker. Over the past 50 years, the coffin industry in the United States has become increasingly centralized. A few manufacturers with large, automated plants now dominate the market. The same phenomenon exists in Canada and the United Kingdom as well. In reaction to this centralization, many small casket makers have recently tried to reach the public directly, selling coffins either through showrooms, by mail, or over the Internet. Some alternatives to the conventional coffin have also arisen. One small manufacturer in England specializes in basket-like coffins made of a traditional willow wicker, while a Swiss entrepreneur advocates the Peace Box, a cardboard coffin made principally of recycled materials.

Raw Materials

Raw materials used in casket making vary greatly. The Peace Box is made from cardboard, and a deluxe coffin for a head of state may be made of solid bronze. Wooden caskets may be assembled from pine boards, or use an expensive hardwood such as cherry or mahogany. The most common American coffin is made from steel. Still others are made of fiberglass.

Most caskets, except for the most simple, contain, in addition to the outer shell, an inner lining. This is typically made of taffeta or velvet. The lining may be backed with a batting material, usually polyester, and cardboard may back the batting.

Other materials used in the manufacture of coffins include steel or other metals for hinges and accessories; rubber, if a gasket is used to seal the coffin; and paint.

The Manufacturing Process

A wooden casket can be manufactured in any woodshop, using cabinet-making tools and techniques. Ambitious consumers can make their own, just as some people make their own bookcases and coffee tables. A typical small casket manufacturer is more often a casket assembler, buying prefabricated parts and putting them together. The three essential elements of the coffin are the shell, the lining, and the handles and acces-

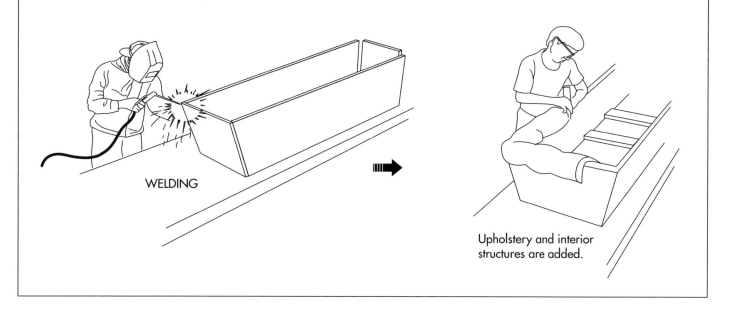

WELDING

Upholstery and interior
structures are added.

sories. A small manufacturer may buy casket shells in a semi-finished state from a casket shell producer, and finished linings from another supplier. The manufacturing process might then consist of painting the shell, stapling or latching the lining into the interior, and then screwing on handles and any additional hardware such as decorative corner pieces or latches. Large casket manufacturers do all the manufacturing and assembling under one roof. The following description is of the process for a typical steel coffin.

Assembling the shell

1 Steel caskets are typically made of 18- or 20-gauge steel, which is delivered to the manufacturer from a steel producer in coils. A small coil may weigh 1,000 lb (454 kg), while the largest may weigh up to 20,000 lb (9,080 kg). The steel coil is first sent through a leveler, which straightens it. Then the metal is cut into large blanks by a blanking machine. The blanks are then fed into a die stamper, which stamps the parts of the shell. The parts are then passed to a welding area. In the welding area, workers feed the parts into an automatic welder, which welds together the body of the coffin. The tops are also welded this way. Then a worker welds by hand any areas the welding machine did not cover.

Painting

2 The shells are then passed to a painting area. Workers apply paint using a spray gun, with a continuous supply of paint piped in through hoses. First the shells are sprayed with primer, next with paint. Then the caskets are baked, to set the paint. Other industries use similar painting processes. The paint used for steel caskets is unique, and specially formulated.

Accessorizing

3 At this point, the lids are ready to be assembled to the shells, and hinges and handles are screwed on. Hinges are usually made of steel. A worker welds these on by hand. Workers then attach handles. These are usually preassembled, either at the casket manufacturer or at a supplier's facility. They attach simply, either by snapping into place or with screws. Next, decorative pieces, such as corner plates, are attached in the same way.

Making the lining

4 The upholstery that lines the casket may be bought in specified dimensions from a supplier, and then simply inserted in to the finished shell. It may also be made on site. Seamstresses take rolls of the lining material, usually taffeta, and feed it through shirring machines. These multi-needle machines gather and stitch the material into a decorative quilted design. Seamstresses working at industrial sewing machines then cut and sew the shirred material into the proper dimensions. Workers also cut and sew a thick batting material, which backs

The metal is cut into large blanks by a blanking machine. The blanks are then fed into a die stamper, which stamps the parts of the coffin shell. The parts are welded together and the fabric liner is added.

Decoration and handles are added to the outside of the coffin.

Handles are screwed on, and decorations are added.

the taffeta. Taffeta and batting are then attached to a cardboard backing. Then workers fit this three-layered upholstery into the finished shell. The upholstery may be glued or stapled to the shell, or it may be designed so it simply snaps into place in the shell.

Packaging

5 After the coffin has passed a final inspection, it is sent to a packaging area. Coffins are prone to scratching, so care is taken to package them well. The finished caskets are first wrapped in large sheets of packing paper to protect the finish from rubs and scratches.

The corners are given additional padding. Then the casket is put in a clear plastic bay. After this, the bag-covered casket is covered with a plastic shrink-wrap. Before shipping, the casket is wrapped in a rug similar to a mover's blanket. Caskets are then taken by truck to warehouses for distribution.

Quality Control

Workers inspect coffins for defects at several points during the manufacturing process. When the steel comes into the factory, it must be inspected to insure it is the proper gauge and quality. Workers check the parts of the shell after they are stamped, and inspect again before the shells go to the painting area. The shells are checked again after painting, as this is particularly important to the final appearance of the casket. The upholstery and accessories have their own quality checks. Then the finished product is examined carefully before it is sent to the packaging area.

The Future

In the United States, cremation is becoming increasingly prevalent, and the demand for coffins is not growing. Future developments in the industry might lie more in the realm of marketing than in the actual manufacture. Traditionally in the United States, coffins are purchased only after a death, usually as part of a burial package offered by a funeral home. Consumers who purchase a coffin directly from the manufacturer are able to reap significant savings by foregoing the middleman service of the funeral home. Since the mid-1990s, many small coffin manufacturers have boldened their efforts to reach consumers. Another growing area is funeral insurance, which covers the cost of a funeral-including casket-for the policy bearer, upon the bearer's death. Though marketing caskets may grow more sophisticated and competitive, the actual technology used in their manufacture is relatively simple, and does not seem prone to quick changes and development.

Where to Learn More

Books

Colman, Penny. *Corpses, Coffins and Crypts*. Henry Holt, 1997.

Periodicals

"Cardboard Coffins." *UNESCO Courier* (March 1993): 26.

French, Howard W. "A Whimsical Coffin? Not Just for Chiefs Anymore." *The New York Times* (December 18, 1995): A4.

Friedman, Dorian. "Caskets: Compare and Save." *U.S. News & World Report* (June 2, 1997): 10.

Lubove, Seth. "Dancing on Graves." *Forbes* (February 28, 1994).

—Angela Woodward

Compass

Global Positioning System (GPS) consists of a system of 24 satellites containing atomic clocks that broadcast extremely accurate time signals to Earth. By analyzing the exact time these signals arrive at a receiver, it is possible to determine position with great accuracy.

Background

A compass is a device used to determine direction on the surface of the earth. The most familiar type of compass is the magnetic compass, which relies on the fact that a magnetic object tends to align itself with Earth's magnetic field. Other types of compasses determine direction by using the position of the Sun or a star, or by relying on the fact that a rapidly spinning object (a gyroscope) tends to resist being turned away from the direction in which its axis is pointing.

The basic parts of a magnetic compass are the needle (a thin piece of magnetic metal), the dial (a circular card printed with directions), and the housing (which holds the other parts in place). Inexpensive compasses, generally used as toys, may have no other parts. Compasses intended for more serious purposes usually have other parts to make them more useful. These other parts may include lids, covers, or cases to protect the compass; sights making use of lenses, prisms, or mirrors to enable the user to determine the direction of an object in the distance; and a transparent baseplate marked with a scale of inches or millimeters so that the compass can be used directly on a map.

An important feature found on many compasses is automatic declination adjustment. Declination, also known as variance, is the difference between magnetic North (the direction to which the needle points) and true North. This difference exists because Earth's magnetic field does not align exactly with its North and South poles. The amount of declination varies from place to place on Earth's surface. If the amount of declination is known for a particular area, automatic declination adjustment allows the compass user to read true direction directly from the compass rather than having to add or subtract the amount of declination every time the compass is used.

History

By 500 B.C., it was known that lodestone, a naturally occurring form of iron oxide also known as magnetite, had the ability to attract iron. No one knows where or when it was first noticed that a freely moving piece of lodestone tended to align itself so that it was pointing North and South. Written records indicate that the Chinese used magnetic compasses by 1100 A.D., western Europeans and Arabs by 1200 A.D., and Scandinavians by 1300 A.D.

Early compasses consisted of a piece of lodestone on a piece of wood, a cork, or a reed floating in a bowl of water. Somewhat later, a needle of lodestone was pivoted on a pin fixed to the bottom of a bowl of water. By the thirteenth century, a card marked with directions was added to the compass. By the middle of the sixteenth century, the bowl of water was suspended in gimbals, which allowed the compass to remain level while being used aboard a ship being tossed by the ocean.

In 1745, the English inventor Gowin Knight developed a method for magnetizing steel for long periods of time. This allowed needles of magnetized steel to replace needles of lodestone. During the early nineteenth century, iron and steel began to be used extensively in shipbuilding. This caused distortions in the operation of magnetic compasses. In 1837, the British Admiralty set up a special commission to study the problem. By 1840, a new compass design using four needles was so successful at overcoming this difficulty that it was soon adopted by navies around the world.

Until the middle of the nineteenth century, navigators used both dry-card compasses, in which the needle pivoted in air, and liquid compasses, in which the needle pivoted in water or another liquid. Dry-card compasses were easily disturbed by shocks and vibrations, while liquid compasses tended to leak and were difficult to repair. In 1862, improvements in the design of liquid compasses quickly made the dry-card compass obsolete for naval use. By World War I, the British Army used liquid compasses on land, and liquid compasses are still the standard for the best hand held magnetic compasses.

Raw Materials

The needle of a magnetic compass must be made of a metallic substance, which can be magnetized for an extended period of time. The most common substance used for compass needles is steel. Steel is an alloy of iron and a small amount of carbon. The raw materials used to produce steel are iron ore and coke (a carbon-rich substance produced by heating coal to a high temperature in the absence of air). Other substances such as cobalt are often added to the steel to produce alloys, which can be magnetized for a very long time.

The housing that holds the needle in place is often made of acrylic plastic. Acrylic plastics are produced from various derivatives of the chemical compound acrylic acid. The most important of these derivatives is methyl methacrylate. Thousands of molecules of methyl methacrylate are linked into a long chain to form polymethyl methacrylate, known by the trade names Lucite and Plexiglas. Polymethyl methacrylate has the advantages of being strong and transparent.

The Manufacturing Process

Making the needle

1 Iron ore, coke, and limestone are heated in a blast furnace by hot pressurized air.

The coke releases heat, which melts the ore, and carbon monoxide, which reacts with iron oxides in the ore to release iron. The limestone reacts with impurities in the ore such as sulfur to form slag, which floats on the molten iron and is removed. The product of this process is pig iron, which contains about 90% iron, 3-5% carbon, and various impurities.

2 To remove the impurities and most of the carbon, oxygen is blasted into the molten pig iron under high pressure. The impurities are released as slag and the carbon is released as carbon monoxide. The remaining molten steel is poured into molds and allowed to cool into ingots weighing thousands of pounds each.

3 The ingots are heated to about 2,200° F (1,200° C) and rolled between grooved rollers to form slabs. The slab is cut with giant shears, reheated, and rolled again until it is the proper thickness for needles. The thin sheet of steel is then stamped with a sharp die in the shape of the needle. The process is repeated to produce many needles from a single sheet of steel.

4 The needles are shipped from the steel manufacturer to the compass manufacturer. At the compass factory, the needles are inserted by hand into holders on an automated turntable. As the turntable spins the "North" end of the needle is sprayed with red paint and the "South" end of the needle is sprayed with white paint. As the needle continues, it is exposed to a strong magnetic field produced by an electronic magnetizer.

5 The magnetized needles are removed from the turntable and the paint is allowed to dry. The needles may also be baked in an oven to dry the paint. They are then placed in storage until needed for assembly.

Making the housing

6 Polymethyl methacrylate is formed by subjecting a solution of methyl methacrylate to light, heat, or various chemical catalysts. The components of the compass housing are then formed by a process known as injection molding. Polymethyl methacrylate is heated until it melts into a liquid. The molten plastic is then injected

A frontal and sideview of a simple pivoted-needle compass.

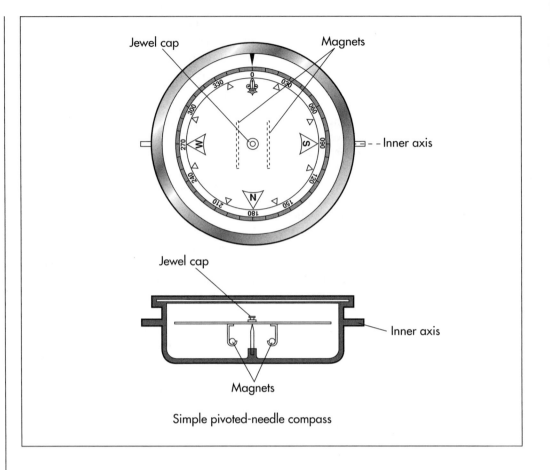

Simple pivoted-needle compass

into a mold in the shape of the desired product. The mold is allowed to cool, opened, and the solid plastic is removed. The various plastic components are shipped from the plastic manufacturer to the compass manufacturer and stored until needed.

Assembling the compass

7 When the compass manufacturer receives an order from a wholesaler, the plant manager arranges for the necessary parts to be issued from storage to workers on an assembly line. As the compass progresses along the assembly line, plastic components are snapped together. Some plastic components move through printers, which stamp them with markings such as a company logo, or with scale markings for use with maps.

8 One of the most critical components of a compass is the vial, which holds the needle. The needle is balanced on a pivot to enable it to move freely. Inexpensive compasses may have a steel pivot, but the best compasses have jeweled pivots in order to resist wear. Jeweled pivots are made of very hard materials such as an osmium-iridium alloy and are capped with a material such as artificial sapphire.

9 The vials are dipped in a liquid that will serve as a dampener. A dampener is a substance that causes the needle to come to rest more quickly when it is disturbed. Various liquids are used for dampeners. These liquids must be transparent and must not react with any of the components of the compass. A typical liquid used for this purpose might be a mixture of ethyl alcohol and water.

10 The vials filled with liquid are sealed using sonic welding. This avoids exposing the needle to heat, which could disturb its magnetism. In this process, ultrasonic waves are used to melt the plastic at the place where the vial is to be sealed. The plastic is then allowed to solidify, forming a tight seal. Assembly of the compass continues as the sealed vial is snapped onto a baseplate.

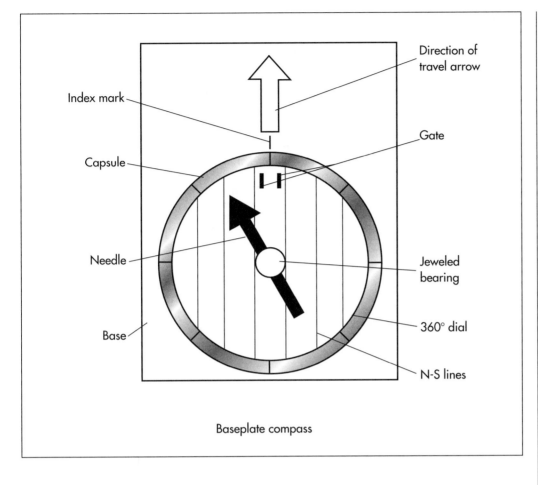

Index mark

Capsule

Needle

Base

Direction of travel arrow

Gate

Jeweled bearing

360° dial

N-S lines

Baseplate compass

11 The completed compasses are packed in ways that protect them from theft and damage. They may be packed in clam packing, in which a plastic container resembling a clam shell surrounds the compass. They may also be packed in blister packing, in which a plastic bubble attached to a flat piece of cardboard surrounds the compass. The packaged compasses are placed in cardboard boxes and shipped to the wholesaler.

Quality Control

At each step of the manufacturing process, the various components which make up the compass are visually inspected and removed if they are defective. Common imperfections include printing errors and bubbles in the dampening liquid. The most important part of the compass, the magnetic needle, is very unlikely to be defective. The few cases in which the needle does not work properly are usually caused by the consumer exposing the needle to a strong magnetic or electric field. In such cases, the needle may be remagnetized so that it points backwards, with the "North" end pointing south.

The most important part of quality control for a magnetic compass is the user's responsibility for learning how to use the compass properly. Compasses are very reliable instruments, but they are useless if the user does not know how to use them correctly. Knowing how to allow for declination is a critical skill in using a magnetic compass. In some parts of the world, failure to allow for declination could lead to an error of several degrees, causing the user to wind up many miles from the intended destination. An excellent way to learn proper use of a compass is to participate in the sport of orienteering. This sport involves using a map and compass to compete with others in finding a path from a starting point to a selected destination.

The Future

During the 1970s, the U.S. Navy began an ambitious project known as the Global Positioning System (GPS). The GPS project was taken over by the U.S. Air Force in the 1980s and completed in June 1993. GPS consists of a system of 24 satellites containing atomic clocks that broadcast extremely accurate time signals to Earth. By analyzing the exact time these signals arrive at a receiver, it is possible to determine position with great accuracy. Devices not much larger than an ordinary compass can determine location within about 100 ft (30 m).

At first glance, it may seem that GPS threatens to make the magnetic compass obsolete. In fact, the exact opposite is true. Because GPS indicates position but not direction, manufacturers of GPS equipment recommend that it be used with a compass. Compasses also have the advantage of requiring no energy supply. Unlike GPS, compasses can be used when heavy tree cover or large buildings block the reception of electronic signals. Although GPS promises to revolutionize navigation, traditional compasses will remain a vital component in how we find our way around.

Where to Learn More

Books

Hogan, Paula Z. *The Compass*. Walker and Company, 1982.

Randall, Glenn. *The Outward Bound Map and Compass Handbook*. Lyons and Burford, 1989.

Periodicals

"Consumer's Guide to Compasses." *Mechanix Illustrated* (April 1978): 9.

Kleppner, Daniel. "Where I Stand." *Physics Today* (January 1994): 9-11.

—Rose Secrest

Concrete Beam Bridge

Background

Nearly 590,000 roadway bridges span waterways, dryland depressions, other roads, and railroads throughout the United States. The most dramatic bridges use complex systems like arches, cables, or triangle-filled trusses to carry the roadway between majestic columns or towers. However, the workhorse of the highway bridge system is the relatively simple and inexpensive concrete beam bridge.

Also known as a girder bridge, a beam bridge consists of a horizontal slab supported at each end. Because all of the weight of the slab (and any objects on the slab) is transferred vertically to the support columns, the columns can be less massive than supports for arch or suspension bridges, which transfer part of the weight horizontally.

A simple beam bridge is generally used to span a distance of 250 ft (76.2 m) or less. Longer distances can be spanned by connecting a series of simple beam bridges into what is known as a continuous span. In fact, the world's longest bridge, the Lake Pontchartrain Causeway in Louisiana, is a pair of parallel, two-lane continuous span bridges almost 24 mi (38.4 km) long. The first of the two bridges was completed in 1956 and consists of more than 2,000 individual spans. The sister bridge (now carrying the northbound traffic) was completed 13 years later; although it is 228 ft longer than the first bridge, it contains only 1,500 spans.

A bridge has three main elements. First, the substructure (foundation) transfers the loaded weight of the bridge to the ground; it consists of components such as columns (also called piers) and abutments. An abutment is the connection between the end of the bridge and the earth; it provides support for the end sections of the bridge. Second, the superstructure of the bridge is the horizontal platform that spans the space between columns. Finally, the deck of the bridge is the traffic-carrying surface added to the superstructure.

History

Prehistoric man began building bridges by imitating nature. Finding it useful to walk on a tree that had fallen across a stream, he started to place tree trunks or stone slabs where he wanted to cross streams. When he wanted to bridge a wider stream, he figured out how to pile stones in the water and lay beams of wood or stone between these columns and the bank.

The first bridge to be documented was described by Herodotus in 484 B.C. It consisted of timbers supported by stone columns, and it had been built across the Euphrates River some 300 years earlier.

Most famous for their arch bridges of stone and concrete, the Romans also built beam bridges. In fact, the earliest known Roman bridge, constructed across the Tiber River in 620 B.C., was called the *Pons Sublicius* because it was made of wooden beams (*sublicae*). Roman bridge building techniques included the use of cofferdams while constructing columns. They did this by driving a circular arrangement of wooden poles into the ground around the intended column location. After lining the wooden ring with clay to make it watertight, they pumped the water out of the enclosure. This

Nearly 590,000 roadway bridges span waterways, dryland depressions, other roads, and railroads throughout the United States.

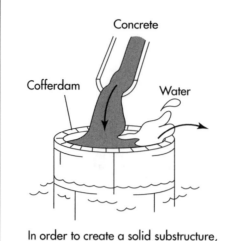

In order to create a solid substructure, cofferdams are used in the construction of the concrete support columns.

allowed them to pour the concrete for the column base.

Bridge building began the transition from art to science in 1717 when French engineer Hubert Gautier wrote a treatise on bridge building. In 1847, an American named Squire Whipple wrote *A Work on Bridge Building,* which contained the first analytical methods for calculating the stresses and strains in a bridge. "Consulting bridge engineering" was established as a specialty within civil engineering in the 1880s.

Further advances in beam bridge construction would come primarily from improvements in building materials.

Construction Materials and Their Development

Most highway beam bridges are built of concrete and steel. The Romans used concrete made of lime and pozzalana (a red, volcanic powder) in their bridges. This material set quickly, even under water, and it was strong and waterproof. During the Middle Ages in Europe, lime mortar was used instead, but it was water soluble. Today's popular Portland cement, a particular mixture of limestone and clay, was invented in 1824 by an English bricklayer named Joseph Aspdin, but it was not widely used as a foundation material until the early 1900s.

Concrete has good strength to withstand compression (pressing force), but is not as strong under tension (pulling force). There were several attempts in Europe and the United States during the nineteenth century to strengthen concrete by embedding tension-resisting iron in it. A superior version was developed in France during the 1880s by François Hennebique, who used reinforcing bars made of steel. The first significant use of reinforced concrete in a bridge in the United States was in the Alvord Lake Bridge in San Francisco's Golden Gate Park; completed in 1889 and still in use today, it was built with reinforcing bars of twisted steel devised by designer Ernest L. Ransome.

The next significant advance in concrete construction was the development of prestressing. A concrete beam is prestressed by pulling on steel rods running through the beam and then anchoring the ends of the rods to the ends of the beam. This exerts a compressive force on the concrete, offsetting tensile forces that are exerted on the beam when a load is placed on it. (A weight pressing down on a horizontal beam tends to bend the beam downward in the middle, creating compressive forces along the top of the beam and tensile forces along the bottom of the beam.)

Prestressing can be applied to a concrete beam that is precast at a factory, brought to the construction site, and lifted into place by a crane; or it can be applied to cast-in-place concrete that is poured in the beam's final location. Tension can be applied to the steel wires or rods before the concrete is poured (pretensioning), or the concrete can be poured around tubes containing untensioned steel to which tension is applied after the concrete has hardened (postensioning).

Design

Each bridge must be designed individually before it is built. The designer must take into account a number of factors, including the local topography, water currents, river ice formation possibilities, wind patterns, earthquake potential, soil conditions, projected traffic volumes, esthetics, and cost limitations.

In addition, the bridge must be designed to be structurally sound. This involves analyz-

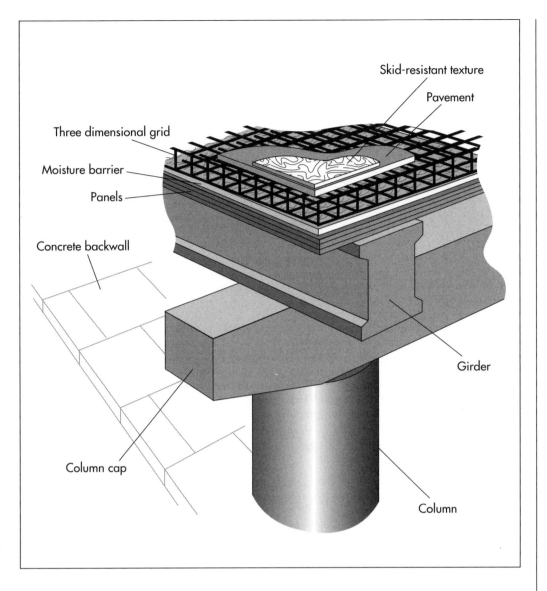

Skid-resistant texture

Pavement

Three dimensional grid

Moisture barrier

Panels

Concrete backwall

Girder

Column cap

Column

ing the forces that will act on each component of the completed bridge. Three types of loads contribute to these forces. Dead load refers to the weight of the bridge itself. Live load refers to the weight of the traffic the bridge will carry. Environmental load refers to other external forces such as wind, possible earthquake action, and potential traffic collisions with bridge supports. The analysis is carried out for the static (stationary) forces of the dead load and the dynamic (moving) forces of the live and environmental loads.

Since the late 1960s, the value of redundancy in design has been widely accepted. This means that a bridge is designed so the failure of any one member will not cause an immediate collapse of the entire structure. This is accomplished by making other members strong enough to compensate for a damaged member.

The Manufacturing Process

Because each bridge is uniquely designed for a specific site and function, the construction process also varies from one bridge to another. The process described below represents the major steps in constructing a fairly typical reinforced concrete bridge spanning a shallow river, with intermediate concrete column supports located in the river.

Example sizes for many of the bridge components are included in the following description as an aid to visualization. Some have been taken from suppliers' brochures or industry standard specifications. Others are details of a freeway bridge that was built across the Rio Grande in Albuquerque, New Mexico, in 1993. The 1,245-ft long, 10-lane wide bridge is supported by 88 columns. It contains 11,456 cubic yards of concrete in the structure and an additional 8,000 cubic yards in the pavement. It also contains 6.2 million pounds of reinforcing steel.

Substructure

1 A cofferdam is constructed around each column location in the riverbed, and the water is pumped from inside the enclosure. One method of setting the foundation is to drill shafts through the riverbed, down to bedrock. As an auger brings soil up from the shaft, a clay slurry is pumped into the hole to replace the soil and keep the shaft from collapsing. When the proper depth is reached (e.g., about 80 ft or 24.4 m), a cylindrical cage of reinforcing steel (rebar) is lowered into the slurry-filled shaft (e.g., 72 in or 2 m in diameter). Concrete is pumped to the bottom of the shaft. As the shaft fills with concrete, the slurry is forced out of the top of the shaft, where it is collected and cleaned so it can be reused. The aboveground portion of each column can either be formed and cast in place, or be precast and lifted into place and attached to the foundation.

2 Bridge abutments are prepared on the riverbank where the bridge end will rest. A concrete backwall is formed and poured between the top of the bank and the riverbed; this is a retaining wall for the soil beyond the end of the bridge. A ledge (seat) for the bridge end to rest on is formed in the top of the backwall. Wingwalls may also be needed, extending outward from the backwall along the riverbank to retain fill dirt for the bridge approaches.

3 In this example, the bridge will rest on a pair of columns at each support point. The substructure is completed by placing a cap (a reinforced concrete beam) perpendicular to the direction of the bridge, reaching from the top of one column to the top of its partner. In other designs, the bridge might rest on different support configurations such as a bridge-wide rectangular pier or a single, T-shaped column.

Superstructure

4 A crane is used to set steel or prestressed concrete girders between consecutive sets of columns throughout the length of the bridge. The girders are bolted to the column caps. For the Albuquerque freeway bridge, each girder is 6 ft (1.8 m) tall and up to 130 ft (40 m) long, weighing as much as 54 tons.

5 Steel panels or precast concrete slabs are laid across the girders to form a solid platform, completing the bridge superstructure. One manufacturer offers a 4.5 in (11.43 cm) deep corrugated panel of heavy (7- or 9-gauge) steel, for example. Another alternative is a stay-in-place steel form for the concrete deck that will be poured later.

Deck

6 A moisture barrier is placed atop the superstructure platform. Hot-applied polymer-modified asphalt might be used, for example.

7 A grid of reinforcing steel bars is constructed atop the moisture barrier; this grid will subsequently be encased in a concrete slab. The grid is three-dimensional, with a layer of rebar near the bottom of the slab and another near the top.

8 Concrete pavement is poured. A thickness of 8-12 in (20.32-30.5 cm) of concrete pavement is appropriate for a highway. If stay-in-place forms were used as the superstructure platform, concrete is poured into them. If forms were not used, the concrete can be applied with a slipform paving machine that spreads, consolidates, and smooths the concrete in one continuous operation. In either case, a skid-resistant texture is placed on the fresh concrete slab by manually or mechanically scoring the surface with a brush or rough material like burlap. Lateral joints are provided approximately every 15 ft (5 m) to discourage cracking of the pavement; these are either added to the forms before pouring concrete or cut after a slipformed slab has hardened. A flexible sealant is used to seal the joint.

Quality Control

The design and construction of a bridge must meet standards developed by several agencies including the American Association of State Highway and Transportation Officials, the American Society for Testing and Materials, and the American Concrete Institute. Various materials (e.g., concrete batches) and structural components (e.g., beams and connections) are tested as construction proceeds. As a further example, on the Albuquerque bridge project, static and dynamic strength tests were conducted on a sample column foundation that was constructed at the site, and on two of the production shafts.

The Future

Numerous government agencies and industry associations sponsor and conduct research to improve materials and construction techniques. A major goal is the development of lighter, stronger, more durable materials such as reformulated, high-performance concrete; fiber-reinforced, polymer composite materials to replace concrete for some components; epoxy coatings and electro-chemical protection systems to prevent corrosion of steel rebar; alternative synthetic reinforcing fibers; and faster, more accurate testing techniques.

Where to Learn More

Books

Brown, David J. *Bridges*. New York: Macmillan, 1993.

Hardesty, E. R., H. W. Fischer, R. W. Christie, and B. Haber. "Bridge." In *McGraw-Hill Encyclopedia of Science & Technology*. New York: McGraw-Hill Book Company, 1987, pp. 49-58.

Troitsky, M.S. *Planning and Design of Bridges*. New York: John Wiley & Sons, Inc., 1994.

Other

"General Information About Concrete Pavement." American Concrete Pavement Association. http://www.pavement.com/general/conc-info.html (24 Feb. 1998).

"Beam Bridge." Nova Online "Super Bridge." November 1997. http://www.pbs.org/wghb/nova/bridge/meetbeam.html (24 Feb. 1998).

—Loretta Hall

Copper

In the United States, the first copper mine was opened in Branby, Connecticut, in 1705, followed by one in Lancaster, Pennsylvania, in 1732. Despite this early production, most copper used in the United States was imported from Chile until 1844, when mining of large deposits of high-grade copper ore around Lake Superior began.

Background

Copper is one of the basic chemical elements. In its nearly pure state, copper is a reddish-orange metal known for its high thermal and electrical conductivity. It is commonly used to produce a wide variety of products, including electrical wire, cooking pots and pans, pipes and tubes, automobile radiators, and many others. Copper is also used as a pigment and preservative for paper, paint, textiles, and wood. It is combined with zinc to produce brass and with tin to produce bronze.

Copper was first used as early as 10,000 years ago. A copper pendant from about 8700 B.C. was found in what is now northern Iraq. There is evidence that by about 6400 B.C. copper was being melted and cast into objects in the area now known as Turkey. By 4500 B.C., this technology was being practiced in Egypt as well. Most of the copper used before 4000 B.C. came from the random discovery of isolated outcroppings of native copper or from meteorites that had impacted Earth. The first mention of the systematic extraction of copper ore comes from about 3800 B.C. when an Egyptian reference describes mining operations on the Sinai Peninsula.

In about 3000 B.C., large deposits of copper ore were found on the island of Cyprus in the Mediterranean Sea. When the Romans conquered Cyprus, they gave the metal the Latin name *aes cyprium*, which was often shortened to cyprium. Later this was corrupted to cuprum, from which the English word copper and the chemical symbol Cu are derived.

In South America, copper objects were being produced along the northern coast of Peru as early as 500 B.C., and the development of copper metallurgy was well advanced by the time the Inca empire fell to the conquering Spanish soldiers in the 1500s.

In the United States, the first copper mine was opened in Branby, Connecticut, in 1705, followed by one in Lancaster, Pennsylvania, in 1732. Despite this early production, most copper used in the United States was imported from Chile until 1844, when mining of large deposits of high-grade copper ore around Lake Superior began. The development of more efficient processing techniques in the late-1800s allowed the mining of lower-grade copper ores from huge open-pit mines in the western United States.

Today, the United States and Chile are the world's top two copper producing countries, followed by Russia, Canada, and China.

Raw Materials

Pure copper is rarely found in nature, but is usually combined with other chemicals in the form of copper ores. There are about 15 copper ores mined commercially in 40 countries around the world. The most common are known as sulfide ores in which the copper is chemically bonded with sulfur. Others are known as oxide ores, carbonate ores, or mixed ores depending on the chemicals present. Many copper ores also contain significant quantities of gold, silver, nickel, and other valuable metals, as well as large quantities of commercially useless material. Most of the copper ores mined in the United States contain only about 1.2-1.6% copper by weight.

The most common sulfide ore is chalcopyrite,$CuFeS_2$, also known as copper pyrite or yellow copper ore. Chalcocite,Cu_2S, is another sulfide ore.

Cuprite, or red copper ore,Cu_2O, is an oxide ore. Malachite, or green copper ore, $Cu(OH)_2 \cdot CuCO_3$, is an important carbonate ore, as is azurite, or blue copper carbonate, $Cu(OH)_2 \cdot 2CuCO_3$.

Other ores include tennantite, boronite, chrysocolla, and atacamite.

In addition to the ores themselves, several other chemicals are often used to process and refine copper. These include sulfuric acid, oxygen, iron, silica, and various organic compounds, depending on the process used.

The Manufacturing Process

The process of extracting copper from copper ore varies according to the type of ore and the desired purity of the final product. Each process consists of several steps in which unwanted materials are physically or chemically removed, and the concentration of copper is progressively increased. Some of these steps are conducted at the mine site itself, while others may be conducted at separate facilities.

Here are the steps used to process the sulfide ores commonly found in the western United States.

Mining

1 Most sulfide ores are taken from huge open-pit mines by drilling and blasting with explosives. In this type of mining, the material located above the ore, called the overburden, is first removed to expose the buried ore deposit. This produces an open pit that may grow to be a mile or more across. A road to allow access for equipment spirals down the interior slopes of the pit.

2 The exposed ore is scooped up by large power shovels capable of loading 500-900 cubic feet (15-25 cubic meters) in a single bite. The ore is loaded into giant dump trucks, called haul trucks, and is transported up and out of the pit.

Concentrating

The copper ore usually contains a large amount of dirt, clay, and a variety of non-copper bearing minerals. The first step is to remove some of this waste material. This process is called concentrating and is usually done by the flotation method.

3 The ore is crushed in a series of cone crushers. A cone crusher consists of an interior grinding cone that rotates on an eccentric vertical axis inside a fixed outer cone. As the ore is fed into the top of the crusher, it is squeezed between the two cones and broken into smaller pieces.

4 The crushed ore is then ground even smaller by a series of mills. First, it is mixed with water and placed in a rod mill, which consists of a large cylindrical container filled with numerous short lengths of steel rod. As the cylinder rotates on its horizontal axis, the steel rods tumble and break up the ore into pieces about 0.13 in (3 mm) in diameter. The mixture of ore and water is further broken up in two ball mills, which are like a rod mill except steel balls are used instead of rods. The slurry of finely ground ore that emerges from the final ball mill contains particles about 0.01 in (0.25 mm) in diameter.

5 The slurry is mixed with various chemical reagents, which coat the copper particles. A liquid, called a frother, is also added. Pine oil or long-chain alcohol are often used as frothers. This mixture is pumped into rectangular tanks, called flotation cells, where air is injected into the slurry through the bottom of the tanks. The chemical reagents make the copper particles cling to the bubbles as they rise to the surface. The frother forms a thick layer of bubbles, which overflows the tanks and is collected in troughs. The bubbles are allowed to condense and the water is drained off. The resulting mixture, called a copper concentrate, contains about 25-35% copper along with various sulfides of copper and iron, plus smaller concentrations of gold, silver, and other materials. The remaining materials in the tank are called the gangue or tailings. They are pumped into settling ponds and allowed to dry.

COPPER PROCESSING

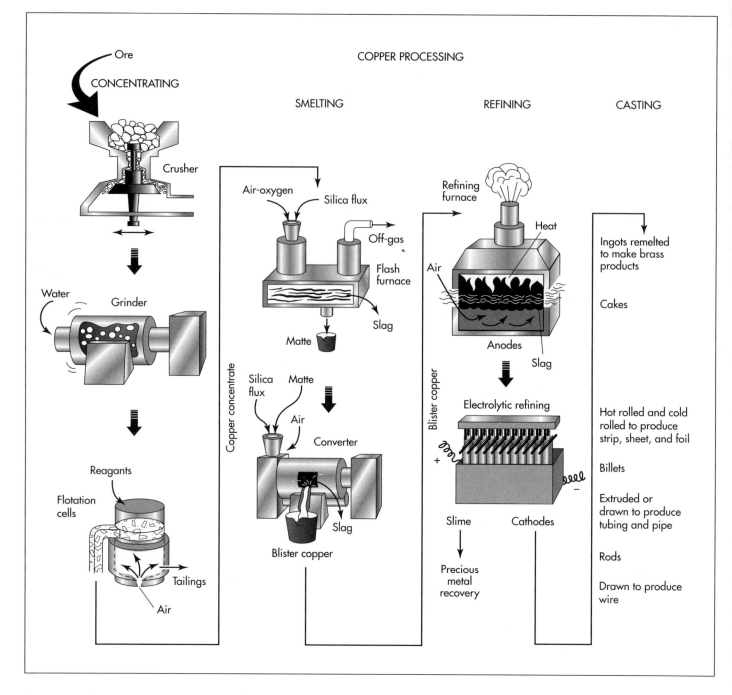

The process of extracting copper from copper ore varies according to the type of ore and the desired purity of the final product. Each process consists of several steps in which unwanted materials are physically or chemically removed, and the concentration of copper is progressively increased.

Smelting

Once the waste materials have been physically removed from the ore, the remaining copper concentrate must undergo several chemical reactions to remove the iron and sulfur. This process is called smelting and traditionally involves two furnaces as described below. Some modern plants utilize a single furnace, which combines both operations.

6 The copper concentrate is fed into a furnace along with a silica material, called a flux. Most copper smelters utilize oxygen-enriched flash furnaces in which preheated, oxygen-enriched air is forced into the furnace to combust with fuel oil. The copper concentrate and flux melt, and collect in the bottom of the furnace. Much of the iron in the concentrate chemically combines with the flux to form a slag, which is skimmed off the surface of the molten material. Much of the sulfur in the concentrate combines with the oxygen to form sulfur dioxide, which is exhausted from the furnace as a gas and is further treated in an acid plant to produce sulfuric acid. The remaining molten material in the bottom of the furnace is called the matte. It is a mixture of copper sulfides and iron sulfides and contains about 60% copper by weight.

7 The molten matte is drawn from the furnace and poured into a second furnace called a converter. Additional silica flux is added and oxygen is blown through the molten material. The chemical reactions in the converter are similar to those in the flash furnace. The silica flux reacts with the remaining iron to form a slag, and the oxygen reacts with the remaining sulfur to form sulfur dioxide. The slag may be fed back into the flash furnace to act as a flux, and the sulfur dioxide is processed through the acid plant. After the slag is removed, a final injection of oxygen removes all but a trace of sulfur. The resulting molten material is called the blister and contains about 99% copper by weight.

Refining

Even though copper blister is 99% pure copper, it still contains high enough levels of sulfur, oxygen, and other impurities to hamper further refining. To remove or adjust the levels of these materials, the blister copper is first fire refined before it is sent to the final electrorefining process.

8 The blister copper is heated in a refining furnace, which is similar to a converter described above. Air is blown into the molten blister to oxidize some impurities. A sodium carbonate flux may be added to remove traces of arsenic and antimony. A sample of the molten material is drawn and an experienced operator determines when the impurities have reached an acceptable level. The molten copper, which is about 99.5% pure, is then poured into molds to form large electrical anodes, which act as the positive terminals for the electrorefining process.

9 Each copper anode is placed in an individual tank, or cell, made of polymer-concrete. There may be as many as 1,250 tanks in operation at one time. A sheet of copper is placed on the opposite end of the tank to act as the cathode, or negative terminal. The tanks are filled with an acidic copper sulfate solution, which acts as an electrical conductor between the anode and cathode. When an electrical current is passed through each tank, the copper is stripped off the anode and is deposited on the cathode. Most of the remaining impurities fall out of the copper sulfate solution and form a slime at the bottom of the tank. After about 9-15 days, the current is turned off and the cathodes are removed. The cathodes now weigh about 300 lb (136 kg) and are 99.95-99.99% pure copper.

10 The slime that collects at the bottom of the tank contains gold, silver, selenium, and tellurium. It is collected and processed to recover these precious metals.

Casting

11 After refining, the copper cathodes are melted and cast into ingots, cakes, billets, or rods depending on the final application. Ingots are rectangular or trapezoidal bricks, which are remelted along with other metals to make brass and bronze products. Cakes are rectangular slabs about 8 in (20 cm) thick and up to 28 ft (8.5 m) long. They are rolled to make copper plate, strip, sheet, and foil products. Billets are cylindrical logs about 8 in (20 cm) in diameter and several feet (meters) long. They are extruded or drawn to make copper tubing and pipe. Rods have a round cross-section about 0.5 in (1.3 cm) in diameter. They are usually cast into very long lengths, which are coiled. This coiled material is then drawn down further to make copper wire.

Quality Control

Because electrical applications require a very low level of impurities, copper is one of the few common metals that are refined to almost 100% purity. The process described above has been proven to produce copper of very high purity. To ensure this purity, samples are analyzed at various steps to determine whether any adjustment to the process is required.

Byproducts/Waste

The recovery of sulfuric acid from the copper smelting process not only provides a profitable byproduct, but also significantly reduces the air pollution caused by the furnace exhaust. Gold, silver, and other precious metals are also important byproducts.

Waste products include the overburden from the mining operation, the tailings from the concentrating operation, and the slag from the smelting operation. This waste may contain significant concentrations of arsenic, lead, and other chemicals, which pose a potential health hazard to the surrounding area. In the United States, the Environmental Protection Agency (EPA) regulates the storage of such wastes and the remediation of the area once mining and processing operations have ceased. The sheer volume of the material involved—in some cases, billions of tons of waste—makes this a formidable task, but it also presents some potentially profitable opportunities to recover the useable materials contained in this waste.

The Future

Demand for copper is expected to remain high, especially in the electrical and electronics industries. The current trends in copper processing are towards methods and equipment that use less energy and produce less air pollution and solid waste. In the United States, this is a difficult assignment because of the stringent environmental controls and the very low-concentration copper ores that are available. In some cases, the production costs may increase significantly.

One encouraging trend is the increased use of recycled copper. Currently over half the copper being produced in the United States comes from recycled copper. Fifty-five percent of the recycled copper comes from copper machining operations, such as screw forming, and 45% comes from the recovery of used copper products, such as electrical wire and automobile radiators. The percentage of recycled copper is expected to grow as the costs of new copper processing increase.

Where to Learn More

Books

Brady, George S., Henry R. Clauser, and John A. Vaccari. *Materials Handbook.* McGraw-Hill, 1997.

Heiserman, David L. *Exploring Chemical Elements and Their Compounds.* TAB Books, 1992.

Hornbostel, Caleb. *Construction Materials.* John Wiley and Sons, Inc., 1991.

Kroschwitz, Jacqueline I. and Mary Howe-Grant, ed. *Encyclopedia of Chemical Technology.* John Wiley and Sons, Inc., 1993.

Stwertka, Albert. *A Guide to the Elements.* Oxford University Press, 1996.

Periodicals

Baum, Dan and Margaret L. Knox. "We want people who have a problem with mine wastes to think of Butte." *Smithsonian* (November 1992): 46-52, 54-57.

Shimada, Izumi and John F. Merkel. "Copper-Alloy Metallurgy in Ancient Peru." *Scientific American* (July 1991): 80-86.

Other

http://www.copper.org.

http://www.intercorr.com/periodic/29.htm.

http://innovations.copper.org/innovations.html.

—*Chris Cavette*

Corn Syrup

Background

Corn syrup is one of several natural sweeteners derived from corn starch. It is used in a wide variety of food products including cookies, crackers, catsups, cereals, flavored yogurts, ice cream, preserved meats, canned fruits and vegetables, soups, beers, and many others. It is also used to provide an acceptable taste to sealable envelopes, stamps, and aspirins. One derivative of corn syrup is high fructose corn syrup, which is as sweet as sugar and is often used in soft drinks. Corn syrup may be shipped and used as a thick liquid or it may be dried to form a crystalline powder.

The use of corn as a food product dates to about 4000 B.C. when it was grown near what is now Oaxaca in Mexico. Because of its natural hardiness, corn was successfully cultivated by people in much of the Western Hemisphere. It was imported to Spain from the West Indies in about 1520 A.D. and soon became a popular food throughout Europe.

As the use of corn as a food product spread, various machines were developed to help process it. Water-powered mills, which had been used to grind wheat and other grains for thousands of years, were adapted to grind dried corn. By the early 1700s, a device to shell corn—remove the dried corn kernels from the cob—had been patented. The refining process used to separate corn starch from corn kernels is called the wet milling process. It was patented by Orlando Jones in 1841, and Thomas Kingsford established the first commercial wet milling plant in the United States in 1842.

The process for converting starches into sugars was first developed in Japan in the 800s using arrowroot. In 1811, the Russian chemist G.S.C. Kirchoff rediscovered this process when he heated potato starch in a weak solution of sulfuric acid to produce several starch-derived sweeteners, including dextrose. In the United States, this acid conversion method was adapted to corn starch in the mid-1800s and the first corn sweeteners were produced in a plant in Buffalo, New York, in 1866. This process remained the principal source of corn syrup until 1967, when the enzyme conversion method for producing high fructose corn syrup was commercialized. At first, this was a batch process requiring several days. In 1972, a continuous enzyme conversion process was developed that reduced the time to several minutes or hours.

Today, corn syrups are an important part of many products. In 1996, there were 28 corn-refining plants in the United States that processed a total of about 72 billion lb (33 billion kg) of corn. Of that amount, about 25 billion lb (11.4 billion kg) were converted into corn syrups and other corn sweeteners. These corn-based products supplied more than 55% of the nutritive sweetener market in the United States.

Raw Materials

There are several thousand varieties of corn, but the variety known as yellow #2 dent corn is the primary source of corn syrup. It is a common variety grown in the Midwestern portion of the United States and elsewhere in the world. It belongs to a family of corn that derive their name from the small dent in the end of every kernel.

Other materials used during the process of converting corn to corn syrup include sulfur

In 1996, there were 28 corn-refining plants in the United States that processed a total of about 72 billion lb (33 billion kg) of corn. Of that amount, about 25 billion lb (11.4 billion kg) were converted into corn syrups and other corn sweeteners.

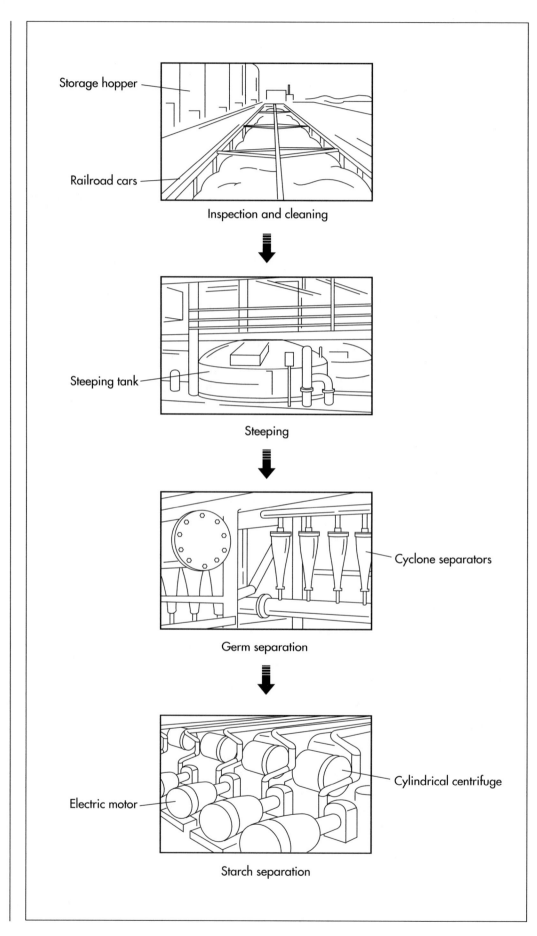

Storage hopper

Railroad cars

Inspection and cleaning

Steeping tank

Steeping

Cyclone separators

Germ separation

Cylindrical centrifuge

Electric motor

Starch separation

dioxide, hydrochloric acid or various enzymes, and water.

The Manufacturing Process

Corn syrup is produced in processing plants known as wet corn mills. In addition to corn syrup, these mills produce many other corn products including corn oil, corn starch, dextrose, soap stock, animal feed, and several chemicals used in other industrial processes.

Separating corn starch from corn

1 Dried, shelled corn kernels are transported to the mill in trucks, railcars, or barges. The corn is unloaded into a storage pit where it is weighed and sampled.

2 The kernels are taken from the pit on conveyors and are passed over a set of vibrating screens or perforated metal grates to remove any sticks, husks, stones, and pieces of cob. A controlled blast of air blows away any chaff and dust, while electromagnets capture any nails, screws, or bits of metal that may have fallen in among the kernels during harvesting, shelling, or shipping.

3 The cleaned kernels are placed in a series of large stainless steel tanks called steep tanks. Each tank holds about 168,000 lb (76,000 kg) of kernels. Warm water with a small amount of sulfur dioxide is circulated through the tanks. The sulfur dioxide reacts with the water to form a weak sulfurous acid solution. This process continues for about 20-40 hours and is used to soften the kernels and make it easier to separate the starch.

4 The softened kernels are passed through coarse grinding mills to remove the inner portion of the kernel, called the germ, which contains most of the corn oil. Each mill has one stationary and one rotating disk. The clearance between the two disks is adjusted to tear the kernel apart without crushing the germ.

5 The resulting pulp is transferred to a set of cyclone separators called germ separators or hydroclones. The germs, which are less dense than the other parts of the kernel, are spun out of the pulp by centrifugal force. The germs are then pumped onto a series of screens and washed several times to remove any remaining starch. The cleaned germs are heated and pressed to extract the corn oil for further processing into food products and soap stock.

6 The remaining material from the germ separators is a slurry composed of starch, protein, and fiber. This slurry passes through another set of mills to tear the starch lose from the fiber. The fiber is then trapped on a set of washing screens and is dried to become animal feed or corn bran fiber for use in cereals.

7 The starch and protein mixture, called mill starch, is pumped into a set of centrifugal separators that spin the mixture at high speeds. Because of a difference in specific gravity between the two materials, the heavier starch can be separated from the lighter protein, which is called the gluten. The gluten is dried and sold as animal feed.

8 The starch is diluted with water before being washed and filtered 8-14 times to remove any remaining protein. It is then rediluted and run through a second set of centrifugal separators. The resulting starch is more than 99.5% pure. Some of this corn starch is dried and packaged for use in food products, building materials, or to produce various chemicals. The rest of it, usually the majority, is converted into corn sweeteners including corn syrup.

Converting corn starch into corn syrup

9 Corn starch is converted into ordinary corn syrup through a process called acid hydrolysis. In this process, the wet starch is mixed with a weak solution of hydrochloric acid and is heated under pressure. The hydrochloric acid and heat break down the starch molecules and convert them into a sugar. The hydrolysis can be interrupted at different key points to produce corn syrups of varying sweetness. The longer the process is allowed to proceed, the sweeter the resulting syrup.

10 This syrup is then filtered or otherwise clarified to remove any objectionable

flavor or color. It is further refined and evaporated to reduce the amount of water.

11 To produce a corn syrup powder, also called corn syrup solids, the liquid corn syrup is passed through a drum or spray dryer to remove 97% of the water. This produces a crystalline corn syrup powder.

Converting corn syrup into high fructose corn syrup

12 Ordinary corn syrup contains dextrose sugar which is about three-quarters as sweet as the sucrose sugar in cane or beet sugar. In many sweetener applications this is an advantage because it does not overpower the other flavors in the food. However, in some applications, such as soft drinks, a sweeter taste is desired. To improve the sweetness of ordinary corn syrup, it undergoes a further process called enzyme conversion. In this process, the dextrose sugars in the syrup are converted into sweeter fructose sugars by the action of an enzyme in a series of steps under carefully controlled temperatures, pressures, and acidity. This produces a high fructose corn syrup with a 42% fructose content. It is used in canned fruits and condiments.

13 To produce corn syrups with a fructose level above 50%, the 42% fructose syrup is passed through a series of fractionation columns, which separate and hold the fructose content. The separated portion is about 80-90% fructose and is flushed from the columns with deionized water. A portion of this is retained and sold for use in "light" foods where only a small amount of liquid sweetener is needed. The remainder is blended with other 42% fructose syrup to produce a 55% fructose syrup, which is used in soft drinks, ice cream, and frozen desserts.

14 Powdered high fructose corn syrups can be produced by evaporating the water from the syrup and then encapsulating the powder grains to prevent them from reabsorbing moisture. Pure fructose crystals may be obtained by further processing the 80-90% fructose syrup. It is used in cake mixes and other food products where a highly concentrated, dry sweetener is desired.

Quality Control

Corn syrup is primarily used as a food product. In the United States, its production and use falls under the control of the federal Food and Drug Administration (FDA), which sets rigid quality standards. The corn refiners, working through the Corn Refiners Association, have developed comprehensive analytical procedures for testing the properties of corn products, including corn syrup. Some of the important properties of corn syrup are dextrose or fructose content, carbohydrate composition, solids content, sweetness, solubility, viscosity, and acidity. In addition to monitoring the materials and processes used to make corn syrup, manufacturers also take frequent samples of the finished product for analysis.

The Future

Because of the ready supply of corn in the United States, it is expected that corn syrup and other corn sweeteners will continue to be used extensively in food products.

Corn is also expected to be a source of many other products in the future. Ethanol can be derived from corn and offers a cleaner-burning fuel than gasoline for use in motor vehicles. Corn starch can be used as a raw material to replace petroleum in the production of chemicals and plastics. Corn products may also find applications in the production of drugs and antibiotics.

Where to Learn More

Books

Considine, Douglas M., ed. *Foods and Food Production Encyclopedia.* Van Nostrand Reinhold, 1982.

Hui, Y.H., ed. *Encyclopedia of Food Science and Technology.* John Wiley and Sons, Inc., 1992.

Matz, Samuel A. *The Chemistry and Technology of Cereals as Food and Feed.* Pan-Tech International, 1991.

McGraw-Hill Encyclopedia of Science and Technology. McGraw-Hill, 1997.

Other

Corn Refiners Association. http://www. corn.org.

"Corn Refining: The Process, The Products." Corn Refiners Association Inc., 1992.

"Nutritive Sweeteners From Corn." Corn Refiners Association Inc., 1993.

"Tapping the Treasure." Corn Refiners Association Inc., 1997.

—*Chris Cavette*

Correction Fluid

Correction fluids were greatly improved during the 1950s when polymer technology was utilized. This allowed production of a product which would adhere better to paper, spread easier, and remain flexible when dry.

Correction fluid is a liquid product designed to cover mistakes made while typing, hand writing, or photocopying markings on paper. Typically, it is applied to paper using a brush. When it dries, it forms a solid film that effectively covers the error and allows the correct mark to be written over it. Correction fluids are composed of pigments, polymeric binders, and solvents that are mixed together in large batch tanks. First developed during the late 1950s, correction fluid formulations have steadily improved over the years.

Background

The need for correcting mistakes made during writing has been around for as long as writing itself. While erasers worked well for pencil marks, they did little to remove mistakes made with a fountain pen, typewriter, or ball point pen. At some point, it was realized that a mistake could be covered using an ink that was the same color as the paper. This led to the development of the first correction fluids. These fluids were typically white inks. These products were inferior because they did not match the paper color very well, took a long time to dry, and were difficult to write over. Correction fluids were greatly improved during the 1950s when polymer technology was utilized. This allowed production of a product which would adhere better to paper, spread easier, and remain flexible when dry. Over the next 40 years, a variety of patents have been granted which show how steady improvements have been made in correction fluid technology.

Correction fluid is a liquid product designed to cover mistakes marked on paper. It is typically sold in a small jar with a brush applicator and works in much the same way as paint. First, the fluid is applied to the paper over the errant mark. Then it forms a film, which bonds to the paper fibers. This film is an elastic polymer that is both strong and flexible. Fixed in this film are pigments, which are supposed to match the color of the paper and cover the incorrect ink mark. When the film is dry it can be written over.

A variety of correction fluid products have been developed for different applications The most common types are those designed to be used on standard, white typing paper. These formulas are typically white and designed to dry relatively quickly. Other fluids are available for special types of paper. For bonded paper, correction fluid formulas are made which give a different texture when they dry. This makes the correction less noticeable. For corrections on paper that is not white, various colored correction fluid are available. Products are also available for photocopying applications. These formulas are made with special additives that reduce the reflection of light off the film.

While the standard product is sold in a plastic jar with an applicator brush built in the cap, this is not the only kind. Some fluids are sold in a pen, which uses a roller ball applicator. These products give better control over the application and the amount of fluid used. Other correction fluid type products are sold as solid films. These products are designed to be placed in front of the typing hammer of a typewriter. When the typewriter hammer hits the film it transfers the correction formula onto the paper in the exact shape as the letter providing a perfect coverup. As computers gradually replace conventional typewriters, this product will be used less frequently.

Design

Before a correction fluid can be made for the first time, a formula must be developed. This is done by trained chemists who are knowledgeable of a variety of raw materials. These scientists begin by choosing what characteristics are required for the fluid. They decide on functional features such as how long the product will take to dry, how strong the film has to be, and how stable it will be during storage. They also consider aesthetic features such as how thick it should be, what color it will be and how it will be delivered from the package. Often consumer testing is employed to help with these determinations.

Preliminary formulae are first prepared in small beakers in the lab so the performance aspects of the formula can be evaluated. Tests for the correction fluid's effectiveness are done on these initial samples. Other tests may be run, including stability tests, safety tests, and consumer acceptance testing. Stability testing is used to detect physical changes in characteristics such as color, odor and thickness over time. It helps ensure that the product on the store shelves will work just like the formula created in the laboratory. Using the information obtained during this testing phase, the formula can be adjusted to produce the best product.

Raw Materials

There are many different types of ingredients that can be used to make a correction fluid formula. In general, the formulas are composed of an opacifying agent, a polymeric film former, a solvent, and other miscellaneous ingredients.

The opacifying agent is a key ingredient in the correction fluid formula. It is the material responsible for covering the errant marking. The most common opacifying agent is titanium dioxide. This is an inorganic material derived from various titanium ores. It is an opaque material, which does not significantly absorb visual light. Since it has a high refractive index, it produces a predominantly white color. By changing the processing method and mixing the titanium dioxide with different materials, a variety of other colors may be produced. These are used for the different colored correction fluids. In general, the opacifying agent makes up from about 40-60% of the formula.

Although the opacifying agent actually covers the error, a polymeric material is used to affix it to the paper. This polymer creates the film that strongly bonds to the paper fibers when it dries, or cures. The film is designed to be strong so it will stay in place, but also flexible so that it will not crack, flake, and fall off under normal conditions. A variety of polymeric resins can be used such as acrylic resins, petroleum resins, chlorinated polyolefin resins and even synthetic rubber. To make the optimal film, often a copolymer is used. One type copolymer system is a latex emulsion. This is made by polymerizing methacrylate with a nitrogen containing monomer in the presence of ethylene vinyl acetate. In a typical correction fluid formula, the polymer resin comprises 5-15% of the formula.

To control the viscosity and dry time of the correction fluid, a solvent is necessary. In general, the correction fluid is made thin so it can be applied evenly and smoothly. The solvent works by diluting the formula and quickly evaporating to leave a dried film. Additionally, the solvent improves stability and helps to make the other materials in the formula more compatible with each other. In developing a correction fluid formula, the solvent must be chosen carefully. On the one hand it must evaporate quickly, so it can be quickly written over. On the other hand, it can not evaporate too fast or the polymer may solidify in the bottle.

Two types of solvents are used including aqueous based and organic based. The aqueous based solvents are used for correction fluids that will cover oil based inks. They are typically a mixture of water and alcohol. Organic based solvents use volatile organic compounds (VOC) and generally dry more quickly than aqueous solvents. They are better for covering water based inks. A variety of organic compounds can be used including acetone, toluene, xylene, ethyl acetate, and methyl ethyl ketone. Some newer formulae include both types of solvents. These "amphibious"

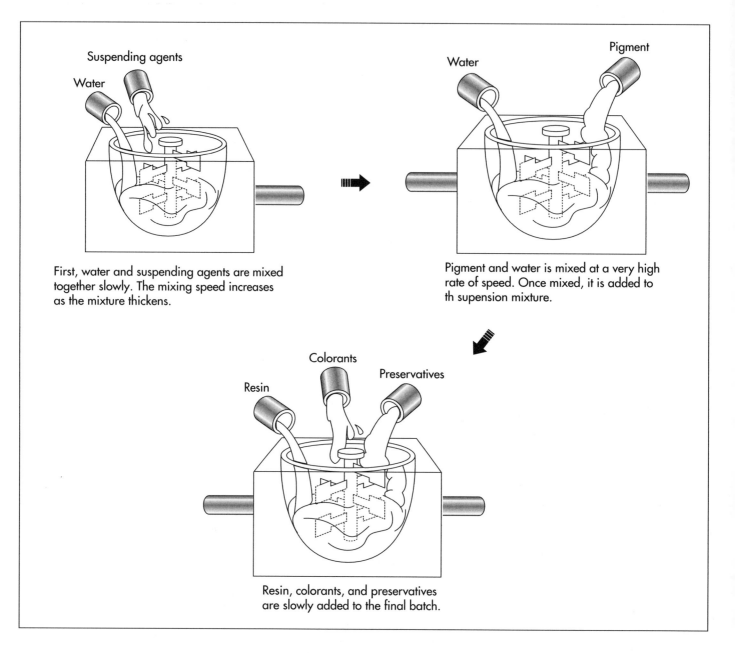

First, water and suspending agents are mixed together slowly. The mixing speed increases as the mixture thickens.

Pigment and water is mixed at a very high rate of speed. Once mixed, it is added to th supension mixture.

Resin, colorants, and preservatives are slowly added to the final batch.

type formulae are useful for all types of inks. Recently, environmental concerns have led to the development of formulae, which use little or no volatile organic solvents. The formula can be composed of anywhere from 25-50% solvent.

A variety of other ingredients are added to the correction fluid formula to optimize stability and performance. Since titanium dioxide is not generally soluble in the solvent it has a tendency to settle out over time. For this reason suspending agents and dispersing agents are added. Examples of the former include hydroxyethylcellulose, xanthan gum or guar gum. Examples of the lat-

ter include phosphate esters, ethoxylated alcohol, and polysorbitans. Sometimes glass or metal mixing beads are included in the container to help re-disperse the titanium dioxide. In this case, the user has to shake before using. Other ingredients added include chelating agents that help protect metal parts in the applicator, defoamers that prevent excessive bubbling, and preservatives that prevent biological contamination.

The Manufacturing Process

The manufacturing process can be broken down into two steps. First, the batch of cor-

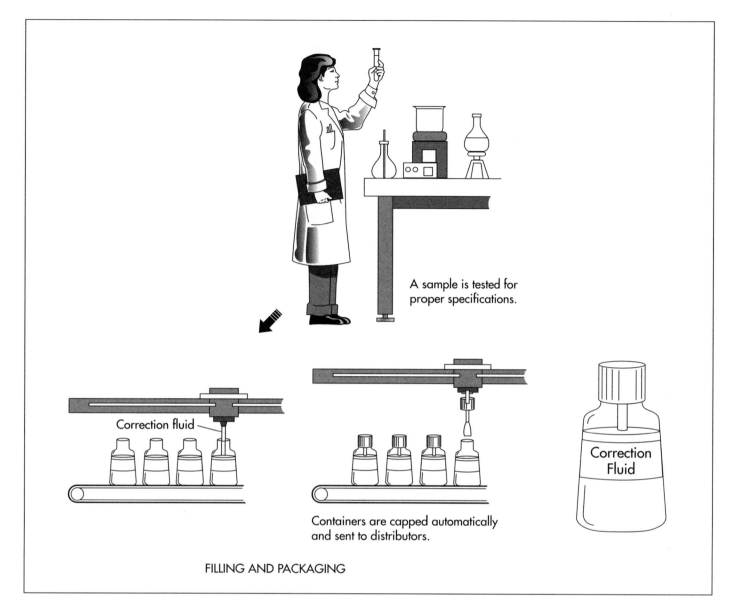

A sample is tested for proper specifications.

Correction fluid

Containers are capped automatically and sent to distributors.

Correction Fluid

FILLING AND PACKAGING

rection fluid is made and then it is filled into its packaging. The following description details the production of an aqueous based correction fluid. Other types are made in a similar manner.

Compounding the batch

1 The batches of correction fluid are made in large, stainless steel tanks which can hold 3,000 gallons or more. These tanks are equipped with mixers and a temperature control system. Workers, known as compounders, follow the formula instructions and add the correct types and amounts of raw materials at specified times and temperatures. Using computer controls, they can regulate the mixing speed and temperature

of the batch. The correction fluid batch is made in three phases.

2 In the first phase, the main batch tank is filled with some of the water. The suspending agents and other miscellaneous ingredients are added at this time. Mixing is done at a low shear rate to get adequate dispersion without incorporating air into the mixture. As the suspending agent is hydrated, it thickens, and the mixing speed is increased.

3 A pigment dispersion is made next. This is done by adding the pigment to an amount of water and dispersing it at a very high shear rate. When the size of the particles is sufficiently small, it is slowly added

to the main batch. In the final phase, the resin is slowly added. Additional ingredients such as colorants and preservatives may also be added at this point.

Quality control check

4 After all of the ingredients are added, a sample of the batch is taken to the quality control lab for approval. Physical and chemical characteristics are checked to make sure the batch falls within specifications outlined in the formula instructions. The quality control scientists run tests such as pH determination, viscosity checks, and appearance and odor evaluations. If the batch does not meet all of the specifications, an adjustment may be made. For example, if the color of the batch is off it can be adjusted by adding more pigment. After the batch is approved, it is pumped to a holding tank where it is stored prior to filling.

Filling and packing

5 The filling operation is dependant on the type of packaging in which the product will be sold. For the typical bottle of correction fluid the process begins with empty containers at the start of the filling line. These bottles are held in a large bin and physically manipulated until they are standing upright. They are then moved along a conveyor belt to the filling heads holding the correction fluid.

6 As the bottles pass the filling heads, they are injected with the exact amount of product needed. The bottles are then moved to a capping machine, which sorts the caps, places them on the bottles, and tightens them. At this point, the bottles may be passed through a labeling machine if necessary. The bottles are then put into boxes and stacked onto pallets for shipping to wholesalers and retailers.

Quality Control

Beyond the quality control tests made during the batching process, other checks are performed during filling. Line inspectors are stationed at various points on the filling line, and they watch the bottles to make sure every thing looks right. They check for things such as label placement or filling weights. They would also see that enough finished bottles are packed into cases. Occasionally, product performance tests are run. For example, the opacity may be checked using a colorimeter. The flexibility and adhesion of the film may also be examined using a fold test. In this test, the fluid is applied to paper and allowed to dry. The paper is then folded numerous times and the film is checked for cracking and flaking. These types of tests are crucial to the production of a quality product.

The Future

There are a variety of challenges facing developers of correction fluids. Many of the correction fluid formulas continue to have certain drawbacks. For example water based correction fluids are still prone to a problem called bleeding when used with water based inks. When this happens, the inks often show through the coating. The new amphibious formulas, which contain both a water based and organic based solvent, help alleviate some of these problems. However, these formulas will be more difficult to produce as governmental regulations require a reduction in the amount of volatile organic solvents used. Other formulation challenges include producing new colors, reducing drying time, reducing the incidence of product dry out in the container, and making the products less poisonous. New and improved forms of product delivery are also expected.

Where to Learn More

Books

Carraher, C. and R. Seymour. *Polymer Chemistry.* New York: Marcel Dekker, 1992.

Kirk Othmer Encyclopedia of Chemical Technology. New York: John Wiley & Sons, 1992.

—*Perry Romanowski*

Cotton Candy

Cotton candy is a light and fluffy sugar confectionery which resembles cotton wool. It is made by melting a sugar composition and spinning it into fine strands. The strands are then collected on a cardboard tube or bundled in a continuous mass. First developed over 100 years ago, cotton candy remains a favorite summertime candy at carnivals, amusement parks, and baseball stadiums. With the development of more efficient, automated machines it is expected that the market for cotton candy will substantially increase in the coming years.

Background

Cotton candy is a popular food at carnivals and amusement parks. Typically, it is sold as a large mass wrapped around a cardboard cone. It has a fibrous texture that makes it unique among sugar confectioneries. This texture is a direct result of the sugar used to make the candy and the method in which it is processed. At the start of manufacture, the sugar is a solid material supplied as individual granules. When it is melted the individual granules become intermixed and form a thick, sticky syrup. This syrup is then spun out to create thin strands that harden. These hardened strands have many of the same characteristics as cotton fibers, which is how cotton candy got its name. When the strands are collected on a cone, they are not packed close together and a certain amount of air gets trapped between them. This increases the volume of the candy, giving it a light and fluffy texture.

History

Sugar confectioneries have been known for thousands of years, however the development of cotton candy is a relatively recent event. Evidence shows that the first sugar confectioneries were used during the time of the ancient Egyptian civilization. True candymaking began only after a sugar refining process was developed during the fourth century. For many years candy was a luxury item available to only the privileged. Eventually, sugar became more widely available and candy could be enjoyed by all.

The modern candy industry developed during the nineteenth century. At this time, special candymaking machinery was invented. These machines were semi-automatic and allowed production on a large scale. The first cotton candy machine was created during the late nineteenth century. This machine consisted of a large pan with a rotating heating core in the middle. Operators could make individual servings, and since it was portable, it became a popular confection at circuses, carnivals, and ball parks.

Prior to the 1970s, cotton candy was only produced on a small scale. This was due to the fact that there were no automated machines that could produce enough product for widespread distribution. Then, in 1972, an automatic cotton candy manufacturing machine was patented. This machine provided an efficient for automatic manufacture and packaging. It led to the mass production of cotton candy.

Raw Materials

Sugar is the most important ingredient used in the manufacture of cotton candy. Chemically, sugar is known as sucrose, which is a disaccharide, made up of glucose and fructose units. It is obtained primarily from sug-

The first cotton candy machine was created during the late nineteenth century. This machine consisted of a large pan with a rotating heating core in the middle. Operators could make individual servings, and since it was portable, it became a popular confection at circuses, carnivals, and ball parks.

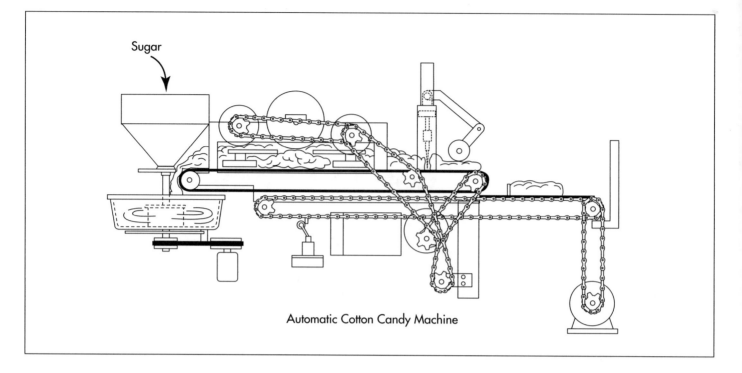

Sugar

Automatic Cotton Candy Machine

After processing the sugar granules into extruded sugar strands, the strands of cotton candy are pulled onto a conveyor belt and transferred into a sizing container. Here, the candy strands are combined into a continuous bundle.

arcane or sugar beets via an extraction process. In cotton candy, sugar is responsible for the candy's physical structure as well as its sweet taste and mouthfeel. The sugar used for cotton candy production, called floss sugar, is specially treated to promote the formation of fibers.

To produce the well-known characteristics of cotton candy, other ingredients such as dyes and flavorings must be added. Since sugar is naturally white, dyes must be added to produce the different colors typical of cotton candy. Usual dyes include Red dye #40, Yellow dye #5, Yellow dye #6, and Blue dye #1. By using only these federally regulated dyes, cotton candy can be made to be almost any color desired. The most popular colors are pink and blue, however purple, yellow, red, and brown cotton candy are also sold.

Cotton candy is available in many different flavors including bubble gum, banana, raspberry, vanilla, watermelon, and chocolate. To produce these flavors, both artificial and natural flavorants may be used. Natural flavors are obtained from fruits, berries, honey, molasses, and maple sugar. Artificial flavors are mixtures of aromatic chemicals produced synthetically via organic reactions. Some important artificial flavoring

compounds include materials such as methyl anthranilate and ethyl caproate.

In addition to the cotton candy ingredients, different packaging raw materials are required. Since moisture can make cotton candy rubbery and sticky, the packaging is designed to inhibit interaction with air. Typically, a plastic bag made out of a high-molecular weight polymer is used.

The Manufacturing Process

There are primarily two types of machines used to produce cotton candy. One of them is semi-automatic and is used to produce the single serve helpings that are immediately sold at carnivals and amusement parks. The other is a fully automated machine that is used to produce large volumes of cotton candy for widespread distribution. Since these machines are very similar, both will be described below.

Sugar processing

1 The first step in making cotton candy is converting the granular sugar into fine filaments. To do this, solid sugar is placed in a large, stainless steel hopper. This hopper has a tapered bottom, which funnels the sugar into the extruder. The extruder is a

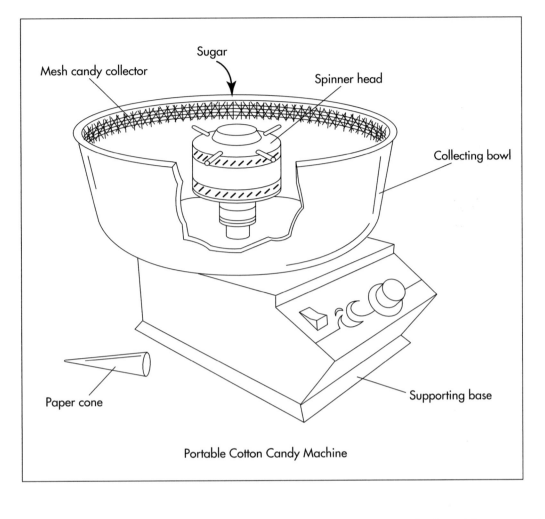

Mesh candy collector

Sugar

Spinner head

Collecting bowl

Paper cone

Supporting base

Portable Cotton Candy Machine

The portable cotton candy machine consists of a large pan with a rotating heating core in the middle. Operators make individual servings at popular venues such as the circus, carnival, and ball park.

rotating metal cylinder, which has holes along its sides and is equipped with a heating element.

2 Inside the extruder, the sugar is heated such that it melts and becomes a molten liquid. The spinning extruder then throws the strands of liquid sugar out in all directions through the holes in its sides. As it exits the extruder, the liquid sugar cools and forms solid strands. These strands, which are the fibers used to make cotton candy, are collected in a large circular pan surrounding the extruder. To prevent coagulation of the strands, moisture is minimized during this phase of manufacture.

Candy collection

3 In machines that produce a small amount of cotton candy, such as those found at carnivals, the strands of cotton candy are then collected by the machine operator. He takes a cardboard cone and passes it around the sides of the collection pan. As the card-

board is passed around, the sticky sugar strands adhere to it. When enough is collected on the cone, the cotton candy is sold to the consumer immediately. The situation is slightly different for automated cotton candy machines. In these machines, the strands of cotton candy are pulled onto a conveyor belt and transferred into a sizing container. Here the candy strands are combined into a continuous bundle.

4 In the sizing container, the bundle of cotton candy is molded into a consistent shape. This is done by rollers that are spaced on the top and sides of the conveyor belt. To prevent the cotton candy from sticking to the rollers, they are typically coated with a non-stick substance such as Teflon. As the candy exits the sizing container, it has the shape of a continuous block with a fixed height and width. This forming process is done with a minimum of force so the candy is not compressed so much that it changes its character or texture.

Cutting

5 After the shaping process, the cotton candy is conveyored to a knife blade where it is cut into segments of a set length. The knife is mounted vertically above the conveyor, and as the candy passes by, it slides down to make the cut. The knife is then retracted and the segmented candy is conveyored away. To help the candy maintain its shape and prevent it from sticking to the knife, it is then passed under another roller immediately after it is cut.

Packaging

6 The cut mass of cotton candy is next transferred to the packaging machine. Here, it is automatically put into a plastic bag or other type of packaging, and sealed shut. It is important that the package is sealed so moisture is prevented from spoiling the candy. The bags are passed by a coding device where they are marked with information related to the date of production, batch number, and other information. The bags are then carefully put into boxes. The boxes are stacked on wooden pallets, transferred to trucks via forklifts, and shipped to the local supermarket. The entire process from loading the sugar to putting the candy in boxes takes only a few minutes.

Quality Control

As in all food processing facilities, quality control begins with a check of the incoming ingredients. These ingredients are tested in a quality control laboratory to ensure they meet specifications. Tests include evaluation of the ingredient's physical properties such as particle size, appearance, color, odor and flavor. Certain chemical properties of the ingredients may also be evaluated. Each manufacturer has their own tests that help certify that the incoming ingredients will produce a consistent, quality batch of cotton candy.

In addition to ingredient checks, the packaging is also inspected to ensure it meets the set specifications. An important property that is routinely examined is the odor of the packaging. Many times plastics can acquire off-odors during processing. These odors can be passed on to the food products and hence must be found before the packaging

can be used. Since excessive water vapor can ruin a bag of cotton candy, the packaging is also checked for its moisture-vapor transmission rate. Other properties that are checked include grease resistance and physical appearance. Correctly produced cotton candy has a shelf life of about six months.

After production, the characteristics of the final product is also carefully monitored. Quality control chemists perform many of the same tests on the final product that they did on the initial ingredients. These include tests of the candy's appearance, flavor, texture, and odor. The usual test method involves comparing the final product to an established standard. For example, to make sure the color is correct, a random sample may be taken and compared to some set standard. Other qualities such as taste, texture and odor may be evaluated by sensory panels. These panels are made up of a group of specially trained people who can determine small differences. In addition to sensory tests, other standard industry instrumental tests may also be performed.

The Future

Cotton candy has changed very little since it was first introduced. Most of the improvements have come in the design of machines that are used to make the candy. It is expected that future improvements will continue to be found in this area. For example, machines will be developed which are more automated with computer controls. These machines will be able to produce the candy more efficiently, economically and safely. In addition to new cotton candy machines, new colors and flavors will also be introduced to make the confection more appealing.

Where to Learn More

Books

Alikonis, J. *Candy Technology.* Westport, CT: AVI Publishing Co., 1979.

Kirk Othmer Encyclopedia of Chemical Technology. New York: John Wiley & Sons, 1992.

Mathlouthi, M. and P. Reiser, ed. *Sucrose: Properties and Applications.* London: Blackie and Sons, Ltd., 1995.

Pennington, N. L. and C.W. Baker, ed. *Sugar, A User's Guide to Sucrose.* New York: Van Nostrand Reinhold, 1990.

—Perry Romanowski

Cotton Swab

The cotton swab was invented in the 1920s by a Polish-born American named Leo Gerstenzang.

Background

A cotton swab is a short spindle with one or both ends coated with an absorbent cotton padding. Such swabs have long been used for various cosmetic and personal hygiene tasks, particularly for cleaning the ear. The cotton swab was invented in the 1920s by a Polish-born American named Leo Gerstenzang. Leo had observed that his wife used a toothpick stuck into a piece of cotton to clean their baby's ears at bath time. Leo was concerned that the wooden toothpick might splinter and cut the baby's ear or that the cotton might come off the stick and become lodged in the ear. He decided to design a ready-made cotton swab that could be used on babies with less risk of injury. Thinking that such a product would appeal to many parents, Leo formed the Leo Gerstenzang Infant Novelty Company to market his swab and other baby related products. It took him several years to solve certain design problems, like how to secure equal amounts of cotton on each end of the swab. Eventually, he not only developed a successful product but he even created special packaging for his swabs. The package was designed to be opened with one hand so a parent holding an infant with one hand could easily open the box and extract a swab with the other. Once Leo had perfected his product, he looked for a commercially viable name. Because he designed his product to keep infants happy as their ears were safely cleaned, he chose the name Baby Gays. In 1926, he changed the name to Q-Tips Baby Gays, claiming that the "Q" stood for quality. Eventually, Baby Gays was dropped from the name and the swabs became known simply as Q-Tips. Today Q-Tips is a registered trademark of Chesebrough-Ponds, Inc.

Design

Cotton swab design has advanced significantly since the 1920s. For most applications, wooden sticks were replaced by paper spindles, which were less likely to splinter and puncture delicate ear tissue. The thin paper rods were made by rolling a heavy gauge paper. More recently, plastic has become a popular choice for spindle material because it offers improved flexibility and imperviousness to water. However, care must be taken to design the plastic shaft so that it does not poke through the cotton mass at the end of the stick. To prevent this from occurring, swabs have been designed with a number of special features. For example, some swabs are made with a protective plastic cap on the end of the spindle, under the cotton coating. Others use a cushioning element, like a dab of soft hot melt adhesive, to protect the end of the stick should it protrude through the body of the tip during manipulation. A third way of circumventing this problem involves a process, which results in a swab with a flared tip. This flared tip can not penetrate too deeply into the ear because of its larger diameter.

While infant care is one of the most popular uses for cotton swabs, they are commonly used in other areas as well. These other applications have their own special design requirements. For example, swabs used to apply color cosmetics can be made with special flocked tips made of non-woven fibers. In addition to personal care, swabs are used for industrial purposes. For example, long handled wooden swabs are designed for sampling microbiological cultures. This is the type of swab used by a doctor to take a throat culture. Other indus-

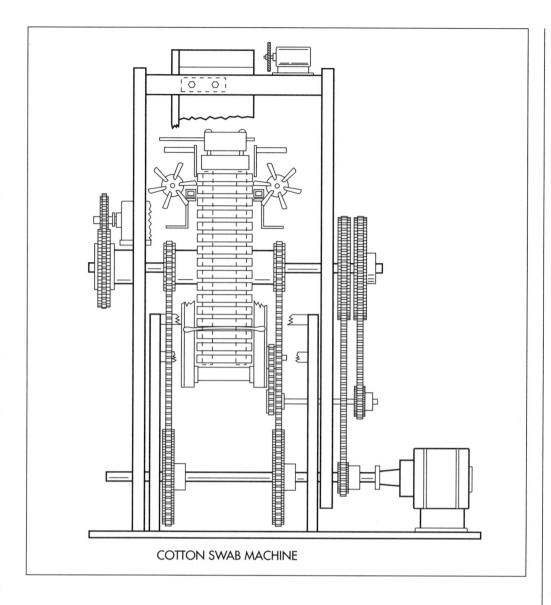

COTTON SWAB MACHINE

trial swabs may be specially designed for cleaning electronic parts; these may have special lint free requirements.

Raw Materials

There are three primary components involved in swab manufacture: the spindle or stick, which forms the body of the swab; the absorbent material coated onto the spindle ends; and the package used to contain the swabs.

Spindle

Spindles can be sticks made of wood, rolled paper, or extruded plastic. They can be made to different specifications depending on the intended use. Personal care products are fairly small and lightweight and are only

about 3 in (75 mm) long. Swabs made for industrial use may be more than twice as long and are typically made of wood for greater rigidity.

Absorbent end material

Cotton is most often used as an end covering for swabs because of its absorbent properties, fiber strength, and low cost. Blends of cotton with other fibrous materials may also be used; rayon is sometimes used in this regard.

Packaging

Packaging requirements vary depending on the application for the swab. Some personal hygiene swabs, like Q-tips, are packaged in a clear plastic shell (known as a blister pack)

Manufacturing processes essential to cotton swab production.

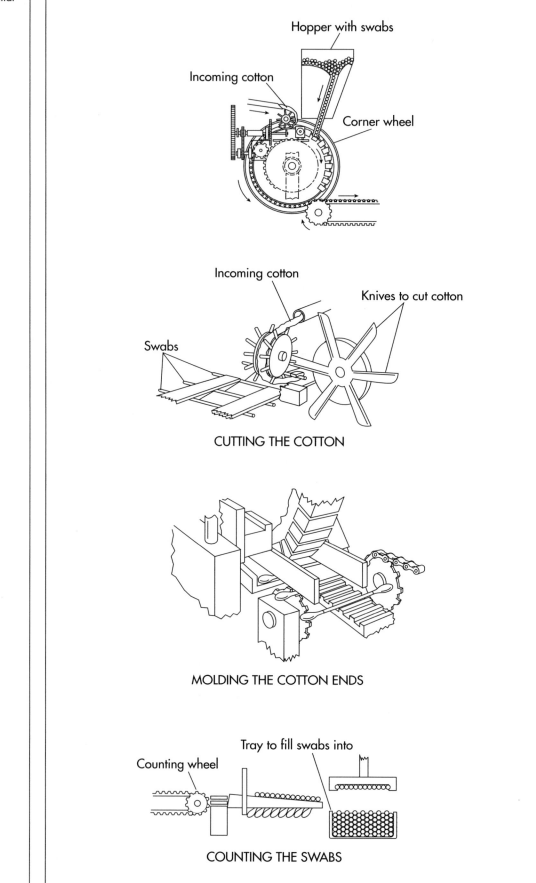

Hopper with swabs

Incoming cotton

Corner wheel

Incoming cotton

Knives to cut cotton

Swabs

CUTTING THE COTTON

MOLDING THE COTTON ENDS

Tray to fill swabs into

Counting wheel

COUNTING THE SWABS

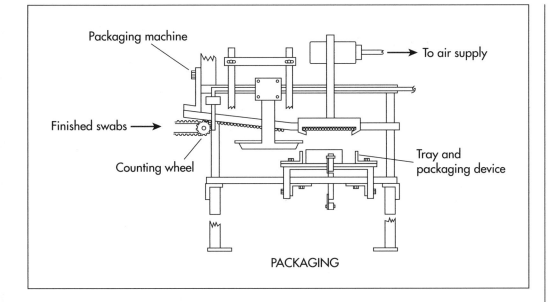

Packaging machine

To air supply

Finished swabs →

Counting wheel

Tray and
packaging device

PACKAGING

A schematic drawing of a typical
packaging machine used in cotton
swab production.

which is attached to a fiberboard backing. Chesebrough-Ponds holds a patent on the design of a self-dispensing package for their Q-tip products. This patent describes a package made of a plastic bubble body with small projections molded into the plastic for the purpose of re-securing the cover onto the body. Other packaging used for swabs includes paper sleeves. This type of packaging is common for swabs used for microbiological applications, which must be kept sterile prior to use.

The Manufacturing Process

Different methods are used in swab manufacture depending on the design of the swab. In general the process can be described in three major steps: spindle fabrication, cotton application, and packaging of the finished swabs.

Spindle fabrication

1 Spindles are made several ways depending on their material of composition. Wooden spindles are shaped by various lathe processes. Paper spindles are made by die-cutting a heavy grade paper and then tightly rolling the paper to form a stick. Plastic spindles are made by an extrusion molding process. In this case, the plastic resin and other additives are blended together, heated, and extruded through a die in a process similar to a the way a pasta maker

squeezes out a strand of spaghetti. After manufacture, the spindles are loaded into a hopper on the manufacturing line in preparation for cotton application.

Cotton application

2 One U.S. patent describes a complex operation involving a series of workstations. At the first station, a hopper full of spindles is vibrated to cause them to fall through a long, thin slot at the bottom of the hopper and onto a rotary stick carrier. This carrier is a drum type mechanism with cross slots placed around its outer rim. The sticks enter these slots one at a time and are carried to the next station.

3 As the wheel moves the sticks along, they contact a rotating friction wheel that rubs against the swab spindles causing them to spin. The carrier advances the sticks through a glue pot station where adhesive is applied to the opposite ends of the sticks.

4 At the next station, incoming cotton in the form of a rope is fed onto another rotating wheel with a series of metal fingers extending from its hub. These fingers direct the rope of cotton to the glue-coated ends of the sticks as they move along the wheel. In the case of Chesebrough Pond's Q-tips the spindles have a series of four to eight indentations, or shallow cuts, along their top and bottom. Presumably these help the shaft grab and hold onto the cotton as it is fed

onto the shafts. The cotton then adheres to the spindle is wound tightly by the spinning sticks. After approximately 0.05-0.1 g of cotton has been deposited on the spindle, a series of rotating blades severs the cotton from the rest of the rope.

5 The ends of the swabs then pass through a series of narrow channels while they are still spinning. These channels compress and shape the cotton on the end of the stick and give it a smooth shape. Chemical coating agents may be introduced to the cotton. These coating agents, which include cellulose polymer solutions, help the tip retain its shape and help prevent spotting and mildewing. Finally, an escapement wheel diverts the sticks from the carrier to a slotted transfer carrier belt and then to an automatic packaging unit.

Packaging

6 On the way to the packaging unit, the finished swabs pass through a counting wheel. A guide rail holds the swabs in place while the counting wheel counts the number of swabs for one tray layer. As the swabs are counted, they are pushed up against an inclined plane. When the counting wheel signals that enough swabs have been collected, it sends a signal to a piston arrangement, which slides the group of swabs onto a paper divider. The divider and the swabs are then dropped into the plastic package. In subsequent operations, a paperboard backing is glued over the filled plastic container and the entire assembly is packed for shipping.

Byproducts/Waste

The swab manufacturing process can generate waste in the form of loose cotton as well as plastic, paper, or wood scrap, depending on what material is used to make the spindles. Some of the cotton can be reclaimed and either incorporated back into the incoming feed path or used elsewhere as scrap. The plastic used in spindles is thermoplastic, which means it can be reground and remelted for later use.

Quality Control

A number of quality control measures are used to ensure cotton swabs are acceptable.

The spindles must be checked to ensure they are straight and free of imperfections, such as stress cracks or other molding defects. The cotton used to coat the ends must be of specific purity, softness, and fiber length. The finished swabs must be free from lose adhesive and sharp edges, and the tips must be tightly wrapped. These measures are particularly critical for swabs designed for infant use. For swabs intended for other applications, other quality requirements may be more important. For example, swabs used for biological purposes must remain sterile until used. For some applications, lack of loose lint maybe imperative. The particular quality control requirements will vary with the application. Of course, each box of swabs must be weighed to make sure the correct number of swabs are packed in each box.

The Future

A more recent innovation used to help prevent the swab from damaging ear tissue is a swab with extra cotton filling the hollow spindle. To achieve the effect, the swab applicator is made by extruding a plastic tube over a resilient mass of cotton. One end of the stick is fitted with a cap and the other end has a more traditional swab-like protrusion of cotton. The cap can be removed and the fiber core filled with any liquid that is desired to be dispensed. This technique could be useful for applying a variety of cleaning fluids or topical medicines. Future developments in swab technology may play a role in space technology was well. The Micro Clean Company, under a technology license from National Aeronautics and Space Administration (NASA), has recently perfected the first cotton swab that has the absorption qualities of cotton yet meets NASA's lint-free, adhesive-free requirement for clean room use. This swab is enclosed in a nylon sheath and the wood handle is enclosed in a shrink film to prevent fiber release or other contamination. The shrink film allows the dowel to absorb more stress, making it easier to use and less likely to slip in the hand. The sheathing and shrink film can be custom designed for special applications or specific solvent compatibility.

Where to Learn More

Books

Feldman, David. *When Did Wild Poodles Roam the Earth?* Harper Collins Publishers Inc., 1992.

Levy, Sidney. *Plastic Extrusion Technology Handbook*. New York: Industrial Press Inc, 1981.

Other

Mourkakos, George M. U.S. Patent 3698040. "Apparatus for Fabrication of Swabs." October 17, 1972.

—Randy Schueller

Credit Card

The VISA Company traces its history back to 1958 when the Bank of America began its BankAmericard program. In the mid-1960s, the Bank of America began to license banks in the United States the rights to issue its special BankAmericards. In 1977 the name Visa was adopted internationally to cover all these cards. VISA became the first credit card to be recognized worldwide.

Background

A credit card allows consumers to purchase products or services without cash and to pay for them at a later date. To qualify for this type of credit, the consumer must open an account with a bank or company, which sponsors a card. They then receive a line of credit with a specified dollar amount. They can use the card to make purchases from participating merchants until they reach this credit limit. Every month the sponsor provides a bill, which tallies the card activity during the previous 30 days. Depending on the terms of the card, the customer may pay interest charges on the amount that they do not pay for on a monthly basis. Also, credit cards may be sponsored by large retailers (such as major clothing or department stores) or by banks or corporations (like VISA or American Express).

Credits cards are a relatively recent development. The VISA Company, for example, traces its history back to 1958 when the Bank of America began its BankAmericard program. In the mid-1960s, the Bank of America began to license banks in the United States the rights to issue its special BankAmericards. In 1977 the name Visa was adopted internationally to cover all these cards. VISA became the first credit card to be recognized worldwide.

The banks and companies that sponsor credit cards profit in three ways. Primarily they make money from the interest payments charged on the unpaid balance, but they also can make money by charging an annual fee for the use of the card. The income from this fee, which is typically only $50 or $75 per customer per year, can be substantial considering that the larger companies have tens of millions of customers. In addition, the sponsors make money by charging merchants a small percentage of income for the service of the card. This arrangement is acceptable to the merchants because they can let their customers pay by credit card instead of requiring cash. The merchant makes arrangements to participate in a credit card program with a merchant bank, which in turn works with a card-issuing bank. The merchant bank determines what percentage of the total purchase value has to be paid by the merchant to the card-issuing bank. The amount varies depending on the volume and type of business, but in general it is between 1-2%. A percentage of that amount is kept by the merchant bank as a transaction-processing fee. For companies like American Express which sponsor cards, the processing fee may be significantly higher. Furthermore, sponsors may generate income by leasing credit card verification equipment to merchants (especially if the merchants can not afford to purchase the equipment themselves.) Finally, sponsors may profit by charging service fees for late payments.

Design

Credit cards are designed with complex security features to prevent the possibility of fraud. These features involve the card's account number, its signature panel, and its magnetic stripe. The card's unique account number is the key piece of information needed to conduct a financial transaction and must be carefully protected. To prevent someone from using a wrong account number, or from making up a phony number, companies rely on the laws of statistics for protection. By using long account numbers they make it unlikely that a number can be faked. For example, the Visa card has 13 digits, American Express has 15, Diners Club 14, and MasterCard has 20. Mathematically, nine digits would provide one billion unique account numbers (000000000, 000000001, 0000000002, and so forth up to 999999999) which would be enough for all the customers of a given company. (The largest companies, Visa and MasterCard, only have about 65 million customers.) If only 65 million numbers are assigned out of a possible 10 trillion possibilities, it is unlikely that anyone will be able to mistakenly use another account number. If an incorrect account number is mistakenly entered by a store clerk, it will almost certainly not be accepted. This statistical security gives companies confidence that someone is not making up a number when conducting business over the phone. Of course, this security measure does not help if someone obtains a real number and uses it fraudulently.

Another security design feature involves the signature panel on the back of the card. The signature is intended to document the owner's handwriting so a forged signature on a receipt can be detected. To prevent criminals from erasing the back panel of a stolen card and putting on their own signature, the panel is printed with a fingerprint design that is difficult to duplicate and that will come off when the original signature is erased. If the signature is erased, this design will disappear too leaving a white spot, which instantly indicates the card has been tampered with. Some card manufacturers imprint the word VOID beneath this panel, which is revealed upon erasure.

The magnetic stripe on the back of the card is a third security feature. The stripe is an area coated with particles of iron oxide that can be encoded with binary information, which identifies the card as authentic. It is difficult to determine exactly what information is coded on the strip because for security reasons companies do not wish to discuss this. However, it is likely that the card's expiration date is one fact recorded on the strip because automatic teller machines (ATMs) will retain cards that have expired. It is unlikely that information like credit limit, address, phone number, employer, is recorded on the stripe because banks do not reissue cards when this type of information changes.

Finally, some cards feature special features that make them hard to duplicate, such as complicated holograms.

Raw Materials

Cards are made of several layers of plastic laminated together. The core is commonly made from a plastic resin known as polyvinyl chloride acetate (PVCA). This resin is mixed with opacifying materials, dyes, and plasticizers to give it the proper appearance and consistency. This core material is laminated with thin layers of PVCA or clear plastic materials. These laminates will adhere to the core when applied with pressure and heat.

A variety of inks or dyes are also used for printing credit cards. These are available in a variety of colors and are designed for use on plastic substrates. Some manufacturers use special magnetic inks to print the magnetic stripe on the back of the card. The inks are made by dispersing metal oxide particles in the appropriate solvents. Additional special printing processes are involved for cards, like VISA, which feature holograms.

The Manufacturing Process

The manufacturing process consists of multiple steps: first the plastic core and laminate materials are compounded and cast into sheet form; then the core is the printed with appropriate information; next the laminates are applied to the core; and finally the assembled sheet is cut into individual cards.

Plastic compounding and molding

1 The plastic for the core sheet is made by melting and mixing polyvinyl chloride acetate with other additives. The blended components are transferred to an extrusion molding apparatus, which forces the molten plastic through a small flat orifice known as a die. As the sheet exits the die, it goes through a series of three rollers stacked on top of each other that pulls the sheet along. These rollers keep the sheet flat and maintain

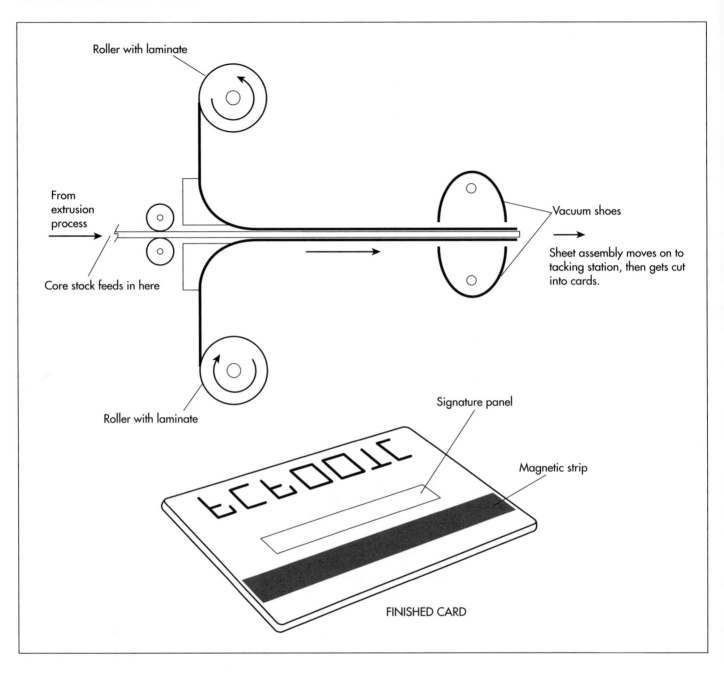

Roller with laminate

From extrusion process

Core stock feeds in here

Roller with laminate

Vacuum shoes

Sheet assembly moves on to tacking station, then gets cut into cards.

Signature panel

Magnetic strip

FINISHED CARD

As the sheet exits the die, it goes through a series of three rollers stacked on top of each other that pulls the sheet along. These rollers keep the sheet flat and maintain the proper thickness. The sheets may then pass through additional cooling units before being cut into separate sheets.

the proper thickness. The sheets may then pass through additional cooling units before being cut into separate sheets by saws, shears, or hot wires. The cut sheets enter a sheet stacker that stacks them into place and stores them for subsequent operations.

2 The laminate films used to coat the core stock are made by a similar extrusion process. These thinner films may be made with a slot cast die process in which a molten plastic film is spread on a casting roller. The roller determines the film's thickness and width. Upon cooling the films are stored on rolls until ready for use.

Printing

3 The plastic core of the card is printed with text and graphics. This is done using a variety of common silk screen processes. In addition, one of the laminate films may also undergo subsequent operations where it is imprinted with magnetic ink. Alternately, the magnetic stripe may be added by a hot stamping method. The magnetic heads used to code and decode the iron oxide particles can only operate if the magnetic medium is close to the surface of the card, so the metal particles must be placed on top of the laminating layer. Upon com-

pletion of the printing process, the core is ready to be laminated.

Lamination

4 Lamination helps protect the finish of the card and increases its strength. In this process, sheets of core stock are fed through a system of rollers. Rolls of laminate stock are located above and below the core stock. These rolls feed the laminate into the vacuum shoes along with the core stock. The vacuum holds the three pieces of plastic together while they travel to a tacking station. At the tacking station a pair of quartz infrared heat lamps warm the upper and lower plastic films. These lamps are backed with reflectors to focus the radiant energy onto a narrow area of the films, which optimizes a smooth bonding of the film to the core stock. The laminate films are then fully bonded to the core stock by pressing with metal platens, which are heated to 266° F (130° C) and applied with a pressure of 166 psi/sq inch. This lamination process may take up to 3 minutes.

Die cutting and embossing

5 After lamination has been completed, the finished assembly is cut and completed by die cutting methods. Each assembly yields a sheet, which is cut into 63 credit cards. This is achieved by first cutting the assembly longitudinally to form seven elongated sections. Each of the seven sections is then cut and trimmed to form nine credit cards. In subsequent operations, the card is embossed with account numbers. The finished cards are then prepared for shipping, usually by attaching the card to a paper letter with adhesive.

Quality Control

Key quality issues are associated with the compounding of plastic and color matching of the inks. The American National Standards Institute has a standard for plastic raw materials (ANSI specification x4.16-1973). As with any compounding procedure, ingredients must be properly weighed and mixed and blended under the appropriate temperature and sheer conditions. Similarly, the molding process must be monitored to avoid defects, which could cause the cards to crack or break. The final quality check is to make sure the correct numbers are stamped on the cards during the embossing process.

The Future

Future credit card manufacturing processes are likely to evolve in three key areas. First, continued improvements in plastic chemistry and molding technology are likely to allow cards to be made increasingly cheaper and easier. Second, breakthroughs in digital technology are likely to improve the way credit cards are kept secure with advanced magnetic coding. One recent advance is the use of a new generation of magnetic stripes which are harder to duplicate. This improvement combats the trend toward duplicating card information and copying it to phony cards. Perhaps even more importantly, new generations of credit cards will carry integrated computer chips, containing a variety of useful information. For instance, these future cards will be able to operate a frequent flyer program on the same card as a debit or credit account. Other services will allow users to participate in frequency or loyalty programs with merchants, including storing hotel reservation preferences. Financial institutions may develop partnerships with local mass transit systems so public transit could be paid for with these "smart" cards in various cities throughout the world. Third, marketing initiatives resulting from these advances in card technology are likely to make credit cards even more pervasive in society. For example, American Express has just launched a new Blue card that is expected to reach new levels of worldwide acceptance.

Where to Learn More

Books

Poundstone, William. *Big Secrets*. New York: William Morrow & Co., 1983.

Sutton, Caroline and Kevin Markey. *More How Do They Do That?* New York: William Morrow & Co., 1993.

Other

U.S. Patent 4,100,011. "Production of Laminated Card with Printed Magnetically Encodable Stripe," issued 1978.

Visa International. http://www.visa.com (May 12, 1998).

—*Randy Schueller*

Cushioning Laminate

The need for efficient, protective packing material has been long recognized. Originally, shredded paper and rags were used for this purpose. Other materials that have been used historically include pulverized mica and corrugated cardboard. As plastics technology matured in the 1950s and 1960s, new and improved packing materials were developed.

Background

Bubble wrap is the trademarked name for a packing material consisting of two plastic sheets laminated together in a way that traps air bubbles in small, uniform pockets. This plastic sheet assembly is used as a flexible cushion to protect fragile objects during storage or shipping. The name Bubble wrap is registered by Sealed Air Corporation of Saddle Brook, New Jersey, however the name has become synonymous with the packaging material itself. Similar materials are known in the industry as cushioning laminates.

The need for efficient, protective packing material has been long recognized. Originally, shredded paper and rags were used for this purpose. Other materials that have been used historically include pulverized mica and corrugated cardboard. As plastics technology matured in the 1950s and 1960s, new and improved packing materials were developed. Foam beads made from polystyrene plastic are one popular example of plastic packing material; these are more commonly known as Styrofoam peanuts. Another innovation based on plastic technology is cushioning laminates, a packing material that relies on air to cushion and protect highly fragile objects. The first use of these laminates dates back to the early 1970s when methods used to process plastics became increasingly sophisticated, allowing cheap and rapid manufacturing. Today, they are made by a number of companies both in the United States and abroad. While a variety of manufacturing methods are used, the basic process involves trapping air bubbles between two laminated sheets of plastic.

Raw Materials

Plastic resin

Cushioning laminate is primarily made of plastic film or thin sheet formed from resins such as polyethylene and polypropylene. These resins are widely used because they perform well and are relatively inexpensive. They can be cast into strong, flexible films, which have the ability to hold air without leaking. Furthermore, these resins are thermoplastic materials, which means they easily can be melted and molded. This is an important property since the plastic sheets may be reheated during processing. Different types of resins may be used to make the top and bottom sheets to give the cushioning laminate special properties. For example, one layer could be made with a more rigid material to give the finished product increased stiffness.

Other additives

The polypropylene or polyethylene films are formed with a variety of additives mixed with the base polymers in order to modify their properties and to facilitate processing. These additives include lubricants and plasticizers which control the flexibility of the resin blend; ultraviolet light absorbers, heat stabilizers, and antioxidants which inhibit different types of degradation; and coupling agents and strength modifiers which improve the bond between the polymer and the filler. Furthermore, antistatic agents are added to reduce buildup of static electricity and biocides may be included to inhibit microbial growth.

The Manufacturing Process

Cushioning laminate is manufactured in a process that consists of three primary steps:

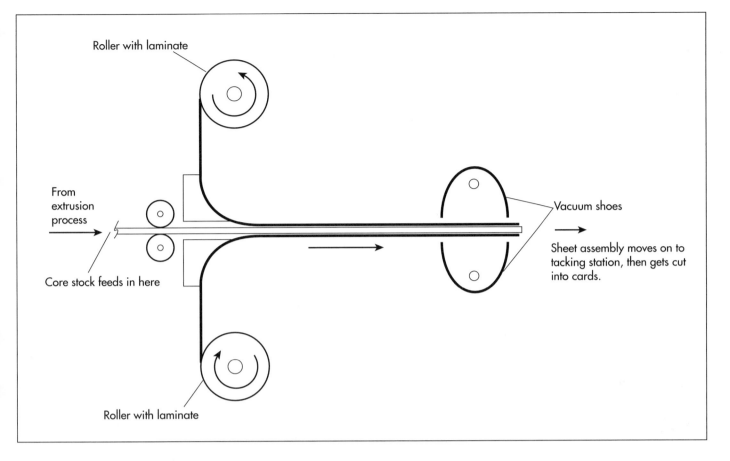

Roller with laminate

From extrusion process →

Core stock feeds in here

Roller with laminate

Vacuum shoes

→ Sheet assembly moves on to tacking station, then gets cut into cards.

plastic compounding and sheet extrusion, lamination, and finishing operations.

Plastic compounding and sheet extrusion

1 Plastic resin that has been compounded to the manufacturer's specifications is purchased in bulk from a supplier. In this compounding process, the polyethylene resin is heated and mixed with the additives described above. This mixture is then melted and formed into small pellets 0.125 in (0.3175 cm) in diameter. At the beginning of the manufacturing process, these pellets are introduced into a molding machine, known as an extruder. At one end of the extruder is a hopper into which the pellets are dumped. This hopper feeds the pellets into a long heated barrel. This barrel is equipped with a screw mechanism, which pushes the plastic forward. At the other end of the barrel is a stainless steel sheeting die that can produce sheets up to 10 ft (3 m) wide.

2 The resin melts as it moves along the heated barrel, and by the time it reaches the end, it can be easily forced out through the opening in the die. As the molten resin is squeezed through the die it is shaped into a sheet which is then processed further. Depending on the process, the sheet can be laminated to another layer immediately while it is still warm or it can be cooled and laminated later. In either case, after being extruded the sheet passes through a series of stainless steel rollers, known as a three roll finisher or a three roll stack. These rollers are 10-16 in (25.4-41 cm) in diameter and are internally cooled with water. As the plastic sheet exits the die, it enters the nip, the point where the top two rollers meet. The sheet is pulled in by the motion of the rollers and is passed through the top, middle, and bottom rollers. These rollers cool the sheet while helping it to maintain the correct size and shape. After passing through the three roll stack, the sheet enters another series of rollers known as pull rolls, which drag the sheet through the rest of the processing.

Lamination

3 Lamination is the process used to seal the two sheets together in such a way

As the molten resin is squeezed through the die, it is shaped into a sheet which is then processed further. After being extruded, the sheet passes through a series of stainless steel rollers, known as a three roll finisher or a three roll stack.

Lamination is the process used to seal the two sheets together in such a way that traps air bubbles.

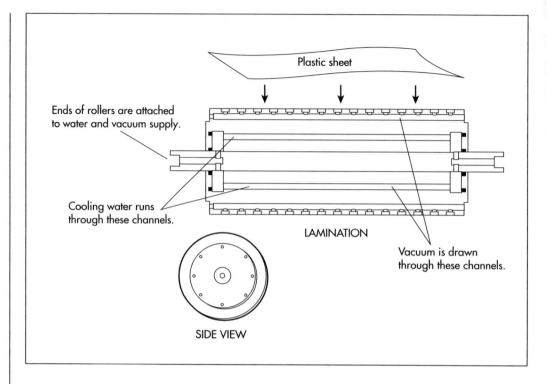

Plastic sheet

Ends of rollers are attached to water and vacuum supply.

Cooling water runs through these channels.

LAMINATION

Vacuum is drawn through these channels.

SIDE VIEW

that traps air bubbles. Uniform placement of these bubbles across the face of the sheet can be achieved by stretching or perforating the substrate sheet in a designated pattern. These uniformly placed deformations in the sheet will retain air and form individual pockets. The process of deforming the substrate sheet requires heat to soften the plastic. As noted, this step can be performed immediately after extrusion while the sheet is still warm or the sheet can be reheated and molded at a later time. Bubbles can then be molded into the softened sheet by exposing it to a forming surface. This surface may be a roller or a plate with protrusions in the desired shape and distribution. When the molten sheet is brought into contact with the forming surface, the plastic is molded in the desired pattern.

4 One method of creating these air pockets uses a rotating belt as the forming surface. This belt has a number of holes spread across it. As substrate sheet moves along the belt, suction is applied from a vacuum source to the holes in the belt. The air pressure differential causes the plastic to stretch down into the holes on the belt, thus creating a series of pockets. Another method employs a molding plate as the forming surface. The plastic sheet is moved into place below this plate through which a vacuum is

drawn. The suction causes the sheet to conform to the bumps in the mold plate and produces a molded sheet having the desired irregular surface. A third method uses a rotating molding cylinder to form the air pockets in the plastic.

5 After the air pockets have been formed by one of the methods described above, the substrate sheet and a second sheet are fed together through a set of laminating rollers. At least one of the sheets must be at the proper temperature to ensure bonding will occur. The pressure and heat seals together the sheets and the air bubbles remain trapped.

Special operations

6 After lamination is complete, the sheets are cooled, if necessary, by open or forced air systems. Air can be blown across from above and below the sheet. Water cooling is sometimes done but this requires extra time for drying and may cause cleaning problems. Depending on the type of cushioning laminate being made, other special processing may be required. For example, some types of cushioning laminate are treated with an adhesive coating on one side. Others are formed into envelopes to hold small fragile objects. Depending on the processing involved, these additional op-

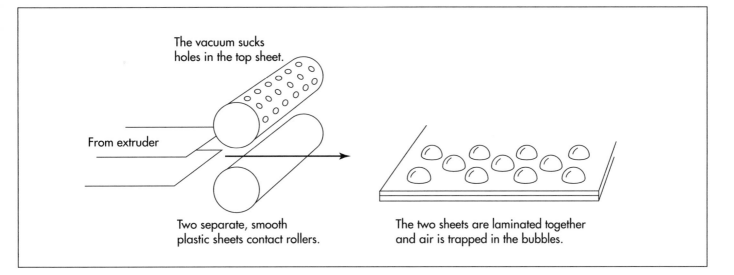

The vacuum sucks holes in the top sheet.

From extruder

Two separate, smooth plastic sheets contact rollers.

The two sheets are laminated together and air is trapped in the bubbles.

erations may be performed before or after lamination process.

Finishing operations

7 After the cushioning laminate is completed, the sheet material is cut to the appropriate size. This may be done as part of the primary processing or the uncut wrap may be stored on large rolls and cut to size later. This cutting process is known as slitting and is accomplished with special knives which can slice cut through the thick layers of plastic. The laminate may be packaged and sold on rolls or in sheet form.

Byproducts/Waste

The major waste product from cushioning laminate manufacturing is the plastic resin. Resin that is contaminated, overheated, or otherwise ruined must be discarded. However, sheets that fail quality checks for reasons related to physical molding problems can be reworked. This recycling process is known as regrinding and shredding the sheets, remelting them, and re-extruding them as new sheets. To ensure the plastic meets physical specifications, regrind may be mixed with virgin resin. This can be done without loss of quality because of the thermoplastic nature of polypropylene.

Quality Control

As with other plastic manufacturing processes there are several key areas that must be closely controlled to ensure a quality product is produced. During the compounding process, the resin and additives must be added carefully to ensure the formula components are blended in the proper ratios. The finished resin may be analyzed to ensure its chemical and physical properties meet specifications before sheet extrusion operations begin. At the start of the extrusion process a small amount may be flushed through the barrel of the extruder. This purging process cleans out the barrel and reveals any problems with the molding systems.

During extrusion, it is critical that the resin is kept at the proper temperature. The flow rate of the polymer will vary according its molecular weight and temperature. If the temperature is too cool, the resin will not move through the die properly. If the temperature is too high, the polymer may undergo thermal degradation. Overheating can cause chemical changes in the resin, making it unusable. Unwanted chemical interactions can also effect the quality of the plastic sheets during the extrusion process. One problem is oxidation, a reaction with air that can negatively affect the plastic. Similarly, interaction with moisture affects the quality of the plastic. If too little moisture is present, certain plastic blends can become too brittle.

After the extrusion process is complete, the extruder must be properly cleaned. Thorough cleaning is necessary before working with a different resin because traces of the previously used resin can contaminate the new batch. Die cleaning is best done while the machine is still warm and left over resin can be easily scraped out.

Other factors must also be monitored. For example, in certain methods of manufacturing it is important that top and bottom plastic sheets respond to heat differently so that during the lamination process one sheet distorts but the other does not. For this type of operation, it is critical that the heat distortion of the two sheets differ by at least 77° F (25° C) or problems will occur during lamination.

After the cushioning laminate is completed, samples may be evaluated to ensure the sheets meet specifications for strength, bubble bursting point and other criteria.

The Future

Improvements in plastics technology continue to occur at a rapid pace. These advances are likely to produce improved plastic compounds that are easier to process, provide better cushioning ability, and are biodegradable. The latter quality is of particular significance considering that packaging material is a disposable product and is used in considerable quantities. Cushioning laminate made of plastic, which could safely breakdown without negatively impacting the environment, would be a great asset to the industry. While improvements in equipment used in the manufacturing process

continue to be made, they may be slow to come to market because replacing existing machines may be prohibitively expensive. One new method of manufacturing circumvents the need for costly forming equipment. Instead, this method uses a plastic substrate sheet as a pattern to form the bubbles without expensive molding equipment. In this process, a thin plastic sheet is first perforated in the desired bubble pattern. This layer is laminated to a substrate sheet and the combination is then passed through heated pinch rolls. Vacuum or gas pressure is applied to draw the film through the perforations in the substrate. This process creates bubbles without the use of a forming surface. It remains to be seen if this, or other new manufacturing methods, will be embraced by the industry in the future.

Where to Learn More

Books

Green, Joey and Tim Nybery. *The Bubble Wrap Book.* Harper Perennial Publishers, 1998.

Other

The Sealed Air Corporation: http://www. sealedaircorp.com/.

US Patent 4,681,648 Process for Producing Cushioning Laminate.

—*Randy Schueller*

Dental Crown

Background

A dental crown is a cap-like restoration used to cover a damaged tooth. Crowns can give support to misshapen or badly broken teeth and permanently replace missing teeth to complete a smile or improve a bite pattern. They may be molded from metal, ceramic, plastics, or combinations of all three. They are cemented in place and coated to make them more natural looking. Historically, a variety of materials have been used as tooth replacements. The ancient Egyptians used animal teeth and pieces of bone as primitive replacement materials. More recently, artificial teeth have been fabricated from substances such as ivory, porcelain, and even platinum. With modern technology, high quality tooth replacements can be made from synthetic plastic resins, ceramic composites, and lightweight metal alloys.

Design

There are several key factors to consider in the design of dental crowns. First, appropriate raw materials with which to make the crown must be identified. These materials must be suitable for use in the oral cavity, which means they must be acceptable for long term contact with oral tissues and fluids. Crown components must have a good safety profile and must be non-allergenic and non-carcinogenic. The American Dental Association/ANSI specification #41 (Biological Evaluation of Dental Materials) lists materials which have been deemed safe for use. In addition to safety considerations, these materials must be able to withstand the conditions of high moisture and mechanical pressure, which are found in the mouth. They must be resistant to shrinkage and cracking, particularly in the presence of water. Metal is preferred for strength but acrylic resins and porcelain have a more natural appearance. Therefore the selection of crown material is, in part, dependent on the location of the tooth being covered. Acrylic and porcelain are preferred for front teeth, which have higher visibility. Gold and metal amalgams are most often used for back teeth where strength and durability are required for chewing but appearance is less critical.

The second factor to consider when designing a crown is the shape of the patient's mouth. Dental restorations must be designed to mimic the bite properties of the original tooth surface so the wearer does not feel discomfort. Since every individual's mouth is different each crown must be custom designed to fit perfectly. Successful crown design involves preparation of an accurate mold of the oral cavity.

Raw Materials

There are four main types of materials used in crown construction: The plasters used to create the mold, the materials from which the crown itself is made (e.g., metal, ceramic, plastic), the adhesives used to cement the crown in place, and the coatings used to cover the crown and make it more aesthetically appealing.

Molding plasters

Plaster molds are made from a mixture of water and gypsum powder. Used for dental applications since the 1700s, gypsum is finely divided calcium sulfate dihydate. Different types of plasters are used depending on application: impression plaster is used to

State of the art crowns can be made with an industrially produced core made of dense-sintered ceramic, and an outer layer of porcelain is added by hand. This futuristic crown material is made by an advanced Computer Aided Design (CAD) process, known as Procera process, which was introduced in the mid-1990s in Switzerland.

An impression of the tooth to be crowned is taken to record its shape. The impression plaster is mixed and then placed in a tray that is fitted over the teeth. The tray is held still in place until the plaster hardens. When the tray is removed from the mouth, it retains a three dimensional impression of the tooth that is to be covered. This impression is a negative, or reverse, image of the tooth.

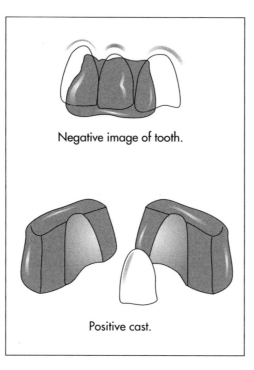

Negative image of tooth.

Positive cast.

record the shape of the teeth, model plaster is used to make durable models of the oral cavity, and investment plaster is used to make molds for shaping metal, ceramics and plastics. Waxes are also sometimes used in this regard.

Crown construction materials

Metals are frequently used in crown construction because they have good hardness, strength, stiffness, durability, corrosion resistance, and bio-compatibility. Metals formulated as mixtures of mercury have been historically used. In fact, one source notes that metal amalgam was used as a dental restorative as early as 1528. Common alloys used in crowns are based on mixtures of mercury with silver, chromium, titanium, and gold. These mixtures form a blend than can be easily shaped and molded, but which hardens in a few minutes.

Ceramics are well suited for use in crowns because they have good tissue compatibility, strength, durability and inertness. They can also be made to mimic the appearance of real teeth fairly closely. However, the tensile strength of ceramic is low enough to make it susceptible to stress cracking, especially in the presence of water. For this reason, ceramic is most often used as a coating for metal-structured crowns. The two primary types of ceramics used in crowns are made from potassium feldspar and glass-ceramic.

The first resin used in denture materials was vulcanized rubber in 1839. Since then, a number of other resins have been developed which are more suitable for dental applications. Today, acrylic polymer resins are commonly used in dentures and crowns. Specifically, polymethyl methacrylate is most often used. This type of resin is made by mixing together chemical entities known as monomers with activating chemicals which cause the monomers to react and link together to form long chains called polymers. Some of these resins harden at room temperature as this reaction progresses. Others require heat or ultraviolet light to catalyze the change.

Special dental adhesives, or dental cements, are used to hold the crown in place. These can be classified as either aqueous or nonaqueous. The aqueous type include zinc phosphates, polycarboxylate cements, glass-ionomer cements, and calcium phosphate cements. The nonaqueous type include zinc oxide-eugenol, calcium chelates, and acrylic resins such as polymethyl methyacrylate.

Coatings are used to make the crown appear more natural. Porcelain is used in this regard, but it is difficult to work with and hard to match to the tooth's natural color. Resins similar to the ones used in tooth construction are also used to create tooth-colored veneers on crowns. These resins have an advantage over other veneers in that they are inexpensive, easy to fabricate, and can be matched to the color of tooth structure. However, acrylic coatings may not adhere to the crown's surface as well as porcelain or other materials. Therefore, the prosethedontist may design the crown's surface with mechanical undercuts to give the coating a better grip. Resin coatings also have relatively low mechanical strength and color stability and poor abrasion and stain resistance as compared to porcelain veneers.

The Manufacturing Process

Creating the mold

1 Before beginning, the dentist may need to prepare the area where the crown is to be

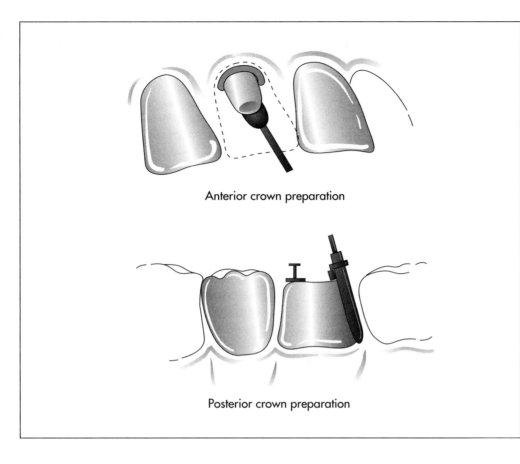

Anterior crown preparation

Posterior crown preparation

installed. This may require the removal of 2-3 millimeters of tooth structure from the four sides and the biting edge. Then, an impression of the tooth is taken to record its shape. This step uses impression plaster which is the softest and fastest setting type of dental plaster. The impression plaster is mixed with a small amount of water until it is fluid. This slurry is placed in a tray that is fitted over the teeth. The tray is held still in place until the plaster hardens. When the tray is removed from the mouth, it retains a three dimensional impression of the tooth that is to be covered. This impression is a negative, or reverse, image of the tooth.

2 The next step is to prepare another type of plaster, known as model plaster. This type of plaster is harder than the impression plaster. Once again the plaster is mixed with the appropriate quantity of water. Then the slurry is poured into the impression mold. In this way a positive model of the tooth can be made. This positive model made from the negative impression mold is called a cast. The cast is used by the dentist for study purposes.

3 The impression is also used to make a mold, called an investment, which is capable of withstanding high temperatures. This is an important consideration because some metals and ceramics require temperatures higher than 2,372° F (1300° C) for molding. These investments are made from calcium phosphate mixed with silica and other modifying agents.

Fabrication

4 Fabrication of the crown is done by filling the investment with the appropriate material. In the case of metals, this is done at a high temperature so the metal is molten. For ceramics and plastics, the mixture is initially fluid but may require the addition of heat to cause the materials to cure and harden. A vertical vise may be used to help pack the casting investment tightly. The process also requires the mold first be treated with a release agent to ensure the crown can be easily removed after it has hardened. Some acrylic resins must be heated for up to eight hours to make sure they are fully cured. After the processing is done and the invest-

ment has cooled, the mold is broken apart and the crown is removed.

Installation

5 After the crown has been successfully completed, it is ready for installation. The prosethedontist applies cement to the inside of the crown surface and then fits it into place over the tooth. Because of the number of processing steps there may be a slight discrepancy in the fit and the crown may require minor grinding and smoothing of its surface to ensure it fits correctly.

Finishing

6 The crown may require a finishing coat to seal it and improve its natural appearance. Such coatings are typically acrylic polymers. The polymer can be painted on as a thin film, which hardens to a durable finish. Some polymers require a dose of ultraviolet light to properly cure.

Quality Control

Good quality control is critical to ensure the crown fits and looks natural in the patient's mouth. Every crown is unique because every person's mouth is different and every crown is custom molded to fit. To ensure appropriate fit and feel, fine details can be added to the crown by hand after the molding process is completed. Even with minor adjustments, quality problems and failures in crowns are likely to occur. Key quality control issues include failures due to biological factors (such as caries, recurrent decay, sensitivity problems, and periodontal diseases), mechanical reasons (including fracture of the crown surface, and poor cementation), aesthetic problems (discoloration of the surface), and damage due to traumatic accidents. In such situations it may become necessary to reposition or remove a crown to allow for either replacement or other dental operations. There are special crown and bridge removal systems that have been developed for easy removal of these prosthedontics. This is accomplished by placing a precision vertical channel in the surface of the crown, then threading the surface until the cement layer has been broken. The crown can then easily be lifted from the underlying tooth without force.

Byproducts/Waste

Denture manufacture generates little waste other than minimal amount resulting from the gypsum and plaster materials used in mold making and the excess acrylic resins used in crafting the teeth and mounts. These materials are not generally in large quantities since crowns are crafted by hand and are not mass produced on a production line.

The Future

Dental technology is constantly advancing and these improvements are already finding application in dental crown manufacturing. State of the art crowns can be made with an industrially produced core made of dense-sintered ceramic, and an outer layer of porcelain is added by hand. This futuristic crown material is made by an advanced Computer Aided Design (CAD) process, known as Procera process, which was introduced in the mid-1990s in Switzerland. This process results in crowns with improved strength and optimal fit. Unlike other crown materials, crowns made by the Procera process can be used anywhere in the mouth due to the strength of its core material and its more natural appearance. Another advance in crown technology involves pre-made and pre-sized stainless steel crowns, which are designed as generic tooth replacements. Usage of this new type of crown is very simple: first the tooth surface is prepared then the selected crown is cemented in place with a standard stainless steel crown adhesive. The crown can be crimped or cut to fit and the epoxy finish will not chip or peel. While this new technology offers increased simplicity, it does not give the same appearance as a custom made crown. Other future advancements are likely to come from new resins, which have improved adhesion in the high moisture environment of the oral cavity.

Where to Learn More

Books

Geering, Alfred H., Martin Kundert, and Charles Kelsey, ed. *Complete Denture & Overdenture Prosthetics.* Thieme Medical Publishers, Inc., 1993.

Goldstein, Ronald. *Change Your Smile.* Chicago: Quintessence Publishing Co., 1997.

Woodforde, J. *The Strange Story of False Teeth*. New York: Universe Books, New York, 1968.

Other

Ellison, Robin E. "Developing the Wax Pattern for Removable Partial Dentures." VHS tape distributed by the National Audiovisual Center, 1980.

—Randy Schueller

Dice

The modern day cubical dice originated in China and have been dated back as early as 600 B.C. They were most likely introduced to Europe by Marco Polo during the fourteenth century.

Background

Dice are implements used for generating random numbers in a variety of social and gambling games. Known since antiquity, dice have been called the oldest gaming instruments. They are typically cube-shaped and marked with one to six dots on each face. The most common method of dice manufacture involves injection molding of plastic followed by painting.

History

Dice have been used for gaming and divination purposes for thousands of years. Evidence found in Egyptian tombs has suggested that this civilization used them as early as 2000 B.C. Other data shows that primitive civilizations throughout the Americas also used dice. These dice were composed of ankle bones from various animals. Marked on four faces, they were likely used as magical devices that could predict the future. The ancient Greeks and Romans used dice made of bone and ivory. The dice of most of these early cultures were made in numerous shapes and sizes. The modern day cubical dice originated in China and have been dated back as early as 600 B.C. They were most likely introduced to Europe by Marco Polo during the fourteenth century.

Dice were typically handcrafted and produced on a small scale up until the twentieth century. As plastic technology emerged, methods for applying it to dice manufacture were developed. This allowed manufacturers to produce mass quantities of dice in a cost effective manner. Over the years a variety of patents for improved methods of dice manufacture have been granted.

Design

The standard die is a six-sided, plastic cube. Each side is typically marked with one to six spots, or dots. These dots are arranged such that opposite sides always total seven. For example, the one dot side is opposite the six dot side and the three dot side is opposite the four dot side. In a two dice game, the dice are shaken and thrown on a surface. The rolled amount is indicated by the sides of the dice that are face up. If the dice are well-balanced and fair, each side has an equal chance of landing face up. Depending on the game, the player will either move her piece or collect money based on this rolled amount. Some popular gambling games that use dice include craps, chuck-a-luck, and poker dice. Board games such as backgammon, Monopoly, and Parcheesi also use dice.

Standard dice are available in a wide variety of sizes and colors. For board games a pair of 12 mm dice are typically used. These dice are considered imperfect because they have rounded corners, which reduce randomness. Since these dice are often used in children's games, they must be designed to meet certain toy safety standards. Casinos use perfect dice that may be hand made. They are generally larger than board game dice with a side measuring 33 mm. These are red, translucent dice which have precision-edges and corners and white dots. With this construction, rolls with these dice have the greatest probability of being fair.

Specialty dice are produced for many different applications. In some cases, the spots on a standard cube die are replaced by words, pictures, or symbols. Divining dice, which are used to predict future events, have different predictive messages on each face. Poker dice have card faces printed on each side. For blind people, Braille dice are available. Some games require dice that have a different number of sides and can provide a greater number of outcomes than standard dice. These polydice can have anywhere from three to 20 sides. They are used extensively in fantasy role playing games.

The key design element of dice manufacture is the mold. A mold is a cavity carved in steel that has the shape of the product that it forms. Typically, a mold is made up of two pieces which are forced together to form the cavity. When a plastic is injected into this mold, it takes on the mold's shape as it hardens. Since dice are solid cubes, using a standard mold is not practical because they would take too long to cool. For mass production of dice a special mold design is used. This mold is made up of separate chambers, which create individual elements of the die. As the individual pieces cool, they can be forced together to create a unified single object. The mold is then opened and the die is ejected. Special release agents are used to help make the plastic easier to remove from the mold. This mold design saves time because the smaller pieces can cool more rapidly.

Raw Materials

Numerous materials have been forged into dice throughout history. This includes such things as bones, glass, wood, seeds, and metals. Today, the most widely used base material for dice manufacture is plastic. Plastics are high molecular weight polymers that are produced through a variety of chemical reactions. For a plastic to be suitable in dice manufacture it must have good impact strength, be easily colored, and heat stable. It is also desirable that it be clear, colorless, and transparent. Most dice are made with a thermoset plastic. One plastic that meets all of these requirements is polymethyl methacrylate (PMMA). Cellulose based plastics are also used.

Since the polymer, which makes up the bulk of the plastic is typically colorless, colorants are added to make the dice more appealing. These may be soluble dyes or comminuted pigments. To produce a white color, an inorganic material such as titanium dioxide may be used. Other inorganic materials such as iron oxides can be used to produce yellow, red, black, brown and tan dice. Organic dyes such as pyrazolone reds, quinacridone violet, and flavanthrone yellow may also be utilized.

A host of other filler materials are added to the plastics to produce a durable, high quality set of dice. To increase the workability and flexibility of the polymer, a plasticizer is included. Plasticizers are nonvolatile solvents and include things such as paraffinic oils or glycerol. To improve the overall properties of the plastic, reinforcement materials such as fiberglass are added. During production the plastic is typically heated. For this reason, stabilizers must be added to protect the plastic from breaking down. Unsaturated oils such as soybean oil may be used as heat stabilizers. Other protective materials that are added include ultraviolet (UV) protectors such as benzophenones to prevent UV degradation and antioxidants such as aliphatic thiols to alleviate environmental oxidation. Finally, compounds are also used during manufacture to aid in processing. This generally includes materials like ethoxylated fatty acids, silicones, or metal stearates, which help with the removal of the plastic from the mold.

The Manufacturing Process

The exact manufacturing process for any type of die depends on the base raw materials used. For mass production of imperfect standard dice an injection molding process is used followed by painting and packaging.

Forming

1 At the beginning of a dice manufacturing line, plastic pellets are transformed into dice via injection molding. The pellets are plastic beads that have all the colorants and fillers already added. They are placed into a large bin known as a hopper and passed through a hydraulically controlled screw. As they travel through the screw, they are heated and melted. At the end of this screw is a spreader which injects the molten material into a cool, closed two-piece mold.

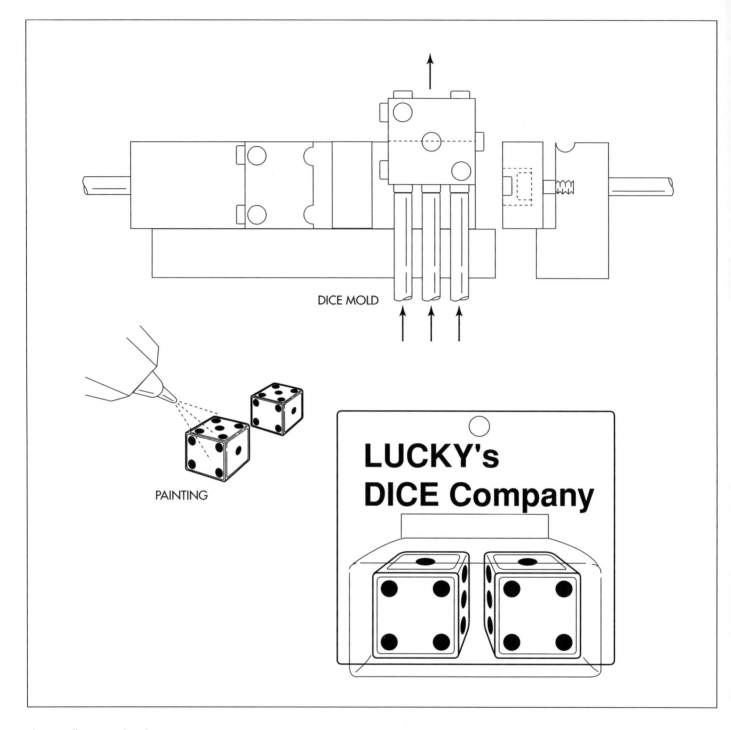

DICE MOLD

PAINTING

LUCKY's DICE Company

Plastic pellets are placed into a large bin known as a hopper and passed through a hydraulically controlled screw. As they travel through the screw, they are heated and melted. At the end of this screw is a spreader which injects the molten material into a cool, closed two-piece mold. The dice is painted and packaged.

2 The mold is made up of several chambers, which create multiple plastic parts. Inside the mold, the plastic is held under pressure and then allowed to cool. As it cools, the plastic pieces harden. The mold is then opened. When the mold is opened, the individual pieces are forced together to form a single solid cube. Because the mold was appropriately designed, this cube has indentations that will become the dots on each side. The cube is then ejected from the

mold, coated and passed to the next phase of production via conveyor. Meanwhile, the two piece mold closes again making it ready to create the next die.

Painting and labeling

3 To complete the production of the die, they may be washed and dried before the final decorations are applied. First, the spots are appropriately painted. For special-

ty dice, words may be printed or images may be applied to each side. Company logos or other advertising may also be applied. These can be stickers or coatings. The coatings used during this step are specially formulated to adhere to the plastic and dry quickly.

Packaging

4 Depending on the final use of the dice, they may be bulk packaged for use in board games, or individually wrapped for consumer sale. For board games, the dice are packed up into boxes and shipped to game manufacturers just like any other component material. When they are sold directly to consumers, dice are typically put into plastic blister packs with a cardboard backing. This package has the dual purpose of protecting the dice during shipping and advertising the product. The finished dice are then placed in cases and shipped by truck to distributors.

Quality Control

To ensure that each die produced meets specified quality standards, a number of quality control measures are taken. Prior to manufacturing, certain physical and chemical properties of the incoming plastic raw materials are checked. This includes things such as molecular weight determinations, chemical composition studies, and visual inspection of the appearance. More rigorous testing may also be done. For example, stress-strain testing can be performed to determine the strength of the plastic. Impact tests help determine the toughness of the plastic. During manufacture, line inspectors are stationed at various points on the production line. Here, they visually check the plastic parts to make sure they are shaped, sized and colored correctly. They also check the integrity of the final packaging. If any defective dice are found, they are removed from the production line and set aside for reforming. Computers are also used to control plastic use, mold retention time, and line speed.

The Future

In the future, dice manufacturers will concentrate on increasing sales and improving the production process. To increase sales, dice marketers will be involved in developing new games that utilize different types of dice. These games will require new types of dice that may have different shapes, sizes, and plastic compositions. From a production standpoint, future improvements will focus on increasing manufacturing speeds, minimizing chemical waste, and reducing overall costs.

Where to Learn More

Books

Beasley, John. *The Mathematics of Games.* Oxford: Oxford University Press, 1990.

Chabot, J. *The Development of Plastics Processing Machinery and Methods.* Brookfield, IL: Society of Plastics Engineers, 1992.

Scarne, John. *Scarne on Dice.* Wilshire Book Co., 1992.

Seymour, R. and C. Carraher. *Polymer Chemistry.* New York: Marcel Dekker, Inc., 1992.

Periodicals

Chabot, J. and R. Malloy. "A History of Thermoplastics Injection Molding. Part I: The Birth of an Industry." *Journal of Injection Molding Technology* (March 1997).

Other

U.S. Patent #4,012,827, 1977.

—*Perry Romanowski*

Drinking Straw

Thermoliquid crystals, a special colorant that responds to changes in temperature, can be added to straws to make them change color when they come in contact with hot or cold liquid.

Background

A straw is a prepared tube used to suck a beverage out of a container. Historians theorize the first straws were cut from dried wheat shafts and they were named accordingly. With the advent of industrial age, methods were developed to mass produce straws by rolling elongated sheets of wax-coated paper into a cylindrical, hollow tubes. This was accomplished by coiling paraffin-coated paper around a rod-shaped form and then securing the paper with an adhesive. The entire straw was then coated with wax to further waterproof it. The wax coating was important since the straw was paper and would eventually absorb some of the liquid being sucked up it. Thus, inevitably these paper straws became soggy and useless. In the 1960s, paper was largely replaced by plastic which were becoming less expensive and increasingly more sophisticated. The explosion of plastic technology led to techniques to manufacture plastic straws via extrusion. Today, straws are made in a wide variety of shapes, colors, and functions.

Raw Materials

Straws are made from a formulated blend of plastic resin, colorants, and other additives.

Plastic

Historically, straws have been made from paper but today polypropylene plastic is the material of choice. Polypropylene is a resin made by polymerizing, or stringing together, molecules of a propylene gas. When a very large number of these molecules are chemically hooked together they form this solid plastic material. Polypropylene was first developed in the mid-1950s and has many properties, which make it suitable for use in straw manufacturing. This resin is lightweight, has fair abrasion resistance, good dimensional stability, and good surface hardness. It typically does not experience problems with stress cracking and it offers excellent chemical resistance at higher temperatures. Most importantly for this application, it has good thermoplastic properties. This means it can be melted, formed into various shapes and, upon reheating, can be melted and molded again. Another key attribute of this plastic is that it is safe for contact with food and beverage. Polypropylene is approved for indirect contact with food and, in addition to drinking straws, is used to make many types of food packaging such as margarine and yogurt containers, cellophane-type wrapping, and various bottles and caps.

Colorants

Colorants can be added to the plastic to give the straws an aesthetically pleasing appearance. However, in the United States, the colorants used must be chosen from a list of pigments approved by the Food and Drug Administration (FDA) for food contact. If the colorants are not food grade, they must be tested to make sure they will not leach out of the plastic and into the food or beverage. These pigments are typically supplied in powdered form, and a very small amount is required to impart bright colors. Through use of multiple colorants, multi-colored straws can be made.

Other additives

Additional materials are added to the plastic formula to control the physical properties of the finished straw. Plasticizers (materials which improve the flexibility of the polypropylene) may be added to keep the resin from cracking. Antioxidants are used to reduce harmful interactions between the plastic and the oxygen in the air. Other stabilizers include ultraviolet light filters, which shield the plastic from the effects of sunlight and prevent the radiation from adversely effecting the plastic. Finally, inert fillers may be added to increase the bulk density of the plastic. All these materials must meet appropriate FDA requirements.

Packaging materials

Straws are typically wrapped in paper sleeves for individual use or bulk packed in plastic pouches or cardboard boxes.

The Manufacturing Process

Straw manufacturing requires several steps. First, the plastic resin and other components are mixed together; the mixture is then extruded in a tube shape; the straw may under go subsequent specialized operations; and finally the straws are packaged for shipment.

Plastic compounding

1 The polypropylene resin must first be mixed with the plasticizers, colorants, antioxidants, stabilizers, and fillers. These materials, in powder form, are dumped into the hopper of an extrusion compounder that mixes, melts, and forms beads of the blended plastic. This machine can be thought of as a long, heated, motor driven meat grinder. The powders are mixed together and melted as they travel down the barrel of the extruder. Special feeder screws are used to push the powder along its path. The molten plastic mixture is squeezed out through a series of small holes at the other end of the extruder. The holes shape the plastic into thin strands about 0.125 inch (0.3175 cm) in diameter. One compounding method ejects these strands into cooling water where a series of rotating knives cut them into short pellets. The pellet shape is preferred for subsequent molding operations because pellets are easier to move than a fine powder. These pellets are then collected and dried; they may be further blended or coated with other addi-

tives before packaging. The finished plastic pellets are stored until they are ready to be molded into straws.

Straw extrusion

2 The pellets are transferred to another extrusion molder. The second extruder is fitted with a different type of die, which produces a hollow tube shape. The pellets are dumped in a hopper on one end of the machine and are forced through a long channel by a screw mechanism. This screw is turned in the barrel with power supplied by a motor operating through a gear reducer. As the screw rotates, it moves the resin down the barrel. As the resin travels down the heated channel, it melts and becomes more flowable. To ensure good movement and heat transfer, the screw fits within the barrel with only few thousands of an inch clearance. It is machined from a solid steel rod, and the surfaces almost touching the barrel are hardened to resist wear. By the time the resin reaches the end of the barrel, it is completely melted and can be easily forced out through the opening in the die.

3 The resin exits the die in a long string in the shape of a straw. It is then moved along by a piece of equipment known as a puller which helps maintain the shape of the straw as it is moved through the rest of the manufacturing process. In some processes, it is necessary to pull the straw through special sizing plates to better control the diameter. These plates are essentially metal sheets with holes drilled in them. Eventually, this elongated tube is directed through a cooling stage—usually a water bath. Some operations run the plastic over a chilled metal rod, called a mandrel, which freezes the internal dimension of the straw to that of the rod. Ultimately, the long tubes are cut to the proper length by a knife assembly.

Special operations

4 Straws with special design requirements may undergo additional processing. For example, so called "crazy" straws, which have a series of loops and turns, may be bent into shape using special molding equipment. Another type of straw with special manufacturing requirements is the "bendable" straw. This type of straw can bend in the middle and is made using a special device that cre-

Plastic drinking straws are extruded through an injection molding machine.

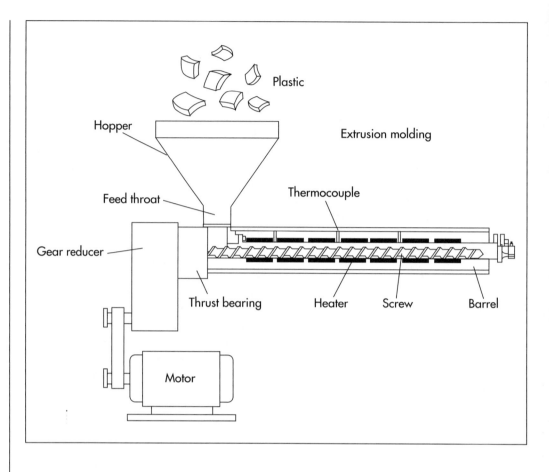

ates a series of grooves that allow the straw to flex. These grooves can be crimped into the straws in a two step process. First, it is first necessary to "pick up" the straw so it can be manipulated. This can be accomplished by spreading the straws across a flat plate, which has slots cut in it. The straws will tend to roll into the slots and remain there. The slots are evenly spaced and are adjacent to a separate metal plate, which has a series of metal pins extending from it. The pins are aligned in a parallel fashion with the slots on the plate. Once the straws have come to rest in the slots, the pins can be easily inserted into the straws. The straws can then be easily lifted up and moved around in any orientation by simply manipulating the plate that holds the pins. The steel pins holding the straws have a series of parallel rings cut into them. As the straws are wrapped around the pin, they are gripped by a pair of semi-circular steel jaws, which have a complementary set of rings. The jaws crimp a series of rings into the straw. The crimp pattern allows the straws to bend without closing off. After these operations, the straws can then by proceed to packaging.

Packaging

5 Straws are typically packaged in individual paper sleeves after manufacturing. This packaging is widely used for applications where each straw must be kept sanitary. One method of packaging involves loading the finished straws into a supply funnel. At the bottom of the hopper is a wheel with straw receiving grooves cut in it around its outer edge. The straws drop out of the hopper and are picked up one at a time by this rotating wheel. As the wheel rotates, it moves the straws along to a second wheel, which has grooves connecting to a vacuum source. Sheets or packaging material (paper wrap) are moved onto this wheel from a supply roller. The vacuum holds the paper in place while the main wheel feeds straws on top of the paper. Another layer of paper is guided over the first and the assembly then passes through a sealing roller. The two layers of paper are then crimped together with the application of pressure or otherwise sealed together. The sealed sheet of straws then travels along the conveyor to a punching region where a die presses down and cuts out indi-

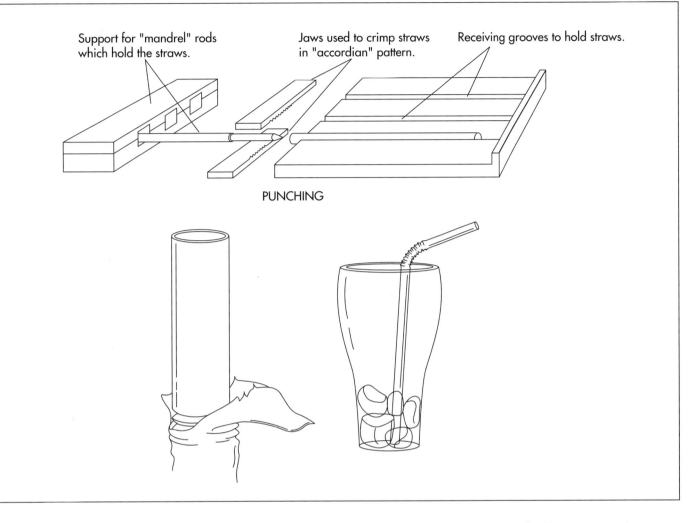

Support for "mandrel" rods which hold the straws.

Jaws used to crimp straws in "accordian" pattern.

Receiving grooves to hold straws.

PUNCHING

vidual straws. The die cut pieces then move along a conveyor to a collection area. The individual straws can then be bundled together and packed in boxes or pouches for shipping.

Quality Control

Drinking straw quality is determined at a number of key steps during the compounding and extrusion phases of the manufacturing process as well as after extrusion is complete. During compounding, the mixing process must be monitored to ensure the formula components are blended in the proper ratios. Before beginning the extrusion process, it is a common practice to purge some resin through the extruder. This purging helps clean out the barrel and acts as a check to make sure all molding systems are operating properly. At this stage, sample straws can be checked to make sure they achieve the proper dimen-

sions. These samples can also be used to ensure manufacturing equipment is operating at the proper line speed.

During the extrusion process, it is critical that the resin is be kept at the proper temperature. Depending on the processing temperature (and the molecular weight of the polymer), plastic can flow as slowly as tar or as quickly as corn syrup. If the temperature is too cool, the viscosity increases dramatically, and the resin will not flow through the die. If the temperature is too high, thermal breakdown can occur. Overheating can cause chemical changes in the resin, weakening the plastic and rendering it unsuitable for use in straw manufacturing. Under certain circumstances, die buildup occurs. When this happens, a glob of plastic gets stuck somewhere in the die. This glob eventually breaks free, becomes attached to the molded straw, and ruins its appearance. Unwanted chemical interac-

Flexible grooves can be cut into the straws in a two step process. As the straws rest in slots, they are gripped by a pair of semi-circular steel jaws, which crimp a series of rings into the straw. The crimp pattern allows the straws to bend without closing off.

tions can also effect the quality of the finished straws during the extrusion process. One problem is oxidation, which results from contact with air. This reaction can negatively impact the plastic. Similarly, the plastic interacts with any moisture that is present; too little moisture can make certain plastic blends too brittle.

After the manufacturing process is complete, it is critical that the extruder be properly cleaned. Thorough cleaning is necessary because different types of different colored plastics can be left behind in the extruder barrel. This residue can cause contamination in the next batch that is made. Die cleaning is done when the machine is still hot and traces of resin can be easily scraped from the metal.

Byproducts/Waste

The major waste product from straw manufacturing is the plastic resin. Resin, which is contaminated, overheated, or otherwise ruined must be discarded. However, straws, which fail for other reasons, can be reworked. This process of reusing plastic is known as regrinding and involves pulverizing the straws and remelting them. This can be done without loss of quality because of the thermoplastic nature of polypropylene.

The Future

There are a number of interesting new developments in straw technology. First, new and improved plastic blends are constantly being evaluated. This is necessary to keep costs down, meet regulatory requirements, and improve quality. In addition, new processing and design methods are being developed. These can expand the straws into new areas. For example, thermoliquid crystals, a special colorant that responds to changes in temperature, can be added to straws to make them change color when they come in contact with hot or cold liquid. Other unique applications include ways of printing straws with the identity of the beverage (e.g., diet, root beer, etc.). The straw can then be used to mark what the drink contains. Other advances include straws made by a blow molding process, which creates faces or other artifacts in the middle of the straw.

Where to Learn More

Books

Richardson, Paul. *Introduction to Extrusion.* Brookfield Center, CT: Society of Plastic Engineers, 1974.

Other

US patent 5,722,219. Method of Making a Drinking Straw.

—Randy Schueller

Drum

Background

A drum is a musical instrument which produces sound by the vibration of a stretched membrane. The membrane, which is known as the head, covers one or both ends of a hollow body known as the shell. Instruments that produce sound by means of a vibrating membrane are also known as membranophones. Drums are part of the larger category of musical devices known as percussion instruments. Percussion instruments other than membranophones are known as idiophones. Idiophones, such as bells and cymbals, produce sound by the vibration of the instrument itself rather than by an attached membrane.

Drums exist in a wide variety of shapes and sizes. The two basic shapes for shells are bowls and tubes. The most familiar bowl-shaped drums in Western music are kettledrums, also known as timpani. Tubular drums may be taller than they are wide, such as conga drums, or shorter than they are wide. Short drums, also known as shallow drums, are the most common tubular drums used in Western music. Shallow drums include snare drums, tenor drums, and bass drums. If a tubular drum is so shallow that the shell does not resonate, it is known as a frame drum. The most familiar type of frame drum is the tambourine.

Drums are usually played by being struck. Some drums, such as bongo drums, are designed to be played by striking them directly with the hand. In modern Western music, most drums are designed to be played by being struck with various devices known as beaters. The most familiar beaters are wooden sticks, generally used to play smaller drums such as snare drums, and padded wooden mallets, used to play larger drums such as bass drums. Sometimes drums are struck with wire brushes or other types of beaters to produce a different sound.

Some drums, particularly in non-Western cultures, are played in ways other than being struck. Rattle drums contain pellets within the shell or knotted cords attached to the head and are played by being shaken. Friction drums are played by being rubbed. Some membranophones have the vibrating membrane set into motion by sound waves coming from a human voice or from another musical instrument. These devices are known as mirlitons. The most familiar mirliton is the kazoo.

Drums are either tunable, so that they produce a particular note, or nontunable. Most drums in Western music are nontunable. The only commonly used tunable drums in Western music are timpani. Idiophones, which exist in an even greater variety than membranophones, may also be tunable, such as a xylophone, or nontunable, such as a rattle.

History

Percussion instruments have been used since prehistoric times. The earliest drums consisted of fish or reptile skin stretched over hollow tree trunks and were struck with the hands. Somewhat later the skins of wild or domesticated mammals were used to make larger drums which were struck with sticks. Besides tree trunks, skins were also stretched over pits dug into the ground to make large drums or over openings in pots or gourds to make small drums.

Frame drums were used by the ancient civilizations of the Middle East about 5,000

An important development in drum manufacturing occurred in the 1950s when drum makers began to experiment with using plastic instead of animal skin to make heads. Although some drummers, particularly timpani players, preferred the sound of heads made with animal skins, plastic heads soon almost completely replaced traditional heads.

years ago. They were later adapted by the ancient Greeks and Romans. The Romans also used tubular drums with skins stretched over both ends of a hollow shell. After the fall of Rome, drums were not commonly used in Western Europe, although they continued to be used by the Arabs. The Crusades brought Europeans in contact with the Arab culture. From the Arabs, Europeans adapted the tambourine (a small frame drum), the naker (a small kettledrum), and the tabor (a small tubular drum). The tabor was often used with a snare, which consisted of thin cords of animal gut stretched across one of the heads in order to produce a rattling sound. The snared tabor is the ancestor of the modern snare drum.

Large kettledrums, long used in the Middle East, were introduced to Western Europe in the fifteenth century. These instruments consisted of calfskin stretched over large copper cauldrons and were used for military and ceremonial purposes. They were first used in orchestras in the late seventeenth century.

The bass drum, a large tubular drum, was rare in Europe until the late eighteenth century. The snare drum and the tenor drum (a somewhat larger version of the snare drum, but without the snare) were used primarily for military purposes until the nineteenth century.

Timpani became an important part of orchestral music during the nineteenth century. During the 1880s, devices were developed which allowed timpani players to change the pitch of the instrument quickly, allowing them to play more complex melodies.

An important development in drum manufacturing occurred in the 1950s when drum makers began to experiment with using plastic instead of animal skin to make heads. Although some drummers, particularly timpani players, preferred the sound of heads made with animal skins, plastic heads soon almost completely replaced traditional heads. A few individual drum makers still make heads from animal skins for musicians who prefer this type of product.

During the twentieth century, percussion instruments of all kinds became important in both orchestral music and in popular music. A modern drum set used by popular musicians such as jazz and rock drummers often

consists of a bass drum struck with a mallet operated by a foot pedal, a snare drum, a series of tubular drums of various sizes, and a set of cymbals.

Raw Materials

Until the late 1950s, the head of a drum was almost always made of animal skin. Modern heads are now almost always made of plastic. Usually some form of polyester is used. Polyesters are plastics in which numerous small molecules are linked together into a long chain using a chemical bond known as an ester group. The most common form of polyester used in the drum industry is known as polyethylene terephthalate, available under trade names such as Mylar. Polyethylene terephthalate has the advantage of being strong and resistant to moisture, heat, sunlight, and many chemicals. Polyethylene terephthalate is made from the chemical compounds ethylene glycol and terephthalic acid. These substances are derived from petroleum

The shell of a drum is usually made of wood. Commonly used woods include maple, birch, and poplar. Some drums have a shell made of metal. Commonly used metals include steel, aluminum, brass, and bronze. Sometimes synthetic materials are used to make shells. These materials are usually strong, hard plastics.

The various hardware components that hold the drum together are usually made of steel. Sometimes other metals such as brass or aluminum are used. In some cases, these components are made of wood or strong plastic.

Optional attachments such as stands to hold the drum in front of the drummer are usually made of steel or aluminum. Straps to hold the drum in place while marching in a band are generally made from leather, plastic, or cloth. The snare of a snare drum consists of thin strands of various materials such as steel, aluminum, plastic, or animal gut.

The Manufacturing Process

Making the hardware components

1 Metal hardware components are made using precision metalworking equipment

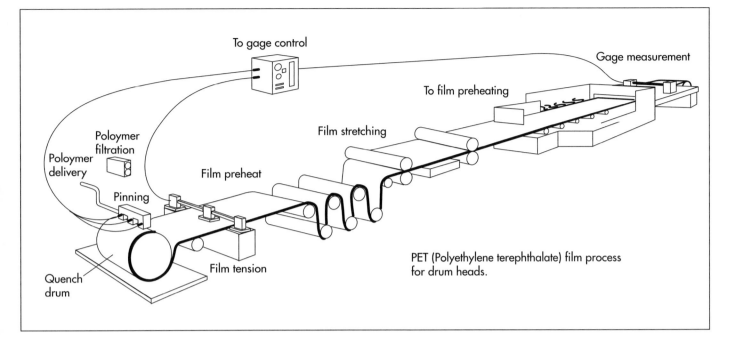

To gage control

Gage measurement

To film preheating

Film stretching

Polymer filtration

Poloymer delivery

Film preheat

Pinning

Quench drum

Film tension

PET (Polyethylene terephthalate) film process for drum heads.

---

such as drills and lathes. Wood hardware components are carved from blocks of wood using various kinds of cutting instruments. Plastic hardware components are often made using a process known as injection molding. This process involves heating the plastic until it melts, injecting the molten plastic into a mold in the shape of the desired component, allowing the plastic to cool back into a solid, opening the mold, and removing the completed component.

Making the head

2 Polyethylene terephthalate is made by combining terephthalic acid (or a derivative such as dimethyl terephthalate) with ethylene glycol. These chemicals are subjected to heat to produce hot, liquid plastic. The liquid is cooled on a large metal roller to form a solid, then stretched between smaller metal rollers to produce a thin film. Various additives may be included to produce films in many colors, which may be either transparent or opaque.

3 The polyethylene terephthalate film is shipped to the drum manufacturer on large rolls. Circles of the proper sizes to make drum heads are cut from the film using precision cutting tools.

4 The edge of the plastic circle is softened by a heating element and allowed to

cool to form a collar around the circumference. A steel ring is placed within the collar and an aluminum ring is placed outside the collar. The ring is then closed on a rolling machine to produce a tubular, secure circle of metal, which serves to keep the plastic skin taut. The completed head is stored until needed for assembly.

Making the shell

5 Metal shells are made using a variety of methods including casting (pouring molten metal into a mold in the shape of the shell) and machining (shaping the metal with various metalworking machines such as drills and lathes). Plastic shells may be made using injection molding.

6 Most shells are made of wood. Wooden shells are made from large, thin panels of wood known as veneer. Veneer is cut from lumber with large saws and shipped from the lumber company to the drum manufacturer.

7 Veneer is cut to the proper size to make the shell by a computer-controlled saw. The cut pieces of veneer are sorted by size and stored under controlled temperature and humidity until needed.

8 Pieces of cut veneer of the proper sizes are moved through a glue press. This device applies glue to the veneer as it passes

Drum head material is made by cooling hot liquid plastic on a large metal roller to form a solid, then stretched between smaller metal rollers to produce a thin film.

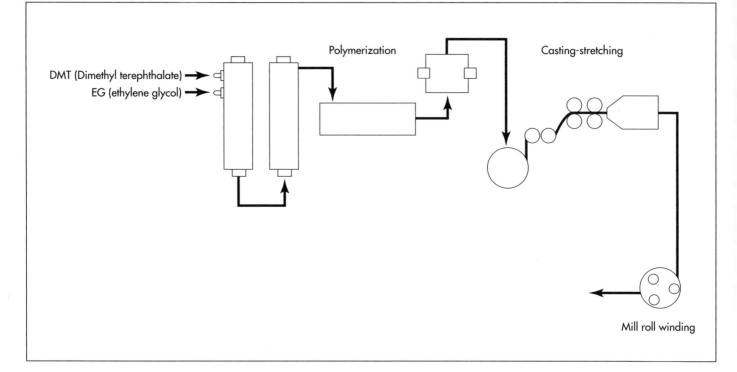

DMT (Dimethyl terephthalate) →
EG (ethylene glycol) →

Polymerization

Casting-stretching

Mill roll winding

Schematic diagram of the formation of the plastic drum heads.

between large metal rollers. The pieces of veneer are then rolled together to form a cylinder consisting of about 7-9 layers of veneer. The rolled veneer is inserted into a metal mold of the proper size to make the desired shell.

9 A variety of methods exist to apply pressure to the glued layers of veneer within the mold. A bag within the mold may be inflated to apply pressure. Water or oil may also be used to apply pressure. The glue may be allowed to dry slowly on its own or an electrical current may be applied to dry the glue quickly.

10 The shell is removed from the mold. It is then cut with a rapidly rotating blade to produce a 45° angle on its edge. In other words, the layers of veneer are cut at slightly different heights, with the innermost layer being the shortest and the outermost layer being the tallest. This sloping edge will allow the head to fit in place.

11 The shell is sanded with high power sanders to produce a smooth surface. It is then either stained or covered with a sheet of shiny, decorative plastic in various colors and patterns. If it is stained, it is rotated while wood stain of various colors is sprayed on it. The stain is then dried quickly with ul-

traviolet light. A clear, shiny topcoat is then applied and dried in a similar way.

Assembling the drum

12 Precision drills are used to drill holes in the shell to allow the hardware components to be attached. Lugs, which hold the head in place, are screwed into the shell. Long, thin metal rods known as tension rods pass through the aluminum ring surrounding the head and into the lugs. The degree of tension on the head can be controlled by using a drum key, which tightens or loosens the tension rods.

13 For a snare drum, a small amount of one edge of the shell is cut away to allow the snare to fit in place so that it touches one of the heads. The snare is attached in such a way that it can either be held tight, producing the snared sound, or held loose, producing the unsnared sound.

Packaging and shipping

14 The drum is placed in a plastic bag. It is then placed in a cardboard box containing pieces of expanded polystyrene foam, a firm, light plastic that prevents the drum from moving during shipping. The boxes are then shipped to musicians, music

stores, orchestras, marching bands, and other consumers.

Quality Control

The most important factor in the quality control of drum manufacturing is the size and shape of the various components. The wooden veneer must be cut to the precise size to allow several layers to fit together to form a cylinder. The plastic head and the metal rings that hold it in place must fit together properly. The lugs and other hardware components must be positioned correctly in exact holes drilled in the proper places in the shell.

The external appearance of the drum is important to drummers. Each drum is visually inspected to ensure that the wood stain or decorative plastic wrapping is free from defects.

The Future

During the 1980s, it seemed that electronic drum machines (flat panels that produce a synthesized sound when struck) might replace traditional drums in popular music. It soon became obvious that drummers preferred playing traditional drums. In the future, small electronic devices may be attached to drums to allow the sound to be manipulated in new ways while allowing the drummer to enjoy the experience of playing traditional drums.

Where to Learn More

Books

Bonfoey, Mark P. *Percussion Repair and Maintenance*. Belwin Mills, 1986.

Holland, James. *Percussion*. Schirmer Books, 1978.

Percussion Anthology. The Instrumentalist Company, 1988.

—*Rose Secrest*

Dulcimer

Background

The origin of the dulcimer is as elusive as its haunting sound. Two types of instrument stake claim to the name—both have different shapes, different methods of being played, and diverse origins. The fretted dulcimer resembles an elongated violin with a limited number of strings (usually three to five) that can be plucked or bowed. In the United States, the fretted dulcimer is better known as the Appalachian or Mountain dulcimer.

The hammered dulcimer is rectangular or trapezoidal in shape and has sets of multiple strings with a range of up to three octaves. The instrument is played with two lightweight beaters called hammers that are shaped like long-handled spoons and are used to strike the strings.

History

The history of both dulcimers is confused because they were developed to play folk music and sprang up independently in a number of locations in Europe and the Middle East. It is not known how or if varieties of dulcimers crossed cultural or topographic barriers.

The hammered dulcimer is considered a member of the zither family and may have origins in Iran as the citar or santir, an instrument used to produce the ancient classical music of Persia. The spice and silk trades that criscrossed the Middle East during the Middle Ages and the Renaissance may have been responsible for the instrument's presence in Spain by the twelfth century and its appearance in China. In China, it is called the *yangqin*, *yang ch'in*, or foreign zither.

The French version of the hammered zither was called the *tympanon*, and it's strings were struck with leather-covered hammers. The instrument experienced great popularity from about 1697-1770 due to an inventor named Pantaleon Hebestreit, who constructed a version with 186 strings and named it a pantaleon. The instrument appeared to fade in popularity with the rise of the piano.

By unknown paths of immigration, hammered dulcimers arrived in the United States. The instrument was sufficiently popular to have been carried by both the Montgomery Ward and Sears Roebuck catalogues in the 1800s and early 1900s. The hammered dulcimer was also reportedly the favorite instrument of Henry Ford and enjoyed a mild revival thanks to his admirers. The harpsichord and pianoforte (or piano) are hammered dulcimers to which keyboards have been attached.

The Appalachian dulcimer or dulcimore is generally hourglass-shaped with three to five strings and frets (low ridges against which the strings are pressed). Its strings are plucked with the fingers, a pick, or a quill, and the player's left hand holds a stick or plectrum on the strings as a stop. Nordic settlers claim to have brought the dulcimer to the New World. The Swedes brought their version called a *humle*, Icelanders imported the *langspil*, and Norwegian immigrants brought the *langleik*. The Germans and Dutch had developed two instruments, the *scheitholt* and the *hummel*, which became the folk instrument of Pennsylvania where so many Germans settled. To this mix of variants, the French added their *Epinette des Vosges*, which is less box-like and more similar in shape to a violin.

Many of the Scotch-Irish who left Ireland in the early 1700s settled first in Pennsylvania before following the frontiersmen to the Appalachian mountains and along the Ohio River. While in Pennsylvania, they may well have heard the scheitholt or humle played by the Pennsylvania Dutch. The German instruments were elongated boxes with frets and a limited number of strings; these either merged with British versions or were developed by the Scotch-Irish settlers to form the hourglass-shaped instrument now known as the Appalachian dulcimer.

Design

As the history of the dulcimer suggests, almost anything goes in selecting the shape of an Appalachian dulcimer. The size and depth of the soundbox must be chosen for the desired sound of the instrument. The deeper and larger boxes produce both louder and lower tones. Perhaps the most eye-catching and apparently ornamental feature of the dulcimer is the sound holes. Some of these are both beautiful and elaborate. The actual shape of the hole does not affect the sound, but the length of slots or elongated holes is important. Like the f holes in a violin, they free longer pieces of the soundboard from the constraining effects of the rigid sides so that the soundboard vibrates more responsively to the strings.

Design of a dulcimer begins with making a pattern and selecting the size of the instrument. Size is determined from the outside in. Strings are available from manufacturers in standard lengths between 25-30 in (63.5-76.2 cm). The soundbar is as long as the strings, but because it is fixed to the soundbox with a nut several inches down from the top of the solid end of the box, the soundbox must be longer than the soundboard. The soundbox is usually 6-8 in (15.2-20.3 cm) wide and up to 2 in (5.08 cm) deep. The maximum measurements have proven to produce the best sound character, including loudness and timbre, or quality of sound.

After the basic dimensions are chosen, a pattern is made on paper or cardboard. One half of the dulcimer is outlined on the paper, which is folded on the axis and cut to produce the mirror image. A form will be made to shape the upper and lower curves of the

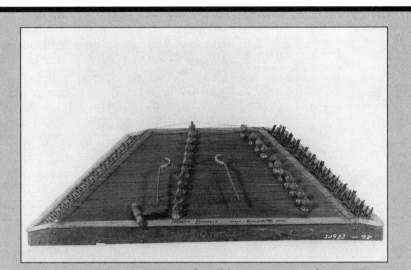

A dulcimer made by Henry Bryant of Wolfeboro, New Hampshire, in 1898. (From the collections of Henry Ford Museum & Greenfield Village, Dearborn, Michigan.)

The hammered dulcimer produces a sweet sound. In fact, the word dulcimer is derived from the Greek meaning sweet (dulce) and song (menos). While the Greeks or Persians may have invented it nearly 1,000 years ago, Europeans were likely introduced to the dulcimer during the Crusades. It soon became a favorite at royal court, but musicians complained about the lack of dynamic range of the instrument. The dulcimer's stringed keyboard eventually led to the development of the pianoforte, which did offer a dynamic range, ultimately eclipsing the dulcimer in popularity. While the dulcimer was no longer a court instrument, it remained a great favorite with street musicians, gypsies, and other plain folk.

This trapezoidal-shaped hammered dulcimer was lovingly made at the turn-of-the-century, Red with gold-painted scrolls and varnished trim, it is proudly emblazed with "Henry Bryant-Maker-Wolfeboro NH 1898." While we don't know much about Henry Bryant, we suspect that he may have been a dulcimer player who typically found it difficult to acquire dulcimers and had to resort to constructing his own instruments. Happily, the sweet song of the hammered dulcimer was rediscovered in the early twentieth century with the revival of early American folk music and interest swelled again in the 1960s. Today, hammered dulcimers are available readily and there is a burgeoning fascination with its soothing sound.

Nancy EV Bryk

dulcimer, so before the pattern is cut, a partial pattern is made of this shape so that top and bottom match.

Raw Materials

Many types of fine wood can be used to make a dulcimer. The outer wood forms the finish and should be selected for its beauty and grain, while the inner wood (usually oak) does not have to be attractive, although the material must respond to steam used to soften and shape the sides. The outer or finish wood used for the soundboard may be walnut, spruce, pine, or yellow poplar and should produce a bright sound due to the presence of hard and soft streaks in the wood. Such striations will give the finished dulcimer visual as well as aural beauty. The body of the dulcimer is usually made of harder wood like cherry, black walnut, or mahogany. The tuning pegs are Brazilian rosewood, although old dulcimers have ebony, rosewood, metal, or crude wooden pegs.

Frets on old dulcimers were also made of a variety of materials including silver, steel, brass, staghorn, ivory, bone, and various woods. Today, commercially-made fretting material is manufactured by specialists in a tee-shaped cross section that is convenient to fit to an instrument. These frets are made of nickel silver or brass. Also made by specialty manufacturers, dulcimer strings are standard, using the same strings as are manufactured for 12-string guitars.

The Manufacturing Process

Master jig assembly

1 After the design is selected and the pattern is made, dies must be constructed to form the sides of the dulcimer. Only two shapes are involved—the curves of the upper and lower sections of the instrument called bouts and the center violin-like cutout called a waist, which may be any simple curve including a circle. Heavy blocks of wood are used to make the dies. The curves are cut with a band saw and sanded to match exactly.

2 A master jig assembly is also an important preliminary. The jig is a piece of plywood that is used to mount and form the dulcimer. Typically, a piece of 0.75 in (1.9 cm) thick plywood that is about 10 in (25.4 cm) longer than the dulcimer and 4 or 5 in (10.2-13 cm) wider is selected. The pattern is mounted on the center of the board, and fourpenny finishing nails are driven into the board at 2 in (5.1 cm) spacings around the

pattern's edge with extra room at the ends. The nails support the inside of the sides of the instrument while it is being shaped.

Sides and internal bracing

3 To shape the sides, 0.125 in (0.32 cm) thick wood is cut in pieces that are as deep as the dulcimer and longer than the curves forming the soundbox. One piece of finish wood backed by a piece of oak is placed in a steam cooker for about 20 minutes to soften the wood. The wood is carefully removed, placed in between the halves of the die, and clamped together. The form with the wood pieces enclosed is then dried in an oven. All the sections are formed this way. Then the outer finish wood piece is glued to the oak backing with polyester or epoxy resin glue. The glued sections are returned to the forming dies and clamped until the glue hardens.

4 When all the curved side sections are formed, the center curves are placed in the master jig assembly. They must be adjusted by eye to be symmetrical. Then, finish nails are driven along the outsides of the sections to hold them in place. The end sections should be fitted to the centerpieces, however all sections were cut long. After the ends are properly fitted, the center sections are removed, cut to their proper lengths, and rasped or sanded for smoothness. The end sections are then refitted. They will overlap the center sections and must be trimmed to mate with the ends of the center sections. Also, the inner oak pieces are trimmed back so they are not exposed on the finished surface. Visual symmetry and tightly fitting joints will produce the most beautiful instrument in sound and appearance.

5 Internal braces are required to make the instrument structurally strong but also to provide the best resonance. Braces are fit to the junctions of the curved sections and also across the centers of the upper and lower curved sections. The braces fit against the back of the dulcimer and the front (soundboard), but they do not extend to the full depth of the instrument. The braces are seated on notches in the side sections. The back braces are glued and clamped into place first. Placement of the front braces of-

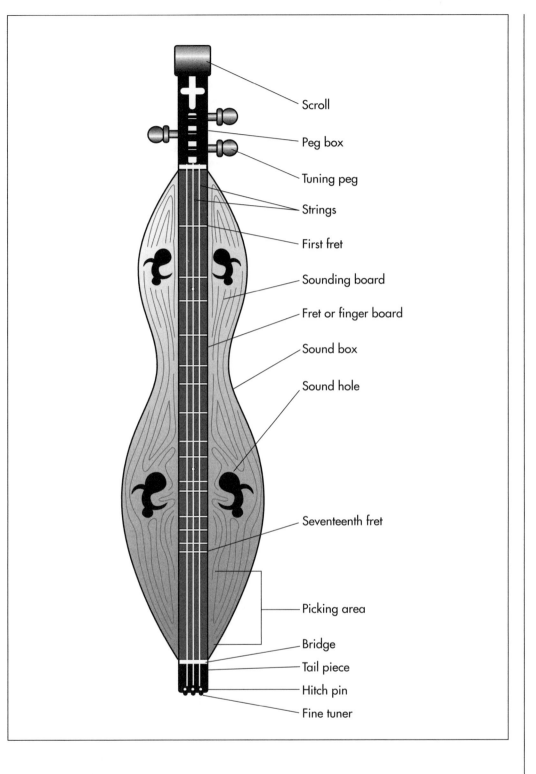

A typical dulcimer.

- Scroll
- Peg box
- Tuning peg
- Strings
- First fret
- Sounding board
- Fret or finger board
- Sound box
- Sound hole
- Seventeenth fret
- Picking area
- Bridge
- Tail piece
- Hitch pin
- Fine tuner

fers the final chance to adjust the shape of the instrument.

Tuning head and pegs

6 Dulcimers often have elaborately scrolled tuning heads at the tops of the instruments. The tuning head has to be large enough to allow 1 in (2.54 cm) spacings be-

tween the pegs. Care is also taken to arrange the pegs so the string from one does not ride on the peg for another. Corresponding peg holes on the head are cut and tapered to match.

7 The head consists of three pieces of finish wood that will be sandwiched together. The overall dimension is about 7 in (18 cm) in length and 4 in (10.2 cm) in width. The pieces are clamped or lightly glued together and cut with a band saw to shape the scroll. The pieces are taken apart, a string slot is cut in the bottom of the center section, and the pieces are then permanently glued together. After the glue has hardened, the peg holes are drilled. The instrument-end of the tuning head must then be mortised to fit the instrument and the taper of the tuning head is shaped by careful woodworking and smooth sanding.

8 Four tuning pegs are required. Pieces of 1 in (2.54 cm) square rosewood are cut to 4 in (10.2 cm) lengths. Two inches of the total length are turned to form a tapered peg, and the heads are flattened and shaped so they are easy to grip. The pegs must be individually fitted to the tuning head because the tapers will vary. A tapered reamer is used to ream the holes that were previously drilled through the tuning head. The tuning head can be fitted to the sides of the sound box by carefully matching and fitting the mortises. A fitted block is made to match the base of the tuning head and reinforce the upper end of the sound box. The joints are then glued and clamped securely. For the lower end of the sound box, a tail section is made with mortises to fit the sides of the soundbox. A similar reinforcing block is made for the tail section.

Soundboard and soundbar

9 The soundboard works as a diaphragm that allows air (and sound) to resonate in the instrument. Grains in the wood should extend the full length of the soundbox. The soundboard is made in two sections with a gap of about 1 in (2.54 cm) between the two halves to allow for the hollow portion of the soundbar. The soundboard is also cut to extend about 0.125 in (0.32 cm) beyond the sides of the dulcimer. The two halves are fitted to the sides with the overhanging lip and glued and clamped in place.

10 The soundbar is the length of the soundbox plus 0.5 in (1.3 cm), extending up into the tuning head. It is constructed of three pieces of wood to be semihollow. That is, two pieces of finish wood form the sides and the top piece forms the fingerboard, leaving a three-sided center tunnel or tube down the middle of the soundboard.

Finish pieces are cut to close the top and bottom ends of the soundbar, and the whole assembly is fitted and glued into place. Edges and ends are checked for squareness and are smoothed. The hole for the nut used to fix the soundbar to the soundbox is drilled several inches below the junction of the soundbar with the tuning head so the two are not acoustically connected. The nut is the only connection attaching the soundbar to the soundbox, again allowing for vibration and resonance.

11 Frets are positioned on the fingerboard of the soundbar based on calculations related to the string length of the instrument. The fret positions are usually tempered by comparing tones with a tuned piano and repositioning the frets before they are permanently placed. Prefabricated metal fret material is hammered into saw cuts at the locations of the frets. The saw cuts are extremely thin and about 0.0625 in 0.16 cm) deep. Overcutting causes the frets to rattle. The edges of the finished frets are filed for smoothness and level. A bridge is also cut from hard maple. It is placed at the tail end of the dulcimer to support the strings and is allowed to float in place, rather than being glued, to permit fine adjustment of the strings and optimal sound.

Soundbox and strings

12 Finally, the back of the soundbox is cut from the finish wood in a single piece and with a 0.125 in (0.32 cm) overlap around the edge. The back is glued to all the edges and braces on the backside of the instrument.

13 The strings are fitted to the instrument before the wood is finished. Holes are drilled in the tailpiece and the tuning pegs. Fine, equally spaced notches are cut in the bridge and the nut to support the strings when they are tightened. Three steel strings that are each 0.012 in (0.03 cm) in diameter and one 0.022 in (0.56 cm) wound steel string are typically used on a four-string dulcimer. The strings are threaded individually through the tailpiece, over the bridge and nut, and through the holes in the tuning pegs. Each peg is turned until the string is tight. The clearance of the string at each position is checked and the strings are lowered at the nut by cutting deeper notches. The bridge is also adjusted to correct the string length.

Finishing

14 The strings, bridge, and tuning pegs are removed before the instrument is finished. The finish that is selected is essentially a part of the design of the instrument because the type of wood, its grain, the desired sound effects, and the final appearance are all considerations. Some dulcimers are left completely unfinished. Rubbing fine wood with linseed oil or applying superb varnish finishes are options. Hand-rubbing and oiling or waxing can take hundreds of hours to fill the wood pores. Application of a sealer followed by wax will produce a similar appearance with less labor.

Quality Control

As with all handmade products, quality control is in the hands of the maker. The end result of the dulcimer maker's craftsmanship is considered from the conception of the instrument forward, and many careful hours are spent in achieving a final product that is as beautiful as intended. Quality control may also be in the ear of the beholder; the craftsman's care will be evident in the sound the instrument produces.

The Future

Renewed interest in folk music has awakened the interest of hobbyists and musicians in dulcimers. The instrument is easier to play than more sophisticated instruments that require long hours of training and practice. Dulcimer-making kits are available from a number of suppliers.

Where to Learn More

Books

Alvey, R. Gerald. *Dulcimer Maker: The Craft of Homer Ledford.* The University Press of Kentucky, 1984.

Hines, Chet. *How to Make and Play the Dulcimore.* Stackpole Books, 1973.

Murphy, Michael. *The Appalachian Dulcimer Book.* St. Clairsville, Ohio: Folksay Press, 1980.

Ritchie, Jean. *The Dulcimer Book.* New York: Oak Publications, 1974.

Other

Charlie Alm's Hammered Dulcimer Book. www.dcwi.com/dulcimer.

The Dulcimer Factory. www.Instar.com/mall/dulcimer/dulcfac2.htm.

The Dulcimer Shop. www.databahn.net/dulcimer.

Frequently Asked Questions About the Hammered Dulcimer. www.dulcimer.com/faq.html.

Hammered Dulcimer Page. www.rtpnet.org/~hdweb/.

Handcrafted Cimbaloms. www.cimbalom.com.

Joe Zsigray's Mountain Dulcimer Page. www.wcnet.org/~jrz100/mountain-d/.

Mike's Hammered Dulcimer. www.halcyon.com/riston/home/dulcimer.htm.

—Gillian S. Holmes

DVD Player

In the early stages of DVD introductions, a single disk will hold as much as 4.7 GB of information. This is roughly equivalent to seven CD-ROMs. In future releases of this technology, a disk may hold up to 17 GB.

Digital video disk or digital versatile disk (DVD) is a type of optical data storage medium capable of holding up to 17 gigabytes (GB) of information. First introduced during the mid-1990s, they were developed as an improved form of compact disk (CD) technology. DVDs can produce such high quality pictures and sounds, they are expected to eventually replace both VCRs and CD players. It is anticipated that the market for DVD players will reach 10 million units by the year 2000.

Background

DVDs work much the same way as conventional CDs. Just like in a CD, the information is coded as a series of tiny pits in the disk. The pits are organized on a spiral track in a structure similar to vinyl records. By using a laser, these pits can be interpreted as binary code. When a smooth surface is read, the machine interprets the data point as a 0. When a pit is encountered, the data point is read as a 1. However, the key innovation that makes DVDs superior to CDs is the laser used to read and create the pits. DVDs use a shorter-wavelength, red laser that can place pits more densely on the surfaces of the disks. This not only allows for more data, it also requires that the disks be only half as thick as conventional CDs. Consequently, two layers can be bonded together to create a double-sided disk which has the same thickness as a CD (1.2 mm).

The DVD system has three features, which make it highly desirable including its high storage capacity, interoperability, and backward compatibility. In the early stages of DVD introductions, a single disk will hold as much as 4.7 GB of information. This is roughly equivalent to seven CD-ROMs. In future releases of this technology, a disk may hold up to 17 GB. This amount of storage space will literally change the way computer programs are developed and will allow for the inclusion of more video clips. The data format and laser used in DVDs will be the same for the computer players as for the television players. This will enable consumers to play the same disks in their computers that they play on their TV. DVD players will also have the ability to play current technology CDs. In this way, consumers will not have to buy replacement products for their current CD collections.

DVDs can be used for a variety of applications including movies, audio systems, computers, and video games. Since the information stored on these disks are electronic, the picture quality is estimated to be three times better than conventional VHS pictures. Additionally, the picture will not degenerate with age or use. Computer programs will also benefit from DVDs. For example, programs, which used to take up multiple CDs can now be condensed onto a single DVD. Video games will also benefit from DVD technology. Since DVDs offer high memory and interactivity possibilities, video clips can be included to enhance the playing experience.

History

Developing the ability to store data for later retrieval has always been important. The first true data storage and retrieval systems were journals and ledgers. While they are still used today, they are slow, inefficient, and bulky. When the computer was being developed during the 1950s and 1960s, one of their main benefits was their ability to store and retrieve data quickly. This has now become one of the cornerstones of information storage and retrieval. Early computerized storage mediums included such things as punch cards, vinyl LPs, magnetic tape, cartridges, and magnetic disks. As computers improved, so did the data storage capabilities. In the late 1970s, the internal hard drive was introduced. Each of these data storage systems were developed to improve on the convenience and efficiency of the best storage methods available. Many of these systems continue to be improved on even today.

The development of DVD began with the introduction by Sony of the CD in the early 1980s. This new storage medium employed a laser to read tiny pits carved in a disk. The first CD audio players were introduced in 1983. They were useful because it was possible to store more than 75 minutes of music on one disk. That was nearly twice what a vinyl LP could hold. While their acceptance was slow, the CD eventually replaced vinyl records as the preferred medium of choice for audio releases.

Video CD players were introduced later in the decade. For various reasons, they never became popular enough to replace VCRs. The use of CDs in computers began during 1987. These devices were useful because they allowed storage of up to 650 megabytes (MB) on a single disk. Until then, the maximum storage on a magnetic disk was 1.3MB. Early computer CDs were slower than typical disk drives and were read only. Data transfer speeds steadily increased as did their capacity to write data. In 1994, 4X speed CD-ROMs (Read Only Memory) were introduced. In the next two years this was doubled. By 1996, 24X speed CD ROMs were available. Recordable CD players were first produced in 1996.

Introduced in 1996, the multiple write CD was developed by Matsushita using a phase-change dual process. This uses a laser to change the reflective properties of the disk. Current CD-RW recorders can use this technology.

As all these advancements in CD technology were occurring, researchers continued to search for ways to improve the storage capacity of these machines. Then, scientists discovered that by using a shorter wave-length laser, much more data could be packed on a single disk. This led to the creation of the DVD. In 1997, the first DVD players were introduced. These machines are slower than the fastest CD players and are not yet recordable. However, the next generation DVDs that are scheduled for release during 1999 will be faster and employ recording technology. Eventually, DVD is expected to replace VCRs and CD players.

Design

A DVD player is designed much like a CD player. For example, computer DVD drives are made the same size and shape as CD-ROM drives. They also have an outer plastic housing and come complete with plastic buttons on the front panel. Some DVD drives have a plastic tray, which extends out from the machine to accept disks. Others have an automatic feed system in which the disk is inserted. Inside the DVD drive, the electronics are also much the same as a CD-ROM drive. Both have sophisticated electronics and include a disk drive mechanism, a printed circuit board, and an optical system assembly. While DVD drive mechanisms come in various designs, each basically consists of a spindle that holds the disk and a motor that spins it. The circuit board contains all of the electronic components, which help convert the data being read into a usable format.

The optical system assembly is the part of the DVD that reads the data from the disk and transmits it to be converted into binary code. In a DVD machine, this consists of a red-laser diode, which has the ability to produce short-wavelength pulses. This is a low noise red laser producing light in the 600-650 nanometer (nm) range. It is much shorter than the 780 nm lasers used in conventional CDs. The other primary component of the optical system assembly is the photodiode, which receives the optical signal from the laser and converts it to an electronic signal. Highly polished lenses and mirrors make up the rest of the optical system assembly.

DVD disks also look like CD-ROMs, but the data is packed together more tightly. The surface of the disk is coated with a reflective silver layer that is protected by a thin, hard coating of lacquer. If a semi-transparent gold layer is put on top of the

A comparison of the amount of data a compact disk and a DVD disk can contain. DVD pit densities are much greater, allowing the disk to store at least seven times more data as a CD.

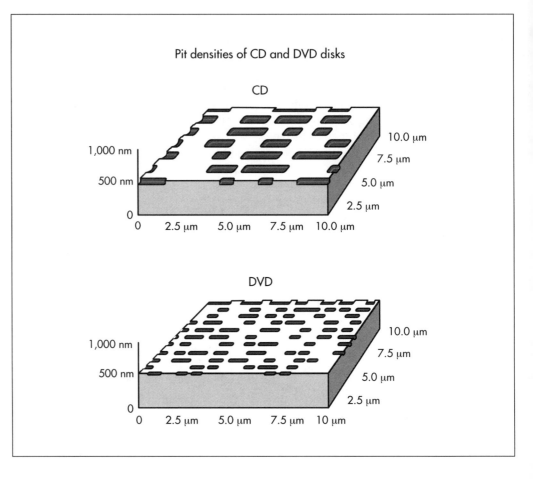

Pit densities of CD and DVD disks

reflective silver layer, the disk can be made to store 2 layers of data on one side. By using less power, the laser can read the data from the gold layer first and then by increasing the power, it can read the silver layer. This nearly doubles the capacity of one side to 8.5 GB. Eventually, a single double-sided disk will be able to hold up to 17 GB of data.

Since DVDs were invented primarily for movies, a compression system is required. To do this, manufacturers have agreed to use the MPEG-2 (Motion Picture Experts Group) compression system. This is a system in which only the elements of the picture that change from frame to frame are stored. For audio, Dolby digital compression is used. Because both of these compression systems are used, a decompressor or decoder must also be included in the DVD player. Currently, this is a separate card that plugs into the computer. The decoder board processes data from the disk and sends it right to the computer's graphics and audio system.

Raw Materials

A variety of raw materials are used in the construction of DVD players and disks. Glass is used to make the laser and other diodes in the system. The primary components on the circuit board are made from silicon. Aluminum metal is used for the housing as well as a hard plastic. The base material of the disks is plastic. They are additionally coated with a silver colored layer and a thin gold layer. The surface of the disk is further coated with a hard layer of lacquer to protect it from damage.

The Manufacturing Process

The components of a DVD machine are typically manufactured by separate companies and then assembled by the DVD manufacturer. The production of the component parts is a highly specialized process, and only a few companies are equipped to supply the entire industry. The main components include the optical system assembly,

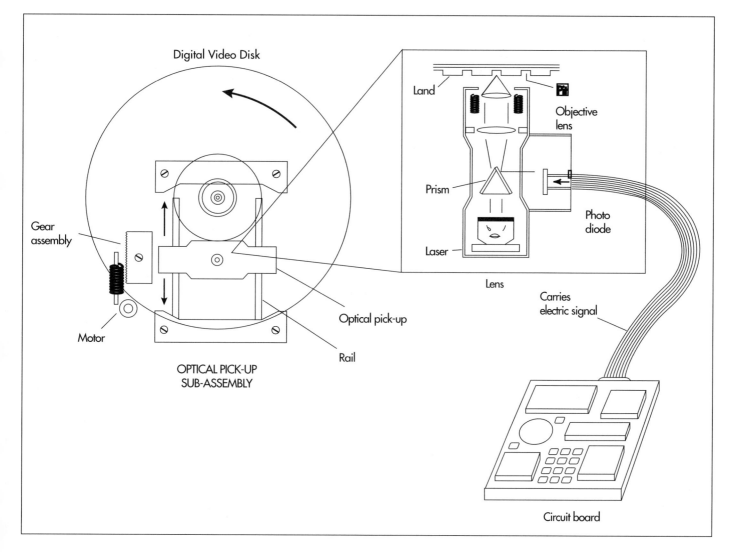

Digital Video Disk

Land

Objective lens

Prism

Photo diode

Laser

Lens

Gear assembly

Motor

Optical pick-up

Rail

OPTICAL PICK-UP
SUB-ASSEMBLY

Carries electric signal

Circuit board

---

internal electronic circuit board and the disk drive mechanism.

Optical system

1 The optical system is made up of a laser, photodetector, prism, mirrors, and lenses. The laser and photodetector are installed on a plastic housing, and the other components are placed in specific places. Great care is taken in the positioning of each of these pieces because without proper alignment, the system will not perform properly. Electrical connections are attached and the optical system is then ready to be attached to the disk drive mechanism.

Disk drive mechanism

2 The optical system is attached to the motor that will drive it. This in turn is connected to the other principle parts of the disk drive including the loading tray (if present) and the spindle motor. Other gears and belts are attached and the entire assembly is placed in the main body.

Internal electronics

3 The electronic components of the DVD machine are sophisticated and use the latest in electronic processing technology. The circuit board is produced much like that of other electronic equipment. The process begins with a board which has the electronic configuration printed on it. This board is then passed through a series of machines, which place the appropriate chips, diodes, capacitors and other electronic pieces in the appropriate places. The process is done in a clean room to prevent damage that can be caused by airborne dust. When completed, it is sent to the next step for soldering.

A DVD player is composed of sophisticated electronics, including a disk drive mechanism, a printed circuit board, and an optical system assembly. It consists of a spindle that holds the disk and a motor that spins it. The circuit board contains all of the electronic components, which help convert the data being read into a usable format. The optical system assembly is the part of the DVD that reads the data from the disk and transmits it to be converted into binary code.

4 A wave-soldering machine is used to affix the electronic components to the board. Before a board is put into the machine however, it is first washed to remove any contaminants. The board is then heated using infrared heat. The underside of the board is passed over a wave of molten solder and through capillary action, the appropriate spots are filled. As the board is allowed to cool, the solder hardens holding the pieces in place.

Final assembly and packaging

5 When all the components are ready they are assembled to produce the final product. The electronic board is hooked up to the rest of the machine and the main cover is attached. The DVD machine is then sent along to a packaging station where they are boxed along with accessories such disks, manuals, and power cords. They are then put on pallets and sent to distributors and finally customers.

Quality Control

To ensure the quality of the DVD machines, visual and electrical inspections are done throughout the entire production process and most flaws are detected. Additionally, the functional performance of each completed DVD machine is tested to make sure it works. These tests are done under different environmental conditions such as excessive heat and humidity. They involve playing a test disk, which will produce specific electronic signals. Since most DVD manufacturers do not produce all of their own parts, they rely heavily on their suppliers for good quality. Most manufacturers set their own quality specifications, which their suppliers must meet.

The Future

DVD technology is relatively new. There are many areas which will be improved in the coming years. Key developments for DVD include greater storage capacity, improved reader capability, and an increase in the number of movies available in DVD format.

Currently, the most intensely studied area of DVD technology is increasing data storage capabilities. While the technology has already been developed to produce 17GB disks, some companies have found ways to store even more. A new encoding technique is being developed that can create a three-fold improvement in DVD storage. In this method, the pits made on the disk will have varying degrees of depth. This will allow the pit to encode for numbers from 0 to 8 instead of just a 0 or 1. It is anticipated that DVD devices using this technology will be available during 1999. Other storage mediums also show some promise. A new technology has recently been demonstrated that can hold up to 30 GB of data. This system uses red lasers and a magnetic field to retrieve the data. The use of blue lasers may allow for even greater storage capacity.

Another area of improvement will be found in the ability of the DVD players to read two layers of information on a single side of the disk. Even though DVD players can theoretically read two layers of information both layers are rarely used because of its high cost. As technology improves however, this obstacle should be overcome and the full potential of DVDs may be realized.

Currently, one the most inhibiting factors in the development of DVDs is the lack of a universal standard for storing and picking up media. This is similar to the problem that developed in the 1980s between VHS and Beta videotape players. In the near future, this problem should be resolved when major DVD manufacturers agree on a format.

Where to Learn More

Books

Williams, G. *Compact Disk Players.* TAB Books, 1992.

Periodicals

Hogan, Dan. "I want my DVD." *Current Science* (October 3, 1997).

Vandendorpe, L. "A Rose by Any Other Name Couldn't Hold This Much Data." *R&D Magazine* (July 1997).

Poor, Alfred. "21st Century Storage." *PC Magazine* (January 21, 1997): 164.

—*Perry Romanowski*

Fire Hose

Background

The term fire hose refers to several different types of hose specifically designed for use in fighting fires. The most common one consists of one or more outer layers of woven fabric with an inner layer of rubber. It is usually manufactured in 50 ft (15.3 m) lengths with threaded metal connections on each end. Unlike other hoses, most fire hose is designed to be stored flat to minimize the space required. For example, the average fire pumper in the United States can carry 1,200 ft (366 m) of 2.5 in (64 mm) diameter fabric-covered, rubber-lined hose in a space about the size of a king-size bed.

The earliest recorded use of fire hose was in ancient Greece. According to the Greek author Apollodorus, one end of an ox's intestine was attached to a bladder filled with water. When the bladder was pressed, the water was forced through the long ox gut and was directed "to high places exposed to fiery darts."

The forerunner of the modern fire hose was invented in 1672 in Amsterdam, Netherlands, by Nicholas and Jan van der Heiden (Heides). Their discharge hose was made of leather with tightly sewn seams. Brass fittings were attached to each end to allow several sections to be coupled together. In 1698, they made a suction hose of heavy sailcloth coated with paint or cement to make it watertight. The hose was reinforced with internal metal rings to prevent it from collapsing under a vacuum.

Early leather hoses leaked badly, and their sewn seams were prone to rupture under pressure. The first riveted leather hose was developed in 1808 in Philadelphia by a group of volunteer firefighters. Their hose had seams held together by 20-30 metal rivets per foot (65-100 rivets per meter) to eliminate leaks. Two members of the group patented this design in 1817 and began manufacturing it. Although woven cotton and linen hoses were also introduced in the early 1800s, and rubber-coated hoses were introduced in 1827, none of these designs was developed enough to replace riveted leather hose until about the 1870s.

Modern fire hoses use a variety of natural and synthetic fabrics and elastomers in their construction. These materials allow the hoses to be stored wet without rotting and to resist the damaging effects of exposure to sunlight and chemicals. Modern hoses are also lighter weight than older designs, and this has helped reduce the physical strain on firefighters.

Types and Sizes of Fire Hose

There are several types of hose designed specifically for the fire service. Those designed to operate under positive pressure are called discharge hoses. They include attack hose, supply hose, relay hose, forestry hose, and booster hose. Those designed to operate under negative pressure are called suction hoses.

Attack hose is a fabric-covered, flexible hose used to bring water from the fire pumper to the nozzle. This hose ranges in nominal inside diameter from 1.5 in (38 mm) to 3.0 in (76 mm) and is designed to operate at pressures up to about 400 psi (2,760 kPa). The standard length is 50 ft (15.3 m).

The forerunner of the modern fire hose was invented in 1672 in Amsterdam, Netherlands, by Nicholas and Jan van der Heiden (Heides).

Supply and relay hoses are large-diameter, fabric-covered, flexible hoses used to bring water from a distant hydrant to the fire pumper or to relay water from one pumper to another over a long distance. These hoses range in nominal inside diameter from 3.5 in (89 mm) to 5.0 in (127 mm). They are designed to operate at pressures up to about 300 psi (2,070 kPa) for the smaller diameters and up to 200 psi (1,380 kPa) for the larger diameters. The standard length is 100 ft (30.6 m).

Forestry hose is a fabric-covered, flexible hose used to fight fires in grass, brush, and trees where a lightweight hose is needed in order to maneuver it over steep or rough terrain. Forestry hose comes in 1.0 in (25 mm) and 1.5 in (38 mm) nominal inside diameters and is designed to operate at pressures up to about 450 psi (3,105 kPa). The standard length is 100 ft (30.6 m).

Booster hose is a rubber-covered, thick-walled, flexible hose used to fight small fires. It retains its round cross-section when it is not under pressure and is usually carried on a reel on the fire pumper, rather than being stored flat. Booster hose comes in 0.75 in (19 mm) and 1.0 in (25 mm) nominal inside diameters and is designed to operate at pressures up to 800 psi (5,520 kPa). The standard length is 100 ft (30.6 m).

Suction hose, sometimes called hard suction, is usually a rubber-covered, semi-rigid hose with internal metal reinforcements. It is used to suck water out of unpressurized sources, such as ponds or rivers, by means of a vacuum. Suction hose ranges in nominal inside diameter from 2.5 in (64 mm) to 6.0 in (152 mm). The standard length is 10 ft (3.1 m).

Another suction hose, called a soft suction, is actually a short length of fabric-covered, flexible discharge hose used to connect the fire pumper suction inlet with a pressurized hydrant. It is not a true suction hose as it cannot withstand a negative pressure.

Raw Materials

In the past, cotton was the most common natural fiber used in fire hoses, but most modern hoses use a synthetic fiber like polyester or nylon filament. The synthetic fibers provide additional strength and better resistance to abrasion. The fiber yarns may be dyed various colors or may be left natural.

Coatings and liners include synthetic rubbers such as styrene butadiene, ethylene propylene, chloroprene, polyurethane, and nitrile butadiene. These compounds provide various degrees of resistance to chemicals, temperature, ozone, ultraviolet (UV) radiation, mold, mildew, and abrasion. Different coatings and liners are chosen for specific applications.

Hard suction hose consists of multiple layers of rubber and woven fabric encapsulating an internal helix of steel wire. Some very flexible hard suction hose uses a thin polyvinyl chloride cover with a polyvinyl chloride plastic helix.

Hose connections may be made from brass, although hardened aluminum connections are more frequently specified because of their lightweight.

Design

A fabric-covered fire hose has one or more layers of woven fabric as a reinforcement material. A hose with one layer is called single jacket hose and is used where lightweight is important or where the hose is expected to have infrequent service. A forestry hose is single jacket for lightweight. An industrial fire hose is single jacket because it sees infrequent use. A hose with two layers is called a double jacket hose and is used where weight is not as critical and where the hose is expected to have frequent, sometimes harsh use, as in urban fire service.

A jacketed hose is usually lined with a thin-walled extruded tube of rubber or another elastomer material that is bonded to the inside of the hose. This prevents the water from seeping through the hose jacket. Some forestry hose is made with a perforated rubber liner to allow it to "weep" a little water through the jacket as a protection against embers that might otherwise burn the hose.

Another type of fabric hose construction is called through-the-weave extrusion. In this design a single fabric jacket is fed through a rubber extruder. The extruder coats both the

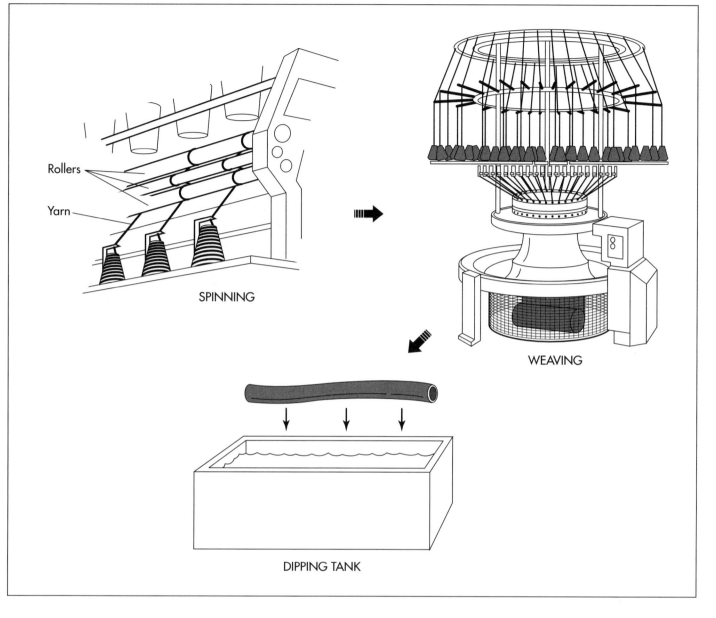

Rollers

Yarn

SPINNING

WEAVING

DIPPING TANK

inside and outside of the fabric with a rubber compound to form both an inner liner and an outer coating at the same time. The extruder forces the rubber into and through the jacket weave to form an interlocking bond. This construction produces a lighter weight hose and is primarily used for larger-diameter supply hoses.

The Manufacturing Process

Fire hose is usually manufactured in a plant that specializes in providing hose products to municipal, industrial, and forestry fire departments. Here is a typical sequence of op-

erations used to manufacture a double jacket, rubber-lined fire hose.

Preparing the yarn

1 There are two different fiber yarns that are woven together to form a hose jacket. The yarns that run lengthwise down the hose are called warp yarns and are usually made from spun polyester or filament nylon. They form the inner and outer surfaces of the jacket and provide abrasion resistance for the hose. The yarns that are wound in a tight spiral around the circumference of the hose are called the filler yarns and are made from filament polyester. They are trapped between the crisscrossing warp yarns and

As the loom starts, the filler bobbins wind the filler yarn in a circle through the warp yarns. The inner and outer jackets are woven separately. If the outer jacket is to be coated, it is drawn through a dip tank filled with the coating material.

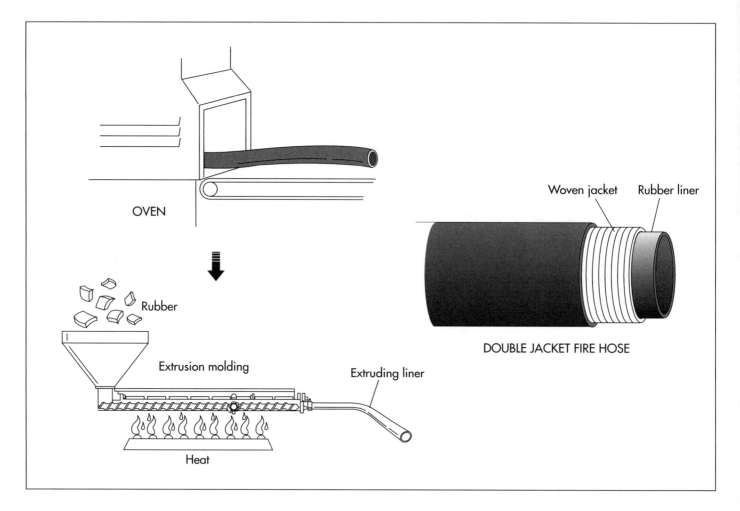

OVEN

Rubber

Extrusion molding

Extruding liner

Heat

Woven jacket   Rubber liner

DOUBLE JACKET FIRE HOSE

Once the outer jacket is coated, it passes through an oven where the coating is dried and cured. The rubber liner is extruded. Jackets and liner are joined to create the hose.

provide strength to resist the internal water pressure.

The spun polyester warp yarns are specially prepared by a yarn manufacturer and are shipped to the hose plant. No further preparation is needed.

2 The continuous filament polyester fibers are gathered together in a bundle of 7-15 fibers and are twisted on a twister frame to form filler yarns. The plied and twisted yarn is then wound onto a spool called a filler bobbin.

Weaving the jackets

3 The warp yarns are staged on a creel, which will feed them lengthwise down through a circular loom. Two filler bobbins with the filler yarn are put in place in the loom.

4 As the loom starts, the filler bobbins wind the filler yarn in a circle through

the warp yarns. As soon as the bobbins pass, the loom crisscrosses each pair of adjacent warp yarns to trap the filler yarn between them. This weaving process continues at a high speed as the lower end of the jacket is slowly drawn down through the loom, and the bobbins continue to wrap the filler yarns around the circumference of the jacket in a tight spiral. The woven jacket is wound flat on a take-up reel.

5 The inner and outer jackets are woven separately. The inner jacket is woven to a slightly smaller diameter so that it will fit inside the outer jacket. Depending on the expected demand, several thousand feet of jacket may be woven at one time. After an inspection, the two jackets are placed in storage.

6 If the outer jacket is to be coated, it is drawn through a dip tank filled with the coating material and then passed through an oven where the coating is dried and cured.

Extruding the liner

7 Blocks of softened, sticky, uncured rubber are fed into an extruder. The extruder warms the rubber and presses it out through an opening between an inner and outer solid circular piece to form a tubular liner.

8 The rubber liner is then heated in an oven where it undergoes a chemical reaction called vulcanizing, or curing. This makes the rubber strong and pliable.

9 The cured liner passes through a machine called a rubber calendar, which forms a thin sheet of uncured rubber and wraps it around the outside of the liner.

Forming the hose

10 The jackets and liner are cut to the desired length. The inner jacket is inserted into the outer jacket, followed by the liner.

11 A steam connection is attached to each end of the assembled hose, and pressurized steam is injected into the hose. This makes the liner swell against the inner jacket and causes the thin sheet of uncured rubber to vulcanize and bond the liner to the inner jacket.

12 The metal end connections, or couplings, are attached to the hose. The outer portion of each coupling is slipped over the outer jacket and an inner ring is inserted into the rubber liner. A tool called an expansion mandrel is placed inside the hose and expands the ring. This squeezes the jackets and liner between the ring and serrations on the outer portion of the coupling to form a seal all the way around the hose.

Pressure testing the hose

13 Standards set by the National Fire Protection Association require that each length of new double jacket, rubber-lined attack hose must be pressure tested to 600 psi (4,140 kPa), but most manufacturers test to 800 psi (5,520 kPa). Subsequent to delivery, the hose is tested annually to 400 psi (2,760 kPa) by the fire department. While the hose is under pressure, it is in-

spected for leaks and to determine that the couplings are firmly attached.

14 After testing the hose is drained, dried, rolled, and shipped to the customer.

Quality Control

In addition to the final pressure testing, each hose is subjected to a variety of inspections and tests at each stage of manufacturer. Some of these inspections and tests include visual inspections, ozone resistance tests, accelerated aging tests, adhesion tests of the bond between the liner and inner jacket, determination of the amount of hose twist under pressure, dimensional checks, and many more.

The Future

The trend in fire hose construction over the last 20 years has been to the use of lighter, stronger, lower maintenance materials. This trend is expected to continue in the future as new materials and manufacturing methods evolve.

One result of this trend has been the introduction of lightweight supply hoses in diameters never possible before. Hoses up to 12 in (30.5 cm) in diameter with pressure ratings up to 150 psi (1,035 kPa) are now available. These hoses are expected to find applications in large-scale industrial firefighting, as well as in disaster relief efforts and military operations.

Where to Learn More

Books

NFPA 1961: Fire Hose. National Fire Protection Association, 1997.

NFPA 1963: Fire Hose Connections. National Fire Protection Association, 1993.

Periodicals

Goldwater, Sam and Robert F. Nelson. "Large-Diameter Super Aquaduct Flexible Pipeline Applications in the Fire Service." *Fire Engineering* (April 1997): 147-149.

—*Chris Cavette*

Fire Hydrant

In order to provide sufficient water for fire fighting, hydrants are sized to provide a minimum flowrate of about 250 gallons per minute (945 liters per minute), although most hydrants can provide much more.

Background

A fire hydrant is an above-ground connection that provides access to a water supply for the purpose of fighting fires. The water supply may be pressurized, as in the case of hydrants connected to water mains buried in the street, or unpressurized, as in the case of hydrants connected to nearby ponds or cisterns. Every hydrant has one or more outlets to which a fire hose may be connected. If the water supply is pressurized, the hydrant will also have one or more valves to regulate the water flow. In order to provide sufficient water for firefighting, hydrants are sized to provide a minimum flowrate of about 250 gallons per minute (945 liters per minute), although most hydrants can provide much more.

The need for fire hydrants developed with the advent of underground water systems. Prior to that time, water was obtained from easily accessible public wells or ponds. During the 1600s, London, England, began installing an underground water system using hollowed-out logs as pipes. When there was a fire, firefighters had to dig up the street and bore a hole in the wooden pipes. Later wooden plugs were inserted into pre-drilled holes at fixed intervals along the log pipes to make it easier for the firefighters to get water. This gave rise to the term fire plug, which is still sometimes used to refer to a hydrant.

As cities grew, so did their water systems. Larger systems meant increased pressures, and cast iron pipes were laid to replace the rotting wooden logs. When Philadelphia's new water system commenced operations in 1801, it not only served 63 houses and sev-eral breweries, but it also had 37 above-ground hydrants for fire protection. The first fire hydrant in New York City was installed in 1817 by George Smith, who was a fireman. He wisely located it in front of his own house on Frankfort Street.

Following the earthquake and fire that devastated San Francisco in 1906, the city installed an extensive emergency water system that is still in use. In addition to more than 7,500 hydrants connected to standard-pressure water mains, the system includes a reservoir and two tanks located on hills to supply nearly 1,400 high-pressure hydrants throughout the city. There are also two salt-water pumping stations to draw water from San Francisco Bay, plus five additional connections along the waterfront to allow the city's fireboats to pump into the hydrant system. As a final line of defense, the city has over 150 underground cisterns connected to unpressurized hydrants. Fire pumpers can connect a rigid suction hose to these hydrants and pull the water out of the cisterns by creating a vacuum.

Today, the size and location of fire hydrants in an area affect not only the degree of fire protection, but also the fire insurance rates. In many urban areas the lowly fire plug is all that stands between the first spark and a multi-million-dollar fire loss.

Types of Hydrants

There are two types of pressurized fire hydrants: wet-barrel and dry-barrel. In a wet-barrel design, the hydrant is connected directly to the pressurized water source. The upper section, or barrel, of the hydrant is always filled with water, and each outlet has its own valve with a stem that sticks out the side of the barrel. In a dry-barrel design, the hydrant is separated from the pressurized water source by a main valve in the lower section of the hydrant below ground. The upper section remains dry until the main valve is opened by means of a long stem that extends up through the top, or bonnet, of the hydrant. There are no valves on the outlets. Dry-barrel hydrants are usually used where winter temperatures fall below 32° F (0° C) to prevent the hydrant from freezing.

Unpressurized hydrants are always a dry-barrel design. The upper section does not fill with water until the fire pumper applies a vacuum.

Raw Materials

The hydrant barrel is usually molded in cast or ductile iron. Some iron wet-barrel hydrants have an epoxy coating on the inner surface to prevent corrosion. Other wet-barrel hydrants are molded in bronze. The hydrant bonnet is usually made from the same material as the barrel. The valve stem in a dry-barrel hydrant design is steel. The valve stems in a wet-barrel hydrant are usually made from silicon bronze.

The hydrant outlets are molded in bronze. If the barrel is cast or ductile iron, the bronze outlets are threaded into the barrel. If the barrel is bronze, the outlets are cast as part of the barrel. The outlet caps may be bronze, cast iron, or plastic.

Valve seats, seals, and gaskets are made from a variety of synthetic rubbers including styrene butadiene, chloroprene, urethane, and butadiene acrylonitrile. Fasteners may be zinc-plated steel or stainless steel.

Hydrants are given a coat of primer paint before they are shipped. When a hydrant is installed, the outer surface is coated with an exterior-grade paint.

Design

The basic design and construction of pressurized fire hydrants in the United States are defined by the American Water Works Association (AWWA), which sets general standards for hydrant size, operating pressure, number of outlets, and other requirements. Unpressurized hydrants may be the same design as the pressurized hydrants within a city or fire district in order to maintain commonality, or they may be a simple capped pipe design with no valves.

The main body of the hydrant is called the barrel or upper standpipe. It may consist of a single piece or it may be made in two pieces. If it is made in two pieces, the upper portion with the outlets is called the head and the lower portion is called the spool. This terminology is not exact and varies from one manufacturer to another, as well as from one city to another.

The hydrant outlets usually have male National Standard Threads (NST) to mate with fire hose couplings. The smaller outlets, sometimes called the hose nozzles or connections, are 2.5-inch NST. The larger outlets, sometimes called the steamer nozzles or connections, are 4-inch or 4.5-inch NST. The outlet caps are secured to the hydrant body with short lengths of chain. The terms hose connection and steamer connection date back to the 1800s. Before the advent of modern fire apparatus, minor fires were often fought by connecting a single hose line directly to the smaller outlet on a pressurized hydrant. If the fire was larger, a steam-powered pumper, called a steamer, took water from the larger hydrant outlet and pumped it into several hose lines.

The hydrant valves are actuated by turning metal stems. The portion of each stem that protrudes from the exterior of the hydrant is pentagonal shaped and is called the operating nut. This five-sided nut requires a special wrench to turn and helps prevent unauthorized use. On some hydrants the operating nut is a separate piece that slips over the stem. This allows the nut to be replaced if it becomes worn from use.

Some dry-barrel hydrants include a breakaway feature to allow easy repair if the hydrant is struck by a vehicle. This design includes a breaker ring on the barrel of the hydrant near the ground and a breakable coupling on the valve stem inside the hydrant. When struck, the upper barrel and stem snap free without disturbing the underground piping or valve.

Although the basic components of all fire hydrants are similar, the shape of hydrants

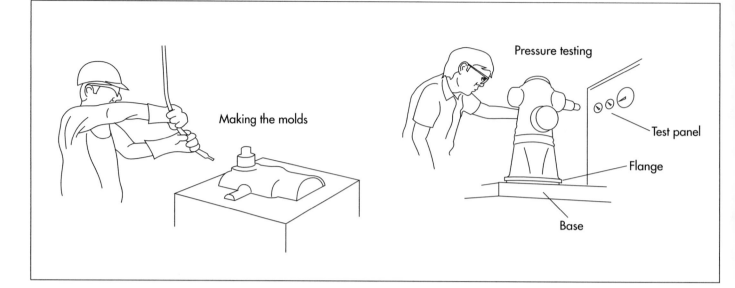

Making the molds

Pressure testing

Test panel

Flange

Base

Fire hydrants are made through a process of metal casting. Once manufactured, each hydrant is filled with water and pressurized to twice the rated pressure to check for leaks.

varies from one manufacturer to another. Some hydrants have the classical round body with a domed bonnet. Others have square or hexagonal bodies. Some areas that are undergoing urban renewal have hydrants that are low and modern looking.

The Manufacturing Process

Making a fire hydrant is primarily a metal-casting process, and most hydrant companies are metal foundries that specialize in manufacturing a variety of municipal water works components.

Here is a typical sequence of operations for manufacturing a wet-barrel fire hydrant.

Forming the molds

1 The outer surface of a mold is formed by a piece called the pattern. To make a hydrant pattern, the hydrant's outer shape is generated in three dimensions on a computer. This data is fed into a stereo lithography machine, which uses laser beams to harden liquid plastic into the shape of the hydrant. This hardened plastic piece is used to make multiple copies of left and right pattern halves out of rigid polyurethane.

2 The inner surface of a mold is formed by a piece called the core. To make a hydrant core, the hydrant's inner shape is machined into two halves of a block of aluminum or cast iron to form a cavity. The

two halves are clamped together, and the cavity is filled with a mixture of sand and a plastic polymer. When the block of aluminum or cast iron is heated gently, the polymer hardens the sand to form the core. The block is then opened, and the core is removed. This process is repeated to make multiple cores.

Casting the barrel

3 When a production run of hydrants is ready to start, the patterns and cores are brought to the mold-making machine. The left and right patterns are pressed into the two halves of a mold filled with sand to form impressions in the shape of the outer surface of the hydrant. Molding sand is a special mixture that holds its shape without crumbling. The hardened sand core is then carefully laid on its side and held with short spacers to form a cavity between the core and the impression in one of the mold halves. The other half of the mold is put in place over the core and the mold is clamped together. This process is repeated for each hydrant.

4 Molten metal is poured into each mold through an inlet passage called a gate. Pouring continues until the metal starts to rise through outlet on the opposite side called a riser. As the molten metal hardens, it cooks the polymer in the core sand. This raises the temperature of the polymer far beyond its initial setting point and causes it to break down and allow the sand to become loose again.

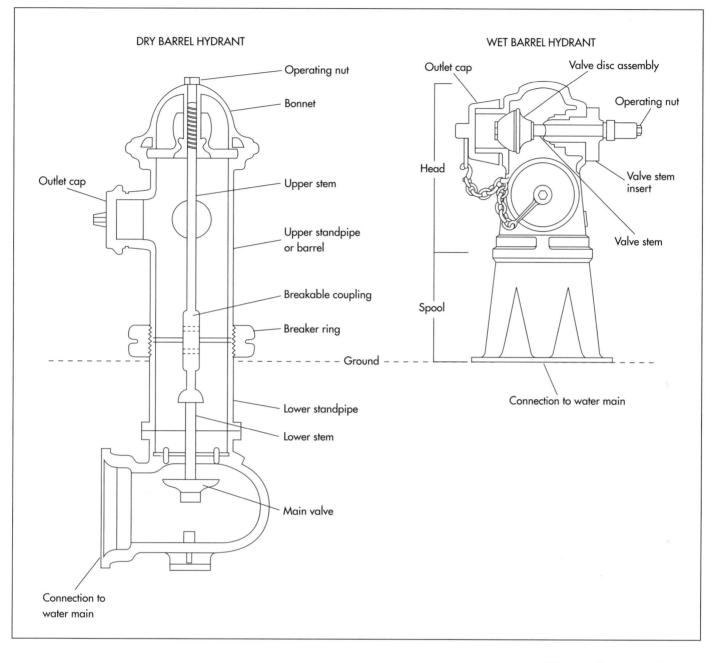

DRY BARREL HYDRANT

- Operating nut
- Bonnet
- Outlet cap
- Upper stem
- Upper standpipe or barrel
- Breakable coupling
- Breaker ring
- Ground
- Lower standpipe
- Lower stem
- Main valve
- Connection to water main

WET BARREL HYDRANT

- Outlet cap
- Valve disc assembly
- Operating nut
- Head
- Valve stem insert
- Spool
- Valve stem
- Ground
- Connection to water main

Sideviews of a dry barrel and wet barrel hydrant.

5 After the casting has completely hardened, the mold is split apart and the core sand is dumped out. The casting is placed in a horizontal cylinder filled with small metal pellets and tumbled to remove any small bits of metal or molding sand that may have adhered to the casting.

6 The cast gates and risers are cut off with an abrasive cut-off saw, and are returned to the furnace. The cast barrel is ground with a handheld power grinder to remove any rough surfaces.

7 If the hydrant has a two-piece barrel, the head and spool are cast, ground, and finished separately. If the hydrant is made from cast or ductile iron, the outlets are cast, ground, and finished separately in bronze.

Machining the barrel and valves

8 The entire hydrant is fixed lengthwise in a lathe, and shallow concentric grooves are cut into the face of the lower flange. This allows the flange to seal against a gasket when the hydrant is mounted. The flange bolt holes may be drilled at this point or they may be drilled just before shipment.

9 If the barrel is a two-piece design, the lower portion of the head has National Pipe Taper (NPT) threads cut on the inside and the upper portion of the spool has NPT threads cut on the outside to allow the two pieces to be joined. The head is drilled and tapped on one side in the area of the NPT threads to hold a locking set screw.

10 The hydrant—or the head, if it is a two-piece design—is repositioned crossways in a lathe along the centerline of the larger outlet. A rotating piece, called a fixture, clamps the hydrant in place and provides a counterbalance as the hydrant is spun. The lathe bevels the inner surface of the barrel around the outlet opening to provide a smooth seating surface for the valve disc. The opening for the valve stem insert is drilled and threaded. Finally the outlet or outlet opening is threaded. This process is repeated for each of the outlets.

11 The valve stems, valve stem inserts, and valve disc holders are machined, and threaded separately.

Assembling the hydrant

12 Starting with the upper valve, an o-ring seal is placed over the valve stem, and the stem is threaded into the stem insert. The inside end of the stem is pushed through the stem insert opening, and the disc holder, rubber disc, and locking nuts are reached up inside the barrel, threaded onto the stem, and locked in place with a set screw. The stem insert is then threaded into the barrel, and the replaceable operating nut is slipped over the outside end of the stem and held in place with a nut. This process is repeated for each of the valves.

13 If the barrel is a two-piece design, an o-ring is slipped over the threaded portion of the spool and the assembled head is screwed down to seal against the o-ring. The threads are locked in place by a set screw.

Testing the hydrant

14 The AWWA standards require that bronze hydrants be rated at 150 psi (1,034 kPa), and ductile iron hydrants be rated at 250 psi (1,723 kPa). Each hydrant is filled with water and pressurized to twice the rated pressure to check for leaks.

Preparing for shipment

15 After the hydrant is pressure tested, the outlet caps and chains are attached, a plastic protector is slipped over the bottom flange, and the exterior of the hydrant barrel is given a coat of primer paint.

Quality Control

All incoming material is inspected to ensure it meets the required specifications. This includes spectrographic analysis of the raw materials used to make the castings. The moisture content of the molding sand is critical to the casting process, and it is checked before every casting run. When a run of castings is machined, the first piece is checked for proper dimensions before the remainder of the castings is machined.

The Future

It is unlikely that the fire hydrant will disappear from the urban landscape anytime in the near future. Water is still the most cost-effective fire suppressant, and the hydrant is still the most cost-effective way to provide a ready supply of water. If anything, the fire hydrant will gain importance as fire departments and taxpayers alike realize that strategically placed, high-capacity hydrants can significantly reduce fire insurance rates.

Where to Learn More

Books

NFPA 291: Fire Flow Testing and Marking of Hydrants. National Fire Protection Association, 1995.

NFPA 1231: Water Supplies for Suburban and Rural Fire Fighting. National Fire Protection Association, 1993.

Periodicals

Long, Germaine R. "Fire Plugs with Personality." *Firehouse* (June 1977): 36-37, 59.

Stevens, Larry H. "Water Works: Get the Most Out of Your Hydrants." *Firefighter's News* (August/September 1996): 32-33, 35-39.

—*Chris Cavette*

Flea Collar

Background

A flea collar is a device used to protect dogs and cats from fleas and ticks. The collar is a plastic strip made by mixing an insecticide with plastic resins and molding the mixture into a thin strip. They are designed to deliver enough pesticide to continually kill fleas for up to 12 months. The pesticide must be safe for prolonged skin contact with animals and non-toxic in the event the animal chews on the collar.

The fleas found on dogs (*Ctenocephalides canis*) and cats (*C. felis*) are ectoparasites, which live off the blood of their host animal. The adult female flea lays eggs on the animal, or in its sleeping place, which hatch into larvae. These larvae pass through a pupae (or intermediate) stage before entering adulthood. In the course of their normal nine-month lifespan, two fleas can produce more than one million offspring. At this rate of reproduction, fleas can easily become a painful annoyance to house pets. They can cause animals to scratch and bite themselves almost constantly. The pets' coats become soiled and roughened and their skin becomes irritated. Dogs in particular may develop severe flea allergies. The need to control these pests is therefore very important to pet owners. Flea control requires killing both the adults and larvae with an insecticide, which can be delivered via sprays, dips, powders, and collars. While collars are somewhat less effective than the other methods because they provide less direct contact, they do offer the long-lasting performance and ease of application.

Design

Flea collars are designed to provide animals with effective protection against parasitic infestations. To achieve this goal, both the insecticide composition and the components used to make the collar must be carefully selected. The most important selection criterion is that the insecticide must efficiently kill fleas without being toxic to pets. It must be effective under a variety of environmental conditions, free from significant taste or odor, and non-staining to fur and surfaces such as carpet and furniture that the pet may contact. Other important considerations are related to the plastic components used to manufacture the collar. These must provide the appropriate release characteristics for the chosen active ingredient. Since collars are designed to be worn for three, six, or even 12 months, the plastic must be durable. The type of animal being treated is another factor to consider. Dogs, being larger, require a relatively higher dose of the active ingredient. Cats require about half the dose as that of a dog. Collars are designed to be a standard size, which is approximately 0.375 inch (0.95 cm) wide and 0.125 inch (0.32 cm) thick. They may be made in different lengths to fit different size animals. Collars are deliberately made slightly longer than necessary to allow the pet owner to trim the collar with scissors to ensure a perfect fit for their pet.

Collars can be designed to deliver insecticide in a solid or liquid form. One patented collar uses a type of insecticide known as a carbamate which can be incorporated into plastic resins in such a way that the carbamate molecules on the collar's surface which are worn away or displaced by contact with the animal fur are replaced by additional carbamate molecules from within the collar. The effect is a continually replenished supply of insecticide at the collar's surface. Another approach is to use a volatile liquid insecticide

In the course of their normal nine-month lifespan, two fleas can produce more than one million offspring.

like Naled (dimethyl 1,2-dibromo-2,2-dichloroethyl phosphate) which is released from the collar as a vapor. Caution must be used in formulating with liquid insecticides as they may be released too quickly and condense on the surface of the collar in small droplets that may be ingested by the animal. In the industry, this droplet formation phenomenon is known as spewing.

Raw Materials

Insecticide

A variety of materials are approved for use against flea and ticks. One type is based on chemicals known as carbamates. Materials of this type include 2-isopropoxyphenyl N-methyl carbamate, 3-(1-methylbutyl) phenyl N-methyl carbamate, 1-napthyl N-isopropyl carbamate, and many others. To provide effective protection, between 3-25% of these materials must be used as a single active or mixture of actives. Other insecticide materials include tetrachorovinphos, which is also known by the trade name Rabon. More accurately this material is 2-chloro-1(2,4,5-trichlorophenyl) vinyl dimethyl phosphate. It is used in commercial brands such as Hartz Mountain.

Collar

The collar itself is a made from a mixture of plastic resins and resin modifiers. The resins used to make collars must be formulated to have appropriate strength and flexibility so it can withstand the shaping operations without cracking or crumbling. The resin must have the appropriate release characteristics such that the active ingredients can escape at the proper rate, but such other inert ingredients remain in the collar. Resins commonly used in flea collars include polyvinyl chloride, polyacrylate, and polymethacrylate esters. These resins may comprise up to 35-70% by weight of the total mixture. Plasticizers must be incorporated into the mixture to make the plastic resin flexible. If these materials are not added to the formulation, the collar would be brittle and would tend to crack or break during the molding operation. Suitable plasticizers include esters made from phosphoric acid (such as tricresyl phosphate) or phthalic acid (such as dioctyl phthalate). Other useful esters are made from adipic, maleic, myristic, palmitic, or oleic acid.

Flea collars also include stabilizers, such antioxidants, which protect the plastic from attack by sunlight and other oxidizing agents. Soybean oil derivatives are useful in this regard. Lubricants, such as stearic acid and low molecular weight polyethylene, are added to improve the physical characteristics of the collar. Inert filler material may also be added to increase the bulk density of the plastic and colorants may be included to give the collar a more pleasing appearance.

Closure

The closure is typically made from a metal buckle or from a solid piece of plastic.

The Manufacturing Process

Blending

1 The polyvinyl chloride and dioctyl phthalate are weighed into a mixing kettle equipped with a high speed, high shear mixer such as a Henschel mixer. They are mixed at approximately 1,800 rpm and the soybean oil, and other plastic agents are added. Mixing is continued until the blend is homogeneous. Then the insecticide, stearic acid, and other components are added. Mixing is continued until the mixture is homogenous.

Collar formation

2 The finished mixture is typically shaped by extrusion molding to form a solid fused product in the desired shape. The plastic enters the extruder through a hopper, which feeds it into a long heated barrel. A screw mechanism pushes the plastic forward toward a die at the end of the machine. The molten plastic is forced through the die at approximately 300° F (149° C) and exits as a flat ribbon.

Final processing and packaging

3 Through subsequent operations, holes are punched in the collar and a buckle is attached. Some collars are designed with plastic closures, which operate by friction and do not require holes to be punched. After final processing, the finished collars are sealed in plastic-laminated foil pouches. This pouch helps prevent the loss of active

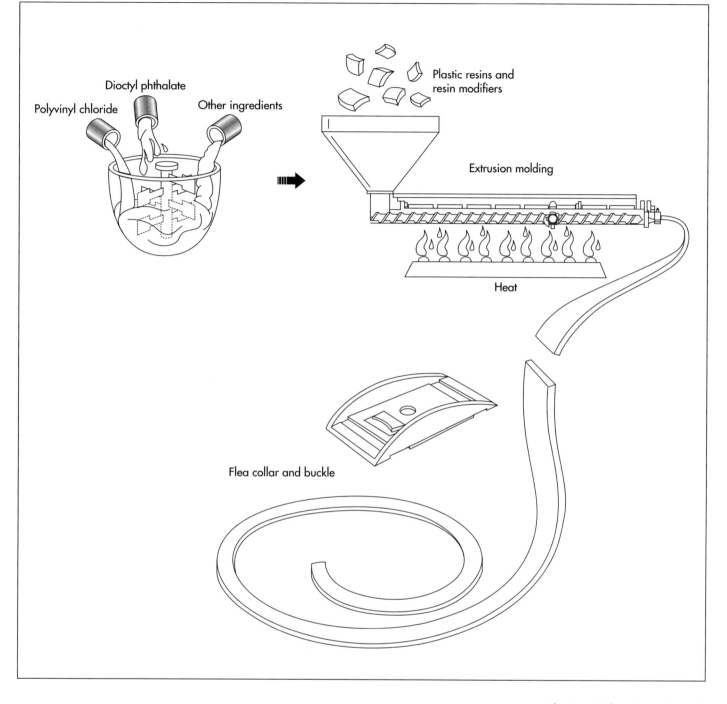

Polyvinyl chloride

Dioctyl phthalate

Other ingredients

Plastic resins and resin modifiers

Extrusion molding

Heat

Flea collar and buckle

ingredients and prolongs the shelf life of the collar. Lastly, they are placed in cardboard boxes and packed for shipping. Individual boxes may be marked with lot numbers as required by Environmental Protection Agency (EPA) regulations.

Quality Control

Depending on the quality control procedures of the manufacturer, a number of collars may be selected for testing to ensure they are functioning properly. Collars can be tested by intentionally infesting a dog with a known number of fleas, usually 50-100. The test collar is then attached to the dog and the animal is placed in a paper-lined wire cage. After 48 hours, the dead fleas, which have collected on the paper, are counted. The collar is removed and the dog is doused with an aerosol insecticide, which should kill the remaining fleas. This is confirmed after six hours by counting the additional dead insects. The animal is then washed and re-im-

The insecticide mixture is mixed with the plastic resins and then the collar is formed through a process called injection molding.

pregnated with more fleas and the process is repeated several times over 13 weeks. The pesticidal efficiency of the collar is measured as a percentage of fleas killed by the collar versus those killed by the aerosol treatment. An effective collar will kill at least 75% as many fleas as the aerosol treatment. Since insecticides are regulated by the EPA, flea collars require EPA registration. Every formula requires testing to ensure specific requirements for safety and performance are met. Furthermore, the EPA mandates that certain information appears on the package. For example, all product labels are required to have appropriate storage instructions. Factors, which must be considered in establishing storage conditions, include conditions that might alter the composition or usefulness of the pesticide (e.g., extreme heat or humidity).

Byproducts/Waste

Since collars use EPA-registered biocidal agents, all waste materials must be processed in accordance with EPA regulations. Such waste may be considered biohazardous and will require special handling and disposal techniques. The disposal techniques for empty containers depend on the type of material. For example, metal containers (non-aerosol) must be triple rinsed before recycling or reconditioning. Glass containers must also be triple rinsed and disposed of in a sanitary landfill or by other approved state and local procedures. Paper and plastic bags must be disposed of in a sanitary landfill or by incineration. Fiber drums with liners must be emptied completely and disposed in a sanitary landfill or by incineration if allowed by state and local authorities.

The Future

Future improvements are likely to come from the development of improved com-

pounds for insect control. For example, one area of current research is focused on finding nonpoisonous ways of controlling fleas. Compounds known as Insect Growth Regulators (IGRs) offer one way to control the population without dangerous side effects. These are hormones, which affect the egg and larva of the flea and make them unable to mature. Therefore, the fleas cannot reproduce. IGRs only affect insects, which go through a four-stage reproductive life cycle. They do not kill the adult flea and must be used in combination with more aggressive pesticides such as the carbamates. They are also non-staining and are practically odor free. Other improvements may come in the design and manufacturing process of the collar. One novel collar design features interlocking pieces which link together to form a continuous band. This method allows collars to uniquely incorporate colors and letters into their design. Another interesting approach involves a kit, which allows the pet owner to add additional liquid insecticide to a refillable collar. Novel designs such as these may hold a place in the future of flea collars.

Where to Learn More

Other

United States Patent 3852416 "Tick and Flea Collar of Solid Solution Plasticized Vinylic Resin-Carbamate Insecticide."

United States Patent 4044725 "Pet Collar."

United States Patent #D264260 "Flea Collar Buckle."

U.S. Department of Agriculture Bulletin on Fleas, February 1997.

—Randy Schueller

Fruitcake

Background

The fruitcake bears the brunt of many holiday jokes in forums as varied as the Sunday funny pages and boxes of greeting cards. One entrepreneur manufactures fruitcakes—and sells them for use as doorstops. Yet the fruitcake has historical associations with the Holy Land, and its internal bounty is said to represent the gifts of the Wise Men. Like many other fruit breads and cakes, it has been venerated since Medieval times when fruit in the wintertime was an extraordinary treat. Modern fruitcakes are based on traditional recipes that are cherished among families, but, thanks to mail-order catalogues and a wide variety of fruitcakes among manufacturers, new twists for this familiar friend abound.

History

History and lore mingle in the retelling of the fruitcake story. The ancient Egyptians made fruitcake for their departed loved ones to carry with them to the afterlife. The dense cake and preserved fruit were thought to withstand the journey, and the riches of the fruits and nuts communicated the wealth of the consumer and the family's esteem for their relative. The Middle East overflowed with the variety of dates, citrus fruit, and nuts that were virtually unknown in Northern Europe until the Crusades. Returning Crusaders brought fruit with them, but the trade that was initiated was frequently interrupted by war, and, of course, the fruit was highly perishable. These dilemmas were partially solved by drying or candying the fruit for travel, and, when the fruit reached Northern Europe, it was shared by mixing it in breads and cakes. Because the fruit came from the Holy Land, it was also revered and saved for feast days, particularly Christmas and Easter.

The Austrians reencountered the bounty of Middle Eastern fruit when the Turks lay siege to Vienna in the seventeenth century. In gratitude for having survived that face-off, the Viennese served German turban cake, or *gugelhupf*, with its filling of raisins, citron, lemon and orange peel, almonds, and spices on Christmas morning. Similarly, the Scandinavians bake fruit breads and cakes variously called *julekage*, *julekakke*, or *julebrod* at Christmas time; like fruitcake, these contain fruits, nuts, and exotic spices and are glazed. The German Christmas bread called stollen and the Italian holiday bread known as panettone are other close kin for the fruitcake. They are characterized by variations in the bread or cake base, choices among fruit and nuts (panettone, for example may be baked with pine nuts), and the optional addition of rum or brandy. Italian panettone is a Milanese tradition surrounded by legend. Supposedly, eggs and fruit were used to make bread for the poor only at holiday times. Panettone became associated with the unification of Italy during the uprisings of 1821 when the traditional raisins were replaced with red cherries and green citron to represent the Italian tricolor flag. Still other similar traditions are Russian Easter bread, known as *kulich* and topped with lemon icing, and Irish fruit bread, which is called *barmbrack*, and accompanies Halloween and All Saints Day festivities.

The English fruitcake or Christmas cake reached its heyday in Victorian times when, with the introduction of the Christmas tree and other festive customs, religious tradi-

Collin Street Bakery makes 1.6 million fruitcakes per year and ships this product, totaling four million lb (approximately two million kg), to customers in all 50 states and 200 countries.

First, the dry and liquid ingredients are mixed separately. Once mixed, they are combined and fruit and nuts are added.

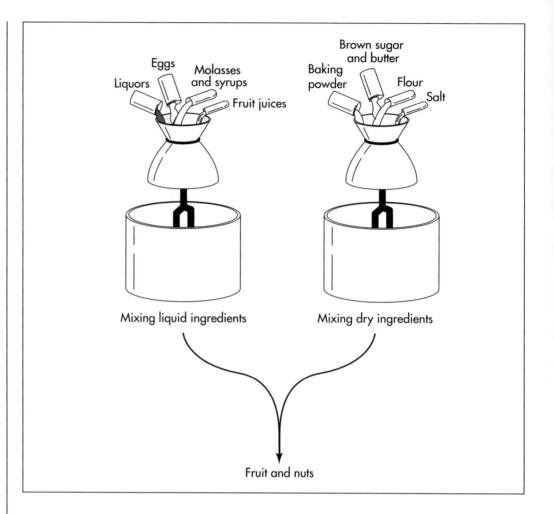

Mixing liquid ingredients

Mixing dry ingredients

Fruit and nuts

tions exploded into colorful, season-long celebrations. Fruitcakes (and other fruit-bearing holiday treats like the plum pudding and Irish plum cake) were made well in advance of the holidays. The cakes were wrapped in cheesecloth that had been soaked in brandy; periodically, the cheesecloth was resoaked and the cakes rewrapped to absorb the liquid. The day before Christmas, the cakes were unwrapped, coated with marzipan or almond paste, further coated with royal icing that dried and hardened, and then glazed with apricot glaze. These Christmas cakes demonstrated such abundance that the same kind of cake is used today in England as wedding cake, and it has the advantage of preserving well for anniversary celebrations.

Raw Materials

Fruitcake character is largely determined by the wealth of fruit and nuts it contains. These can include a whole range or be limited to selected fruit or nuts, depending on the recipe, taste, or market. Fruit can include lemon and orange peel, raisins, dates, currants, figs, apricots, cherries, citron (the preserved rind of the citron fruit, which is similar to a lemon), and pineapple. These fruits are all preserved, dried, candied, or glazed so that much of their natural moisture is removed, and they will keep longer. The cherries and pineapple in particular may also be colored with food coloring. Nuts include walnuts, pecans, almonds, and pine nuts; broken pieces are incorporated in the cake, but walnut or pecan halves may be used to decorate the outside. Most fruitcake bakers purchase fruit and nuts from specialty manufacturers or suppliers.

Spices are other key ingredients that harken back to the Middle Eastern heritage of the fruitcake. Cinnamon, cloves, allspice, and nutmeg are most typical of fruitcakes. Because the blend of spices greatly influences the cake flavor, these are carefully guarded secrets. Liquids include eggs, molasses, other syrups, fruit juices, and liquors of

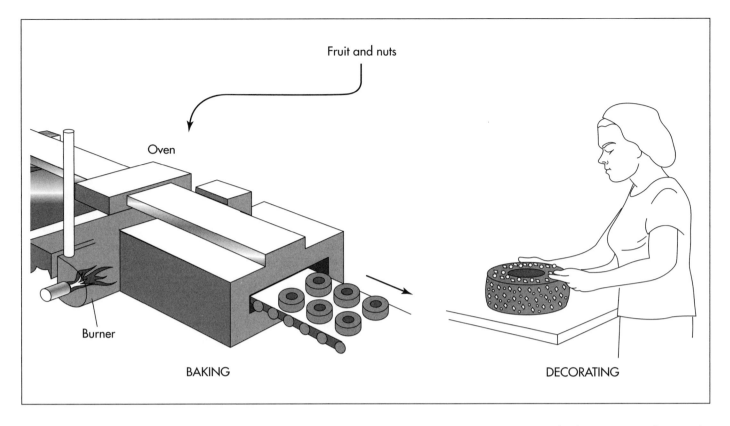

Fruit and nuts

Oven

Burner

BAKING

DECORATING

which rum and brandy are the most popular. The cake itself is made of high-quality flour, salt, baking powder, brown sugar, and butter. Again, these ingredients are purchased directly from suppliers.

Design

The choice of fruit and nuts to be included is subject to availability and the taste of the baker. The spice blend for most manufacturers is carefully guarded, and the proportion of cake to fruit is also a design choice. Ideally, the cake is delicious by itself, but its molasses and brown sugar ingredients (not common to other cakes) are added to help the fruit stick together with the cake as a minimal matrix. Rum and brandy leave their flavor but no alcoholic content because the alcohol is driven off during baking. While these potent potables are flavorful, the choice of making fruitcake intoxicating rests with the consumer who can adjust the cake's moisture level by wrapping it in soaked cheesecloth. If the consumer chooses to do this, any favorite liquor flavor, such as wine, fruit juice, liqueurs, or the traditional brandy or rum can be used. The designer may also select the shape of the fruitcake. Collin Street Bakery, the largest

producer of fruitcake in the world, prefers a ring shape, while circular and loaf-shaped cakes are also manufactured.

The Manufacturing Process

1 In the bakery, the liquid and dry ingredients are mixed separately and then blended together. In large bowls, a generous proportion of fruit and nuts is mixed with the cake dough until the surfaces of the fruit are coated.

2 The rich mixture with its sparkling cherries is scooped into baking pans lined with greased paper that prevents the sugary fruit from burning against the sides of the metal pans. The pans and their contents are weighed to produce uniform products, and the pans are placed on conveyor belts that carry them past a team of inspectors.

3 The inspectors watch for excessive variations and arrange the nuts and cherries so they show to best advantage. The fruit and nuts will not move in the cake once they are in place, so the appearance of the cake before it is baked will be much the same as the finished product.

The batter is poured into cake pans and baked slowly.

4 The cake pans that have passed inspection are placed on large trays that are loaded into a 5 ft (1.5 m) tall industrial baking rack. The entire rack is rolled into a rotating convection oven for the baking process, which is also a well-guarded secret. Because this type of cake is dense rather than airy and because of the fruit and sugar contents, fruit cakes are baked at relatively low temperatures for a long time to drive off the moisture without singeing the fruit. The racks are wheeled out of the convection oven, and the cakes are left in their pans to cool.

5 After the cakes are cooled, they must be wrapped and packaged quickly before they begin to reabsorb moisture from the air. The cakes are removed from their pans, and the paper lining is striped off the cakes. Some manufacturers decorate the finished surfaces with sugared nuts or extra fruit, and some apply a syrup glaze. Following any decorative steps, the fruitcakes are packaged, usually in an inner wrap of decorative cellophane that is seated on a lace doily and a piece of fruitcake-sized cardboard. This inner set of packing is placed in a box or ornamental tin and sealed. That container may also be sealed in an outer box for store display, mailing or shipment.

Quality Control

Quality control begins with the selection of ingredients including the most beautiful candied fruit that should be free of discolorations and tough pieces and should shine like the colors in a stained glass window. Spices and liquors are the other ingredients that give flavorful character to the fruitcake, and they must be fresh and true to their flavors with no sharp twangs or bitterness. During the mixing processes, the ingredients, mixtures, and machinery are monitored carefully. The fruitcake bakers take tremendous pride in their product and scrutinize its quality. The inspection team along the assembly line prior to baking performs the most detailed check while arranging the surficial nuts and fruit. The cakes are inspected again as they are packed, but, after the fruitcakes have been baked, they are virtually indestructible and will last years if kept in cool, dark storage.

The Future

Even in our age of cholesterol-consciousness and health concerns, the fruitcake has an upwardly trending future. If a diet is to be broken, it is most likely to fall by the wayside during the holidays. Fruitcakes may bear the brunt of many jokes, but they are still firmly implanted in our collective holiday traditions. Fruitcakes are also an adult and acquired taste, not only because of the liquors but because of the complex blend of spices and other flavors that usually appeal more to grownups. Collin Street Bakery makes 1.6 million fruitcakes per year and ships this product, totaling four million lb (approximately two million kg), to customers in all 50 states and 200 countries.

Several monasteries in the United States, including the Assumption Abbey in the Ozark Mountains of Missouri, the Abbey of Gethsemani in Trappist, Kentucky, and the Our Lady of Guadalupe Trappist Abbey in Carlton, Oregon, are recent entrants in the race to produce the best fruitcake. The 14 monks of Assumption Abbey create 23,000 fruitcakes per holiday season to finance their less worldly vocations. Furthermore, fruitcake baking doesn't require much conversation, so the monks can maintain their vows of silence while fully supplying customers by mail order and e-mail. Fruitcakes, therefore, may also help us maintain a connection with the religious origins of holidays in these commercial times.

Where to Learn More

Books

Bailey, Adrian, ed. *Mrs. Bridges' Upstairs, Downstairs Cookery Book.* New York: Simon and Schuster, 1974.

Cosman, Madeleine Pelner. *Fabulous Feasts: Medieval Cookery and Ceremony.* New York: George Braziller, 1976.

Field, Carol. *The Italian Baker.* New York: Harper & Row Publishers, 1985.

General Mills, Inc. *Betty Crocker's International Cookbook.* New York: Random House, 1980.

McGrath, Jean. *Butte's Heritage Cookbook.* Butte, MT: Butte-Silver Bow Bicentennial Commission, 1976.

Morris, Sally. *British and Irish Cooking: Traditional dishes prepared in a modern way.* New York: Garland Books, 1972.

Other

Abbey of Gethsemani. http://www.monks. org/abbey_pg.htm.

Collin Street Bakery. http://www.collin streetbakery.com/.

Harry and David. http://www.harryand david.com/.

Southern Supreme Nutty Fruitcakes & Gourmet Confections. http://sosupreme. com/.

Sunshine Hollow Bakery. http://www.sun-shinehollow.com/.

—*Gillian S. Holmes*

Furniture Polish

The most popular form in the United States today is aerosol furniture polish, which sells over 80 million units per year.

Background

Furniture polishes are pastes, creams, or lotions used to clean, protect, and shine wooden furniture. These products were originally made from natural waxes, which were hard to apply and tended to leave a heavy buildup over time. Today these formulations combine natural waxes and oils with petroleum based ingredients and synthetic polymers. These modern formulations can clean the film residue and lay down new polish in a single step so periodic stripping of old polish layers is not necessary. The most popular form in the United States today is aerosol furniture polish, which sells over 80 million units per year. However, these aerosol products are coming under scrutiny as new legislation regulates propellants that can be used in these products.

Wood has been used for ages for making furniture such as tables, bed frames, and sofas. As a natural material, wood is vulnerable to the effects of aging which means it can become dried out, cracked, or stained. Since biblical times, and probably before, people have recognized the usefulness of coating wooden surfaces with oils, balms, and unguents. Early historical accounts have been found with instructions for using linseed or cedarwood oil to treat wood surfaces. Other natural oils used for polishing wood include tung, and Perilla oils. In twelfth-century Italy, these oils were commonly used to polish wooden floors. By the fourteenth century, beeswax was being used to treat inlaid wood and parquetry floors in France. Beeswax became a very popular wood polish but had to be applied with hot irons and then hand buffed. Despite this drawback, beeswax, sometimes mixed with hard animal fats, remained the predominant

form of polish until the late eighteenth century. In 1797, a natural plant wax, called carnauba wax, was discovered on the leaves of the Brazilian cerara palm. Carnauba wax is tough, high melting and, when properly compounded, imparts a fine shine without all the buffing required by beeswax. By the late nineteenth century, other waxes were discovered and polishes were developed that utilized blends of carnauba with ouricui, candelilla, esparto, sugar cane, cotton fiber, flax, palm, hemp and raffia waxes.

By the early 1900s, petroleum chemistry yielded a number of raw materials, which were useful in polish formulations. These included paraffin waxes (which can be varied in melting point and hardness) and inexpensive solvents (like kerosene and naphtha). Similarly, mineral waxes, like montan wax, became commercially available and were incorporated into polish products. By 1929 chemists had prepared a suspension of carnauba wax in a soap and water base and marketed it as the first self-polishing wax emulsion. This formulation was an improvement over its predecessors because it required less buffing but it had significant drawbacks because it caused streaking and the soap tended to make it more easily removed upon contact with water. In the last few decades, synthetic polymer emulsions have been introduced which offer significant improvements over wax systems. The most widely used polymers are based on silicone oils, which provide lubricity and good gloss. Aerosol sprays are the most popular delivery system for these products because they offer easy application over a large surface area. In addition to ease of application, today's products offer excellent gloss, wearability and water resistance.

Raw Materials

The primary ingredients used to prepare furniture polishes are polishing agents, solvents, and emulsifiers. Auxiliary materials include preservatives, colorants, and fragrance.

Polishing agents

The waxes, polymers, and oils are used to improve the condition of the furniture surface can be loosely grouped together and labeled as polishing agents. The waxes employed can be of vegetable, animal, or mineral origin. Common examples of vegetable waxes are carnauba (from palm leaves) and candelilla (from the Mexican plant of the same name). Sugar cane wax, cotton wax, and many others are also used. The primary animal wax (or more accurately, insect wax) is beeswax which is useful for its unique physical and chemical properties. Shellac is another popular insect waxes, which comes from the lac insect of the genus *Ficus religiosa*. Spermaceti wax, from the sperm whale, was popular at one time but ecological concerns have forced development of synthetic replacements. Lanolin fractions from sheep may be used as animal waxes. Mineral waxes, although they are not true waxes by definition, have similar chemical properties. These can be categorized as ozokerite, paraffin waxes, microcrystalline waxes, oxidized microcrystalline waxes, Fischer-Tropsch waxes, and montan waxes. In addition to these naturally derived waxes, synthetic resins are also commonly used in polishes. These include a multitude of polymers, some of which were originally developed for use in the paint and coating industry. These are materials like methyl acrylate, ethyl acrylate, butyl acrylate, vinyl acetate, styrene, vinyl chloride, acrylonitrile. Finally, oils derived from vegetable, petroleum, or silicone sources are added to formulations to enhance shine.

Solvents

Solvents are used to help dissolve or soften some of the water insoluble materials used in polishes. Common solvents include mineral spirits, turpentine, and naphtha. In addition to solvency, factors to consider during solvent selection include flammability and toxicology.

Emulsifiers/surfactants.

Proper blending of oil and water-soluble ingredients requires special chemicals known as surfactants (short for surface active agents). These surfactants (which may also act as emulsifiers) have the ability to bridge water and oil to create a stable cream, paste, or lotion called an emulsion.

Propellants

Propellants are liquified gases, which are used to dispense aerosol products as a spray. The most common propellants are short chain hydrocarbons such as propane or butane, both of which are highly flammable.

Other ingredients

In addition to the ingredients listed above, polishes may contain abrasives, colorants, fragrance, and preservatives. Still other ingredients are added to limit the chance of corrosion of the metal can. These are often nitrogen containing materials that raise the pH of the solution.

These ingredients can be formulated into pastes, creams, liquids, and aerosol (including non-aerosol pump sprays).

Design

Furniture polishes are designed with a blend of waxes and oils because no one single ingredient provides all the desired properties. For example, in theory, a 20% paste of carnauba wax should produce the best gloss but in reality this mixture is gritty and hard to spread. It is beneficial to add different wax materials (e.g., some of the mineral waxes) that may not add appreciable gloss, but which will modify the spreading properties of the waxes with the more desirable characteristics. Of course the solvents and other materials play an important role in the product's consistency as well. Factors to consider when formulating furniture polish include hardness, buffability, flexibility and mechanical strength, water proofing, stain resistance. Cost and ease of manufacture are important considerations was well. When designing such products, one should also take into consideration the type of surface the product is targeted toward. Some polishes are designed for specific types of wood, others are primari-

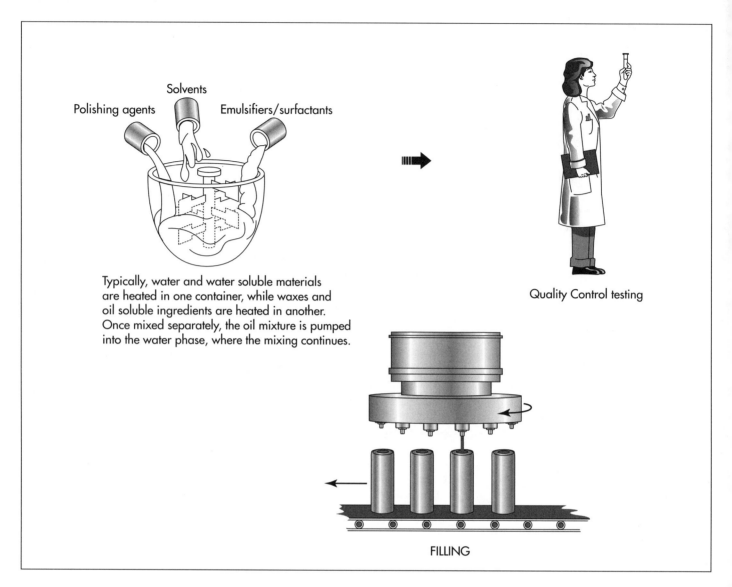

Polishing agents

Solvents

Emulsifiers/surfactants

Typically, water and water soluble materials are heated in one container, while waxes and oil soluble ingredients are heated in another. Once mixed separately, the oil mixture is pumped into the water phase, where the mixing continues.

Quality Control testing

FILLING

ly intended to add protective shine, and still others are made to also to clean and remove dust. The formulator must recognize what kind of surface finish the wood has and consider its attraction for dust and resistance to water spills and grime. Safety and toxicological concerns can not be overlooked and there may be regulatory issues, which affect polish formulation as well.

The Manufacturing Process

The manufacturing procedure for furniture polish varies depending on the type of product being made. The following is a discussion of the mode of manufacture used for aerosol polishes. The production of aerosol polishes requires four important operations: compounding the wax emulsion, filling the

primary container, pressurizing/gassing the can, and finishing operations.

Compounding the wax emulsion

1 The type of emulsions used in furniture polishes can be made with a variety of mixing techniques. One common method is to heat the water and water-soluble materials together in one vessel and the waxes and oil soluble ingredients in a separate vessel. These mixing tanks are typically constructed of stainless steel and are equipped with a jacketed shell that allows steam and cold water to be circulated around the tank. This provides a way to heat and cool the batch without letting it come in contact with external water. The mixing kettle is also configured with temperature controls, and inlet and outlet plumbing for adding ingredients and

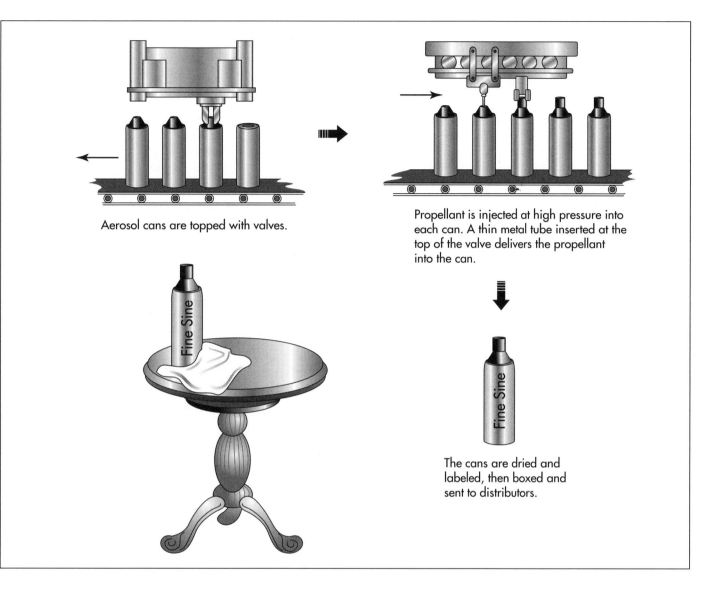

Aerosol cans are topped with valves.

Propellant is injected at high pressure into each can. A thin metal tube inserted at the top of the valve delivers the propellant into the can.

The cans are dried and labeled, then boxed and sent to distributors.

pumping out the finished product. When the water and oil phases are at the appropriate temperature 158-176° F (70-80° C) they are mixed together using a turbine type mixer that provides relatively high shear. Typically the oil is pumped into the water phase. Heating and mixing continues until the batch is homogeneous at which point cooling is initiated. As the batch cools, other ingredients such as preservatives, dyes, and fragrance are added. When the batch is complete, it is assayed to insure it meets quality control standards for solids, pH, etc. The batch may be pumped to a filling line or stored in tanks until it is ready to be filled.

Filling the primary container

2 Aerosol furniture polish is packaged in metal cans, which are capable of with-

standing the pressures required by an aerosol product. Typical can construction may be tin-coated steel or aluminum. When the product is ready to be filled into the package, the emulsion is pumped to a filling line outfitted with a conveyor, which carries the cans to the liquid filling equipment. At the filling head, there is a large hopper which holds the polish emulsion and discharges a controlled amount, usually set by volume, into the can. The filled can then proceeds down the conveyor line where, depending upon the production method, the valve may be inserted and sealed in place immediately before gassing.

Pressurizing/gassing the can

3 After the can is filled with the polish concentrate, propellant is added in a process known as gassing. Aerosol cans may be gassed before or after the valve is crimped into place. In order to fill the can, the propellant is shot into the can around the circular metal cup that forms the base of the valve. This method, which is preferred for economy and speed, is known as undercupping. The other gassing method fills the propellant under high pressure through the stem of the valve after it is crimped in place. This method, known as pressure filling, is much slower because all the gas has to enter the can through a very small opening. Both operations are conducted with special pneumatically operated equipment, which is properly grounded to limit the chances of electrical spark, which can trigger ignition of the highly flammable propellants.

Final operations/finishing steps

4 At some point during, or immediately after, the filling operations, the cans are coded with the date and other batch information to allow traceability. This is useful because if there is a problem with a specific batch of products, for example a spoiled raw material does not allow the emulsion to be stable, then the finished goods made with that lot of raw material can be traced. If necessary, this number can even be used to issue a product recall, although this is a rare occurrence. The can is also usually capped with a plastic overcap, which prevents the valve from being accidentally triggered causing leakage of the slippery contents. After final capping and coding and after the appropriate quality checks, the cans are packed, usually in cardboard shipping cartons, and palletized. They are then sent to a warehouse or distribution center to await final shipping instructions.

Quality Control

Quality of furniture polish is assayed at various points in the manufacturing procedure. Before manufacture, the raw materials are checked to ensure they conform to specifications. After the product is batched, it is analyzed to make sure it was correctly prepared. Key formulation parameters include water content, pH, solids level, and preservative activity. After the product is filled into aerosol cans and charged with propellant the spray characteristics of the valve are checked. The can is passed through a heated water bath to ensure it does not leak. Before filling, a representative number of cans are tested to make sure they are of the appropriate strength. The United States sets limits for bursting strength of aerosol cans.

The Future

As with any technology driven product, improvements will be made as the advances are made in underlying technology. For example, new silicone polymers are constantly being developed and some of these are likely to be incorporated into future furniture polish formulations. Perhaps the most significant changes in store for the future of the polish industry are likely to be driven by regulatory concerns. Air pollution legislation is limiting the types of propellants and solvents used in furniture polishes. In the mid-1970s a similar situation occurred in the antiperspirant industry. Safety concerns caused aerosol antiperspirants to almost vanish from the market where they were originally the most popular type. Whether the industry will respond to regulatory challenge with improved aerosol formulations, non-aerosol pump products, or some new delivery system altogether remains to be seen.

Where to Learn More

Books

Chalmers, Louis. *Household and Industrial Chemical Specialties.* Chemical Publishing Co Inc., 1979.

Flexier, Bob. *Applying Finishes.* Rodale Press Inc, 1996.

Lawrence, David. *Stripping and Polishing Furniture.* International Speciality Book Services, 1987.

—Randy Schueller

Gasoline Pump

Background

A gasoline pump is used to dispense gasoline into motor vehicles. The gasoline pump evolved from a simple mechanism into a more elaborate, specialized one as automobiles grew popular. When cars were rare, drivers usually filled a canister of gas from a barrel or tank at a hardware store, and then tipped the canister into the opening of the car's gas tank. This process was inconvenient and messy, and possibly dangerous, as the gas could easily leak or splash from the container. The first pump specifically marketed for gasoline was adapted from a kerosene pump designed by Sylvanus Bowser of Fort Wayne, Indiana. Bowser had come up with his kerosene pump in 1885, and when he brought out the gasoline version—the "Self-Measuring Gasoline Storage Pump"—in 1905, he was still somewhat in advance of consumer demand. Bowser's invention operated with a manual suction pump, which dispensed the gasoline into the car through a flexible hose. The 50-gallon metal storage tank housed in a wooden cabinet could be set up at the curbside in front of a store.

A slightly earlier invention was John Tokheim's 1901 glass-domed pump for gas or kerosene. The gas was pumped from a storage tank up into the dome. The consumer could measure the amount visually, and then release a valve to let the gas down a tube and into the gas tank. One advantage of this pump was that the consumer could inspect the gas, and make sure that the vendor had not adulterated it. Watering down was apparently a common problem, especially when the gas was stored in an underground tank, out of sight. By the 1920s, many manufacturers were producing gas pumps similar to Tokheim's and Bowser's. Common features were hand-operated pumps, glass-dispensing areas, dial gauges-often of questionable reliability-and a globe-shaped head on top of the pump, bearing a logo. Gasoline itself was often not brand name, but supplied by small dealers and distributors. The pumps were gaudy and decorative, boldly declaring their brand name, and drivers seemed to pick their preferred filling station by the pump and not by what it dispensed. This reversed by the end of the 1920s, when gas became a branded item sold by large companies such as Shell and Gulf.

The first electric gas pump came out in 1923. Further refinements to pump technology mostly concerned how the amount of pumped gas was indicated. As the glass globe was abandoned, consumers needed some other way to tell how much they were buying. One model used a dial with revolving hands similar to a clock to indicate gallons dispensed. In 1933, a Fort Wayne, Indiana, manufacturer brought out a pump with a mechanical calculator called a variator. The variator used a revolving number wheel to show how much gas was being pumped, while a second wheel displayed the price. This was the forerunner of the system in use today, which makes it easy for consumers to buy five dollars worth of gas without having to calculate fractions of gallons. The inventor of the variator, the Wayne Oil Tank & Pump Company, licensed its technology to most other pump manufacturers, and by the end of the 1930s, the revolving wheels were standard across the country.

The Tokheim Company, which had brought out one of the very first gas pumps, intro-

The first pump specifically marketed for gasoline was adapted from a kerosene pump designed by Sylvanus Bowser of Fort Wayne, Indiana. Bowser had come up with his kerosene pump in 1885, and when he brought out the gasoline version—the "Self-Measuring Gasoline Storage Pump"—in 1905, he was still somewhat in advance of consumer demand.

duced the variator's successor in 1975-electronic measurement. Instead of wheels that turned, and interior electronic device calculated the amount and price, and displayed this information on a small screen. Today's pump uses virtually the same system, except that many pumps can also handle other sophisticated transactions, such as debiting the user's bank account.

Raw Materials

There are three basic systems in every gas pump: the hydraulics portion, which includes the actual pumping device; the electronics; and the frame or housing. The hydraulics are generally made from cast iron or cast aluminum. Synthetic rubber may also be used in the hydraulic segment, for seals and gaskets. The electronics portion may use printed circuit boards and plastic parts. The outer housing of the pump is generally made of sheet steel, or stainless steel.

Design

Gas pumps are generally manufactured on a semi-custom basis. That is, though many parts may be virtually the same from pump to pump, the manufacturer designs some aspects of the pump to accord with particular customer wishes. The number of hoses and where they are placed may vary, as may the type and sophistication of the electronics and the design of the housing. Before the manufacturing process begins, engineers must draw up the specifications for the particular order. Customized sub-assemblies may have to be ordered from suppliers, and machines may have to be re-set to cut parts according to the design.

The Manufacturing Process

Gas pump manufacturing involves putting together three basic units: the hydraulics, the electronics, and the housing. One gas pump factory may not manufacture all three units. Typically, some parts are bought from a nozzle manufacturer, and the hose from a company that makes only hoses. The electronics may be sub-contracted, and the gas pump manufacturer only assembles and installs this portion. Assuming a large man-ufacturer makes most of its parts in-house, this might be a typical process:

Making the hydraulics

1 The electromechanical devices that actually bring the gas from its storage tank to your car are usually made from either cast aluminum or cast iron. The iron or aluminum is melted and poured into molds, then allowed to cool. Workers then remove the parts from the molds and clean them by abrasion. Other parts may be stamped in a die. Sheets of metal are fed into a machine that punches out a piece in the desired shape. Metal tubing is placed in machines that bend it according to specifications. Workers using welding irons assemble smaller pieces into larger ones. In making the hydraulics, groups of workers trained in a few specialized skills work together in a cell—a unit of perhaps four to 50 workers—to produce parts that are consistently high quality. Each cell makes specific parts, and the parts are then passed on to other cells that specialize in assembling the parts in order.

Electronics assembly

2 The electronics for a gas pump control the display that tells consumers how much gas is being pumped and what the cost is. Many pumps also have electronic scanners that can read credit cards and debit the customer's account. These devices are micro-processors similar to those used in computers and calculators. The electronics manufacturer produces these by running small, stiff pieces of non-conducting material—typically cardboard or ceramic-through a solder printer, which imprints a pattern of circuit paths on the board. Other parts are placed by hand or automatically, and the board is heated in an oven. The oven melts the solder paste put down by the solder printer, forging electrical connections. The gas pump manufacturer might not do any of this, but buy the circuitry according to specifications. Workers at the gas pump plant may assemble the electronics by snapping or soldering pieces together. These workers, like the hydraulics workers, typically would be assigned to a cell responsible for the head of the pump.

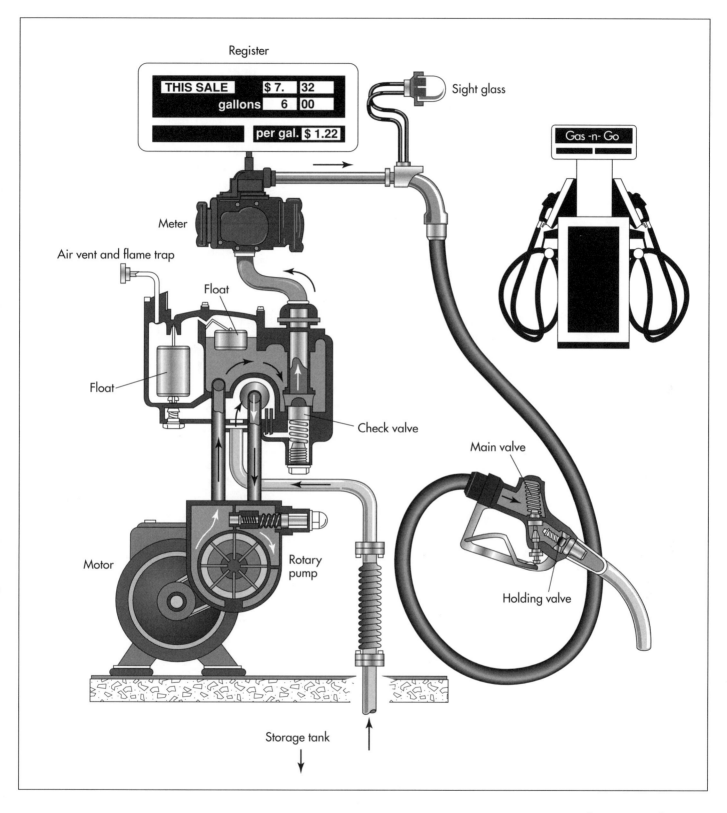

Register

THIS SALE	$ 7.	32
gallons	6	00
per gal.	$ 1.22	

Sight glass

Gas -n- Go

Meter

Air vent and flame trap

Float

Float

Check valve

Main valve

Motor

Rotary pump

Holding valve

Storage tank

A typical gas pump mechanism.

Housing

3 After the hydraulics and electronics are completed, workers bring the parts over to the housing area of the plant. Here, doors, panels and outer walls are cut from sheet steel. These are then sent to a painting area. Workers apply a high-quality, corrosion-resistant paint to the housing, according to a design specified by the customer. The painted parts are sent through an oven to dry, and then cleaned. Next, the hydraulics are fitted to the electronic head, and the whole pump is encased in the housing. Hoses are attached, and gaskets or seals applied. The sections of the housing are welded together, or attached with hinges. The unit is cleaned, and the paint may be touched up. Next, the unit is inspected and tested, then sent to the customer for installation.

Quality Control

Workers inspect gas pumps for quality at many steps along the manufacturing process. If parts are ordered from subcontractors, these are tested and inspected on arrival at the factory. Raw materials, such as the steel for the housing must also be inspected to make sure it is the proper gauge, and free from irregularities. Quality-control inspectors visually check pieces as they come off the assembly line out of the cells. When the entire hydraulics portion of the gas pump has been assembled, it is tested for leakage. Water may be run through it, or pressurized air. Any leakage would be extremely detrimental to the operation of the pump, so great care is taken at this point in its construction that it is absolutely leak free. Similarly, when the electronics portion of the pump is fully assembled, it is tested for accuracy. The weights and measures in the pump must be calibrated exactly. After the entire pump is assembled, it is tested again. One test is a running test, where the machine is put to use for a certain number of hours or days, and checked for any faults, particularly leaks. Quality control inspectors also administer what is called a dielectric withstand test. In this case, a high voltage current is run through the wiring. Any flaws in the electrical system are made evident by this test. Each gas pump maker tests its own products, because the manufacturer wants to assure its customers of a high quality product. Outside inspectors working for Underwriters Laboratory, an agency responsible for a variety of electrical products, also run tests on gas pumps. Pumps that use radio frequencies to identify vehicles also come under the jurisdiction of the Federal Communications Commission (FCC), which administers its own tests.

The Future

Current innovations in gas pump technology focus on three main areas. One is meeting increasingly stringent environmental standards, especially by detecting leaks and containing vapors. A second area where new technology is being applied is data collection. This means automated systems that can check a vehicle's odometer, and record how much fuel it uses, providing convenience for vehicle fleets such as those owned by trucking firms, police forces and bus companies. Perhaps the most fascinating new gas pump technology is the robotic pump that fills the gas tank while the consumer sits in the car.

A variety of devices are being developed to recover vapors lost when gas is dispensed. Some are placed in the hose and others in the nozzle. The challenge is to prevent vapor leaks without slowing down the rate at which the gas can be pumped. The newest high-tech nozzles contain electronic sensors, which record vapor and liquid leaks, and automatically shut off the pump if the gas tank becomes full.

New technology is coming into play for owners of fleets of vehicles. An experimental system is in use in some parts of the country, where the gas pump nozzle can read information from a device installed in a vehicle's fuel tank. Two loop antennas, one in the nozzle and one in the tank, communicate by low-frequency magnetic induction when they come in contact. As soon as the driver begins fueling the car, the antennas connect, and begin sending information from an electronic unit mounted in the car to a terminal at the fueling station. The system can record which vehicle in the fleet is being filled, how much gas it is getting, how much the gas costs, how many hours the car has been driven, and what its odometer reads. This can cut down significantly on the paperwork needed to maintain a large commercial fleet. Similar technology is also being tried out for individual consumers. The consumer uses a small radio-like device, which can be hung on a key chain, carried like a card in a wallet, or mounted on a window. The gas pump carries a similar device that emits a low-frequency radio signal. When the consumer drives up, the gas pump reads the card or tag, identifying the customer and debiting an account for the gas bought.

The first fully robotic gas pumps were tested on consumers in 1997. To use the robotic pump, the consumer needs a transmitter on the vehicle's dashboard and a specially modified gas tank cap. The driver parks the car by the pump and enters the amount of gas needed on a keypad similar to the keypad of an automatic teller machine (ATM). The pump reads the vehicle's make and model from the transmitter, and a robotic arm then finds the gas tank and opens it with suction. Then the robot inserts a nozzle through a flap in the tank cap and dispenses gas. The whole fill-up process is supposed to take only two minutes. Because this process requires drivers to have two modifications to their cars—the dashboard transmitter and the special tank cap—use of the robotic system may evolve slowly.

Where to Learn More

Books

Witzel, Michael Karl. *The American Gas Station*. Motorbooks International, 1992.

Periodicals

Emond, Mark. "Drive-by Dispensing." *National Petroleum News* (November 1996): 47-50.

LeDuc, Doug. "Indiana Gas Pump Maker Invests in Advanced Leak-Detection System." *Knight-Ridder/Tribune Business News* (August 26, 1997): 826B0903.

Shook, Phil. "Pumping Out Innovation." *National Petroleum News* (January 1994): 32-36.

Witzel, Michael Karl. "Gas Pumps." *American Heritage of Invention & Technology* (Winter 1997): 58-63.

—Angela Woodward

Globe

Gerardus Mercator (1512-1594) is best known for developing the type of map, now called a Mercator projection, in which all the meridians and longitudinal lines are parallel and the lines of latitude intersect the longitudinal lines at right angles.

Background

Globes fall into two broad categories: terrestrial and celestial. Terrestrial globes are spherical maps of the world, and celestial globes use the earth as an imaginary center of the universe to map the stars in spherical form. A globe is the only "true" map of the world because there is no distortion in relationships of areas, directions, or distances. The actual flattening of the true earth at its poles and "fattening" around the equator are such small, real distortions that they don't appear at the scale of most globes. The sphere constituting the globe is mounted on an axle and stand so it can be rotated like the earth. The axle's tilt (23.5°) is the same as Earth's rotation on its axis (relative to the plane in which it orbits the Sun).

There are many types of globes within the classification of terrestrial globes. A physical globe depicts Earth as the astronauts see it (except that they also see the intervening clouds and the shadows cast by the sun). Although physical globes emphasize natural land features (sometimes showing them in relief), the features of the bottom of the sea can also be shown. A political globe shows the nations of the world in a variety of colors as well as other features of civilization like locations of cities. Varieties of celestial globes extend to globes of the planets and the moon. Thanks to satellite imagery and other technological advances, the physical features of the world are now available in globe form on CD-ROM as the digital globe.

History

The ancient Greeks never gave credence to "flat earth" theories. They knew the world was spherical and made the first globes to depict their understanding of it. A Greek named Crates is credited with making the first globe in about 150 B.C. Our ancient ancestors were quick to adapt the principle of the globe to mapping the skies. The Romans made a celestial globe called the Farnese globe in 25 A.D. Because they used local marble for this feat, the globe survives today.

German geographer Martin Behaim made the earliest terrestrial globe that has survived. Behaim's accomplishment was timely; he made his globe in 1492, and Christopher Columbus was almost certainly aware of it and strengthened by it in his conviction to sail West to find the Orient. Today's globes would not be the same without the Flemish geographer Gerhard Kremer who is better known by the Latin form of his name, Gerardus Mercator. Mercator lived from 1512-1594 and was also a cartographer, mathematician, astronomer, and engraver. He is best known for having developed the type of map, now called a Mercator projection, in which all the meridians and longitudinal lines are parallel and the lines of latitude intersect these at right angles and are also parallel to each other. The Mercator projection simplified map reading; for instance, a navigator can plot a ship's course between any two points in a straight line and follow that course without changing compass direction. Mercator also widely influenced all other aspects of mapmaking; the world atlas is also his invention. He made Louvain, Belgium, the center of the world of cartography and scientific instruments; and, there, he and Myrica Frisius constructed terrestrial and celestial globes in 1535-1537.

Raw Materials

In the past, globes were generally solid and made of a variety of materials including glass, marble, wood, and metal. Hollow globes, including those made in Mercator's day, were produced from thin metal sheets including copper. Today, globes are almost always hollow and can be made of any material that is both strong and lightweight. Cardboard, plastic, or metal can be used. A three-dimensional jigsaw puzzle with paper pieces backed with foam rubber is manufactured for puzzle fanatics, plastic globes with snap-on continents and other features are learning tools for children, decorative globes of Waterford crystal can ornament desk tops, and inflatable globes (both terrestrial and celestial) are useful tools and toys.

The George F. Cram Company and Replogle Globes Inc. are the only two manufacturers of traditional globes in the United States. The George F. Cram Company has made maps since 1867 and globes since 1929. The company's manufacturing processes for producing the two basic types of globe remain largely unchanged in 70 years. One type is made of fiberboard or cardboard, and the illuminated globe is made of plastic that will withstand the heat from a light bulb that is placed inside the sphere to light it from the inside out. Recycled cardboard is used for the cardboard globes. Injection molding plastic is also used to partially fill the plastic globe. Specialty manufacturers produce all other parts for the globe. These include tape required to join the two globe hemispheres ("Equator tape"); the axis, stand, base, or other mounting; and electrical wiring and the bulb socket for the illuminated version.

Design

Globes are made in two standard sizes. The 12in (30.5 cm) diameter globe (roughly the size of a basketball) is the most popular globe sold to schools and retailers, and the second most popular size is 16 in (40.6 cm) in diameter. Of all the globes sold, 80% of them are 12 in (30.5 cm) globes. Apart from distinctions like terrestrial, political, relief, celestial, etc., globes are made in a variety of color schemes because they are made as ornamental as well as informative objects to decorate homes and offices. Interestingly, children prefer globes with blue oceans, while adults like non-blue globes, of which the antique or off-white color is favored.

Globe manufacturers decide on new product lines based on constant input from the marketplace. Teachers may be the most important source of new globe concepts because they request changes in globes as the curriculum is modified. Globe makers also watch design and fashion trends because many globes are spontaneous purchases made because of appearance, and purchasers expect ornamental globes to be available in designs to match their decors. The globes themselves don't necessarily change for reasons of fashion, but stands and display pedestals do. Obvious choices include selections in dark and light wood; current trends toward Southwestern-style decor and wrought iron work have made globes mounted in these styles popular.

Manufacturers also produce new globes as changes in our world occur. Each manufacturer's research staff monitors changes in data that may require artwork adjustments. Physical globes tend to change little simply because geologic processes are slow and small and don't appear at the scale of most globes (660 miles to the inch on a 12-inch globe). Political changes occur more rapidly but are still not frequent. In the past five years, only three political changes have affected world globes, with two in Africa and one in Europe. By making computerized changes to the artwork printed on the globe, corrections can be made almost instantaneously. Sources for political changes in the world include the Office of Geographic Names (part of the U.S. Department of the Interior), the State Department, and the embassies of various governments. Globe-makers in the United States do not change political names until the State Department has officially recognized that a name change has occurred. For physical changes, the embassies are again sources, as is the Library of Congress. In the United States, the respective states are sources for information about changes within their boundaries. For example, the State of Louisiana provides data about the changing configuration of the Mississippi River Delta.

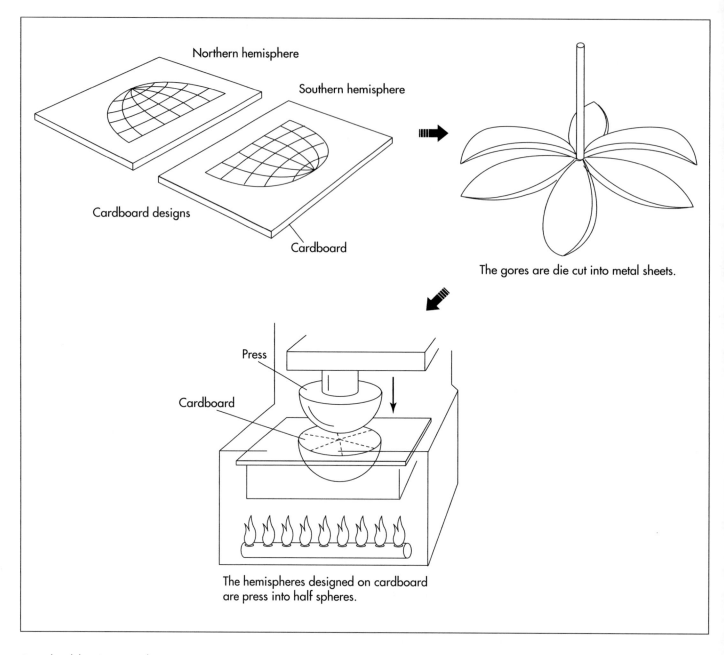

Northern hemisphere

Southern hemisphere

Cardboard designs

Cardboard

The gores are die cut into metal sheets.

Press

Cardboard

The hemispheres designed on cardboard are press into half spheres.

Once the globe pieces are die cut, they are pressed together to form half spheres, one for each hemisphere.

The Manufacturing Process

1 The world is flat when the process of making a globe begins. Highly detailed and informative artwork prepared by a staff of researchers and cartographers is printed on sheets of cardboard. The Southern Hemisphere is printed on one sheet, and the Northern Hemisphere is reproduced on a second sheet of cardboard.

2 Gores, or tapering triangles, are then die-cut into the printed sheets by a specialized machine; the half globe with cut gores looks like a pinwheel or a banana peel with

a pole at the center and the parts of the peel forming segments of the world.

3 The artwork is designed and the gores are located in such a way that adjacent segments will match correctly when joined.

4 The cardboard hemispheres are then subjected to heat and pressure in a forming press to shape them into half spheres. The forming press works much like a curling iron and heats each hemisphere to about 300° F (148.9° C) for 90 seconds. In the joining process, the two halves are glued together to produce the round ball, and Equator tape is placed to cover the seam.

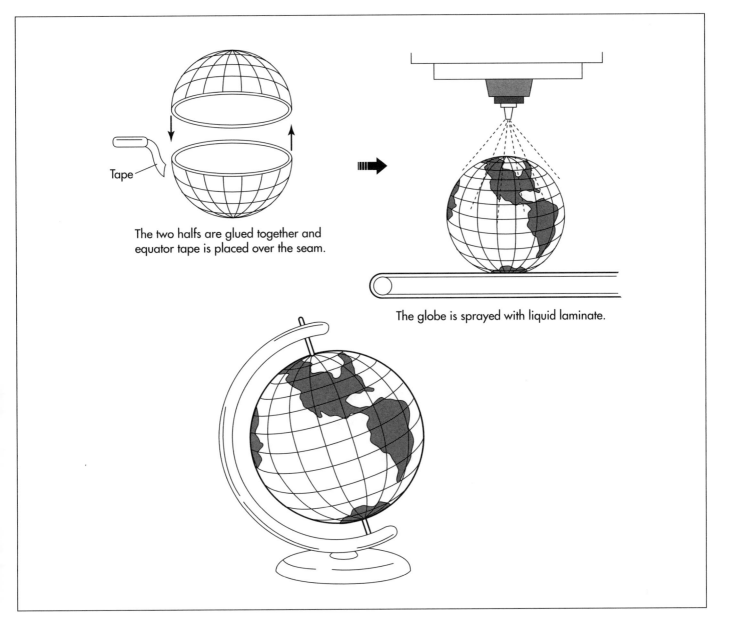

The two halfs are glued together and equator tape is placed over the seam.

The globe is sprayed with liquid laminate.

5 The completed ball is then sprayed with a liquid laminate to make it durable, fingerprint-proof, and glossily attractive.

6 After manufacture of the spheres is completed, they are fitted to any of the wide variety of mountings from inexpensive plastic to brass. The illuminated globes are equipped with light bulbs and electrical sockets, switches, and cords. The completed globes are packaged for sale or shipment.

Illuminated globes

Illuminated globes are made in a very similar manner except that the basic material is different. Artwork is printed on flat sheets of plastic substrate, this time with both hemispheres on the same sheet. The substrate is vacuum-formed into hemispheres by a one-of-a-kind machine that heats the plastic to thousands of degrees and sucks it into shape by applying a vacuum to the pliable plastic. The formed hemispheres are shipped offsite to an injection-molding factory where plastic is injected into them to harden the product. Space remains inside for the illumination source, and a hole is cut in Antarctica so the light bulb and socket can be inserted later. The two hemispheres are glued and taped together. The finished globe is so tough that it can actually be dribbled like a basketball on a concrete floor for five or six dribbles before it will break. The

The two halves are glued together to form a globe, which is then laminated for durability.

main advantages of owning an illuminated globe are that it is easier to read and it is more durable. The disadvantage is that cardboard used to make globes can be formed into a greater variety of products, including globes with topographic relief, and the vacuum-forming process for making the illuminated globes can only produce a smooth surface.

Quality Control

Technicians who manufacture globes are ISO 9000 certified and trained to ensure that each production step is consistent with established standards. Each production step is also a quality station. The technicians are responsible for rejecting products for any flaws, not just those occurring during their particular step of the process.

Byproducts/Waste

There are no byproducts from globe manufacture, although globe-makers often produce maps and related items. Waste is very limited. When the fiberboard is die-cut, the triangles that are removed are scrap; however, the cardboard is again recycled.

The Future

The globe's future is assured as a method of better understanding the changing face of the world we live in. Like manufacturers of many products, globe-makers face the challenge of identifying new ways of catching the public's fancy. Globes are often given as gifts to be used in specific settings (that is, by a student of a particular age or for business reference when the globe's mounting should match the office decor), and they must also be easy to use. Globes illustrate a tremendous amount of information, and manuals that are purchased with them need

to be useful tools. The newest types of globes are becoming interactive and speak the names of countries, as they are touched. They are also designed to present certain information for users in the "global village"; time zone information, for example, can help corporate leaders communicate with their international counterparts in a timely manner. One of the latest and most significant advances in globe making has already occurred, thanks to digital technology. The development and manufacture of the digital globe is described in a companion article.

Where to Learn More

Periodicals

Mickle, Linda, ed. *Map Report.* Kankakee, IL: International Map Trade Association.

Sell, Colleen T., ed. *Mercator's World: The Magazine of Maps, Exploration & Discovery.* Eugene, OR: Aster Publishing Company.

Other

Captain's Globes. http://www.finest1.com/globes/.

George F. Cram Company. http://www.georgefcram.com.

International Map Trade Association. http://www.maptrade.com.

Mercator's World. http://www.mercator-mag.com.

Motion Globes. http://www.motionglobes.com.

National Geographic Society. http://www.nationalgeographic.com.

—*Gillian S. Holmes*

Golf Club

Background

A golf club is used to strike the ball in the game of golf. It has a long shaft with a grip on one end and a weighted head on the other end. The head is affixed sideways at a sharp angle to the shaft, and the striking face of the head is inclined to give the ball a certain amount of upward trajectory. The rules of golf allow a player to carry up to 14 different clubs, and each one is designed for a specific situation during the game.

History

The origins of golf are shrouded in history and probably evolved from other games in which a small object was struck with a stick. The Romans had a game called *Paganica*, which involved hitting a stone with a stick. The French had a similar game called *chole*, while the English had cambuca, which used a ball made of wood. Possibly the strongest claim to golf comes from the Dutch, who were known to play a game called *kolf* as early as 1296. In its original form, *kolf* was played on any available terrain including churchyards, highways, and frozen lakes. The object was to hit a succession of targets by striking the ball with a long-handled wooden club. To allow a clear shot, the ball was slightly elevated on a pile of sand called a *tuitje*, from which we get the modern term tee.

The Dutch claim to the origin of the game is hotly disputed by the Scots who point out that they had been playing golf for as long or longer than the Dutch. Whatever the origin, there is no dispute that it was the Scots who popularized the game. It became so popular that in 1467 the Scottish Parliament passed an act banning golf because it was taking time from archery practice necessary for national defense. The ban was widely ignored. Ironically, the first manufactured golf club was made by a Scottish bow maker named William Mayne, who was appointed Clubmaker to the court of King James in 1603.

Early golf clubs were made entirely of wood. Not only was this material easy to shape, but it was also soft enough not to damage the stuffed leather golf balls that were used until the mid-1800s. With the introduction of the hard rubber gutta-percha golf ball in 1848, golfers no longer had to worry about damaging the ball and began using clubs with iron heads. Because iron heads could be formed with sharply inclined striking faces without losing their strength, iron-headed clubs, called irons, were most often used for making shorter, high-trajectory shots, while wooden-headed clubs, called woods, were used for making longer, low-trajectory shots.

Until the early 1900s, all golf clubs had wooden shafts whether they had iron heads or wooden heads. The first steel-shafted golf clubs were made in the United States in the 1920s. It was about this time that some club makers started using the current numbering system to identify different clubs, rather than the old colorful names. The woods were numbered one through five, and the irons were numbered two through nine. The higher the number, the more inclined the surface of the striking face. The putter rounded out the set of clubs and retained its name instead of being assigned a number. The sand wedge was developed in 1931 to help golfers blast their way out of traps. In time, the sand wedge was joined by several other specialty golf clubs.

In the early 1970s, manufacturers introduced golf clubs with shafts made from fiber-reinforced composite materials originally developed for military and aerospace applications. These shafts were much lighter than steel, but they were expensive and some golfers felt the new shafts flexed to much. Later, when ultrahigh-strength fibers were developed to control the flex, composite shafts gained more acceptance.

The first metal-headed drivers were developed in 1979. In 1989, they were followed by the first oversize metal-headed drivers. The oversize heads were cast with a hollow center and filled with foam, which made them the same weight as smaller wood heads. When combined with a longer, lightweight composite shaft, the oversize metal woods achieved a greater head velocity at impact and drove the ball further. The oversize club heads also had larger striking faces, which made them more forgiving if the ball was struck off-center.

Today, the design and manufacture of golf clubs is both an art and a science. Some club makers use the very latest computer-aided design and automated manufacturing techniques to build hundreds of thousands of clubs a year, while others rely on experience and hand-crafting skills to build only a few dozen custom-made clubs a year.

Raw Materials

Golf clubs are manufactured from a wide variety of materials, including metals, plastics, ceramics, composites, wood, and others. Different materials are chosen for different parts of the club based on their mechanical properties, such as strength, elasticity, formability, impact resistance, friction, damping, density, and others.

Club heads for drivers and other woods may be made from stainless steel, titanium, or graphite fiber-reinforced epoxy. Face inserts may be made from zirconia ceramic or a titanium metal matrix ceramic composite. Oversize metal woods are usually filled with synthetic polymer foam. Traditionalists can even buy woods that are made of real wood. Persimmon, laminated maple, and a host of exotic woods are used. Wood club heads are usually soaked in preserving oil or coated with a synthetic finish like polyurethane to protect them from moisture.

Club heads for irons and wedges may be made from chrome-plated steel, stainless steel, titanium, tungsten, beryllium nickel, beryllium copper, or combinations of these metals. Heads for putters may be made of all of the same materials as irons, plus softer materials like aluminum or bronze, because the velocity of impact is much slower when putting.

Club shafts may be made from chrome-plated steel, stainless steel, aluminum, carbon or graphite fiber-reinforced epoxy, boron fiber-reinforced epoxy, or titanium. Grips are usually made from molded synthetic rubber or wrapped leather.

Design

The rules of the United States Golf Association (USGA) have only a few brief paragraphs regarding the design of golf clubs. There are no restrictions on weight or materials, and only a few restrictions on dimensions. Shafts must be at least 18 in (457 mm) long. The distance from the heel to the toe of the head must be greater than the distance from the face to the back of the head. The cross-sectional dimension of the grip must not be greater than 1.75 in (45 mm) in any direction. Of all the rules, however, the most important one requires that the club "shall not be substantially different from the traditional and customary form and make."

It is this last rule that sometimes gives club designers the fits. It means, for example, that club heads may not have features like aiming fins or holes to reduce aerodynamic drag. Shafts may not have flexible joints, and so forth. In short, anything that is not "traditional and customary" is not allowed. All new club designs must be submitted to the USGA for review and approval before they may be used in tournament play.

Within the USGA guidelines, many new features have been incorporated into golf clubs. Using computer-aided design programs and mathematical models of club and ball dynamics, designers have learned to utilize new materials, redistribute weight, and

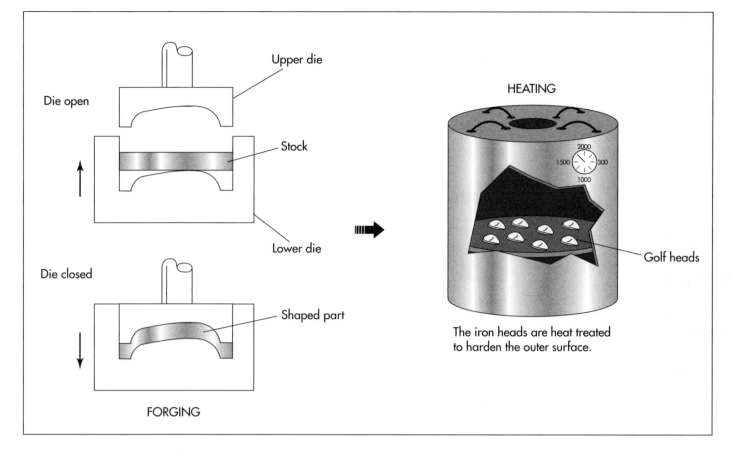

Die open

Die closed

FORGING

Upper die

Stock

Lower die

Shaped part

HEATING

Golf heads

The iron heads are heat treated to harden the outer surface.

alter the general shape of the club in an attempt to help both professional golfers and weekend duffers improve their games.

One common feature of modern irons is perimeter weighting, which places most of the club head weight around the edges, leaving the center with less material. This added mass reduces the amount of club twist when the ball is struck towards the edge of the club, rather than in the center. The effect is to increase the size of the effective hitting area, or the "sweet spot" as golfers call it. The hollow oversize metal heads on some drivers have the same effect.

Another design feature of some modern clubs is the offset head, where the striking face is located to the rear of the centerline of the shaft. This places the golfer's hands slightly ahead of the ball at impact, which tends to square the club face and give better direction control.

Other design features help golfers make cleaner shots from uneven terrain, get the ball up in the air from grassy lies, and correct their tendency to hit to one side or the

other. As with any product, some features offer more psychological help than physical help. Despite three decades of golf club design improvements, the driving distance of the best professional golfers increased only 12 yd (11 m) between 1968 and 1995, and the average winning score fell at a rate of only one stroke every 21 years.

The Manufacturing Process

Every golf club maker uses a slightly different manufacturing process. The largest companies use highly automated machinery, while the smallest companies use hand tools. Some parts of the manufacturing process may be unique to one company and regarded as trade secrets.

Here is a typical sequence of operations used to produce a machine-made, perimeter-weighted golf iron.

Forming the head

1 The head is formed by a process called investment casting. A master die of the

The golf club head is molded in a process called investment casting. Once cast, the head is heat treated to harden the iron.

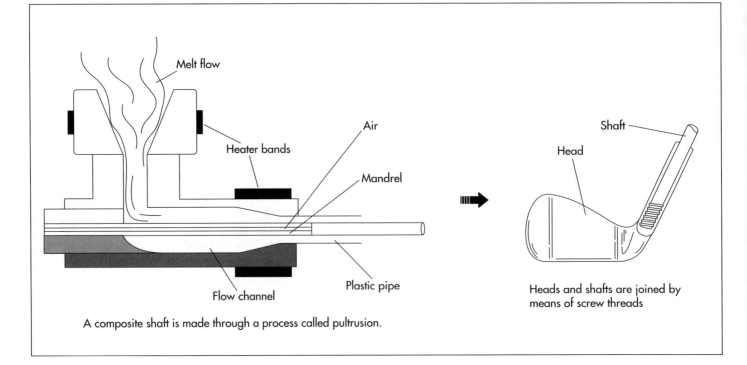

Melt flow

Heater bands

Air

Mandrel

Plastic pipe

Flow channel

A composite shaft is made through a process called pultrusion.

Shaft

Head

Heads and shafts are joined by means of screw threads

If the shaft is made of steel or stainless steel, it is formed by a process called tube drawing. The shaft is connected to the golf club head with screw threads.

club head is made from metal. The die consists of two halves with a hollow cavity that is the exact shape and size of the desired club head. Molten wax is poured into the die cavity and allowed to harden.

2 When the wax is hard, the die is opened, and the wax pattern is removed. This process is repeated several times. Several wax patterns are attached to a central wax column, called a sprue, to form a tree.

3 The tree is dipped into a liquid mixture of powdered ceramic material, various chemicals, and a gelling agent. It is set aside until the coating dries. The tree is then placed in a container, and the container is filled with a liquid molding slurry, which is allowed to harden.

4 The hardened mold is heated to about 1,000-2,000° F (550-1,100° C) in an oven to melt the wax patterns. The melted wax runs out the bottom and any wax residue is vaporized. The mold is then inverted.

5 Molten metal for the club head is poured into the hot mold and allowed to harden. When the metal has cooled, the mold material is broken away from the tree, and the individual cast heads are cut off the sprue. The investment casting process pro-

duces parts with an excellent surface finish and no flash or parting lines to remove. The parts can be made from a wide range of metals and their weight is uniform from one part to another.

6 Most iron heads are heated treated to harden the outer surface. The head is heated either with a flame or an induction coil, and then quickly cooled. This causes the steel near the surface to form a different grain structure that is much harder than the rest of the head.

Forming the shaft

7 If the shaft is made of steel or stainless steel, it is formed by a process called tube drawing. A tube of the desired length is pulled part way through an opening in a die slightly smaller than the tube diameter, which causes the drawn portion of the tube to neck down in diameter. This process is repeated several times. Each time the die diameter is made slightly smaller, and the length of tube pulled through the die is several inches less. The result is a tube that decreases in diameter from about 0.50 in (13 mm) to about 0.37 in (9.5 mm) in seven or eight small steps spaced along the length of the shaft. If the shaft is made of steel, it is chrome plated after it is formed.

8 If the shaft is made of graphite fiber-reinforced composite material, it is formed by a process called pultrusion. A bundle of graphite fibers is pulled through a circular opening in one or more heated dies while epoxy resin is forced through the opening at the same time. The graphite fibers become imbedded in the epoxy and the heat makes the epoxy harden to form the shaft. The shaft is then cooled by air or water and cut to length. Graphite fiber-reinforced shafts are the same diameter along the entire length.

Assembling the club

9 There are several ways to fasten the head to the shaft. With some metal shafts, the shaft is inserted into the socket on the head and a small hole is drill crossways through both the socket and the shaft. A small metal pin is then pressed into the hole and held in place with an epoxy adhesive. With graphite shafts, the head is bonded to the shaft with an adhesive. This second process is becoming more common for all shaft materials, including metal shafts.

10 The other end of the shaft is placed in a hollow die and a rubber grip is molded around its upper portion. The shaft may then be labeled with an adhesive sticker to show the manufacturer, brand name, degree of flex, or other information.

11 The raised metal parts are polished to give the club a finished appearance. As a final step, any recessed lettering or logos on the club head may be filled with paint or another color finish. Adhesive stickers or adhesive-backed metal plates may be affixed to the club head for identification or decoration as well.

Quality Control

Golf clubs are treated with almost as much attention to specifications as components for aircraft. In fact most golf club manufacturers emphasize their specifications as a means of differentiating their clubs from the competition. Swing weight, lie angle, shaft torque, and a host of other specifications are not only important to the club designers, but are also important to the company's customers. In addition to dimensional checks and process controls, clubs are randomly tested for a variety of specifications that affect performance.

The Future

The popularity of golf is expected to continue to grow. As the number of recreational players increases, there will be an emphasis on designing clubs that make the game more enjoyable for the average golfer. Despite objections from purists, oversize club heads and other game-improving features will continue to be offered.

Where to Learn More

Books

Plumridge, Chris. *The Illustrated Encyclopedia of World Golf.* Exeter Books, 1988.

Zumerchick, John, editor. *Encyclopedia of Sports Science.* Simon & Schuster MacMillan, 1997.

Periodicals

Crecca, Donna Hood. "Fore!" *Popular Science* (February 1995): 56-60, 84.

Sauerhaft, Rob. "Easier than Ever." *Golf Magazine* (October 1994): 56-57.

Sauerhaft, Rob. "Iron Wars." *Golf Magazine* (April 1998): 168-169.

Other

Callaway Golf . http://www.callawaygolf.com.

Cobra Golf, Inc. http://www.cobragolf.com.

Karsten Manufacturing Corporation. http://www.pinggolf.com.

United States Golf Association (USGA). http://www.usga.org.

—*Chris Cavette*

Hair Remover

The earliest recorded use of hair removers is found in ancient India where hair removal was highly desirable. Abrasive pastes and resinous plasters were frequently used to physically remove hair.

Hair removers, or depilatories, are products designed to chemically or physically remove undesirable hair from areas on the body. Hair removers are made by mixing together the appropriate raw materials in large stainless steel tanks and then filling them into individual packages. In use for thousands of years, they continue to be an important part of many people's everyday hygiene. Currently, new hair removers are being investigated which are less irritating, more effective, and longer lasting.

Background

Epilatories were the first type of hair removers. The most common of these products is an epilating wax. This product is heated and spread on the skin in the desired area. It is then allowed to cool and harden. The mass of wax is then rapidly removed, pulling with it about 80% of the hairs. It is a slightly painful procedure and a mild antiseptic is typically applied to protect against skin irritation. Epilatories have not been widely used by individual consumers, however they are popular in beauty salons.

While epilatories continue to be an important method of hair removal, depilatories are much more common in personal care. Depilatories rely on a chemical reaction between materials in the formula with components of the hair. When the depilatory is applied to the skin, a component in the product, such as thioglycolic acid, reacts with the protein in the hair and weakens it. The hair can then be removed from the skin by gentle wiping, scraping, or rinsing. This is effective on any part of the hair structure that is above the level of the skin.

The compounds in the depilatories, which react with hair, also react with protein in the skin, albeit at a much slower rate. For this reason, depilatories must be left on the skin for only a short while. The manufacturers of depilatories realize this and strive to develop formulas, which have only minimal negative effects on skin. Typically, if a consumer follows the directions as stated on the package, no problems will arise. Epilatories generally will not have a negative effect other than physical irritation on skin since they do not rely on a chemical reaction to function properly.

The first step in producing a hair remover, or any personal care product for that matter, is developing a formula. Cosmetic chemists use their knowledge of standard cosmetic ingredients, consumer research information and various other types of information to construct their formulas. Since hair removers can be sold in many different forms including creams, gels, lotions, and aerosols, the formula must be adapted to the product form. The formulas are first prepared in small beakers in the lab so aspects of the formula can be evaluated. Tests for product effectiveness, stability, and safety are all completed at this point. Other studies such as consumer acceptance testing may also be completed.

History

Hair removal from various parts of the body has been an important part of beauty for thousands of years. The earliest recorded use of hair removers is found in ancient India where hair removal was highly desirable. This society frequently used abrasive pastes and resinous plasters to physically remove hair. In the Middle East, a lime mixture was used for a similar purpose. Other materials, such as antimony and arsenic compounds, were also used; however, it is now known that these materials are quite toxic, and their use has been discontinued.

The earliest hair removers used a method of hair removal known as epilation, or physically pulling hair out. Common procedures included using devices like tweezers to pull hair out selectively or waxes which pulled hair out in large masses. Since hair removal by physical means was often a painful experience, scientists worked on developing formulas, which would chemically remove hair. Little progress was made in this area until about 100 years ago when it was first reported that barium sulfide was used for this purpose. A few years later a similar idea was patented in the United States. Cream depilatories were first introduced in the 1920s and many more patents were issued during the 1930s. Thioglycolate depilatories, which were first introduced in 1938, have become the most important hair removers.

Raw Materials

There are many different materials that have been used in hair remover formulas. Some of these materials are responsible for the hair removing properties of the product while others are needed to improve the product's aesthetics.

As suggested, depilatories and epilatories remove hair in a very different manner. Obviously, they then require different compounds to function. A standard epilatory may be composed of a wax such as beeswax and a sticky, polymeric resin. The wax provides the setting action needed for peeling the product off the skin and the resin helps bind the material to the hair. The active ingredients used for depilatories include thioglycolate salts and sulfides. Thioglycolate salts include materials such as calcium thioglycolate and potassium thioglycolate. In an aqueous solution at the proper pH, they are converted to an acid, which then affects the hair. Sulfides such as barium sulfide or strontium sulfide are also used because they react more rapidly than thioglycolates however, they have other characteristics which

make them less appealing. Since pH is critical to the proper performance of depilatories, ingredients such as sodium hydroxide or calcium hydroxide, which adjust the pH, are included.

In addition to the hair removing ingredients, other compounds are necessary to complete formulation. This includes diluents, emollients, thickeners, fragrances, and colorants. Water is used most often as a diluent for depilatories because it is compatible with a large range of raw materials, non-irritating, and inexpensive. Since epilatories are waxes, they are not compatible with water so mineral oil is typically used as the diluent. Emollients are included in formulations to reduce the harshness of the formula and improve the feel. Materials like oils, silicones, and esters are all examples of commonly used emollients. Depending on the product's form, a thickening agent may be required. These materials are typically polymers, surfactants, or modified clays. For aerosol products, a propellant is needed.

To improve the aesthetics of the formula, fragrances are included. These fragrances must be specially designed to overcome the generally offensive odor of the hair removal ingredients. For cream or lotion products, emulsifiers are needed and dyes are used to modify the color. Various other ingredients such as preservatives, antioxidants, extracts may also be included.

Beyond the ingredients that go into the hair remover formula, packaging components are another important raw material. Bottles are primarily used and are made of plastics such as polyvinyl chloride (PVC) or high-density polyethylene (HDPE). For aerosol products, a steel or aluminum can is used. The outer graphics can be either directly silk screened on to the package or an adhesive label can be applied.

The Manufacturing Process

The process for making a hair remover can be divided into two steps. First, a large batch of the product is made, and then it is filled into the individual containers. While there are many different product forms the hair remover may take such as creams, aerosols, or waxes, the following descrip-

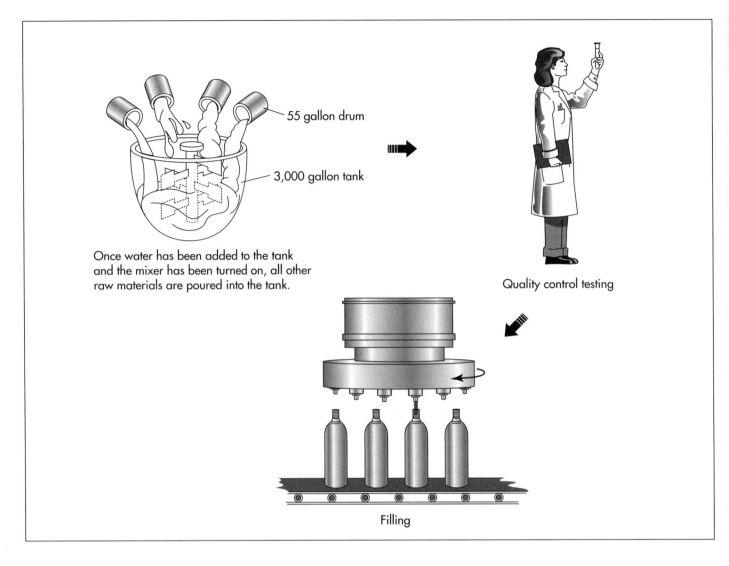

55 gallon drum

3,000 gallon tank

Once water has been added to the tank
and the mixer has been turned on, all other
raw materials are poured into the tank.

Quality control testing

Filling

tion will only outline the method for making
a lotion depilatory.

Compounding

1 The bulk batches of depilatories are produced in a designated area of the manufacturing plant. Plant workers follow a
standard formula to make batches, which
can be over 3,000 gal (11,355 l). The tanks,
which are stainless steel, are equipped with
a large mixer and a heating and cooling system. The temperature and the mixer speed
are both computer controlled, so the compounder can modify them who is making
the batch.

When the controls are set, the raw materials
can be added. In most formulas, water is
added first by being pumped in at the appropriate volume. The mixer is then turned on

and the other raw materials are added in the
order called for in the formula. These raw
materials are either poured into the batch
from bags or 55 gal (208.2 l) drums. As
each goes into the tank, it is thoroughly
mixed. Depending on the formula, the batch
is heated and cooled as necessary to help the
raw materials combine more quickly. A
typical 3,000 gal (11,355 l) batch of depilatory may take anywhere from two to five
hours to make.

Checking quality

2 After the batch is finished, the quality
control department must test it before it
can be sent along for filling. Physical characteristics are examined to ensure the batch
conforms to the specifications outlined in
the formula instructions. Typical tests done
on a depilatory batch include viscosity

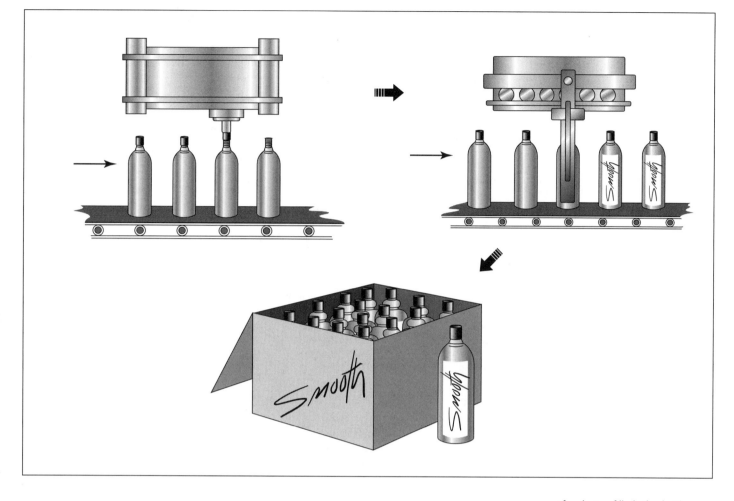

checks, pH determination, and appearance and odor evaluations. They may also check the activity of the thioglycolate. If the batch does not meet the ranges set for the specifications, sometimes adjustments to the formula can be made. For instance if the pH is too low, a depilatory will not function properly. Therefore, a certain amount of a base such as sodium hydroxide could be added to adjust the pH. Salt can be added to increase the thickness of many of these products. Fragrance and color may also be adjusted at this point. After quality control approves the batch, it is pumped out of the main batch tank into a holding tank where it can be stored until it gets filled. From the holding tank it may be pumped into a carousel-style, piston head filler.

Filling

3 The depilatory is filled on a filling line, which is a series of machines that connected, by a conveyor belt system. At the start of the filling line, empty bottles are put in a hopper. This hopper is a large bin, which contains a device that can physically manipulate the bottles so they are standing upright when they come out. From this bottle-sorting hopper, they are moved along a conveyor belt to the filling machine.

4 The filling machine contains a finished batch of depilatory. It is a carousel is made up of a series of piston filling heads that are programmed to deliver the correct amount of product into the bottles. When a bottle passes under the filling machine, product is pumped into it.

5 After being filled, the bottles are moved to the capping machine. Just like the bottles, the caps are stored in a hopper that is designed to physically align them in the right order. As the bottles move by, the capping machine automatically attaches and tightens the caps.

6 Next, the bottles move to the labeling machines (if necessary) and then on to

After being filled, the bottles are capped, labeled, and boxed for shipment.

the boxing area. In the boxing area, the products are lined up and put into boxes. The boxes are stacked onto pallets and hauled away in large trucks to distributors. The entire filling process can produce more than 500 bottles per minute or more.

Quality Control

To produce a consistent product, quality control inspections are done throughout the manufacturing process. At the start, the raw materials are checked to ensure that they meet the manufacturer's specifications. Typically, quality control chemists sample incoming raw materials and run numerous tests before qualifying them. These tests may include checks for appearance, pH, odor, or viscosity. More complex testing may also be performed. During manufacture, the batch of hair remover is periodically tested to make sure that a functional product will be produced.

While the product is being filled, quality inspectors are stationed along the entire filling line. These workers watch the containers as they pass by and pull off any which are defective. This includes those that are inadequately filled, have misplaced labels, or otherwise damaged. Regulations also require that samples be periodically checked for microbial contamination during filling.

The Future

The market for hair removers is relatively small compared to other personal care products such as shampoos or conditioners, so only minimal research is currently being pursued. The focus of this research has been on making products that are less irritating and more moisturizing to the skin, lower in odor and more effective. Irritation is likely to always be a problem for all depilatories that chemically alter proteins so compounds, which weaken hair in other ways, may be developed. Beyond depilatories and epilatories, new drugs could be developed which can inhibit the growth of hair from follicles. This might represent a kind of permanent hair remover.

Where to Learn More

Books

Knowlton, John and Steven Pearce. *The Handbook of Cosmetic Science and Technology.* Oxford: Elsevier Science Publishers, 1993.

Umbach, Wilfried. *Cosmetics and Toiletries Development, Production, and Use.* New York: Ellis Horwood, 1991.

Periodicals

Breuer, Hans. "Depilatories." *Cosmetics & Toiletries* 105 (April 1990): 61-66.

—*Perry Romanowski*

Hammer

Background

A hammer is a handheld tool used to strike another object. It consists of a handle to which is attached a heavy head, usually made of metal, with one or more striking surfaces. There are dozens of different types of hammers. The most common is a claw hammer, which is used to drive and pull nails. Other common types include the ball-peen hammer and the sledge hammer.

The concept of using a heavy object to strike another object predates written history. The use of simple tools by our human ancestors dates to about 2,400,000 B.C. when various shaped stones were used to strike wood, bone, or other stones to break them apart and shape them. Stones attached to sticks with strips of leather or animal sinew were being used as axes or hammers by about 30,000 B.C. during the middle of the Old Stone Age.

The dawn of the Bronze Age brought a shift from stone to metal in the toolmaker's art. By about 3,000 B.C., axes with bronze or copper heads were being made in Mesopotamia, in what is now Iraq. The heads had a hole where a handle could be inserted and fastened. Nails made of copper or bronze were being used in the same area during the same period, suggesting that hammers with metal heads may have also existed. By about 200 B.C., Roman craftsmen used several types of iron-headed hammers for wood working and stone cutting. A Roman claw hammer dating from about 75 A.D. had a striking surface on one side of the head, and a split, curved claw for pulling nails on the other side. It's appearance is so much like a modern claw hammer that you

might expect to find it in a hardware store, rather than a museum.

With the development of commerce and the specialization of trades, many different hammer designs evolved. Coachbuilders, wheelwrights, blacksmiths, bricklayers, stone masons, cabinetmakers, barrel makers (coopers), shoe makers (cobblers), ship builders, and many other craftsmen designed and used their own unique hammers. In 1840, a blacksmith in the United States named David Maydole introduced a claw hammer with the head tapering downwards around the opening for the handle. This provided additional bearing surface for the handle and prevented it from being wrenched loose when the hammer was used to pull nails. His hammer became so popular that his blacksmith shop grew into a factory to keep up with the demand. Most claw hammers made today use this same design.

Modern hammers come in a variety of shapes, materials, and weights. Although some specialty hammers are no longer used, there is still a wide array of hammer configurations as new designs are developed for new applications.

Types of Hammers

In general, hammers have metal heads and are used to strike metal objects. The curved claw hammer used to drive nails into wood is one example. Other hammers include the framing hammer with a straight claw that can be driven between nailed boards to pry them apart. It is often used in heavy construction where temporary forms or supports must be removed. The ball peen hammer has a semi-spherical end and is used to

A Roman claw hammer dating from about 75 A.D. had a striking surface on one side of the head, and a split, curved claw for pulling nails on the other side. It's appearance is so much like a modern claw hammer that you might expect to find it in a hardware store, rather than a museum.

251

shape metal. A tack hammer is one of the smallest hammers. It is used by upholsterers to drive small tacks into wood furniture frames. A sledge hammer is one of the largest hammers. It usually has a long handle and is used for driving spikes and other heavy work. Other modern hammers include brick hammers, riveting hammers, welder's hammers, hand drilling hammers, engineer's hammers, and many others.

A related class of hammer-like tools are called mallets. They have large heads made of rubber, plastic, wood, or leather. Mallets are used to strike objects that would be damaged by a blow from a metal hammer. Rubber mallets are used to assemble furniture or to beat dents out of metal. Wood and leather mallets are used to strike wood handled chisels. Plastic mallets have smaller heads and are used to drive small pins into machinery. A very large wooden mallet is sometimes called a maul.

Design

The two major components of a hammer are the head and the handle. The design of these two components depends on the specific application, but all hammers have many common features.

The striking surface of the head is called the face. It may be flat, called plain faced, or slightly convex, called bell faced. A bell-faced hammer is less likely to bend a nail if the nail is struck at an angle. Another face design is called a checkered face. It has crosshatched grooves cut into the surface to prevent the hammer from glancing off the nail head. Because it leaves a checkered impression on the wood, it is usually only found on framing hammers used for rough construction.

The surface of the head around the face is called the poll. The poll is connected to the main portion of the head by the slightly tapered neck. The hole where the handle fits into the head is called the adze (adz) eye. The side of the head next to the adze eye is called the cheek.

On the opposite end of the head, there may be a claw, a pick, a semi-spherical ball peen, or a tapered cross peen depending on the

type of hammer. There may also be a second face, as in a double-faced sledge hammer.

Hammers are classified by the weight of the head and the length of the handle. The common curved claw hammer has a 7-20 oz (0.2-0.6 kg) head and a 12-13 in (30.5-33.0 cm) handle. A framing hammer, which normally drives much larger nails, has a 16-28 oz (0.5-0.8 kg) head and a 12-18 in (30.5-45.5 cm) handle.

Raw Materials

Hammer heads are made of high carbon, heat-treated steel for strength and durability. The heat treatment helps prevent chipping or cracking caused by repeated blows against other metal objects. Certain specialty hammers may have heads made of copper, brass, babbet metal, and other materials. Dead-blow hammers have a hollow head filled with small steel shot to give maximum impact with little or no rebound.

The handles may be made from wood, steel, or a composite material. Wood handles are usually made of straight-grained ash or hickory. These two woods have good cross-sectional strength, excellent durability, and a certain degree of resilience to absorb the shock of repeated blows. Steel handles are stronger and stiffer than wood, but they also transmit more shock to the user and are subject to rust. Composite handles may be made from fiberglass or graphite fiber-reinforced epoxy. These handles offer a blend of stiffness, light weight, and durability.

Steel and composite handles usually have a contoured grip made of a synthetic rubber or other elastomer. Wood handles do not have a separate grip. Steel and composite handles may also be encased in a high-impact polycarbonate resin. The addition of this material around the handle increases shock absorption, improves chemical resistance, and offers protection against accidental overstrikes. An overstrike is when the hammer head misses the nail and the handle takes the impact instead. This is a common cause of handle failure.

There are several materials and methods used to attach the head to the handle. Wood handle hammers use a single thin wood wedge driven diagonally into the upper end

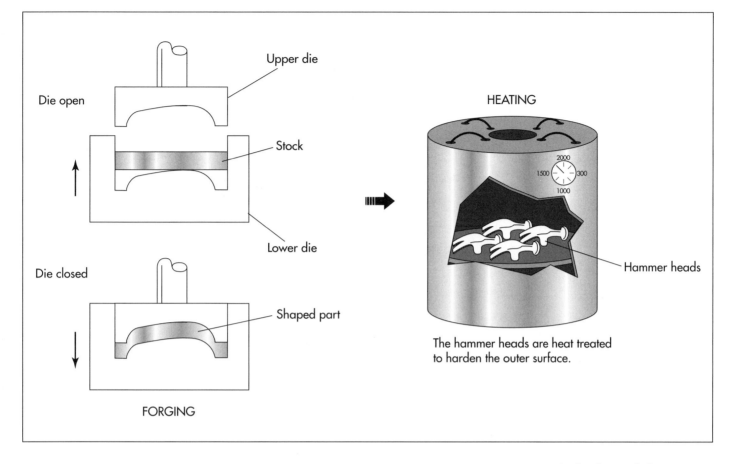

Die open

Upper die

Stock

Lower die

Die closed

Shaped part

FORGING

HEATING

Hammer heads

The hammer heads are heat treated to harden the outer surface.

of the handle, with two steel wedges driven through the wood wedge at right angles to secure it in place.

The Manufacturing Process

The manufacturing process varies from one company to another depending on the company's production capacity and proprietary methods. Some companies make their own handles, while others purchase the handles from outside suppliers.

Here is a typical sequence of operations for making a claw hammer.

Forming the head

1 The head is made by a process called hot forging. A length of steel bar is heated to about 2,200-2,350° F (1,200-1,300° C). This may be done with open flame torches or by passing the bar through a high-power electrical induction coil.

2 The hot bar may then be cut into shorter lengths, called blanks, or it may be fed continuously into a hot forge. The bar or blanks are positioned between two formed cavities, called dies, within the forge. One die is held in a fixed position, and the other is attached to a movable ram. The ram forces the two dies together under great pressure, squeezing the hot steel into the shape of the two cavities. This process is repeated several times using different shaped dies to gradually form the hammer head. The forging process aligns the internal grain structure of the steel and provides much stronger and more durable piece.

3 During this process, some of the hot steel squeezes out around the edges of the die cavities to form flash, which must be removed. As a final step the head is placed between two trimming dies, which are forced together to cut off any protruding flash. The head is then cooled, and any rough spots are ground smooth.

4 In order to prevent chipping and cracking of the hammer head in service, the face, poll, and claws are heat treated to harden them. This is done by heating those

The head is made by a process called hot forging. A length of steel bar is heated to about 2,200-2,350° F (1,200-1,300° C) and then die cut in the shape of the hammer head. Once cut, the hammer head is heat treated to harden the steel.

In order to make a wooden handle, the wood is cut to the desired length and then shaped into a handle on a lathe.

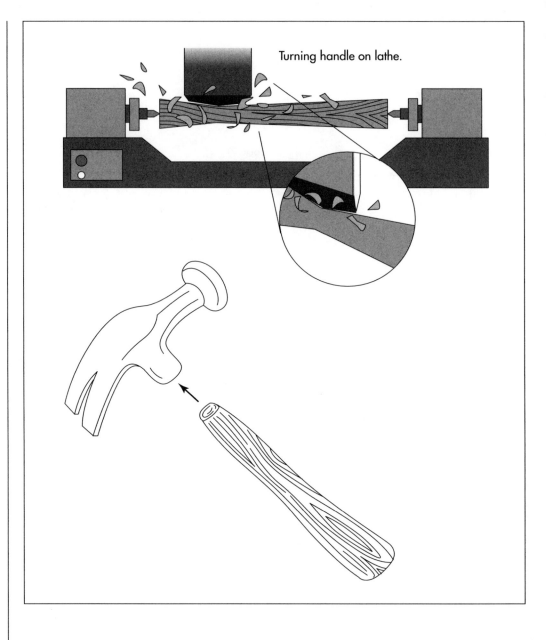

Turning handle on lathe.

areas, either with a flame or an induction coil, and then quickly cooling them. This causes the steel near the surface to form a different grain structure that is much harder than the rest of the head.

5 The heads are cleaned with a stream of air containing small steel particles. this process is called shot blasting. The head may then be painted.

6 The face, poll, claws, and cheeks are polished smooth. This removes the paint in those areas. As part of this operation, the v-shaped slot in the claws is smoothed using an abrasive disc.

Forming the handle

7 If the hammer has a wood handle, it is formed on a lathe. A piece of wood is cut to the desired length and secured at each end in the lathe. As the wood spins around the long axis of the handle, a cutting tool moves in and out rapidly to cut the handle profile. The position of the cutting tool is driven by a cam that has the same shape as the finished handle. As the cutting tool moves down the length of the handle, it follows the shape of the cam and cuts the handle to match it. The finished handle is clamped in a holding device and a slot is cut diagonally across the top of the handle. The handle is then sanded to give it a smooth surface.

8 If the hammer has a steel-core handle, the core is formed by heating a bar of steel, until it becomes plastic, and forcing it through an opening that has the desired cross-sectional shape. This process is called extrusion. If the hammer has a graphite fiber-reinforced core, the core is formed by gathering together a bundle of graphite fibers and pulling them through an opening that has the desired cross-sectional shape while epoxy resin is forced through the opening at the same time. This process is called pultrusion. In either case, the core may then have a protective plastic jacket molded around it.

Assembling the hammer

9 If the hammer has a wood handle, the handle is inserted up through the adze eye of the head. A wood wedge is tapped down into the diagonal slot on the top of the handle to force the two halves outward to press against the head. This provides sufficient friction to hold the head on the handle. The wood wedge is secured in place with two smaller steel wedges driven through it crossways. The handle may then be stenciled with ink or labeled with an adhesive sticker to show the manufacturer, brand name, or other information.

10 If the hammer has a steel or graphite fiber-reinforced core, the handle is inserted up through the adze eye of the head. Liquid epoxy resin is then poured through the top of the hole to bond the handle in place. The handle is placed in a hollow die and a rubber grip is molded around its lower portion. The handle may then be labeled with an adhesive sticker to show the manufacturer, brand name, or other information.

Quality Control

In addition to the normal visual inspections and dimensional measurements, various steps in the manufacturing process are monitored. Probably the most important step is the heat treatment used to harden portions of the head. The temperatures and rate of heating and cooling are critical in forming the proper hardness, and the entire operation is closely controlled.

The Future

Having survived for thousands of years, it is unlikely that the hammer will disappear from civilization's toolbox anytime soon. It does have some serious competition though. The most formidable competitor is the gas-driven nail gun. This device uses a compressed gas, usually air, to drive a nail into wood with a single shot. Although nail guns are heavier and more expensive than hammers, they are also significantly faster. This is especially true in repetitive nailing operations such as installing floor or roof sheathing for new home construction. Nail guns are also favored in areas where noise is a concern. Because a nail gun can drive a nail in a single shot, it produces much less overall noise than the five or six hammer blows it takes to drive a nail.

Where to Learn More

Books

Salaman, R.A. *Dictionary of Tools.* Charles Scribner's Sons, 1975.

Vila, Bob. *This Old House Guide to Building and Remodeling Materials.* Warner Books, Inc., 1986.

Periodicals

Capotosto, Rosario. "Hammer Basics." *Popular Mechanics* (October 1996): 104-107.

Neary, John. "When rules and drills drive you just plane screwy." *Smithsonian* (February 1991): 52-60, 62-65.

Other

Stanley Tools. http://www.stanleytools.com.

—Chris Cavette

Helium

As of 1993, there were about 35 billion cubic feet (1.0 billion cubic meters) of helium in government storage.

Background

Helium is one of the basic chemical elements. In its natural state, helium is a colorless gas known for its low density and low chemical reactivity. It is probably best known as a non-flammable substitute for hydrogen to provide the lift in blimps and balloons. Because it is chemically inert, it is also used as a gas shield in robotic arc welding and as a non-reactive atmosphere for growing silicon and germanium crystals used to make electronic semiconductor devices. Liquid helium is often used to provide the extremely low temperatures required in certain medical and scientific applications, including superconduction research.

Although helium is one of the most abundant elements in the universe, most of it exists outside of Earth's atmosphere. Helium wasn't discovered until 1868, when French astronomer Pierre Janssen and English astronomer Sir Joseph Lockyer were independently studying an eclipse of the Sun. Using spectrometers, which separate light into different bands of color depending on the elements present, they both observed a band of yellow light that could not be identified with any known element. News of their findings reached the scientific world on the same day, and both men are generally credited with the discovery. Lockyer suggested the name helium for the new element, derived from the Greek word *helios* for the sun.

In 1895, English chemist Sir William Ramsay found that cleveite, a uranium mineral, contained helium. Swedish chemists P.T. Cleve and Nils Langlet made a similar discovery at about the same time. This was the first time helium had been identified on Earth. In 1905, natural gas taken from a well near Dexter, Kansas, was found to contain as much as 2% helium. Tests of other natural gas sources around the world yielded widely varying concentrations of helium, with the highest concentrations being found in the United States.

During the early 1900s, the development of lighter-than-air blimps and dirigibles relied almost entirely on hydrogen to provide lift, even though it was highly flammable. During World War I, the United States government realized that non-flammable helium was superior to hydrogen and declared it a critical war material. Production was tightly controlled, and exports were curtailed. In 1925, the United States passed the first Helium Conservation Act which prohibited the sale of helium to nongovernmental users. It wasn't until 1937, when the hydrogen-filled dirigible Hindenburg exploded while landing at Lakehurst, New Jersey, that the restrictions were lifted and helium replaced hydrogen for commercial lighter-than-air ships.

During World War II, helium became a critical war material again. One of its more unusual uses was to inflate the tires on long-range bomber aircraft. The lighter weight of helium allowed the plane to carry 154 lb (70 kg) of extra fuel for an extended range.

After the war, demand for helium grew so rapidly that the government imposed the Helium Act Amendments in 1960 to purchase and store the gas for future use. By 1971, the demand had leveled off and the helium storage program was canceled. A few years later, the government started storing helium again. As of 1993, there were about 35 billion cubic feet (1.0 billion cubic meters) of helium in government storage.

Today, the majority of the helium-bearing natural gas sources are within the United States. Canada, Poland, and a few other countries also have significant sources.

Raw Materials

Helium is generated underground by the radioactive decay of heavy elements such as uranium and thorium. Part of the radiation from these elements consists of alpha particles, which form the nuclei of helium atoms. Some of this helium finds its way to the surface and enters the atmosphere, where it quickly rises and escapes into space. The rest becomes trapped under impermeable layers of rock and mixes with the natural gases that form there. The amount of helium found in various natural gas deposits varies from almost zero to as high as 4% by volume. Only about one-tenth of the working natural gas fields have economically viable concentrations of helium greater than 0.4%.

Helium can also be produced by liquefying air and separating the component gases. The production costs for this method are high, and the amount of helium contained in air is very low. Although this method is often used to produce other gases, like nitrogen and oxygen, it is rarely used to produce helium.

The Manufacturing Process

Helium is usually produced as a byproduct of natural gas processing. Natural gas contains methane and other hydrocarbons, which are the principal sources of heat energy when natural gas is burned. Most natural gas deposits also contain smaller quantities of nitrogen, water vapor, carbon dioxide, helium, and other non-combustible materials, which lower the potential heat energy of the gas. In order to produce natural gas with an acceptable level of heat energy, these impurities must be removed. This process is called upgrading.

There are several methods used to upgrade natural gas. When the gas contains more than about 0.4% helium by volume, a cryo-genic distillation method is often used in order to recover the helium content. Once the helium has been separated from the natural gas, it undergoes further refining to bring it to 99.99+% purity for commercial use.

Here is a typical sequence of operations for extracting and processing helium.

Pretreating

Because this method utilizes an extremely cold cryogenic section as part of the process, all impurities that might solidify—such as water vapor, carbon dioxide, and certain heavy hydrocarbons—must first be removed from the natural gas in a pretreatment process to prevent them from plugging the cryogenic piping.

1 The natural gas is pressurized to about 800 psi (5.5 MPa or 54 atm). It then flows into a scrubber where it is subjected to a spray of monoethanolamine, which absorbs the carbon dioxide and carries it away.

2 The gas stream passes through a molecular sieve, which strips the larger water vapor molecules from the stream while letting the smaller gas molecules pass. The water is back-flushed out of the sieve and removed.

3 Any heavy hydrocarbons in the gas stream are collected on the surfaces of a bed of activated carbon as the gas passes through it. Periodically the activated carbon is recharged. The gas stream now contains mostly methane and nitrogen, with small amounts of helium, hydrogen, and neon.

Separating

Natural gas is separated into its major components through a distillation process known as fractional distillation. Sometimes this name is shortened to fractionation, and the vertical structures used to perform this separation are called fractionating columns. In the fractional distillation process, the nitrogen and methane are separated in two stages, leaving a mixture of gases containing a high percentage of helium. At each stage the level of concentration, or fraction, of each component is increased until the separation is complete. In the natural gas

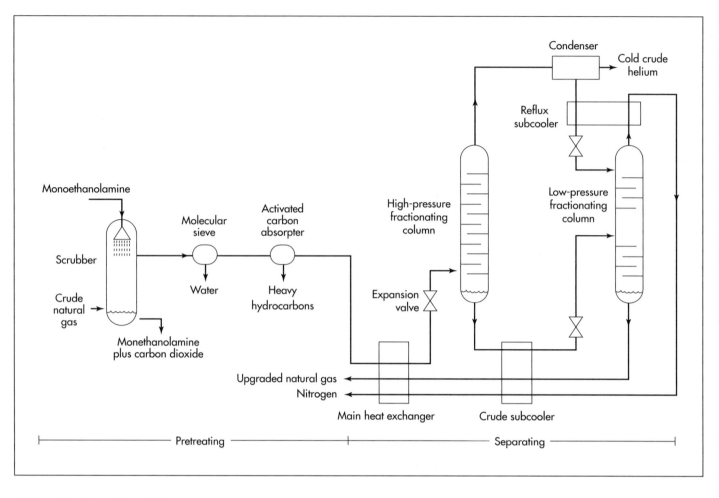

All impurities that might solidify and clog the cryogenic piping is removed from the natural gas in a pretreatment process. After pretreatment, the natural gas components are separated in a process called fractional distillation.

industry, this process is sometimes called nitrogen rejection, since its primary function is to remove excess quantities of nitrogen from the natural gas.

4 The gas stream passes through one side of a plate fin heat exchanger while very cold methane and nitrogen from the cryogenic section pass through the other side. The incoming gas stream is cooled, while the methane and nitrogen are warmed.

5 The gas stream then passes through an expansion valve, which allows the gas to expand rapidly while the pressure drops to about 145-360 psi (1.0-2.5 MPa or 10-25 atm). This rapid expansion cools the gas stream to the point where the methane starts to liquefy.

6 The gas stream—now part liquid and part gas—enters the base of the high-pressure fractionating column. As the gas works its way up through the internal baffles in the column, it loses additional heat. The

methane continues to liquefy, forming a methane-rich mixture in the bottom of the column while most of the nitrogen and other gases flow to the top.

7 The liquid methane mixture, called crude methane, is drawn out of the bottom of the high-pressure column and is cooled further in the crude subcooler. It then passes through a second expansion valve, which drops the pressure to about 22 psi (150 kPa or 1.5 atm) before it enters the low-pressure fractionating column. As the liquid methane works its way down the column, most of the remaining nitrogen is separated, leaving a liquid that is no more than about 4% nitrogen and the balance methane. This liquid is pumped off, warmed, and evaporated to become upgraded natural gas. The gaseous nitrogen is piped off the top of the low-pressure column and is either vented or captured for further processing.

8 Meanwhile, the gases from the top of the high-pressure column are cooled in a

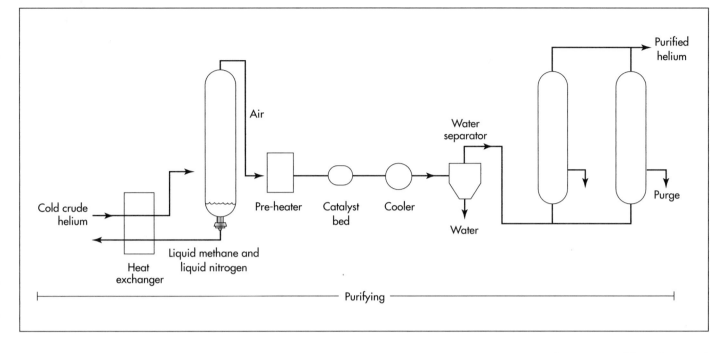

Purifying

condenser. Much of the nitrogen condenses into a vapor and is fed into the top of the low-pressure column. The remaining gas is called crude helium. It contains about 50-70% helium, 1-3% unliquefied methane, small quantities of hydrogen and neon, and the balance nitrogen.

Purifying

Crude helium must be further purified to remove most of the other materials. This is usually a multi-stage process involving several different separation methods depending on the purity of the crude helium and the intended application of the final product.

9 The crude helium is first cooled to about -315° F (-193° C). At this temperature, most of the nitrogen and methane condense into a liquid and are drained off. The remaining gas mixture is now about 90% pure helium.

10 Air is added to the gas mixture to provide oxygen. The gas is warmed in a preheater and then it passes over a catalyst, which causes most of the hydrogen in the mixture to react with the oxygen in the air and form water vapor. The gas is then cooled, and the water vapor condenses and is drained off.

11 The gas mixture enters a pressure swing adsorption (PSA) unit consisting of several adsorption vessels operating in parallel. Within each vessel are thousands of particles filled with tiny pores. As the gas mixture passes through these particles under pressure, certain gases are trapped within the particle pores. The pressure is then decreased and the flow of gas is reversed to purge the trapped gases. This cycle is repeated after a few seconds or few minutes, depending on the size of the vessels and the concentration of gases. This method removes most of the remaining water vapor, nitrogen, and methane from the gas mixture. The helium is now about 99.99% pure.

Distributing

Helium is distributed either as a gas at normal temperatures or as a liquid at very low temperatures. Gaseous helium is distributed in forged steel or aluminum alloy cylinders at pressures in the range of 900-6,000 psi (6-41 MPa or 60-410 atm). Bulk quantities of liquid helium are distributed in insulated containers with capacities up to about 14,800 gallons (56,000 liters).

12 If the helium is to be liquefied, or if higher purity is required, the neon and any trace impurities are removed by passing the gas over a bed of activated carbon in a

Once separated from the natural gas, crude helium is purified in a multi-stage process involving several different separation methods depending on the purity of the crude helium and the intended application of the final product.

259

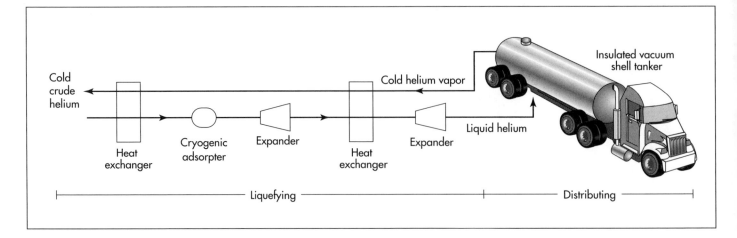

Cold crude helium

Cold helium vapor

Insulated vacuum shell tanker

Heat exchanger

Cryogenic adsorpter

Expander

Heat exchanger

Expander

Liquid helium

Liquefying

Distributing

Helium is distributed either as a gas at normal temperatures or as a liquid at very low temperatures.

cryogenic adsorber operating at about -423° F (-253° C). Purity levels of 99.999% or better can be achieved with this final step.

13 The helium is then piped into the liquefier, where it passes through a series of heat exchangers and expanders. As it is progressively cooled and expanded, its temperature drops to about -452° F (-269° C) and it liquefies.

14 Large quantities of liquid helium are usually shipped in unvented, pressurized containers. If the shipment is within the continental United States, shipping time is usually less than a week. In those cases, the liquid helium is placed in large, insulated tank trailers pulled by truck tractors. The tank body is constructed of two shells with a vacuum space between the inner and outer shell to retard heat loss. Within the vacuum space, multiple layers of reflective foil further halt heat flow from the outside. For extended shipments overseas, the helium is placed in special shipping containers. In addition to a vacuum space to provide insulation, these containers also have a second shell filled with liquid nitrogen to absorb heat from the outside. As heat is absorbed, the liquid nitrogen boils off and is vented.

Quality Control

The Compressed Gas Association establishes grading standards for helium based on the amount and type of impurities present. Commercial helium grades start with grade M, which is 99.995% pure and contains limited quantities of water, methane, oxygen, nitrogen, argon, neon, and hydrogen. Other higher grades include grade N, grade P, and grade G. Grade G is 99.9999% pure. Periodic sampling and analysis of the final product ensures that the standards of purity are being met.

The Future

In 1996, the United States government proposed that the government-funded storage program for helium be halted. This has many scientists worried. They point out that helium is essentially a waste product of natural gas processing, and without a government storage facility, most of the helium will simply be vented into the atmosphere, where it will escape into space and be lost forever. Some scientists predict that if this happens, the known reserves of helium on Earth may be depleted by the year 2015.

Where to Learn More

Books

Brady, George S., Henry R. Clauser, and John A. Vaccari. *Materials Handbook,* 14th Edition. McGraw-Hill, 1997.

Heiserman, David L. *Exploring Chemical Elements and Their Compounds.* TAB Books, 1992.

Kroschwitz, Jacqueline I., executive editor, and Mary Howe-Grant, editor. *Encyclopedia of Chemical Technology,* 4th edition. John Wiley and Sons, Inc., 1993.

Stwertka, Albert. *A Guide to the Elements.* Oxford University Press, 1996.

Periodicals

Powell, Corey S. "No Light Matter." *Scientific American* (March 1996): 28, 30.

Other

http://www.intercorr.com/periodic/2.htm (This website contains a summary of the history, sources, properties, and uses of helium.)

—*Chris Cavette*

Hockey Stick

In 1892, Canada's Governor General Lord Stanley introduced the hockey's first national title, the "Lord Stanley's Dominion Challenge Trophy," now simply called the Stanley Cup.

History

The origin of the game of hockey is a hazy and contentious issue. Various forms of field hockey were played in Scotland, Ireland, and France as early as the sixteenth century. A game involving sticks and balls played on ice called kolven was first recorded in Holland in the seventeenth century. The French called their game *hocquet*, which likely translated into the modern name of the game. Native Americans in eastern Canada had been playing *baggataway*, now known as lacrosse, for hundreds of years by the time the French arrived. A fresco from the Athenian acropolis shows two men holding sticks and battling for possession of a ball.

The modern game is generally accepted as having originated somewhere in Canada in the mid to late 1800s. Claims are laid to the official birthplace of the game from Kingston (Ontario), Montreal, and Halifax. The first know codified rules were produced in 1879 at McGill University in Montreal. In 1892, Canada's Governor General Lord Stanley introduced the game's first national title, the "Lord Stanley's Dominion Challenge Trophy," now simply called the Stanley Cup. Regional leagues prevailed across Canada and the United States for several years. In 1917, the National Hockey League (NHL) was founded in Montreal, and by 1926, the NHL had absorbed most of the competing leagues and took sole possession of the Stanley Cup.

Until a Montreal company began manufacturing hockey sticks in the late 1880s, most players made their own. A player would cut down an alder or hickory sapling, cut out 3 ft (91.44 cm) sections of trunk with branches attached, and file the wood into the desired shape. These first sticks had short handles and small, rounded blades, much like field hockey sticks. Even as play became more organized and stick manufacturing moved from the woodshop to the factory, stick development was slow and evolutionary. First the blades grew longer and squarer, allowing better control of the puck. Then, the shaft grew longer so that players lost the hunched over stance of early games. However, the stick was still fashioned from a single piece of wood, which made it heavy and made the thin blades prone to splitting. In 1928, the Hilborn Company, a stick manufacturer in Montreal, produced what is acknowledged by some to be the first two-piece hockey stick. The new design, with separate pieces for the blade and shaft, freed stick makers from having to find appropriately shaped lumber and allowed blades to be replaced when they cracked. Separating the blade from the shaft also gave manufacturers new latitude to experiment with the shape and thickness of blades. It was not until the late 1960s that blades took their largest jump in shape change when they began to take on a curve. One story, possibly apocryphal, attributes the design revolution to Chicago Blackhawks star Stan Mikita. The story tells that Mikita, frustrated during practice, was trying to break his stick by jamming it between the player's bench gate and the boards. The stick would not break, but its blade did bend remarkably. The new curve gave Mikita exceptional control and speed when shooting. Today, every player has his own preference about the amount and placement of the curve, but all blades have it. The next change came in the 1970s, primarily as a response to increased competition for wood supplies. During that time, foreign demand for ash wood raised prics beyond what stick makers could afford. Typically, only 10% of the wood in a shipment would be of acceptable quality for use in a stick, so manufacturers needed an enormous supply. They began to experiment with lamination as a way to use less and varied types of wood. Ultimately, the cost-cutting measure produced stronger, lighter, more responsive hockey sticks. Today, most high quality wood sticks are laminated.

Design

The three qualities any player is seeking in a hockey stick are stiffness, lightness, and responsiveness. Lack of flex is supremely important in the blade, where any twist can deflect a shot from its intended path. Most stick makers still use ash wood for their blades because ash is hard and durable but can be easily curved. Blades are wrapped with fiberglass to increase their stiffness and their abrasion resistance. While rotation or twisting of the shaft will cause the same problems as they would in the blade, a certain amount of flexibility along its length is necessary. When a player makes a slap shot, he lifts the stick behind his body and swings it down toward the puck. The first thing the blade contacts is not the puck, but the ice. When a baseball player swings a bat or a golfer swings a club, they are taking advantage of a physical property known as rotational inertia. Inertia means that an object in motion will tend to stay in motion until encountering an opposing force and an object at rest will tend to stay at rest until acted upon by an opposing force. With rotational inertia, that tendency to stay in motion is multiplied by the weight at the end of the diameter of an object's arc or swing. The greater the weight and the further it is from the center of the swing, the greater the rotational inertia. However, hockey sticks are light, especially on the end, precisely because a player needs to be able to quickly change the stick's direction. A hockey stick needs another way to make power. In driving the stick into the ice, the player flexes the shaft slightly and stores energy in it. As the blade comes closer to the puck, the player lifts the stick, unflexing it and releasing its stored energy onto the puck.

Stick makers achieve this strength while maintaining the stick's lightness and ultimately its responsiveness increasingly through the use of materials such a fiberglass, graphite, and aluminum. The quality of lightness has two functions. The first is fairly straightforward: players have to carry their stick up and down the ice as they skate. The second contributes, along with stiffness, to a less concrete issue—responsiveness. Responsiveness is the stick's ability to translate contact with the puck and with the ice to the player's hands. It is also the stick's sensitivity to input from the player. A lightweight stick will carry less of its own inertia, so it is able to transmit more of the inertia or feeling of both the puck and the ice (the ice has a sort of permanent inertia). Its lack of inertia also allows it to respond more quickly to changes in force and direction given by the player. Lightweight and stiffness are not the total sum of responsiveness. Materials must also be able to transmit minute sensations to a player's hands, which players translate as the "feel" of the stick. This particular quality is likely the reason that no one material has emerged as dominant in the industry.

Raw Materials

The greatest variety comes in the materials used to make the shaft of the stick. The three primary materials are wood, aluminum, and composite.

Wood, long the traditional material, is still widely used today. Many companies still make solid wood, two-piece sticks for their junior and budget lines. And professional players use laminated wood sticks across the world. Laminated wood sticks come in four varieties. The first is made of 19-21 layers or plies of a variety of hardwoods. These sticks are stiff but relatively heavy. The next is called an aircraft veneer stick and is made with a core of aspen and 8-10 plies of birch or maple. It is lighter than the first type. The third kind of stick is a combination of plies of hardwood and fiberglass. It can be finished either as an "epoxy exposed" stick, which has the layers of fiberglass on the outside or as a veneer finish stick, with strips of wood covering the fiberglass. These sticks are durable, stiff, and lightweight. The fourth type of stick is a tradi-

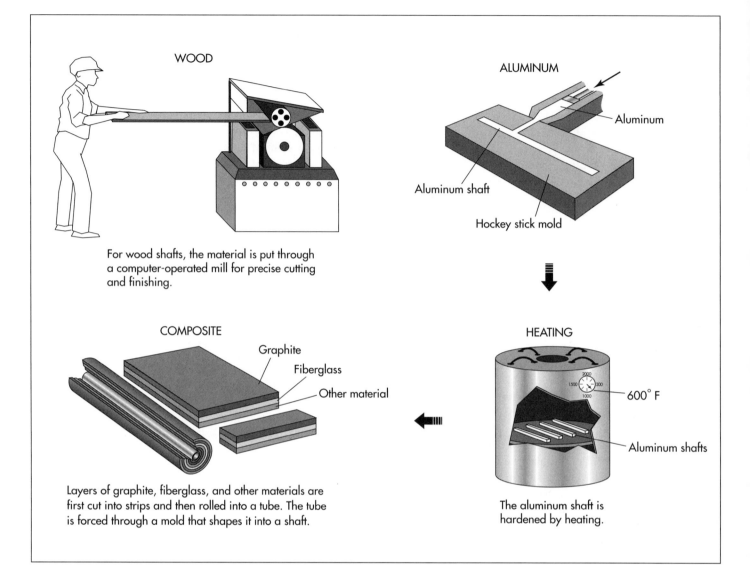

WOOD

For wood shafts, the material is put through a computer-operated mill for precise cutting and finishing.

ALUMINUM

Aluminum

Aluminum shaft

Hockey stick mold

COMPOSITE

Graphite

Fiberglass

Other material

Layers of graphite, fiberglass, and other materials are first cut into strips and then rolled into a tube. The tube is forced through a mold that shapes it into a shaft.

HEATING

600° F

Aluminum shafts

The aluminum shaft is hardened by heating.

Hockey stick shafts can either be made out of wood, aluminum, or composite material. If aluminum is used, it must be heat treated to harden the aluminum.

tional wood laminate with graphite fiber wound around its core. Graphite adds little weight to the stick and is extremely rigid.

Aluminum is used by itself to make a hollow shaft. Properly formed, it has substantial rigidity and extremely low weight.

Composites are comprised of reinforcing fibers, such as graphite and kevlar, and binders, such as polyester, epoxy, or other polymeric resins that hold the fibers together. Composites are used both by themselves to form hollow shafts and to reinforce wood sticks, much in the same way as fiberglass is used. These materials are generally stiffer and lighter than fiberglass.

Fiberglass is used both as a layer in laminated shafts and as a reinforcement around the

outside of sticks. Use in combination with wood as a laminate, fiberglass adds substantial stiffness to a shaft at the same time reducing its overall weight. As reinforcement on the outside of a stick, fiberglass mesh both contributes to stiffness and increases a stick's abrasion resistance.

The Manufacturing Process

A hockey stick is produced in two parts—the blade and the shaft. Today, most stick makers purchase these pieces in unfinished form from subcontractors and then customize them to their own specifications. Although some manufacturers use blades made of graphite or laminate the traditional ash wood over materials such as ABS plas-

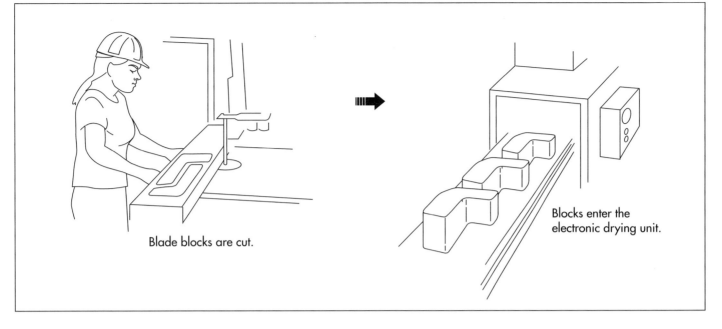

Blade blocks are cut.

Blocks enter the electronic drying unit.

tic, most blades are still made from pure ash. Blades are replaceable, and as they suffer inevitable wear from constant contact with the ice, the puck, and other players' sticks, it is generally more economical not to make them from expensive materials.

Wood

1 The lumber used to make wood shafts is first dried and cured in building-sized kilns to prevent the wood from warping after manufacture. Logs are then fed through a multi-bladed saw that cuts the wood into thin sheets. The sheets of wood and layers of fiberglass (if it is to be used) are coated with adhesive and pressed together in a heated hydraulic mold. The finished laminate is then cut into the rough shape of a stick and shipped to the stick maker.

2 The stick maker uses a saw with splayed teeth to cut the rough wood into the desired shape. Each pass on the saw cuts two corners of the shaft, so after two passes the shaft has four beveled corners and has reached its final shape.

3 A block of ash is glued onto the end of the stick. This will form the joint between the shaft and the blade. A groove is cut into the center of the block to accept a tongue shaped into one end of the blade. The tongue and groove are glued, fit togeth-

er, and placed in a heated hydraulic press to cure for about half an hour.

4 The dried assembly is then shaped and smoothed on a large drum sander. The fiberglass is pulled over this assembly like a sock and is dipped in resin and allowed to dry. Once dry, the stick is again sanded to remove any rough edges.

5 Graphics are silk-screened on and parts of the stick may be painted, and the stick is finished with a clear gloss varnish.

Aluminum

6 Aluminum shafts begin as flat sheets, which are folded and compressed into a long block. The advantage of this is that the walls of the final shaft will be made up of numerous extremely thin layers, each reinforcing the next and making a much stronger material than one layer.

7 The long block is cut into billets roughly the size of the intended stick. The billets are then fed into a machine that heats it and pushes it under enormous pressure through a hole the shape of a hollow hockey stick shaft in a process called extrusion. The extruded metal is cut off in lengths about the measurement of a stick.

8 These pieces are then drawn through a series of smaller and smaller dies to compress the metal and bring it to its final size.

Hockey blades begin as wood blocks that are first cut and shaped into blades, and finally, they are dried.

9 Finally, the metal is heated to approximately 600° F (315.5° C). This final step is called heat treating and is designed to strengthen the metal.

10 Finishing an aluminum shaft is simpler finishing a wood one. Aluminum can be painted. The metal itself may also be colored or anodized. To fit the blade, one end of the shaft is left open, a peg on the heel end of the blade is coated with hot glue, and the two pieces are clamped and left to dry. A rubber plug is inserted in the other end of the shaft for grip and safety.

Composite

11 Composite materials begin as a synthetic cloth just as fiberglass does, but most are far lighter, stiffer, and sometimes more durable than fiberglass. The primary reinforcing fiber used in composite hockey sticks is graphite. Kevlar, used in bulletproof vests, and Nomex, used in racecar drivers' fire suits, are used in small amounts but both are expensive and somewhat more difficult to use. Most makers of composite shafts use pre-preg composite, which has been saturated with the epoxy resin that will eventually bond it.

12 Several layers of the composite material are wrapped around a mold in the shape of the finished shaft and then heated and pressed through one of three methods. The first method utilizes a traditional hydraulic press. The composite-wrapped mold is placed inside another split mold carrying its mirror image, the outer mold is closed, and hydraulic pressure compacts the composite material. The second method uses vacuum pressure to force the composite against the mold and take its shape. The third method uses an inflatable bladder as the inside mold. The composite-wrapped bladder is placed into the outer mold and then is inflated to force the material into shape. The molded shaft is finished and attached to the blade in the same fashion as an aluminum shaft.

Blades

Ash blades also arrive at a stick maker's factory in unfinished form. Most manufacturers have a catalog not only of the blade shapes of their own models, but of the favored shapes of all the professional players who use their sticks. The NHL requires that a blade be 12.5 in (31.75 cm) long, between 2-3 in (5.08-7.62 cm) high, with a maximum curve of less than 0.5 in (1.27 cm) (goalies' blades have slightly different dimensions). Within these parameters, endless variables are possible. Stick makers can change the angle of the blade to the stick, called the lie. They can curve the blade at the heel or at the toe, and they can make a high toe and a low heel or just the opposite.

13 Blades are placed on a pattern or jig to be cut into their final shape. They are steam-heated or boiled then clamped in a hydraulic press to curve them.

Quality Control

Every piece of wood used in a hockey stick is inspected before it enters the assembly process. Lumber with irregular grain, knots, or mineral deposits is rejected. The NHL sets guidelines for every dimension of a stick. Strangely, those dimensions are only enforced during a game by protest of the opposing team. If a complaint is lodged and the stick meets specifications, the protesting team receives a delay-of-game penalty.

The Future

Several stick makers are experimenting with sticks made entirely of aluminum or composite material. New, lighter, more durable composites are always in development, but these materials have been available for some time and still many players choose wood-based sticks.

Where to Learn More

Books

Hockey Hall of Fame. *Hockey Hall of Fame Legends.* Viking/Opus Productions, 1993.

Hughes, Morgan. *The Best of Hockey: Hockey's Greatest Players, Teams, Games, and More.* Publications International, 1997.

Periodicals

Wilkins, Charles. "Sapling to Slapshot-that all — Canadian symbol-the hockey stick-has come a long, long way." *Canadian Geographic* (February 1989): 12-21.

Other

"Sports A to Z." http://test.olympic-usa/sports/az (5/7/98).

"Hockey Hall of Fame Home Page." http://www.hhof.com (5/7/98).

—*Michael Cavette*

Home Pregnancy Test

The presence of hCG in urine has long been a factor used by doctors to determine pregnancy. During normal gestation, hCG level doubles approximately every two days and it can be detected in urine as early as seven days following conception.

A home pregnancy test is a self-diagnostic tool that allows women to quickly and easily determine if they are pregnant. These tests measure Human chorionic gonadotrophin (hCG), a hormone that is secreted in urine during pregnancy. Human chorionic gonadotrophin is measured using a technique known as an immunoassay, which involves a complex reaction between the hormone and various protein antibodies. When urine, containing hCG, is applied to the test strip, a reaction occurs which causes a portion of the stick to change color thus signaling the hormone is present and that the woman is pregnant.

Background

The presence of hCG in urine has long been a factor used by doctors to determine pregnancy. During normal gestation, hCG level doubles approximately every two days and it can be detected in urine as early as seven days following conception. Originally, hCG measurement was a complicated test which had to be administered by a doctor. The test involved taking a urine sample from the woman and injecting it into a frog or rabbit; the animal was subsequently dissected and examined. If the woman was pregnant, interactions between hCG in her urine and the animal's reproductive system could be observed. The cliche "the rabbit died" is actually a reference to this type of biological assay. Such testing has been replaced by immunoassay testing techniques, which are more sensitive and easier to administer. An immunoassay is a type of test that measures protein molecules, which interact with the pregnancy hormone in blood or urine. Early immunoassays were done with blood because this type of test is the most sensitive and can detect minute quantities of hCG. A positive test can be obtained within days of conception and before a missed period. However, tests done with blood still require the involvement of a doctor or nurse.

Further improvements in immunoassay techniques led to the development of a urine-based test, which is simple and foolproof enough for virtually anyone to self-administer at home. In home test kits, a protein called a monoclonal antibody (MAb) reacts with any hCG present in the urine. This reaction causes a color change if the level of hCG is consistent with known pregnancy levels. This kind of test can be positive two weeks after conception or several days after a missed menstrual period. To administer the test, a woman applies urine to a latex-coated test strip, which has been treated with different antibodies. The antibodies are placed in three distinct zones or bands along the test strip. The first band contains two types of antibodies: "anti-a" hCG antibody, which will combine with any hCG in the urine, and Immunoglobin G (or IgG), which is a control to determine that the test strip is working properly. When the urine flows past this first antibody band, two things happen. The hCG in the urine reacts with the "anti-a" hCG antibody and forms a complex. At the same time the urine suspends the IgG and carries both the IgG and the urine/"anti-a" antibody complex along the strip to the second antibody band. The second band contains anti-b hCG antibody which reacts further with the hCG to create a chemical "sandwich" which turns a bright color. As a result, a colored line appears in the test window of the strip. If hCG is absent, no binding takes place and no colored line is formed. In either case, the IgG from the first antibody band continues to move along the test strip until it contacts the third antibody band. This band contains an antibody, which will react with the IgG, forming another colored line under the control window. If two colored lines appear (one in the test window and one in the control), the result is positive and the user is pregnant. If only one line appears (in the control window), the result is negative. If no line appears in the control window, the test was not conducted proerly and should be repeated with a fresh test strip.

By the mid-1990s, this immunoassay technique had been commercialized in over two dozen retail products. According to the October 1994 issue of *Pharmacy Times*, the home pregnancy test market had grown to nearly $200 million in sales per year. Major brands include EPT, Advance, ClearBlue Easy, and Answer Plus.

Design

The test kits on the market differ primarily by the way the user must collect and process the urine. One type of kit requires the user to urinate directly onto an absorbent area of the test kit by holding onto one end of the stick and urinating on the other. The urine is absorbed by the pad on the end of the test stick and travels along the test strip by capillary action. This design is used in products like Advance, Confirm, Clear Blue Easy, Answer, and EPT. The other type of test requires the woman to collect urine in a separate vessel. She then utilizes a dropper to deliver a precise amount of urine into a well on the test unit. Products using this design include Fact Plus, Precise, and Be Sure. Some brands, like Q Test and Answer Plus, require the woman to mix the urine with other test components before adding it to the test strip.

Components

A typical home pregnancy test contains the following components:

Immunoassay strip

The immunoassay strip is formed by compressing nonwoven fibers into a narrow strip and coating them with reactive antibodies. The antibodies combine with the pregnancy hormone in a series of steps, ultimately resulting in a color change.

Absorbent pad

The direct application type of test contains an absorbent pad that extends from the test chamber and is used to contact the urine stream. The pad absorbs the liquid and draws it into contact with the immunoassay strip.

Plastic housing

The test strip and absorbent pad are contained within a two piece housing that allows the unit to be handheld and protects the strip from environmental contaminants. A leak-proof, clear plastic window on the side of the housing prevents urine from accidentally splashing on the test strip and allows the test and control zone portions of the strip to be viewed.

Urine collection cup/vials and reagents

A plastic collection cup is included with test kits that require collection of urine in a separate step. They may also include plastic vials with pre-measured amounts of reagents that must be mixed with the urine before application to the test strip.

The Manufacturing Process

Forming the nonwoven fiber strip/pad

1 The immunoassay strips used in home pregnancy tests are manufactured by pharmaceutical supply firms. They are formed by compressing nonwoven fibers into a narrow strip and coating them with reactive antibodies. These strips are made from fibers such as rayon polyester blends, which can be mechanically or thermally formed into pads. This type of nonwoven fabric pad is similar to that used in disposable diapers and other applications which require highly absorbent material. The nonwoven pad first may be coated with latex then treated with the liquid assay agents. As many as four different antibodies may be coated onto the strip in three different zones.

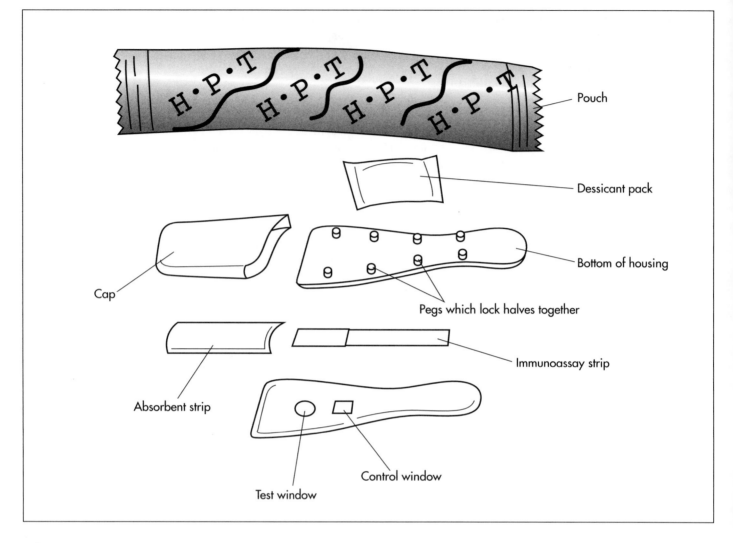

Pouch

Dessicant pack

Bottom of housing

Cap

Pegs which lock halves together

Immunoassay strip

Absorbent strip

Control window

Test window

A home pregnancy test is comprised of two basic components: the immunoassay strip/pad and the housing. The immunoassay strip is coated with a protein called a monoclonal antibody (MAb), which will react with any hCG present in the urine. This reaction causes a color change if the level of hCG is consistent with known pregnancy levels.

After drying, the pads may be cut into appropriately sized strips for the test kit.

Forming the plastic housing

2 The housing used to contain the test kit is made by injection molding. In this process, plastics, colorants, resin modifiers, and plasticizers are mixed together, heated, and injected into a mold under pressure. After cooling, the mold is opened and two halves of the plastic housing are ejected. One side is molded with a series of pins, which fit into corresponding holes on the other half to lock the pieces together. The housing is also designed with contours to hold the immunoassay strip and the absorbent pad in place. In a separate molding operation a clear plastic window is created. This window prevents urine from accidentally splashing on the test strip and allows the user to view the test and control areas of

the test strip. A plastic cap, which covers the absorbent pad, is made in another separate molding operation.

Assembly, packaging, and shipment

3 The individual pieces are transferred to an assembly line where the immunoassay strips are first inserted into the housing along with the absorbent pad. The clear plastic window is affixed with adhesive to the upper half and the two portions are snap-fitted together. Finally, the cap is locked into place.

4 The assembled test kit is packed in a foil/plastic laminate pouch along with a small sachet containing a silica gel desiccant. The desiccant absorbs excess moisture in the package and helps prolong the shelf life of the test kit. The pouch is then crimped tightly shut to minimize exposure

to air. The filled pouch and a detailed instruction booklet are inserted in a cardboard box. One or two complete test kits are packed per box depending on the price and manufacturer. The filled box proceeds down the assembly line where an online ink jet printer is used to encode each box with critical information, such as the date the product was made, the expiration date, and the lot number. The box travels further down the assembly line where it is shrink wrapped for added protection. The finished box is then packed in a case for shipping.

Quality Control

All pregnancy tests are designed with a built in quality control check—they have a control area that is designed to show if the test is working properly. This window will reveal a colored line if the test is functioning. During the manufacturing process, a random number of test kits may be activated to determine if they are operating acceptably. To limit the potential for false or misleading reading the reaction pack is wrapped in a foil wrapper. It should not be removed from the wrapper for more than five minutes before the test is conducted, or a false result may be obtained. After the test is complete, the colored line of a pregnant test result will last at least 24 hours. Interesting to note that a positive result will not change several days after the test is completed. However, some reddish background might be noticed several hours after the test is performed. A negative result should not be read in the test window more than five minutes after the test is performed. After 30 minutes, some negative tests might appear to be weakly positive.

The Future

Advances in biotechnology have created the modern home pregnancy test. Further breakthroughs in immunology and related fields may result in tests which are more sensitive and which can detect pregnancy even quicker after conception. Improvements in biotechnology processes are also likely to lead to tests which are easier to manufacture and therefore less expensive. For example, improvements in the manufacturing processes of the antibodies used in these tests could result in lower prices for home pregnancy tests.

Where to Learn More

Books

"Biopolymers in Immunoassay." *Kirk Othmer Encyclopedia of Chemical Technology,* p. 197.

Periodicals

Lewis, Ricki. "Replacing Test Animals, Improving Diagnoses." *FDA Consumer* (January/February 1993): 15-18.

—Randy Schueller

Hot Dog

Americans consume over 16 billion hot dogs each year.

Hot dogs are a processed meat product made by mixing chopped meat with various curing ingredients, flavorants, and colorants. The meat is then stuffed in casings, cooked, removed from the casing, and put in the final package. Although the technology for hot dog making was developed thousands of years ago, these meat products continue to be a popular summertime food. In fact, in America alone over 16 billion hot dogs are consumed each year.

Background

The typical raw hot dog is a pink, cylindrical-shaped piece of meat. It is about 1.6 oz (45.36 g) on average and contains anywhere from 0.175-0.245 oz (5-7 g) of protein. It also contains about 0.455 oz (13 g) of fat, 450 mg of sodium and 150 calories. Since hot dogs are meat products, they are an excellent source of nutrients including iron, zinc, niacin, riboflavin, and B vitamins. When hot dogs are made using pork meat, they are good source of thiamin. Since they are a pre-cooked food, they are less prone to spoilage than other types of meat products. This makes them one of the safest meat products available.

Hot dogs are known by many different names including frankfurters, franks, red hots, and wieners. While there are many varieties of hot dogs, one of the most famous is the Kosher hot dog. These hot dogs are prepared in a manner, which follows 3,000 year old traditions that comply with Jewish religious practices. Specially trained Rabbis oversee the entire kosher hot dog making process. Ultimately, the main difference between a kosher and a regular hot dog is that kosher hot dogs do not contain pork.

History

The technology for making hot dogs and sausages was developed thousands of years ago. This makes these products one of the oldest forms of processed food. The earliest record of a hot dog type product dates back to 1500 B.C. in Babylonia. Sausages were mentioned in Homer's *Odyssey* written during the ninth century B.C. These early forms of hot dogs were made by grinding up meat, stuffing it in animal intestine and cooking it over a fire.

The exact origin of the product we call a hot dog is debated. Some claim that it was first developed in Frankfurt, Germany, in 1484. Others claim that it was developed in Vienna, Austria, and suggest that the term wiener reflects this point. Still others suggest that it was not developed until the late 1600s when Johann Georghehner (who was from Coburg, Germany) produced a sausage product known as the dachshund sausage.

In 1852, a butchers' guild in Frankfurt produced a spiced, smoked sausage product which they named frankfurter after their hometown. It was slightly curved in shape and was often called the dachshund sausage. The product was brought over to America by Charles Feltman and Antoine Feuchtwanger. Feltman sold frankfurters and sauerkraut from a pushcart in New York's Coney Island. He opened up the first Coney Island hot dog stand in 1871. Shortly thereafter, he started selling the frankfurters with milk rolls, which were the precursors to hot dog buns. The buns that we use today were probably first introduced in St. Louis by Feuchtwanger in 1904. He was a sausage concessionaire who loaned white gloves to his customers to hold the hot sausages. Since most of his customers did not return his gloves he worked with a baker to develop a bun, which people could use to hold their sausages.

In 1893, sausages became a popular food at baseball parks. They were first introduced in the St. Louis Browns ballpark and then spread to the rest of baseball. The term hot dog was coined in 1901 by a sports cartoonist named Tad Dorgan. He was at the New York Polo Grounds, where he had heard some vendors selling red hot dachshund sausages. This prompted him to draw a cartoon of a real dachshund covered with mustard on a bun. Since he did not know how to spell dachshund he wrote on the caption "get your hot dogs." The cartoon was a hit and the name persisted.

Raw Materials

The primary ingredient in hot dogs is the meat. The U.S. Department of Agriculture requires that meats used for hot dogs must be the same type of quality ground meat sold in supermarkets. While pork is most often used, other types may be used such as beef, chicken or turkey. Sometimes variety meats like livers are used however, the hot dog producers must clearly label the product with the statement "with variety meats" or "with meat by-products." The proteins and fats of which meats are composed are responsible for meat characteristics. For example myofibrillar proteins give meat its texture and structure. Myoglobin and hemoglobin proteins create the natural color of the meat. Fats in the meat give the characteristic flavor.

During processing, the meat is mixed with a curing solution to improve the taste and increase the shelf life. A major ingredient in this curing solution is salt. It is used to make the meat easier to work with, improve flavor, and inhibit bacterial growth. Water is another component of the curing solution. It has a variety of functions including helping create the necessary meat emulsion and adding to the meat's juiciness. Sodium nitrite is included in the curing solution to retard the development of rancidity and stabilize the meat color. Curing accelerators such as sodium ascorbate or sodium erythor-

Constructed in 1936, the original Oscar Mayer Wiener mobile was a 13 ft (4 m) hot dog on wheels used to transport the hot dog chef known as Little Oscar. (From the collections of Henry Ford Museum & Greenfield Village, Dearborn, Michigan.)

Product promotion requires a willingness to "take a chance." The Oscar Mayer Wienermobile, used to promote Oscar Mayer hot dogs, was an audacious promotion that was successful. The original Oscar Mayer Wienermobile was constructed in 1936 by the company founder's nephew, Carl. It was a 13 ft (4 m) hot dog on wheels used to transport the hot dog chef known as "Little Oscar," featured open cockpits at the rear and center and was primarily driven in Chicago.

The Wienermobile proved very popular and was modified over the next sixty years. The car, currently on exhibit in Henry Ford Museum in Dearborn, Michigan, was made between 1950 and 1954 and is one of five second-generation Wienermobiles. It is made by the Gerstenslager Company, a custom truck and trailer manufacturer that modified a sheet bodies on a customized Dodge chassis. This Wienermobile is 22 ft (6.7 m) long, has glass over the cockpit, and weighs about 8,000 lb (3,632 kg). This, along with the other four made around 1954, were stationed near key distribution plants and deployed for promotional events along with hot dog chef Little Oscar. The company continued to modify its successful promotional car and produced versions of it in 1958, 1969, 1988, and 1995. Kids still delight in spotting the Wienermobile (sightings of it often makes the local paper) and love to play kazoos in the form of Wienermobiles to the tune of "I'd love to be an Oscar Mayer Wiener."

Nancy EV Bryk

bate may also be added to preserve the color of the meat during storage.

In addition to the meat and curing ingredients, other ingredients are important in hot dog manufacture. Sugar and corn syrup are

After the meat passes inspection, it is cut into small pieces and placed in a stainless steel mixing container. The container is equipped with high speed choppers, which can reduce the size of the meat pieces even further. The other raw materials including the curing ingredients, flavorings, and ice chips are blended in this container until a fine emulsion, or batter, is produced. The batter is pumped into an automatic stuffer/linker machine, which stuffs the batter into tube-shaped, cellulose casings.

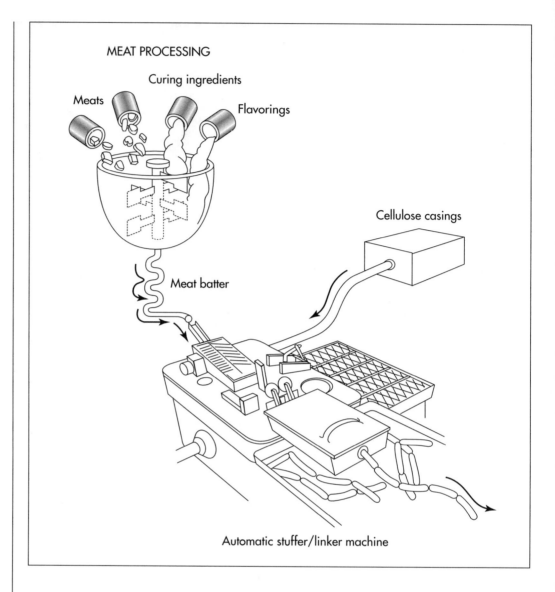

MEAT PROCESSING

Curing ingredients

Meats

Flavorings

Cellulose casings

Meat batter

Automatic stuffer/linker machine

used to give hot dogs a sweeter taste. Flavoring is added to give hot dogs their characteristic taste. The flavorants may be natural or artificial, but typically a mixture of the two is used. Natural flavorants include herbs and spices such as pepper, nutmeg, ginger, cumin, and dill. Artificial flavoring compounds include organically synthesized esters, ketone and amino acids. Monosodium glutamate is an artificial flavor that is often used to intensify the flavor of the meat.

Hot dogs can contain extenders, which are non-meat ingredients that increase the number of hot dogs that can be made from a set amount of meat and improve the nutritive value. Extenders come from plant and animal sources and include things such as non-fat milk, cereal, soy protein and whole milk. In the United States, all of the ingredients

that are used in hot dog manufacture must be clearly labeled on the package.

The Manufacturing Process

Meat Processing

1 The production of hot dogs begins with the preparation of meat. After it passes inspection, the incoming meat is cut into small pieces and placed in a stainless steel mixing container. The container is equipped with high-speed choppers, which can reduce the size of the meat pieces even further. The other raw materials including the curing ingredients, flavorings and ice chips are blended in this container until a fine emulsion, or batter, is produced. This

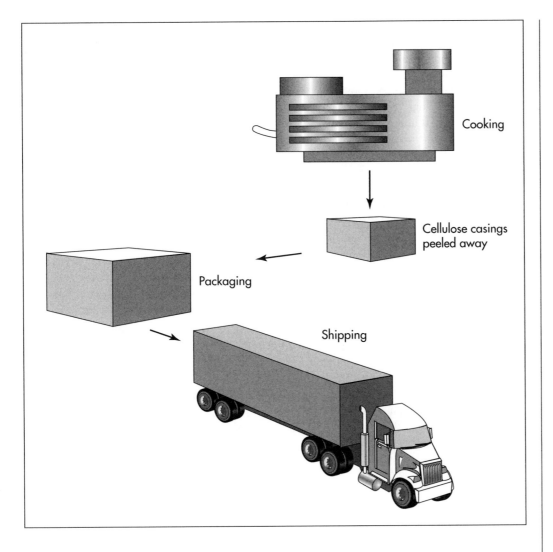

Cooking

Cellulose casings
peeled away

Packaging

Shipping

Once stuffed, the hot dogs are cooked. Their casings are removed and the hot dogs are vacuum-packed to preserve their flavor.

batter has a smooth paste-like consistency, which makes further processing easier.

Linking

2 After the batter passes quality control checks, it is pumped into an automatic stuffer/linker machine. In this machine, batter is put into tube-shaped, cellulose casings. These casing are then twisted at precise points to produce a long linked strand of equally sized hot dogs. Most casing are removed later in the process however, some manufacturers continue to use natural casings, which remain on and are eaten along with the hot dog. This more traditional method of hot dog making is done by smaller manufacturers and tends to cost more.

Cooking

3 The linked hot dog strands are then conveyed to a large smokehouse. Here, they are thoroughly cooked under controlled conditions. The manufacturer has the opportunity at this point to impart a different flavor on the hot dogs by using a variety of smoke sources. The cooking times vary depending on the recipe however, typically it takes about an hour.

Final processing

4 When the cooking is done, the hot dog links are moved via a conveyor to an automatic peeler. During their trip, they are showered with water to help equalize their internal temperature. In the peeler, the cellulose casings are cut away leaving only the bare hot dogs. It should be noted that this step is skipped by manufacturers who use natural casings.

5 From the peeler, the individual hot dogs are transported to the packaging machinery. Here, they are lined up and placed on a plastic film. The films are folded and vacuum-sealed to preserve the hot dog's flavor and increase shelf life. Printed on the films are all of the graphics and required text needed for marketing. The sealed packages are moved to a stamping machine, which prints on a freshness date. They are next transported to boxing devices, put on pallets and shipped in refrigerated trucks to local supermarkets. The entire process of hot dog making from receiving the meat to boxing up the hot dog takes only a few hours.

Quality Control

Quality control is an extremely important factor in any food processing facility. For health and safety reasons, the government regulates all of the raw materials that are used in the hot dog making process. The meat in particular is heavily regulated because the use of poor quality meat represents a significant health risk. Most manufacturers use only high quality meats to assure that their hot dogs are of similar quality. Upon receipt of the raw materials, they are checked for things such as pH, % moisture, odor, taste, and appearance to ensure they meet the previously set specifications. Additionally, the processing equipment is sterilized and checked before any processing can begin. During manufacture, the meat emulsion is continuously checked to assure that all the ingredients are put in at proper proportions. Since hot dogs are eaten, steps must be taken to ensure that they will have an appealing taste and be free from contamination. For this reason, tests similar to the ones run on the initial raw materials are performed on the final product.

The Future

Hot dogs are a well-established product and the technology for their production has changed little over the last century. However, hot dog marketers are continually looking for ways to increase sales. A recent trend is the introduction of more nutritious hot dogs. These products may use poultry meat, which has inherently less fat, or meat substitutes, which have no fat at all. They are also made with less sodium. New varieties of hot dogs are also being produced such as the cheese-containing dog, a product which is injected with a cheese sauce during manufacture. From a production standpoint, hot dog making of the future should be faster. Each year designers of production equipment develop faster, more efficient machines. This will help to make the process more automated and increase the yearly output of hot dogs.

Where to Learn More

Books

LeMaguer, M. and Jelen, T., ed. *Food Engineering and Process Applications.* London: Elsevier Applied Science Publishers, 1986.

Macrae, R., et. al., ed. *Encyclopedia of Food Science, Food Technology and Nutrition.* San Diego: Academic Press, 1993.

Periodicals

Burg, James. "Making More Healthful Meats." *Food Product Design* (March 1998): 32-58.

National Hot Dog and Sausage Council Publications.

—Perry Romanowski

Instant Lottery Ticket

Background

During the last few decades, lottery tickets have become an increasingly popular form of legal gambling in the United States. One popular game is the instant win, or scratch off lottery which features tickets that have the winning (or losing) numbers concealed on the game card itself. The winning numbers are typically hidden by a coating, which is removed by rubbing. By removing this coating, the owner of the ticket can instantly determine the ticket's winning status instead of waiting for a matching number to be drawn. Since the cash value of the ticket is determined at the time of printing, the tickets must be designed and manufactured with extraordinary security precautions to avoid ticket fraud.

Design

The design of instant lottery tickets varies from game to game. To entice potential purchasers, games may be thematically linked to popular interests such as sporting events, television shows, or even other gambling games like poker card or horse racing. Some states have even allowed customers to participate in the design process. For example, in 1993, the Oregon State Lottery held a "Designer Scratch-it Contest" for the general public. Winners were judged based on theme, style of play, graphics, and originality. Regardless of the design type, instant lottery tickets are designed to be played by scratching off a concealing coating to reveal numbers, letters, or symbols that will (hopefully) match the designated winning symbol located somewhere on the ticket. These games are all designed with multiple security features to prevent tickets from being counterfeited or tampered with.

There are several techniques used to breach game security, which must be taken into consideration during the design process. One method of defrauding the lottery is to decode the relationship between the serial number on the ticket and the ticket's lottery number. Each ticket contains an individual serial number composed of a series of digits or alphanumeric characters. This number is used by the game operator to track the distribution of tickets from the operator to the selling agents and for accounting of sold and unsold tickets. It may also include information that shows the ticket is only valid for certain games or dates. These numbers are especially helpful in case tickets are lost or stolen and can be used to track tickets to make sure they are not inappropriately claimed.

By understanding the relationship between the serial number and the lottery number, one could try to locate lots or batches of tickets that are more likely to be winners. Other methods to breach ticket security attempt to directly view the lottery number without scratching off the ticket covering. One way this is done is by candling, which involves shining a bright light on the ticket in an effort to read the lottery number through either the front or back covering. Another technique, known as delamination, involves separating the different layers of the ticket to make the numbers visible. This technique can even be used after the owner has uncovered a winning number and turned in the ticket for redemption. In this scenario, individuals with access to winning tickets could separate the front layer of the ticket that contains the winning number and glue it onto a new back layer that has a different name and address for the winner. Still another way of circumventing

Since the cash value of an instant lottery ticket is determined at the time of printing, the tickets must be designed and manufactured with extraordinary security precautions to avoid ticket fraud.

lottery security, called wicking, uses solvents (e.g., alcohols, ketones, acetate, or esters) to force the lottery number to bleed through the concealing coating.

The design features employed to prevent these security breaches vary from game to game. In general, these features involve the serial number and the concealing coating. One key to controlling game security is to select serial numbers, which do not reveal any information about the winning status of the ticket. This is done by randomly encoding tickets with a series of computer-generated numbers or symbols. Each lottery game uses a specific algorithm, or mathematical process, to randomize the relationship between the serial and lottery numbers. This prevents anyone from discovering the connection between the two numbers. When properly encoded, the serial number cannot be deciphered by the ticket purchaser but still provides useful information to the ticket agent. Printing matching, coded numbers on the front and back of each ticket can help ensure winning tickets have not been tampered with. The security features used to prevent candling, delamination, and wicking involve the coating used to conceal the lottery number. A heavy foil coating can be used over the numbers to prevent light from passing through the ticket and illuminating the numbers. However, this foil is expensive to add and it does not prevent delamination. A better way to prevent the numbers from being read through the coating is to use an opaque covering in conjunction with confusion patterns imprinted on the back and front of the ticket. These confusion patterns are random designs used to obscure the image when light is shined through the ticket. This method can also be used to prevent wicking by utilizing dyes in the coating which are responsive to solvents. If anyone attempts to dissolve the concealing coating, the ink bleeds and obscures the lottery numbers.

Raw Materials

The basic materials required for ticket manufacture are the same as those employed for any similar ticket or card printing. The main component is paper stock of appropriate stiffness, but aluminum foil is also used as a component of some multilayer tickets. Other important raw materials include the suitable inks, adhesives to laminate multi-part tickets, and the scratch-off coating materials used to conceal the number. These coatings are most often made using acrylic resins.

The Manufacturing Process

Printing

1 Depending on the technique employed, tickets may be printed in a single stage (where all the information is printed at once) or in two stages (where the game information and secure numbers are printed separately). Game information includes details such as ticket prices, gaming rules, validation dates, prize information, and graphics to entice the consumer. Tickets can be printed by either a continuous feed process or by a sheet press process. In the continuous feed process a strip of paper ticket stock is fed into the computer controlled printer. In the sheet press method, ticket sheets are imprinted by engraved plates. After printing, the sheets are sliced into individual tickets. These tickets are then stacked and shuffled to ensure the lottery numbers are not in consecutive order. Winning tickets are randomly intermingled so there is a wide and preferred distribution of winners.

Serial number coding

2 Coded serial numbers and corresponding lottery numbers can be added to the tickets in several ways. The continuous printing process uses sequencers, which advance the serial number for each ticket when it passes through the press. These serial numbers go through a complex mathematical transformation (known as algorithmic conversion) to intermediate numbers, which are in a consecutive order. A second algorithmic converter operates on the intermediate number and generates the actual lottery number.

In the sheet printing process, after the tickets have been cut and shuffled, they are fed into a serializer which adds a computer generated serial number. Depending on the process employed, these numbers can be conveniently printed with fast, computer operated ink jet printers. Sensors attached

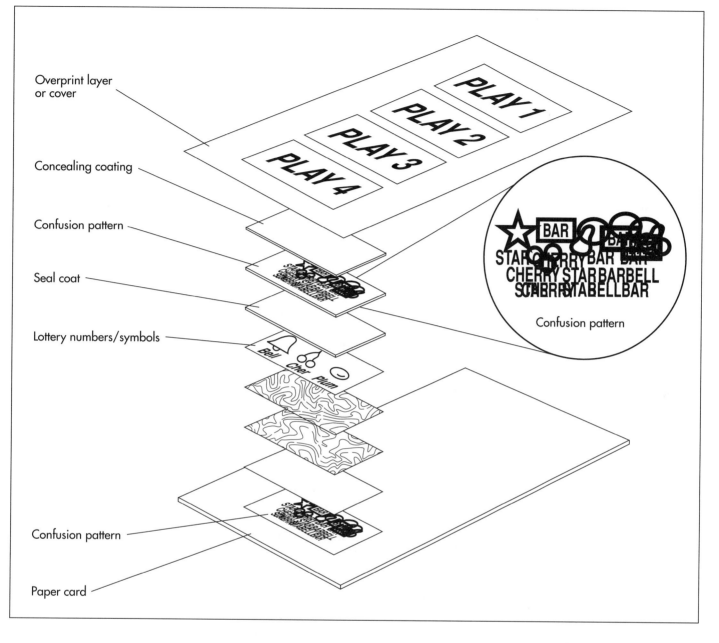

Overprint layer
or cover

Concealing coating

Confusion pattern

Seal coat

Lottery numbers/symbols

PLAY 1
PLAY 2
PLAY 3
PLAY 4

Confusion pattern

Confusion pattern

Paper card

to the printers transmit information on speed and relative position of individual tickets as they are being printed. The computer then generates print control signals that cause the ink jet printers to simultaneously print book numbers and lottery numbers. By printing both these numbers at the same time from opposite facing printers, the front and back of the ticket can be printed at the same time. Similarly, tickets can be printed with a single printer which has a print head located on the top and bottom of the ticket face. After the printing process is complete and both the lottery and serial numbers have been added to the cards, the lottery number can be concealed.

Lottery number concealment

3 The tickets are sent through a cover applicator, which treats the ticket with one of several concealing coatings. One type of coating is an opaque metal foil, which is applied with pressure. A paper pulltab may also be used to conceal the number, although this method is not frequently used today. The preferred concealment method uses an acrylic film to cover the number. This type of coating is applied in several layers. First, the paper card surface is treated with a primer coating which prepares the surface to accept the other layers. Next, a confusion pattern, consisting of random

In order to prevent illegal tampering of a lottery ticket, an opaque covering in conjunction with confusion patterns is imprinted on the back and front of the ticket. These confusion patterns are random designs used to obscure the image when light is shined through the ticket.

lines or symbols the same color as the lottery numbers, is printed on top of the primer coat. The lottery numbers are printed on top of this layer, and a seal coat is applied over the numbers. The uppermost concealing covering is added next. This concealing coating contains highly opaque materials such as carbon black pigment or aluminum paste mixed with acrylic resins and appropriate solvents such as methyl ethyl ketone. The resulting coating effectively hides the numbers beneath but can be easily removed by rubbing. An additional confusion pattern is printed on top of this layer. Finally, an overprint layer featuring additional graphics or instructions may be printed on top.

Conversion and packaging

4 After the printing and coating processes are complete, additional converting operations may be performed on the tickets. These operations include slicing the tickets into rolls or perforating them for ease of dispensing. Finally, the tickets are boxed and readied for shipment to distributors. After delivery, the tickets are sold for use. When the ticket is purchased and the owner scratches off the covering, he or she reveals the ticket's winning status. The ticket is then taken to a ticket vendor to claim the prize, and the game operator inputs the serial number in their computer to decode it and confirm that the ticket is a winner. The ticket vendor then pays the customer and is subsequently reimbursed by the lottery operator.

The Future

Printing technology is continually evolving and this evolution is likely to lead to new methods for lottery ticket production. For example, improved encryption technology could result in the creation of more secure lottery numbers. Likewise, newly developed chemical methods of concealing lottery numbers could produce less expensive tickets than those currently used. Perhaps of more interest for the future are alternate ways in which games may be played. One method under consideration by the lottery commission involves a video terminal instead of a paper ticket. It is conceivable that at some time in the future, an instant lottery game could even be played over the Internet on a personal computer.

Where to Learn More

Books

Johnson, Ben E. *Getting Lucky - Answers to Nearly Every Lottery Question You Can Ask.* Chicago: Bonus Books, 1994.

Periodicals

Lotto Magazine

Other

The Lottery Collectors' Society. 1007 Lutrell St., Knoxville, TN 37917.

—*Randy Schueller*

Insulated Bottle

Background

Since the invention of the insulated bottle, there has never been a question of how to keep hot liquids hot and cold liquids cold. The bigger question has been how does it do that? The answer is this: by using foam or vacuum packing. There is more to an insulated bottle than meets the eye. What we see as one heavy-duty container is actually a container within a container, with either foam or a vacuum between the inside wall of the outer container and the outside wall of the inner container. Foam-packed containers keep cold liquids cold and vacuum-packed containers keep hot liquids hot. This method has proven highly efficient since the early 1900s, which explains why insulated bottles are popular among anyone who needs or desires to drink liquids on the go, like athletes, travelers, campers and hikers, and just the ordinary busy person who gets thirsty once in a while. Many baby bottles are also insulated.

History

The first known bottles, made of glass, were produced in about 1500 B.C. by the Egyptians. The bottles were formed by placing molten glass around a core of sand and clay. The core was then dug out once the glass cooled. The process of making bottles was time-consuming and complex, so they were considered a luxury item in ancient Egypt. By 200 B.C., glass bottles were being made in China and Persia, as well as Egypt using a method whereby molten glass was blown into a mold. The Romans later adopted this same method and the technique spread throughout Europe during the 1400s and 1500s. The first bottle- and glass-making factory in the United States was established in Virginia in 1608.

New variations of the glass bottle began to surface by the 1800s. The baby bottle, for instance, was patented in 1841. The concept was not a new one, however. In ancient times, babies were fed using an urn with two openings. One opening allowed liquid to be poured into the bottle and the second opening was put in the baby's mouth. The sixteenth-century baby bottle resembled a duck; the baby was fed through the beak. Glass-blower John L. Mason devised the first glass jar with a screw-on cap "the now-famous Mason jar" in 1858. Also during the mid-1800s, Dr. Hervey Thatcher devised the glass milk bottle. The Coca-Cola company introduced the first soda pop bottle in 1915. Several other brands followed, each using a bottle with its own distinctive shape. Soda bottle shapes were standardized after 1934, when technology enabled companies to fire permanent color and, thus, company names and logos, onto bottles.

The bottle-making process first became automated in 1865 with the introduction of a pressing and blowing machine. The first fully automatic machine for making various types of glass bottles and jars did not appear until 1903, however, when an employee of a Toledo, Ohio, lamp-chimney company named Michael J. Owens put the Owens Bottle Machine into commercial use. The Owens Bottle Machine revolutionized the industry by enabling the inexpensive, large-scale production of glass bottles. Along with the Crown bottle cap, it helped spur the large-scale carbonated beverage industry as well. By 1920, most bottles were produced on Owens machines or those modeled after Owens' invention. In the early 1940s, manufacturers began to use blow-molding machines to produce plastic bottles. Blow

The first plastic bottles were squeezable and made from polyethylene. Nat Wyeth, a relative of American artist Andrew Wyeth, devised the first plastic bottle strong enough to hold carbonated beverages for the Du Pont Corporation.

Glass-lined aluminum vacuum-sealed bottle made by the American Thermos Bottle Co. of Norwich, Connecticut, in 1915. (From the collections of Henry Ford Museum & Greenfield Village, Dearborn, Michigan.)

This 1915 glass-lined aluminum vacuum-seal bottle was an early product of the American Thermos Bottle Company, founded in Norwich, Connecticut around 1907 (later renamed King Seeley Thermos or KST after 1959). The name Thermos is synonymous with the insulated bottle. Outdoor laborers, particularly construction crews, have been taking bottles like this one to work since KST was founded. Reminiscences of workers sometimes reveal that they often felt like they were really part of the workforce when they purchased, or were ceremoniously handed, a vacuum-seal bottle for their coffee.

School children plead for new lunch boxes and thermos bottles each year, too. Lunch box maker Aladdin and competitor KST wanted to sell to kinds, so their artists began a regular drill of changing the artistry and graphics on their school lunch boxes (first in metal, later in plastic) and vacuum-seal bottles. From 1950-1970 over 120 million kids' lunch boxes and accompanying vacuum seal bottles were sold, from Hopalong Cassidy to Peanuts. That's a lot of chocolate milk in those bottles! Of course, it could be hot chocolate, too—the joke is that no one can figure out how the vacuum-seal bottle could be so intelligent—how does a vacuum-seal bottle figure out when to keep cold liquids cool and hot liquids hot?

Nancy EV Bryk

drew Wyeth, devised the first plastic bottle strong enough to hold carbonated beverages for the Du Pont Corporation.

The first insulated bottle was likely designed by English scientist Sir James Dewar in 1896. In 1892 Dewar had invented a special flask still in use today and attributed to him by its name. Dewar created his insulated bottle by sealing one bottle inside another and pumping out the air between them. This created a vacuum, which is an effective insulator. Dewar never patented his invention, however. It was German glassblower Reinhold Burger and his partner Albert Aschenbrenner, who made bottles for Dewar, who decided to market Dewar's invention. Burger and Aschenbrenner held a contest to name Dewar's device. A resident of Munich suggested the name thermos from the Greek word *threm*, which means hot. Along with Gustav von Paalen, Burger and Aschenbrenner formed a company to manufacture Dewar's invention and called it Thermos GmbH.

Paalen, Burger, and Aschenbrenner did not register the now well-known Thermos name until 1906, the same year they met American businessman William B. Walker in Berlin. Walker learned of their invention and obtained exclusive manufacturing and marketing rights to it in the United States. The American Thermos Bottle Company was incorporated on January 31, 1907, in Portland, Maine, and set up production in Brooklyn, New York. The Thermos insulated bottle quickly gained popularity across the United States. Famous users include President Taft; explorers Lieutenant E.H. Shackelt, who took his to the South Pole and Lieutenant Robert E. Peary, who took his to the Arctic; Colonel Roosevelt on an expedition to Mombassa; Richard Harding Davis on a trip to the African Congo; Count Zeppelin, who took his up in his hot air balloon; and the Wright Brothers, who took theirs to the sky as well in the first airplane.

Raw Materials

The primary raw materials used in the manufacture of the insulated bottle are either plastic or stainless steel, which form the outer cup, and glass, which often forms the inner cup (the inner cup may also be formed

molding is a process in which tiny pellets of plastic resin are heated and forcefully shot into a mold in the shape of the product. When the product cools, it takes the shape of the mold. The first plastic bottles were squeezable and made from polyethylene. Nat Wyeth, a relative of American artist An-

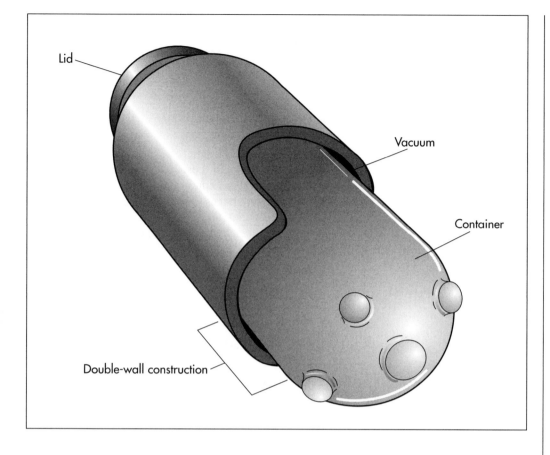

Lid

Vacuum

Container

Double-wall construction

In an assembly line process, the formed outer cup is fitted with its inner liner. A glass filter, made outside the factory, or a stainless still filter, pounded from a sheet of stainless steel, is placed inside the outer cup.

from stainless steel). Insulated bottles for cold drinks are often lined with foam.

The Manufacturing Process

The foam

1 The foam arrives at the factory in the form of liquid chemical balls. These balls are placed together, forming a heat-generating chemical reaction.

2 The liquid mixture is heated slightly, to 75-80° F (23.9-26.7° C).

3 Once the reaction is complete, the mixture is allowed to cool slightly and liquid foam is formed.

The bottle

4 The outer cup is formed. If the outer cup is made of plastic, it is made through a process known as blow molding. In blow molding, small pellets of plastic resin are heated and forcefully blown into a mold in the desired shape. If the outer cup is made of

stainless steel, it pounded into shape from a sheet of stainless steel.

5 In an assembly line process, the formed outer cup is fitted with its inner liner. A glass filter, made outside the factory, or a stainless still filter, pounded from a sheet of stainless steel, is placed inside the outer cup.

6 Next, the insulation is added. For a cold-insulated bottle, foam in a liquid form is sprayed into the space between the cups and allowed to harden. For a hot-insulated bottle, a large vacuuming machine sucks the air from the space between he two cups.

7 A silicone seal coating is sprayed on the cups, holding them together to form a single unit.

8 Steel bottles are then painted.

9 The company logo is pad-printed onto each bottle or a sticker with the company name is placed on the bottle. Pad-printing creates a permanent stamp in the material, rather than on it.

10 After the insulated bottles pass quality inspections, they are wrapped and prepared for shipping.

The top

11 Like the bottle, the top of the insulated bottle is blow molded. If the top has a stopper (many insulated bottled have stoppers that can be pulled upward from the cap. Liquid flows through a small hole in the stopper), that is also blow-molded.

12 A tiny hold is punched in the stopper to allow liquids to flow through it.

Quality Control

Many insulated bottles undergo stringent and thorough quality control inspections. In addition to on-line quality inspections, the company may test the finished product for cosmetic defects and also apply heat retention and leakage tests prescribed by Military Standard 105E . Companies may also employ a special lighting system, called a Macbeth lighting system, which aids them in visually inspecting paint colors to ensure that they match company prototypes. Prior to making the bottle, the foam itself may also undergo strict testing for amount of rise, density, temperature when mixed, viscosity, voids, and discoloration.

The Future

New insulated bottle designs are always being introduced. Designs are adapted to meet the needs and interests of the populations and cosmetic innovations, such as color schemes, change to meet consumer preferences. Many companies not in the insulated bottle business sell or give away water bottles with their names on them, knowing users will provide free advertising every time they carry their insulated bottle down the street. Insulated bottle companies are also expanding farther into international markets, targeting such areas as Asia, Mexico and South America, Africa, and the Middle East.

Where to Learn More

Books

Travers, Bridget, ed. *World of Invention.* Detroit: Gale Research Inc., 1994.

—*Kristin Palm*

Jigsaw Puzzle

Background

A jigsaw puzzle is a picture, which is adhered to a thin and stiff background, like wood or cardboard, and then cut into multiple pieces. The pieces are assembled by the user to reform the original picture. Although the origin of the word puzzle is unknown, it is known that the first jigsaw puzzles were made in the 1760s by European cartographer John Spilbury. In 1762, Spilbury hit upon the idea of gluing maps onto thin mahogany and cedar panels and cutting them up with a fine marquetry saw. He marketed the results of efforts and they became quite popular. Before his death in 1749, Spilbury sold hundreds of puzzles. In the mid-1780s, the next generation of puzzle makers expanded their craft to reach consumers who were not interested in maps. They made puzzles from broadsheets, tabloid size magazines printed with humorous poems or stories. However, broadsheet puzzles were not profitable because their subject material quickly became outdated and new ones had to be printed. Nonetheless, broadsheet puzzles proved that there was a market for puzzles other than maps. Puzzle makers experimented with new images including the alphabet and multiplication tables, Biblical passages, and pictures of historical events and people.

Puzzle popularity increased in England during the following decades, and there is evidence that puzzles arrived in the New World sometime before 1800. Around the same time, the process of color lithography was developed which allowed better quality pictures to be produced more efficiently. This improved the quality and variety of puzzles; some clever manufacturers even made double puzzles with a different scene on each side. In the 1860s, puzzle sales continued to boom as two major companies flooded the market with a variety of puzzle types. These key players were Milton Bradley and the McLoughin Brothers. The 1890s saw the development of die cutting methods, which eliminated the need to cut puzzle pieces by hand. This process allowed puzzles to be mass produced and made them much cheaper. The next few years brought two more significant innovations. First, Parker Brothers, another famous game manufacturer, introduced custom shaped figure pieces into its Pastime brand puzzles. These figure pieces where shaped like recognizable objects, such as dogs or birds. The second innovation was the development of irregular, interlocking pieces. The interlocking format became the standard design because they held the puzzle together and reduced chances that the puzzle would be disturbed during assembly.

Although puzzle sales flagged somewhat in the early 1900s, by the late 1920s and the onset of the Great Depression, there was resurgence in popularity. In 1933, sales peaked at an astounding 10 million per week. With lack of steady employment, people turned to puzzles and other forms of home entertainment instead of outside entertainment like restaurants and nightclubs. Many unemployed architects, carpenters, and other craftsmen made their own jigsaw puzzles for sale or rent. As the puzzle craze of the 1930s continued, drugstores and circulating libraries offered puzzles for rent; they charged 3-10 cents per day depending on the size of the puzzle. For a brief time in 1932, retail stores offered free puzzles with

During the Depression, jigsaw puzzle sales peaked at an astounding 10 million per week in 1933.

the purchase of toothbrushes, flashlights, and hundreds of other products.

By the time World War II ended in the late 1940s, the sales of wooden jigsaw puzzles went into a sharp decline. This was because rising wages increased the labor costs of hand cutting the pieces. At the same time, improvements in the lithography and die cutting (processes which had been introduced decades earlier) made the cardboard puzzles more attractive. The Springbok Company, one major manufacturer, began making puzzles based on high quality reproductions of fine works of art. Hundreds of thousands of Americans struggled to assemble Jackson Pollock's "Convergence," when Springbok introduced it in 1965. By the late 1960s, wooden puzzles had practically vanished. However in the mid-1970s, Stave Puzzles was founded on the belief that there was still an audience for high quality wooden puzzles. Their success has proven them correct, and in the last 25 years, a number of small custom wooden puzzle manufactures have helped re-popularize wooden puzzles.

Raw Materials

Graphics/artwork

Virtually any artwork can be used for puzzle making but most major manufactures use lithographic prints because they are high quality, inexpensive, and easily mass produced. Many of the pictures used in puzzles are based on famous photographs or paintings, but some custom puzzle makers let the customer supply their own photographs or pictures.

Backing material

Mass market puzzle manufacturers use cardboard (also known as chipboard) as a backing material because it is cheap and easy to cut. Higher quality custom-made puzzles still use wood, usually in 5-ply birch. In both cases, adhesive is used to bond the artwork to the backing material.

Cutting equipment

The original wood puzzles were cut with jigsaws, also known as scroll saws, and customized wooden puzzles are still made that way today. These saws have a vertical blade that goes up and down through a fixed horizontal table. The puzzle sheet is guided through the blade by hand to cut the desired pieces. The blades used today are very fine, about 0.016 in (0.041 cm) thick. This allows intricate cuts to be made, which take out very little wood, so the puzzles fit together well. It also leaves a very smooth edge surface, with only a minimum of chipping and fuzzing on the back, which can be sanded off. The majority of puzzles today, however, are the cardboard-backed types and these are mass-produced with die cutting equipment.

Design

Puzzle design varies depending on the type of artwork and the style of puzzle desired. The design of the cuts is hand drawn by artists and, consequently, no two puzzles are alike. Quality puzzles are designed to artfully combine the picture with the design of the cut pieces to enhance the enjoyment of the user. Puzzle artists are cautious in their design not to cut through major features of the artwork such as a person's face. The artists control the puzzle's skill level by varying the number of pieces and the complexity of the cuts. Typically, the more pieces the puzzle is cut into, the more difficult it is to assemble. Some puzzle makers make their puzzles even harder to assemble by avoiding straight-edged border pieces. The lack of a straight border makes the edge pieces harder to locate.

The Manufacturing Process

Nearly 2,000 hours are required to produce a puzzle from start to finish. This process typically stretches over about 12 months. The key steps include printing and laminating the artwork, cutting the pieces, and packaging the finished puzzle.

Printing

1 The first step is to select the artwork and print it in a suitable format. The most common process used to print artwork for puzzles is lithography. Lithography uses a plate, which is specially treated to absorb either water or oil. The portion of the plate, which is not to be printed, is wetted with

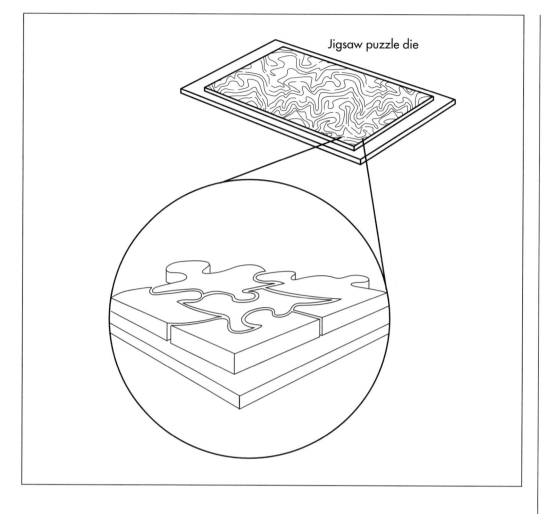

Jigsaw puzzle die

Puzzle pieces are mass produced in a process known as die cutting. A die cutting press uses a sharp, flat metal ribbon to stamp out the individual pieces. The artist's drawings of the cuts are sent to rule-bend experts who bend razor sharp steel rules into the shape of the puzzle pieces. The metal rules are then pounded into a wood mounted die. One side of this metal ribbon is fixed in a wooden block. When this block is pressed with sufficient force onto the softer cardboard backing, the backing surface is cut into the desired shape.

water while the printable portions are coated with grease, which attracts the oil-based ink. When ink is applied to the plate, it sticks only to the grease coated image. As the plate is brought into contact with paper, the image is transferred. Many puzzle pictures may be prepared on the same lithography sheet to save paper and minimize press time. After printing, the litho sheets are laminated onto 0.087 in (0.22 cm) thick chipboard. They are allowed to dry for several days before they are sent to die cut press.

Cutting

2 Puzzle pieces today are mass-produced in a process known as die cutting. A die cutting press uses a sharp, flat metal ribbon to stamp out the individual pieces. The artist's drawings of the cuts are sent to rule-bend experts who bend razor sharp steel rules into the shape of the puzzle pieces. For a 500-piece puzzle of average complexity, it may take 400 hours to make a die.

Three or four such dies maybe made for puzzles of the same size and shape. The metal rules are then pounded into a wood mounted die. One side of this metal ribbon is fixed in a wooden block. When this block is pressed with sufficient force onto the softer cardboard backing, the backing surface is cut into the desired shape. When the laminated artwork is sent through the die cut press, the die is forced down under high pressure. When the die is extracted, the artwork and underlying cardboard are left with cuts in the shape of die.

Packaging

3 After leaving the die press, the sheets go through a breaker, which separates the puzzle pieces and drops them into their package, typically a cardboard box. Today, it is standard for the box to feature a picture of the completed puzzle as a guide. Manufacturers began offering this feature in the mid-1930s. These boxes then go through

final packaging, shrink wrapping, etc. Finally, they are shipped to retail stores.

The Future

While the artwork used in puzzles is constantly changing to keep pace with current consumer tastes, there have been few manufacturing innovations in recent years. Nonetheless, there are areas from which future developments are likely to come. As noted above, quality customized wooden puzzles are gaining in popularity. One company, J.C. Ayer & Co. has developed novel computer-controlled water jets to automate the cutting of wooden puzzles. One new type of puzzle takes two-dimensional jigsaw puzzles and transforms them into three-dimensional puzzles. These puzzles feature die cut pieces which, when assembled, form a three dimensional sculpture. This approach is so novel it has been granted a United States patent (U.S. Patent # 5251900). Lastly, jigsaw puzzles of the future may be electronic without either cardboard or wood. These virtual puzzles are constructed by computer, and exist only on monitor screens. Special software allows puzzle aficionados to continue to enjoy the challenge of reassembling the scrambled pictures without the need for a physical construct.

Where to Learn More

Books

De Cristofor, R.J. *The Jigsaw/Scroll Saw Book.* Blue Ridge, PA: Tab Books, 1990.

Sabin, Francene and Louis. *The One the Only, the Original Jigsaw Puzzle Book.* Chicago: Henry Regnery Company, 1977 .

Williams, Anne D. *Jigsaw Puzzles An Illustrated History and Price Guide.* Radnor, PA: Wallace-Homestead Book Company, 1990.

—*Randy Schueller*

Kazoo

A kazoo is a type of instrument known as a mirliton, which uses a resonating membrane to amplify sound. It belongs in the percussion family of instruments and can be made in a number of ways. Derived from the ancient African mirliton, the kazoo was first manufactured during the 1800s. Today, it is primarily a plastic, toy instrument, which is fun and relatively easy to play.

The primary method of modern manufacture involves injection molding with subsequent assembly of the various pieces.

Background

A kazoo is a wind instrument unlike conventional brass and woodwind instruments. It is typically an open-ended, short tube, which is tapered at one end. Near the wider, front of the instrument there is another opening on top called the turret section, which houses a thin, resonating membrane. It has no valves or buttons typical of other instruments. In addition to these structural differences, a kazoo is also played different than conventional instruments. In brass instruments, the buzzing sound of the lips is amplified to produce the notes. In woodwind instruments, musical notes are created by the vibration of the reed. The kazoo relies on the voice of the musician to produce the sound. When a musician plays a kazoo, he hums into it and that causes the thin film to vibrate. This vibration changes with the sound of the voice giving it the buzzing quality, which is unique to the kazoo.

History

The kazoo was most likely derived from an African instrument called the mirliton. Just like the kazoo, the mirliton has an internal membrane that creates a buzzing sound when it is played. African tribes used it as a voice disguiser during religious ceremonies. In America, a modified version of the mirliton was a popular African-American folk instrument during the 1800s.

The modern day kazoo was invented by Alabama Vest during the 1840s. He drew up the plans for the instrument and had it made by a clockmaker named Thaddeus Von Clegg. In 1852, they demonstrated their kazoo at the Georgia State Fair, and it became a popular instrument in that region. In the early 1900s, a method for large-scale kazoo manufacture was developed by Emil Sorg and Michael McIntyre. McIntyre later started selling kazoos in 1914 and received a patent on the process in 1923. He went on to found the Original American Kazoo Company which is still in operation today. Later, the plastic kazoo was developed and it is now the standard material from which most kazoos are made.

Raw Materials

Kazoos are available in a variety of shapes and sizes. The body of the instrument can be made from numerous materials including plastic, metal, wood, and glass. Of these materials, plastic is most commonly used. Plastics are high molecular weight polymers produced through various chemical reactions. Most kazoos are made from thermoplastics, which are more rigid and durable. Different types of plastics used to make the body of a kazoo include polypropylene and high-density polyethylene (HDPE). To make these materials easier to work with, fillers, which change the properties of the

Derived from the ancient African mirliton, the kazoo was first manufactured during the 1800s. Today, it is primarily a plastic, toy instrument, which is fun and relatively easy to play.

plastic, are often added. These fillers can make the plastic more rigid, more lightweight, and make them less prone to breaking. For decorative purposes, colorants may be added to the plastic to change the appearance of the kazoo. Typically, the plastic is supplied to the kazoo manufacturers as pellets complete with all the filler and colorants already added.

Although plastic is the most often used material, higher priced kazoos are manufactured with metal. These kazoos are usually stainless steel, but silver and gold kazoos have also been produced. While wood and glass kazoos were popular during the late 1800s, they are rarely made today due to their relatively higher cost and difficulty of manufacture.

The resonating membrane is made from materials, which have special characteristics. The materials are thin and flexible to help maximize resonance. They are also resistant to tearing. Wax paper was the material of choice when the first kazoos were made. However, wax paper proved to be unreliable because it often breaks over time. Currently, an advanced silicone plastic is used.

Design

The most critical part in the kazoo manufacturing process is designing the mold. A mold is a cavity carved into steel. When liquid plastic or molten metal is poured into the mold, it takes on the mold's shape when it cools. During manufacture, the mold cavity is highly polished because any flaw on the surface will be reproduced on the plastic. A two piece mold is used to make kazoo parts. The mold pieces are joined together to form the cavity in which the plastic part is formed and opened to release it. Special release agents help make the parts easier to remove. Also, a slight taper in the mold design aids in removal. Steel molds are highly precise and can produce exact parts each time. When molds are designed however, they must be made slightly larger to compensate for the fact that the plastic part shrinks as it cools.

The typical kazoo is composed of three parts including the main body, the turret section, and a resonating membrane. The kazoo's main body is responsible for pro-

viding a playing surface and amplifying the sound. While the main body is typically made in the shape of a tapered tube, it can actually take on almost any shape as long as it is opened on both ends. There are kazoos shaped like bugles, while others look like trumpets or saxophones. The turret section houses the resonating membrane. This section is typically a round piece, which fits into a hole on the main body. The resonating membrane is the heart of the kazoo. It is a thin film, which vibrates as sound passes by. The vibration results in a buzz, which changes the sound emitted from the instrument.

The Manufacturing Process

The manufacture of kazoos has changed very little since the early 1900s. They are still made by a step by step procedure which involves injection molding of the pieces, fitting the pieces together, decorating the kazoo, and putting the finished product in packaging. The major change has been in the type of materials that are used.

Molding

1 At the start of kazoo manufacture, plastic pellets are transformed into kazoo parts via injection molding. The pellets are first put into the hopper of the injection-molding machine. They pass through a hydraulically controlled screw and are melted. As the screw turns, the melted plastic is funneled through a nozzle and physically injected into the mold. Before injection, the two halves of the mold are brought together to form a cavity which has the shape of the kazoo part. Inside the mold, the plastic is held under pressure for a set amount of time and then allowed to cool. As it cools, the plastic inside hardens. After a short while, the mold halves are opened and the kazoo part is pushed out. The mold then closes again and the process begins again. Metal kazoos are stamped from metal sheets and then the two halves are assembled together.

2 After the kazoo parts are ejected from the mold, they are manually inspected to ensure that no significantly damaged parts are used. The damaged parts are set aside to be remelted and reformed into new kazoo parts.

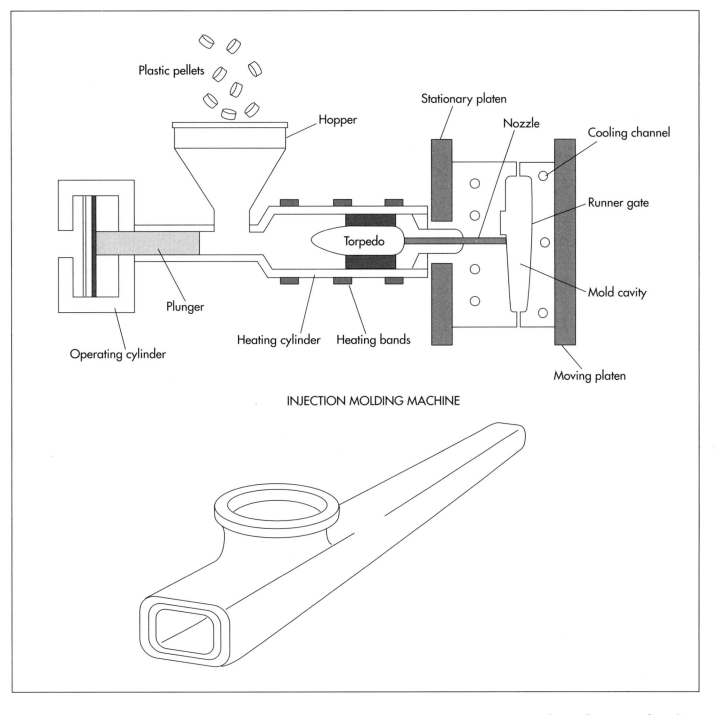

INJECTION MOLDING MACHINE

Assembly of parts

3 The kazoo parts are then transported to an assembly line. In this phase of production, the resonating membrane is first glued to the turret section. This membrane may be supplied to the manufacturer as a pre-cut film or as a large, continuous sheet. After the turret section and membrane are attached, they can be snapped or glued onto the appropriate hole of the main body. At this point, the kazoo is in working order.

Labeling

4 Since kazoos are often sold as toys or advertising specialty items, they are typically decorated with a logo or design. Decoration is applied by either direct labeling or silk screening. Direct labeling involves simply sticking a label on the body of the kazoo. Silk screening is a process, which uses a machine to print colored ink right on the kazoo. The machine consists of a series of rollers and plates, which are dipped in

Plastic pellets are transformed into kazoo parts via injection molding. The pellets are first put into the hopper of the injection molding machine. They pass through a hydraulically controlled screw and are melted. As the screw turns, the melted plastic is funneled through a nozzle and physically injected into the mold. Before injection, the two halves of the mold are brought together to form a cavity which has the shape of the kazoo.

ink. When a kazoo is passed through the machine, the ink is transferred from the plates to the body of the kazoo. The ink is specially formulated so that it dries before the kazoo exits the machine.

Packaging

5 After all the decorations have been set, the kazoos are put in their final packaging. This can be a small box with an instruction booklet included or a plastic blister pack with a cardboard backing. The package serves the dual purpose of protecting the kazoo from damage caused by shipping and advertising the product. The kazoo packages are put into cases and transferred to a pallet. The pallets are then loaded on trucks and the kazoos are shipped to local distributors.

Quality Control

The quality of the kazoo parts are checked during each phase of manufacture. Since thousands of parts are made daily, complete inspection is very difficult. Consequently, line inspectors may randomly check the plastic parts at fixed time intervals and check to ensure they meet size, shape, and consistency specifications. This sampling method gives a good indication of the quality of the overall kazoo production run. Visual inspection is the primary test method. Things that are checked for include deformed parts, improperly fitted parts, and inappropriate labeling. In addition to these checks, more rigorous measurements can also be performed. Measuring equipment is used to check the length, width and thickness of each part. Typically, devices such as a vernier caliper, a micrometer, or a microscope are used. Each of these differ in accuracy and application.

The Future

The future improvements in kazoo manufacture will focus on improving quality,

growing sales, and increasing output. To improve quality, future kazoos will be made with better plastics. For example the resonating film will be more durable and less prone to breakage. To grow sales, manufacturers will likely develop kazoos that have unique designs. For example, they may design a kazoo to tie in with a popular character from an animated movie. In manufacturing, improvements may be made which would reduce the number of parts required to make a kazoo. One process that has potential application in this area is the use of ultraviolet (UV) curable plastics. These plastics rely on UV light to solidify and can be used to make parts without the use of injection molding. Other improvements will focus making the production process more automated and increasing production speeds.

Where to Learn More

Books

Seymour, R. and C. Carraher. *Polymer Chemistry.* New York: Marcel Dekker Inc., 1992.

Chabot, J. F. *The Development of Plastics Processing Machinery and Methods.* Brookfield, IL: Society of Plastics Engineers, 1992.

Stewart, Barbara. *How to Kazoo.* Workman Publishing Co., 1983.

Other

Berghash, R. and D. Jachimowicz. U.S. Patent #4832653.

The Original American Kazoo Company, 8703 South Main Street, Eden, New York 14057. (716) 992-3960.

—*Perry Romanowski*

Kite

Background

A kite is an unpowered, heavier-than-air flying device held to the earth by a line. The kite flies because wind resistance causes the air pressure under the kite to be greater than the air pressure above the kite, making the kite rise. The word kite is derived from the name of a type of bird belonging to the hawk family which is know for its graceful, soaring flight.

A kite consists of three basic parts: the body, the line, and the bridle that attaches the line to the body. To enable the user to control the movement of the kite, the bridle must be attached to the body of the kite in at least two places.

History

Kites were first developed in ancient China. Written references to kites in China date back to 200 B.C., but they were probably invented at a much earlier time. Kites were probably derived from cloth banners, similar to modern flags, which streamed out in the wind while attached to cords or flexible wooden rods. The first use for kites was probably for signaling at a distance. The Chinese later used kites for numerous purposes, ranging from religious ceremonies to warfare. The earliest kites were built of wood and cloth. Paper was invented around the year 100 A.D. and was soon adapted for use in kites.

Kitemaking soon spread from China to Japan, Korea, Burma (now, Myanmar), and Malaysia, regions where kite flying is still an important part of the local culture. From there it spread to Indonesia, India, and the islands of the Pacific. Eventually, the kitemaking technology was adapted by the Arabs, who in turn brought it to North Africa and Europe.

Written references to kitemaking in Europe date back to 1430 A.D. Early European kites were made of cloth or parchment and sometimes had a long slit with a piece of silk sewn into it to help the kite soar. A pair of diagonal sticks were attached to the cloth to hold it in place. A cord was attached to the kite by a ring sewn into the cloth.

The first description of kitemaking in English appeared in 1654 in a book by John Bate entitled Mysteries of Nature and Art. His instructions are not unlike the methods still used to make homemade kites today. "You must take a piece of linen cloth of a yard or more in length; it must be cut after the form of a pane of glass; fasten two light sticks cross the same, to make it stand at breadth; then smear it over with linseed oil, and liquid varnish tempered together...then tie a small rope of length sufficient to raise it unto what height you shall desire."

European kites existed in a variety of shapes, ranging from lozenges to rectangles. They all required tails for stability, and many homemade kites still have such tails. Commercial kites are usually made in such a way that no tail is required.

Kites were used in meteorology as early as the eighteenth century, when two students at the University of Glasgow named Alexander Wilson and Thomas Melville attached thermometers to kites to study the temperature of the air. Kites were used extensively for studying the weather in the 1830s and

The box kite also influenced the design of early aircraft, including the airplane invented by Orville and Wilbur Wright in 1903.

1840s, and continued to be used for this purpose until the middle of the twentieth century, when they were replaced by weather balloons and later by weather satellites.

Innovations in kite design began to appear in the late nineteenth century. In 1891, William A. Eddy, inspired by a Japanese design, invented a diamond-shaped kite, which did not need a tail. In 1893, Lawrence Hargrave invented the box kite, resembling two or more open-ended boxes connected to a wooden frame. Like the diamond kite, the box kite flew well without a tail. Both designs are still commonly used by kitemakers today. The box kite also influenced the design of early aircraft, including the airplane invented by Orville and Wilbur Wright in 1903.

In November 1948, Gertrude and Francis Rogallo applied for a patent on a revolutionary new kind of kite. The patent was issued in March 1951, for the "flexible kite," now usually known as a para-wing. This seemingly simple kite consists of a square of light material (cloth at first, now usually plastic) without any sticks or other parts to hold it in place. Proper length and placement of the cords which make up the bridle enable the para-wing to fly with great stability despite the limpness of its body. Designs similar to the para-wing have been used in parachutes and hang gliders. Military experiments have shown that large versions of this design could be used to carry weapons or vehicles over otherwise impassable terrain. A 4,000 sq ft (372 sq m) para-wing has been used to lift a load of 6,000 lb (2,724 kg).

Raw Materials

Homemade kites are usually made of wood and paper or cloth. Homemade para-wing kites are usually made of Mylar, a tradename for thin sheets of a plastic known as polyethylene terephthalate. This material is extremely strong and very light. The raw materials used to make polyethylene terephthalate are the chemical compounds glycol and dimethyl terephthalate.

Commercial kites are generally made of a strong, light plastic such as nylon. Nylon is the common name for certain types of plastic known as polyamides. Polyamides can be made from a variety of chemical compounds. Nylon-6,6 is the most common form of nylon and is made from the chemical compounds adipic acid and hexamthylenediamine. Another common type of nylon is known as nylon-6 and is made from the chemical compound caprolactam.

The lines attached to the body of the kite are usually made of nylon or cotton. For some large kites, the line is held on a fishing reel, which is made of steel.

The Manufacturing Process

Making nylon

1 The chemicals used to manufacture the various forms of nylon are obtained from a variety of sources. The most common source for these chemicals is petroleum. Crude oil (unprocessed petroleum) is obtained from oil wells. Crude oil contains a mixture of many different substances known as hydrocarbons. The crude oil is pumped into tanks carried by trucks or trains and shipped to refineries.

2 The function of a refinery is to separate crude oil into its various components. During this process known as fractional distillation, the crude oil is pumped into a tall steel furnace shaped like a cylinder. The furnace is heated at the bottom to a temperature between 600-700° F (315-370° C). The heated crude oil boils into a vapor. Any remaining unvaporized residue is removed from the bottom of the furnace as a liquid.

3 As the vapor rises through the furnace, it slowly becomes cooler and cooler. The different hydrocarbons that make up the vapor cool into liquids at different temperatures. Because of this difference in boiling points, each hydrocarbon can be removed as a liquid from a different position within the furnace. Any remaining vapor that does not cool into a liquid is removed from the top of the furnace as a gas.

4 Some hydrocarbons are much more useful than others. In order to maximize the efficiency of refining crude oil, less useful hydrocarbons are chemically transformed into more useful hydrocarbons. This process is known as cracking. Formerly, cracking was accomplished by heating the

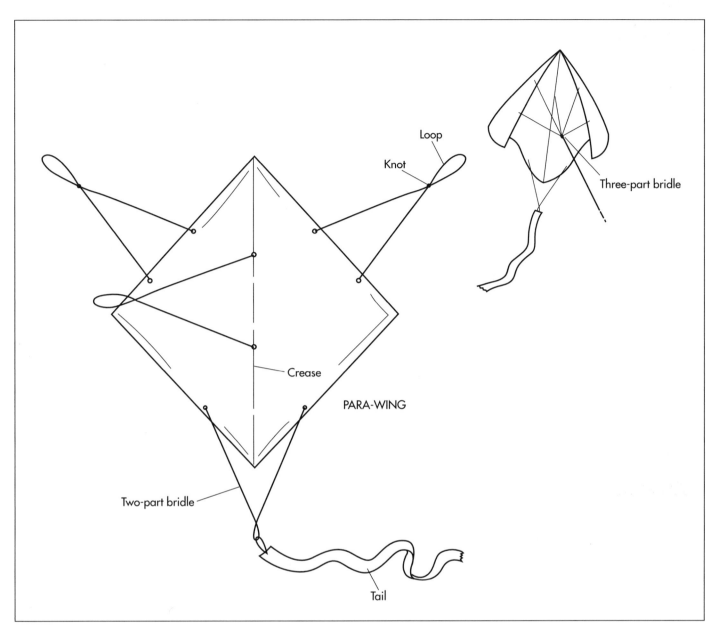

Loop

Knot

Three-part bridle

Crease

PARA-WING

Two-part bridle

Tail

hydrocarbons to a very high temperature under very high pressure. Modern cracking technology makes use of catalysts. A catalyst is a substance that speeds up the rate of a chemical reaction without taking part in it. Catalysts such as natural and artificial clays enable cracking to take place at a much lower temperature and pressure. After cracking is completed, the result is a mixture of various hydrocarbons. These hydrocarbons are separated by once again applying the technique of fractional distillation.

5 Hydrocarbons are shipped from the refinery to the plastics manufacturer. The hydrocarbon needed to manufacture nylon-

6,6 is known as cyclohexane. Cyclohexane is converted into both adipic acid and hexamethylenediamine by subjecting it to a variety of chemical reactions.

6 Adipic acid and hexamethylenediamine (or other chemical compounds needed to produce other forms of nylon) are transformed into nylon-6,6 through a process known as polymerization. This term refers to any process by which hundreds or thousands of small molecules are linked together to form a long chain. Polymerizaiton of nylon combines numerous molecules of an organic acid (such as adipic acid) with numerous molecules of an organic amine (such

The para-wing kite is a seemingly simple kite, consisting of a square of light material (cloth at first, now usually plastic) without any sticks or other parts to hold it in place. Proper length and placement of the cords which make up the bridle enable the para-wing to fly with great stability despite the limpness of its body.

as hexamethylenediamine). For some kinds of nylon, numerous molecules of a single chemical, which contains both an acid group and an amine group, are polymerized. This type of chemical (such as caprolactam, which is polymerized into nylon-6) is known as an amino acid. Polymerization occurs by subjecting the acid and the amine or the amino acid to heat and pressure.

7 The resulting hot liquid nylon is sprayed onto a cool rotating metal drum. This transforms the nylon into a thin, solid sheet. The sheet is cut by sharp metal knives into small chips. The chips can then be processed into many different forms.

Making nylon fabric

8 For some purposes, nylon can be extruded (forced through dies under pressure) or subjected to injection molding (forced into molds as a hot liquid and allowed to cool into a solid). In order to make a kite, nylon must be transformed into fabric. Chips of solid nylon are heated until they melt into a liquid. The liquid nylon is then forced under high pressure through numerous small holes in a steel device known as a spinneret. As the jets of liquid nylon emerge from the spinneret, they are cooled by a blast of cold air. The liquid nylons cool into thin filaments. These filaments are twisted together into fiber. The fiber is woven into fabric and shipped to the kite manufacturer.

Making the kite

9 Large pieces of nylon fabric arrive at the kite factory and are inspected for flaws. Sharp knives and razors are used to cut through several layers of nylon at once in order to produce many pieces of fabric which are all the same shape.

10 The cut pieces of nylon fabric are sewn together using ordinary sewing machines. By efficient cutting and sewing, as little as 3% of the fabric is wasted.

11 In order to hold the limp nylon body of the kite in place, the fabric is sewn around a solid rim, which outlines the shape of the kite. This rim is made of light, rigid tubes of polyethylene. These tubes are made by a plastics manufacturer by injection molding. Solid polyethylene is heated until it melts. The hot, liquid polyethylene is forced into molds in the shape of the tubes and allowed to cool into a solid. The molds are opened, the polyethylene tubes are removed, excess polyethylene is trimmed away, and the tubes are shipped to the kite manufacturer.

12 The bridle lines for the kite are cut to the proper length from spools of cotton or nylon fiber. They are then sewn to the body of the kite in the proper places. For large kites, the line is wrapped around a steel fishing reel. The kite industry is the largest user of fishing reels other than the fishing industry itself. The completed kites are packaged in cardboard boxes and shipped to the retailer or consumer.

Quality Control

The first step in the quality control of kite manufacturing is inspection of the nylon fabric. It must be free from holes and tears, which would damage the ability of the kite to stay aloft. After it is cut, the fabric is inspected to ensure that all pieces have been cut to the proper size and shape. Experienced sewing machine operators inspect the kite at every step of the sewing process to ensure that every piece is sewn into place properly. The position of the bridle line attachments is particularly critical; if they are not properly placed, the kite will be unstable and will fly erratically. Each kite is given a final visual inspection before it is packaged.

Where to Learn More

Books

Eden, Maxwell. *Kiteworks: Exploration in Kite Building and Flying.* Sterling Publications, 1989.

Hart, Clive. *Kites: An Historical Survey.* Frederick A. Praeger, 1967.

Roberts, Keith. *Kiteworld.* Arbor House, 1986.

Wagenvoord, James. *Flying Kites.* Macmillan, 1968.

Yolen, Jane. *World on a String: The Story of Kites.* William Collins and World Publishing, 1968.

—*Rose Secrest*

Krypton

Background

Krypton is chemical element number 36 on the periodic table of the elements. It belongs to the group of elements known as the noble gases. The other noble gases are helium, neon, argon, xenon, and radon. Under normal conditions, krypton is a colorless, tasteless, odorless gas. Its density at normal temperature and pressure is about 0.5 oz per gallon (3.7 g per liter), making it nearly three times heavier than air. At extremely low temperatures, krypton may exist as a liquid or a solid. The boiling point of krypton is –243.81° F (-153.23° C), and its freezing point is only slightly lower at –251.27° F (-157.37° C).

Natural krypton is a mixture of six stable isotopes. Isotopes are atoms which have the same number of protons but which have different numbers of neutrons. The number of protons (the atomic number) determines which element is present, while the total number of protons and neutrons determines the atomic weight of the atom. The isotopes of krypton all have 36 protons and are named for their atomic weights. Krypton-84, which has 48 neutrons, is the most common isotope and makes up 57% of natural krypton. The other stable isotopes of krypton are krypton-86 (50 neutrons, 17.3%); krypton-82 (46 neutrons, 11.6%); krypton-83 (47 neutrons, 11.5%); krypton-80 (44 neutrons, 2.25%); and krypton-78 (42 neutrons, 0.35%).

Krypton can also exist as an unstable, radioactive isotope. These isotopes are created during nuclear reactions. About 20 radioactive isotopes of krypton have been produced. All of these isotopes except krypton-85 are very unstable, with half-lives of a few hours or less. (The half-life of a radioactive substance is the time required for half of the atoms in a sample of the substance to undergo radioactive decay.) Krypton-85, which has 36 protons and 49 neutrons, is much more stable, with a half-life of 10.73 years.

Krypton is used with argon in fluorescent lights to improve their brightness and with nitrogen in incandescent lights to extend their lifetime. It is also used in flashbulbs to produce a very bright light for a very short period of time, for use in high-speed photography. Radioactive krypton-85 can be used to locate small flaws in metal surfaces. The gas tends to collect in these flaws and its radioactivity can be detected.

History

The noble gases were completely unknown to humanity until fairly recently. The first hint of their existence came in 1785, when the English chemist Henry Cavendish discovered that air contained a small amount of an unknown substance that was less reactive than nitrogen. Nothing else was known about this substance until the late nineteenth century.

Meanwhile, the British astronomer Joseph Norman Lockyer discovered a new element in 1868. By analyzing light from the sun, he detected an unknown element that he named helium, from the Greek word *helios* (sun). Helium was not known to exist on Earth for more than a quarter of a century.

In 1894, the English physicist Lord Rayleigh (John William Strutt) and the Scottish

Krypton played an important role in science from 1960-1983, when the length of the meter was defined as 1,650,763.73 times the wavelength of the orange-red light emitted by krypton-86.

Filtered air is compressed under high pressure, raising its temperature. The compressed air is then cooled by rapidly expanding within a chamber. This sudden expansion absorbs heat from the coils, cooling the compressed air. The process of compression and expansion is repeated until most of the gases present in the air are transformed into liquids.

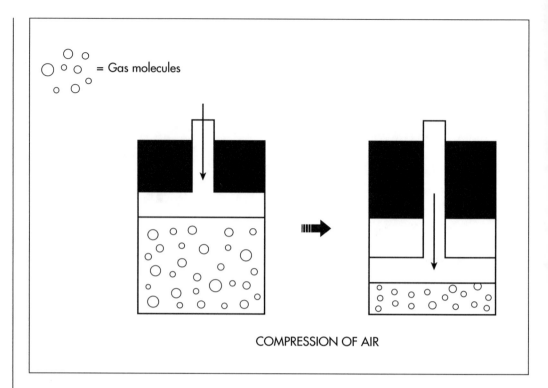

COMPRESSION OF AIR

chemist William Ramsay discovered a difference in the density of nitrogen obtained from the air and nitrogen obtained from ammonia. They soon discovered that the atmospheric nitrogen was mixed with a small amount of an unknown substance. By using magnesium to absorb the nitrogen, they were able to isolate the substance, which they named argon, from the Greek word *argos* (inactive), because it did not react with other substances.

In 1895, Ramsay and his assistant Morris William Travers discovered that the mineral clevite released argon and helium when heated. This was the first time helium was detected on Earth. In 1898, Ramsay and Travers obtained three new elements from air, which had been cooled into a liquid. They named these elements krypton, from the Greek word *kryptos* (hidden); neon, from the Greek word *neos* (new); and xenon, from the Greek word *xenos* (strange).

In 1900, the German chemist Friedrich Dorn noted that the radioactive element radium released helium and an unknown radioactive gas as it decayed. In 1910, Ramsay and his assistant Robert Whytlaw-Gray determined the density of this unknown gas and named it niton, from the Latin word *nitere* (to

shine), because its radioactivity caused it to glow when cooled to a liquid. Niton, later known as radon, was the last noble gas to be discovered. In 1904, Ramsay was awarded the Nobel Prize in Chemistry for his research of noble gases.

The noble gases were formerly known as the rare gases or the inert gases. It was later shown that some were quite common and that some were not completely unreactive. Helium is the second most common element in the universe and argon makes up about 1% of Earth's atmosphere. In 1962, Neil Bartlett created xenon platinum hexafluoride, the first chemical compound of a noble gas. Compounds of radon were created in the same year and compounds of krypton in 1963. No longer thought of as rare or inert, these elements came to be known as the noble gases. Like the so-called noble metals (gold, silver, platinum, etc.), they did not react with oxygen.

Krypton played an important role in science from 1960-1983, when the length of the meter was defined as 1,650,763.73 times the wavelength of the orange-red light emitted by krypton-86. The meter was later defined in terms of the speed of light in a vacuum, but krypton continues to be used in scientific research.

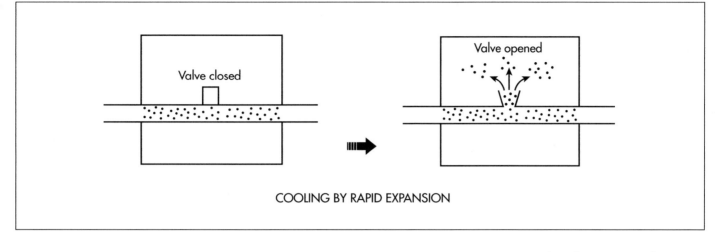

COOLING BY RAPID EXPANSION

Raw Materials

Although traces of krypton are found in various minerals, the most important source of krypton is Earth's atmosphere. Air is also the most important source for the other noble gases, with the exception of helium (obtained from natural gas) and radon (obtained as a byproduct of the decay of radioactive elements). At sea level, dry air contains 78.08% nitrogen and 20.95% oxygen. It also contains 0.93% argon, 0.0018% neon, 0.00052% helium, 0.00011% krypton, and 0.0000087% xenon. Other components of dry air include carbon dioxide, hydrogen, methane, nitric oxide, and ozone.

Krypton can also be obtained from the fission of uranium, which occurs in nuclear power plants. Unlike air, which contains only the stable isotopes of krypton, this process produces both stable isotopes and radioactive isotopes of krypton.

The Manufacturing Process

Making liquid air

1 Air is first passed through filters to remove particulate matter such as dust. The clean air is then exposed to an alkali (a strongly basic substance), which removes water and carbon dioxide.

2 The clean, dry air is compressed under high pressure. Because compression raises the temperature of the air, it is then cooled by refrigeration.

3 The cooled, compressed air passes through coils winding through an empty chamber. A portion of the air, which is compressed to a pressure about two hundred times greater than normal, is allowed to expand into the chamber. This sudden expansion absorbs heat from the coils, cooling the compressed air. The process of compression and expansion is repeated until the air has been cooled to a temperature of about –321° F (-196° C), at which point most of the gases in the air are transformed into liquids.

Separating the gases

4 Gases with very low boiling points are not transformed into liquids and can be removed from the others directly. These gases include helium, hydrogen, and neon.

5 A process known as fractional distillation separates the various elements found in liquid air. This process relies on the fact that the different substances will be transformed from liquid to gas at different temperatures.

6 The liquid air is allowed to warm slowly. As the temperature increases the substances with the lowest boiling points become gases and can be removed from the remaining liquid. Argon, oxygen, and nitrogen are the first substances to be transformed into gases as the liquid air warms. Krypton and xenon have higher boiling points and remain in the liquid state when the other components of air have become gases.

In order to separate krypton, as well as the other gases, from the liquid air, the air is slowly warmed in a process called fractional distillation. Operating under the assumption that each liquid has its own distinct temperature at which it changes to a gas, fractional distillation separates the gases within air one at a time.

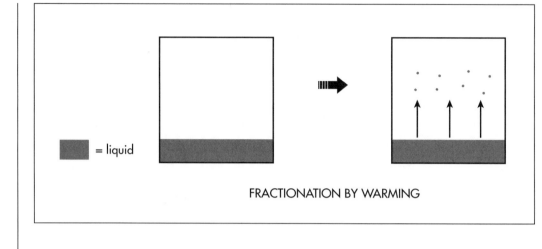

FRACTIONATION BY WARMING

= liquid

Separating krypton from xenon

7 The liquid krypton and xenon are absorbed onto silica gel or onto activated charcoal. They are then once again subjected to fractional distillation. The liquid mixture is warmed slowly until the krypton is transformed into a gas. The xenon has a somewhat higher boiling point and remains behind as a liquid.

8 The krypton is purified by passing it over hot titanium metal. This substance tends to remove all elements except noble gases.

Separating the isotopes of krypton

9 For most purposes, the krypton is now ready to be packaged. For some scientific purposes, however, only one of the six stable isotopes of krypton is desired. To separate these isotopes, a process known as thermal diffusion is used. This process depends on the fact that the isotopes have slightly different densities.

10 The krypton gas is placed in a long vertical glass tube. A heated wire runs vertically through the center of this tube. The hot wire sets up a convection current within the tube. This current of hot air tends to carry the lighter isotopes to the top of the tube, where they can be removed.

Packaging and shipping

11 Krypton gas is packed in bulbs of a strong glass such as Pyrex at normal pressure or in steel canisters at high pressure. Because it is a highly unreactive substance, krypton is very safe. It is nontoxic, nonex-plosive, and nonflammable, so it requires no unusual precautions during shipping.

Quality Control

The most important factor in the quality control of krypton production is ensuring that the final product contains only krypton. The process of fraction distillation has been developed to the point where it produces very pure products from air, including krypton.

Random samples of krypton are tested for purity by spectroscopic analysis. This process involves heating a substance until it emits light. The light then passes through a prism or a grating in order to produce a spectrum, in the same way that sunlight produces a rainbow. Spectroscopic analysis is particularly well suited to studying gases, because heated gases tend to produce sharp, bright lines on a spectrum of pure krypton, it is possible to tell if any impurities are present.

Byproducts/Waste

Krypton is only one of many valuable elements produced by the fractional distillation of liquid air. More than three-quarters of air is made up of nitrogen. Nitrogen is used to produce a wide variety of chemical compounds, particularly ammonia. Because it is much less reactive than oxygen, nitrogen is used to protect many substances from oxidation. Liquid nitrogen is used in freeze-drying and refrigeration.

About one-fifth of air consists of oxygen. The steel industry is the largest consumer of pure oxygen. Oxygen is used to remove excess carbon from steel in the form of carbon dioxide. Oxygen is also used to treat sewage and to incinerate solid waste. Liquid oxygen is used as rocket fuel.

The noble gases obtained from air other than krypton are argon, neon, and xenon. Argon is used in certain types of light bulbs. Passing an electric current through a glass tube containing neon under low pressure produces the familiar neon sign. Xenon is used in strobe lights to produce intense, short bursts of light.

The Future

The future production of krypton is likely to be influenced by the future of nuclear power production. Because krypton can be produced as a byproduct of nuclear fission, nuclear power plants may become an important source of krypton in the future. On the other hand, if nuclear fission is largely replaced by nuclear fusion or by other forms of energy production, krypton is likely to remain almost entirely a product of the atmosphere.

Where to Learn More

Books

Asimov, Isaac. *The Noble Gases*. Basic Books, 1966.

Atkins, P.W. *The Periodic Kingdom: A Journey Into the Land of the Chemical Elements*. Basic Books, 1995.

Compressed Gas Association. *Handbook of Compressed Gases*. Van Nostrand Reinhold, 1990.

—Rose Secrest

Lace Curtain

Background

The word lace is derived from the Latin word *lacques*, meaning loop or snare. The term lace extends to any openwork fabric that is created by looping, twisting, or knotting of threads either by hand or machine. Lace may be made of any fiber—silk, linen, cotton, polyester, rayon, etc. Most handmade laces today are made from linen, cotton, or silk. Machine-made laces are most frequently made from polyester (which performs well on industrial lace machines), cotton, or a combination of both of these fibers.

Lace includes areas of openwork juxtaposed with areas in which fabric or thread has been filled in, thus creating a textile which includes both an open mesh as well as opaque sections. The resulting fabric is airy and light, allowing light and air to filter in through the lace curtain, or permitting a colorful clothing fabric to peek out from behind lace used in clothing decoration.

The first use of lace curtains is unknown, but it is unlikely that anyone but the very wealthy could have afforded to have put handmade lace in their windows where they could be ravaged by sun or rotted by rain. It is more likely that they were used first in the mid-nineteenth century when machine-made laces made such curtains affordable. Furthermore, as heavy curtains are coupled with lighter curtains to shield the privacy of the Victorian house, lace curtains were the logical choice for these filmy barriers. Today, the lace curtain is only made on large lace-making machines that produce thousands of yards each year. While lace curtains of other centuries would have been extraordinarily expensive, the price of mass-produced curtains is very reasonable, and curtains may be purchased of synthetics such as polyester which require little care and are available in a variety of colors.

History

It is difficult for us to imagine the value that our ancestors placed on handmade lace. Always highly prized for its extraordinary beauty and intricate patterns, lace was considered quite a dear commodity until lace-making machines largely destroyed the market for handmade lace. Ancient Egyptian art depicts lace hairnets at about 2,000 B.C. Ancient Babylonian and Assyrian costumes include knotted ornamental braiding and knotting. Lacy fabrics were used nearly 2,000 years ago. By the Middle Ages, ecclesiastical clothing and textiles included lace, as well as exquisite and expensive clothing from the fifteenth century until the early nineteenth century. Particularly complicated and expensive, clothing laces were made by hand in the 1600s and 1700s. Laces made from fine Flemish linen were most highly prized and fortunes were spent on the acquisition of exquisite clothing laces. Sumptuary Laws, which restricted the wearing of gold, silver, jewels, an silk, boosted the popularity of lace, which was often made of plain white linen thread.

By the early nineteenth century, the British were successfully producing machine-made laces with the production of a knitted net. As the machine-made laces became more common, the hand lace-maker could not compete with the low prices of the new laces and the craft waned in popularity. Astonishingly, some old machine made lace imitated the handmade laces to a remarkable degree and sometimes can only be distinguished from the handmade lace by its relentless regularity of pattern (handmade laces incorporate human flaws).

This machine net could then be embroidered or appliquéd by hand. By 1870, several other machine-made lace machines were in production, supplying Americans as well as Europeans with relatively inexpensive lace, including lace curtains. Nottingham lace curtains, with their characteristic square mesh ground, were imported into the United States by at least 1870. By the 1880s, it was affordable and considered a mark of good taste to purchase curtains for the Victorian parlor. By the early 1900s, lace curtains had peaked in popularity and fell from favor—they were a commodity that many had tired of and were associated with those of lesser means who wanted to appear ostentatious.

Today, lace curtains are popular once again. Still prized for their airy beauty, lace curtains permit light to filter through the window, while still providing privacy. Some lace curtain companies offer patterns that have been in machine production for 140 years.

Raw Materials

Lace curtains are made either of polyester, cotton, or a combination of both fibers. Thus, raw materials are simply the yarn used in constructing the curtain. When purchased by the spools, the yarn is generally purchased in 5 lb (2.3 kg) spools. The yarn is generally a beige or natural color; curtains to be sold in colors are dyed after they are knit.

Design

Lace curtain design still generally begins with a sketch on paper. The design is then scanned onto the computer (CAD/CAM). Each scanned design is designated and sized for production on a specific piece of lace-making equipment as the machines vary in width and number of needles used per inch.

Then, it's time to draft the pattern and program the pattern onto a computer disk. Designers draft the pattern using a graphing system on the computer. They determine each machine stitch to be used for each part of the pattern, i.e. "heavy stitch," "lace stitch," and "no stitch." Thus, the pattern is completely filled in, stitch by stitch, for a single repeat. Special software records the pattern and it is saved to the disk. The disk is then inserted into computerized lace-making machines, which are driven by these disks.

The Manufacturing Process

Making the curtains

1 Jacquard Raschel lace-making, computer-driven machines produce many American-made lace curtains. These machines perform admirably with the synthetic yarns, such as polyester, frequently used in modern lace curtains, but can also function well with cotton yarn. (Handmade laces are still generally made of linen or silk.)

The standard Jacquard Raschel machine is 230 in (584.2 cm) long and has at least 12 needles per inch on which four curtains may be produced at one time across the breadth of the loom. Each needle has two threads attached to it that together knit a lace curtain. One thread is attached to the warp, one to the creel. The warp thread makes a chain stitch (the looped background) and the creel yarn is the one that makes the pattern. Patterning is accomplished by varying the interlocking of the rows of loops and then by inserting these creel threads in the loops on the surface of the fabric. Pattern repeats are machine driven and the pattern is repeated again and again. Several panels can be run on a wide loom at time—four curtain panels at a time and three lace tablecloths at a time. (The panels are all attached and are taken off the machine attached to one another. They are cut apart at a later time.) It takes between 4 and 5 minutes to produce one lace curtain but four curtains may be woven horizontally across the breadth of the wide machine at one time.

Dyeing

2 Large rolls of lace curtains are taken off the loom, each roll containing dozens of lace panels. Lace curtain manufacturers may not dye their curtains at their plant but may send it out for chemical dyeing. Generally, when the rolls of curtains are taken from the loom they are of the color of the

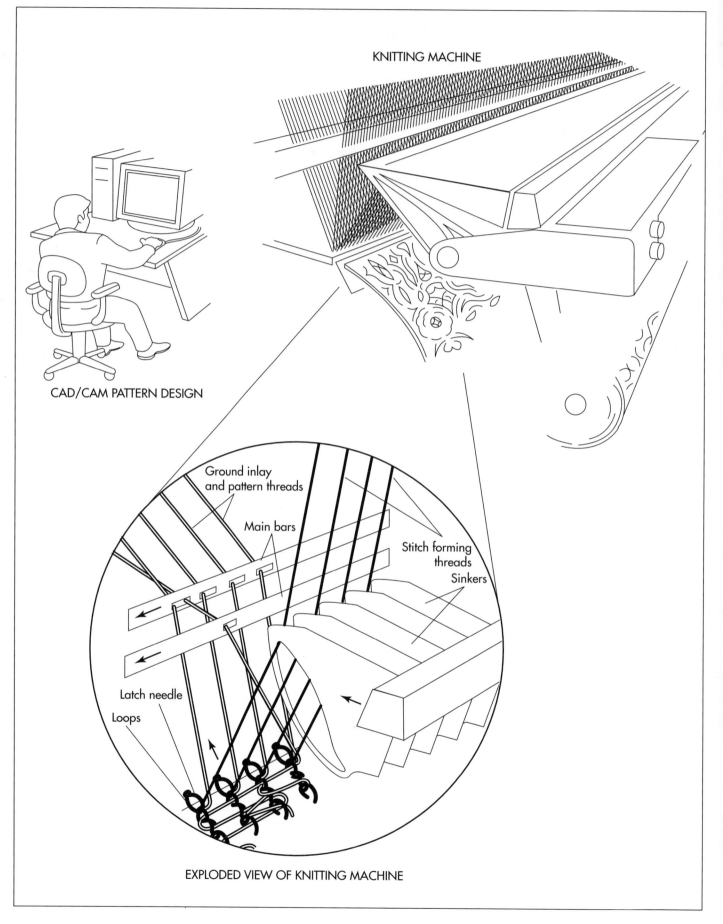

CAD/CAM PATTERN DESIGN

KNITTING MACHINE

Ground inlay and pattern threads

Main bars

Stitch forming threads

Sinkers

Latch needle

Loops

EXPLODED VIEW OF KNITTING MACHINE

original spool of yarn—often a beige or natural color. Curtains that will be sold as "white" must then be dyed white; those that are to be colored are put into dye vats.

3 After dyeing, the roll is placed in a tenter frame, where the lace is secured under clips, dried, and heat set. This drying process makes sure the curtain is framed to the right size (that the repeat is shaped to the correct size) and is then passed under an oven to dry.

4 When dry, the curtains are just the right size and shape for easily cutting the curtain repeats off the roll. The curtain panels are cut apart from the larger roll using automatic cutting equipment, hand scissors, or perhaps a hot pen. The hot pen cuts lace in a specific pattern or configuration (perhaps around a petal or flower) and seals the polyester yarn at the same time so that the knitted pattern will not unravel.

5 Curtains that are cut must be hemmed at top and bottom. Some curtains have a wide "rod pocket" sewn into the top of the curtain to accommodate a wooden or metal rod on which the curtain is hung and this rod pocket is generally 3 in (8 cm) or 5 in (13 cm) wide. The curtains are sent to industrial sewing machines where they are hemmed at top and bottom. The curtains are then folded by hand or machine and sent on a belt to be packaged and sent from the plant.

Quality Control

Each roll of curtain goods has a ticket attached or associated with it that is essentially a report of its production. Dropped stitches or any problems in the production are detailed on the ticket. The company then has a record of damages in order to address quality problems. In addition, the computerized looms may detect and note dropped stitches with an electric eye. Supervisors are responsible for checking inspection reports and use their eyes, experience, and knowledge of the process to monitor quality as well. Dyed curtains are inspected for color, and those that are improperly dyed are sent back to the dye house for re-dye. After cutting and sewing, the curtains are inspected to ensure proper size and shape and may be re-shaped and re-sewn if necessary. Occasionally improperly shaped or sewn curtains may be sold as "seconds" at discount houses.

Byproducts/Waste

Lace curtains made with cotton yarn have a lint problem that must be dealt with in order to provide a safe environment for workers' respiration. In plants with older machines, the company must vacuum the machines daily in order to rid the air of lint. Newer plants have vacuum or suction systems built into them to reduce levels of lint. Unfortunately, cotton lint is not recyclable into new cotton thread and must be discarded as garbage. Polyester yarn does not create extraneous lint. The bits of polyester yarn left on unused spools are bundled into bales and may be recycled into new yarns. Also, dyeing of any fabric includes chemicals suspended in water, which must be neutralized and filtered according to federal standards before the effluvia is released into any water supply.

Where to Learn More

Books

Jerde, Judith. *Encyclopedia of Textiles*. NY: Facts on File, 1992.

Ponde, Gabrielle. *An Introduction to Lace*. NY: Charles Scribner's Sons, 1973.

Warnick, Kathleen and Shirley Nilsson. *Legacy of Lace*. NY: Crown Publishers, 1988.

—Nancy EV Bryk

Opposite page:
A Jacquard Raschel machine is 230 in (584.2 cm) long and has at least 12 needles per inch on which four curtains may be produced at one time across the breadth of the loom. Patterning is accomplished by varying the interlocking of the rows of loops and then by inserting these creel threads in the loops on the surface of the fabric. Pattern repeats are machine driven and the pattern is repeated again and again. Several panels can be run on a wide loom at time—four curtain panels at a time and three lace tablecloths at a time.

Lava Lamp

Credit for the creation of the lava lamp is given to English engineer Craven Walker who, in the late 1940s, saw a prototype of the lamp in a pub in Hampshire, England. By the time Walker left the business in 1990, he had sold over seven million of his creations.

Background

A lava lamp is a decorative tube-shaped light fixture containing a colored, oily fluid that flows up and down throughout the lamp chamber in a manner reminiscent of molten lava. As the fluid rises and sinks in the lamp chamber it changes shape and breaks into globules of various sizes, giving a psychedelic effect of constantly shifting patterns. Credit for the creation of the lava lamp is given to English engineer Craven Walker who, in the late 1940s, saw a prototype of the lamp in a pub in Hampshire, England. This early version, according to the Walker legend, was made of "a cocktail shaker, old tins and things." He purchased the liquid-filled fixture and set out to make his own. Walker formed the Crestworth Company in Dorset, England, and over the next 15 years tried to build a better lava lamp. When first marketed under the name Astro Lite in British stores in the early 1960s, it was not an immediate success. Then, at a 1965 German trade show, two American entrepreneurs saw an early model on display, and bought the rights to manufacturer the lamp in North America. In the United States, they changed the name from Astro Lite, to the infinitely hipper Lava Lite Lamp and began manufacturing operations in Chicago. With the advent of psychedelia and pop-art later that decade, Walker's gimmicky contraption became a major fad. By the time Walker left the business in 1990, he had sold over seven million of his creations. Today, the company ships 400,000 lamps a year to shops around the world. Currently, Haggerty Enterprises is the only U.S. manufacturer of Lava Lite Lamps and they distribute them nationally through a number of retail and mail order outlets.

Design

The lava effect is due to the interaction between the fluids used in the lamp. These fluids are selected on the basis of their density so one tends to barely float in the other. In addition, they are chosen based on their coefficient of expansion, so as they are heated one tends to rise or sink faster than the other. When heat from the light bulb warms the heavier liquid sitting on the bottom, it gets hotter and, due to its lower density, rises to the surface. By the time the "lava" reaches the top of the lamp, it begins to cool, becomes denser, and sinks to the bottom. As the lava sinks, it gets closer to the light bulb, heats up again, and the process is repeated over and over. Therefore, the key to successful lava lamp design is the selection of appropriate immiscible fluids. The exact composition used in lava lamps is a proprietary secret, but in general terms, one fluid is water based and the other is oil based. The aqueous phase may be water mixed with alcohol or other water-soluble solvents. The second fluid must meet a number of design criteria: it must be insoluble in water, heavier and more viscous, non-reactive and non-flammable, and reasonably priced. It must also be non poisonous, unchlorinated, not emulsifiable in water, and must have a greater coefficient of expansion than water.

While fluid selection does not change from lamp to lamp, there are design changes to be considered because lamps are available in different colors, sizes, and styles. The original Century model, which is still manufactured today, was the most popular model during the 1960s and 1970s. Its gold base is perforated with tiny holes which simulate starlight and its 52 oz (1.46 kg) globe is filled with red or white lava and yellow or blue liquid. A number of interesting variations on the Century have been manufactured in past years, although not all of them are still made today. For example, the Enchantress Planter Lava Lite lamp came equipped with plastic foliage. The Continental Lava Lite lamp which, was the only cordless, non-electric model, featured a candle to warm the lava. The Consort Lava Lite lamp, according to the company's 1970s catalog, was designed with a more masculine look "perfect for the study or den, so right for the executive suite." There was also the Mediterranean Lava Lite lamp, which was decorated with black wrought iron. In addition, Haggerty offers so-called giant lamps, which range in size up to 27 in (68.6 cm) tall.

Raw Materials

As noted above, the actual ingredients used in Lava Lite Lamps are proprietary but there are several liquid ingredients, which can be combined to give a lava effect.

Liquid components

Lava-type lamps can be made with water mixed with isopropyl alcohol as one phase and mineral oil as the other. Other materials, which may be used as oil phase ingredients include benzyl alcohol, cinnamyl alcohol, diethyl phthalate, and ethyl salicylate.

Other additives

Other additives used in lava lamp fluids include various oil and water-soluble colorants. The specific gravity of the aqueous phase can be adjusted through the addition of sodium chloride or similar materials. In addition, a hydrophobic solvent may be added to the mixture to help the lava coalesce. Turpentine and similar paint solvents are said to work well in this regard. Antifreeze ingredients can also be added to increase the rate at which the lava warms.

Container

A clear glass cylinder is used to house the fluids and forms the body of the lamp. The classic lava lamp shape is an hourglass about 10 in (25.4 cm) high.

Heat source

A regular incandescent bulb is used as the source for both light and heat in a lava lamp. The type of light bulb is critical to ensure the lava is not over or under heated. Haggerty Enterprises lists several bulb types that are appropriate for their appliances, depending on which the model: 40 watt frosted bulb, 100 watt reflector bulb with inside frost, 7.5 watt bulb 40 watt candelabra type. Although it does not generate at much heat, a florescent bulb is used in their Pacifica model.

Hardware

Other items used in lava lamp production include the base plate, which houses the electrical components, 16-gauge lamp wire, and an electrical plug. Quarter inch (0.635 cm) thick foam rubber may be used as a gasket material to seal the chamber. Miscellaneous hardware, such as screws, is also used. Optional equipment, such as a light dimmer switch or a small fan, may be used for temperature control.

The Manufacturing Process

The manufacturing process of the lamp consists of several steps, both automated and manual. According to a representative of Haggerty Enterprises, the company has the capacity to produce up to 10,000 lamps per day on their assembly line.

Container assembly

1 The glass cylinder is fastened to the ceramic lamp fixture, which forms the base. The lamp is attached to the appropriate wiring and the bulb is screwed into place. The gaskets are glued into place to prohibit leakage. The containers are assembled and checked to make sure they are leak proof.

Compounding liquid phase

2 The liquid phases are mixed and added separately. Isopropyl alcohol is used to lower the specific gravity of the water phase so the mineral oil floats appropriately. By mixing water and alcohol in the correct proportions, the mineral oil can be made to float. In 90% alcohol, the mineral oil will

A lava lamp is made by mixing alcohol and water and mineral oil and dyes, combining each separately. By mixing water and alcohol in the correct proportions, the mineral oil can be made to float. The correct ratio is about six parts 90% isopropyl alcohol to 13 parts of 70% isopropyl alcohol. Dyes, salt, etc. are then mixed into the water phase, and the oils and waxes are added to the second liquid.

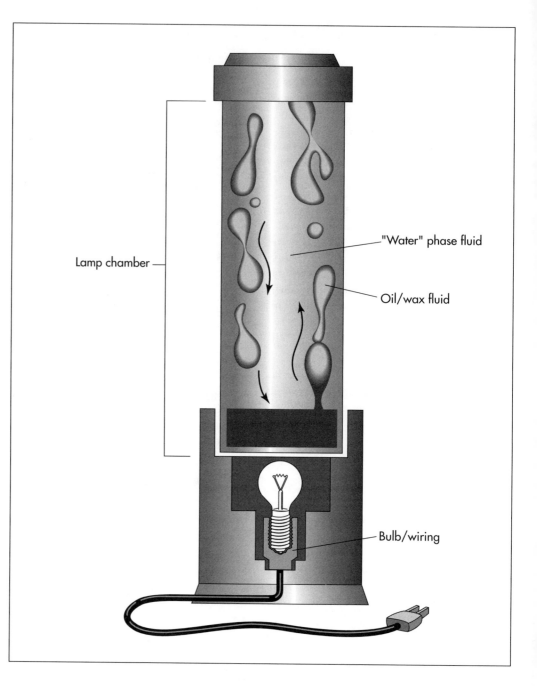

Lamp chamber

"Water" phase fluid

Oil/wax fluid

Bulb/wiring

sink to the bottom. The addition of 70% alcohol will make the oil seem lighter until it is about to "jump" off the bottom. The correct ratio is about six parts 90% isopropyl alcohol to 13 parts of 70% isopropyl alcohol. Dyes, salt, etc. are then mixed into the water phase, and the oils and waxes are added to the second liquid. Some heating may be required to melt the waxy materials.

Filling

3 The lamps are moved along a conveyor line and first filled with the oil/wax phase, then the water phase. A small airspace of about 1 in (2.54 cm) is left at the top to allow for expansion of the hot liquids. This important because the amount of airspace can influence the size of the bubbles formed by the lava. After filling, the cylinder is capped with either a screw type cap or a bottle cap type, which is crimped into place.

Quality Control

During the batching and filling processes, the liquids are checked to ensure they were correctly manufactured. The proper ratio and composition of the two liquid phases is critical to ensure that the lava effect will be achieved. An incorrect ratio may allow the oil and water phase to mix together, separate into bubbles which are too small, rise and fall as one continuous mass, or become mixed with the water and not separate at all. It is critical that all electrical connections are good and seals are tight to ensure safety and that there are no leaks. Improper gasket alignment or poor seals can result in leakage of the fluids. After filling, each lamp is checked to ensure the light bulb is completely centered and tightened. The bulb and socket may move slightly during shipping. If so, the owner is instructed to gently push the socket back to the center of base. Instructions on how to change the bulb are provided inside the lamp socket base. Variation in lamp size or wattage may yield unsatisfactory lava flow and may increase the risk of fire.

During initial use, the lava may not flow properly or may float to the top of the globe. If this happens, the lamp should be allowed to heat up for four hours or more to allow the lava material to become completely molten. Excessive agitation of the lamp may cause the fluids to intermingle and become cloudy, or may even result in permanent malfunction. The lamp should not be stored in direct sunlight as this may cause the colors to fade.

The Future

Given the proprietary nature of lava lamps, it is difficult to speculate on future improvements in the manufacturing process. However, it is interesting to note that computer technology has spawned its own version of the lava lamp—the virtual lava lamp. Also known as the Javalamp after the popular computer language, this virtual lamp is a computer animation that mimics the appearance of the real item.

Where to Learn More

Books

Popular Electronics Hobbyists Handbook. Gernsback Publications Inc, 1992.

Other

Haggerty Enterprises Inc. Chicago, IL.

—*Randy Schueller*

Lead Crystal

Glass making is a 2,000-year-old process that has changed remarkably little in that time.

Background

Ordinary glass has been made for thousands of years and was a product of most ancient cultures. The ancients also began using crystal in its native form of rock crystal for beads, figurines, and dishes. In attempts to imitate nature, man began making glass that was termed crystal by adding metals to change the character of the glass, and lead was found to be the most successful of these additives. Lead crystal produces a product with a ringing sound (without the tin quality of ordinary glass), it is strong and durable, and it has a curious warmth to the touch. Best of all, lead crystal has a brilliant, silvery appearance that is enhanced by cutting.

History

The great glass- and crystal-making countries of Europe include the Netherlands, Czechoslovakia, England—and the City of Venice. The history of the art form in England began with the Roman occupation of Britain, and it had a number of high points including the manufacture of stained glass during the ages of cathedral building. As an industry, it reached a new level in the mid 1500s when several leading glassblowers from Venice moved to London and found the favor of Queen Elizabeth I who promoted the art form. The Venetian influx and the Queen's support made discovery of lead glass possible in the next century.

George Ravenscroft established his own glasshouse in London in 1673 and, shortly after, patented a process for making "flint glass" or lead crystal. Ravenscroft found that the addition of lead to glass during the melting process improved the quality of the glass. Early defects included the introduction of a bluish tinge and "crizzling" of the glass. Increasing the lead content in the crystal eliminated such flaws. He continued experimenting with the chemical composition of glass, and eventually eliminated the imperfections. The practice of cutting glass came into prevalence during Ravenscroft's time (previously, unadorned glass was thought to be beautiful on its own), and his invention was the perfect medium for this kindred art form. In the 1700s, the number of glasshouses in England grew tremendously, however a government tax on glass began to hurt the business.

Manufacturers escaped the Excise Tax by moving their factories to Ireland, and it was during this period that Ireland became the new center for production of lead crystal, notably in the port city of Waterford. There, George and William Penrose founded the Waterford Glass House in 1783, and, by 1851, the house won worldwide attention at the aptly named Crystal Palace Exhibition (one of the first world's fairs) in London. The profitability of the Irish glass houses also caught the attention of the tax authorities, and the tax on glass that was not instituted in Ireland until 1825 finally forced the closure of the Waterford factory during the year of its great Crystal Palace triumph, 1851.

Glass houses elsewhere in Europe thrived throughout the 1800s when Baccarat in France, Orrefors in Sweden, and Swarovski in Austria, to name only three, became leading lead crystal manufacturers. The Irish tradition did not resurface until after World War II when a resurgence of interest in the

Irish arts encouraged a group of businessmen to resurrect Waterford. Today, all of the name glass houses have flocks of admirers and collectors worldwide, and they often market each others' products as a means of boosting international interest and protecting that small brotherhood of artists in glass.

Raw Materials

The raw materials for glass making are a chemical "cocktail" of silica-sand (also called silver sand), potash, and red-lead. A yellow oxide of lead called lethargy was used when lead crystal was first developed, and it is produced from red-lead oxide when some of the oxygen is driven off . Silica occurs in nature as the sand found on beaches (although sand from inland sandstone deposits is used in glassmaking) and the pure form of quartz that produces hexagonal crystals. Each glassworks factory concocts its own formula that produces the qualities needed for its particular manufacture of glass. A typical comparison of the quantities of materials that make the differences between ordinary, or table, glass and crystal follows: ordinary glass with 63% silica-sand, 22% soda, and 15% limestone; and lead crystal with 48% silica-sand, 24% potash, and 28% red-lead.

Colored glass is made by adding other metals to the glass mixture. Manufacturers may also add tiny amounts of saltpeter (a nitrate of potash), borax, and arsenic to their glass recipe. Standards have been devised for the quality of crystal in which the percent of lead or other oxides, the density of the glass, the refractive index of the glass, and its surface hardness are established. Crystal glass, pressed lead crystal, lead crystal, and full lead crystal are defined differently based on these standards.

Design

Although they may be similar among crystal makers, each firm typically has its own designs, just as it has its own mixture of raw materials (an element that makes particular designs possible). Approaches to design vary depending on the purpose of the product. If the crystal-maker is designing a line of glasses and decanters, customers will want to add or replace pieces over the years. Designs must then be chosen to be enduring over the years, and contemporary pieces, particularly, must be carefully made to age as well as having trendy appeal. Collectibles are made to reflect the best characteristics of the crystal house, but they can truly change with fashion and taste more often than pieces in a matched set. Emphasis in design for most houses is on preserving the high standards associated with that name.

Actual design is accomplished by skilled artisans who are knowledgeable in the techniques of glass making, the thicknesses required for particular objects, the depth of cuts that can be made in those thicknesses, and the skills of the glass cutters and engravers who will do that work. The patterns or designs are classified according to flat, hollow (rounded), or miter (v-shaped) cuts, and the motifs developed from these basic cuts are an encyclopedia of combinations with names like hollowed or strawberry diamond, flute, fans and splits, alternate panels, and hobnail. Sizes and types of pieces that will be part of that design line are important, and even the names are selected to evoke heritage or modernity as appropriate.

The Manufacturing Process

Glass making is a 2,000-year-old process that has changed remarkably little in that time. Raw materials are essentially the same, although experiments over the years with the addition of lead to crystal have improved the product.

1 The crystal "cocktail" is mixed and made into molten crystal in a furnace heated to 2,192° F (1,200° C). The red-lead oxide is introduced into the furnace on a rapidly moving air current in a difficult procedure that can create different grades of purity. The furnace conditions must also be carefully controlled so the lead fully oxidizes and doesn't leave metallic lead, which not only discolors the glass but also attacks the fireclay of the furnace. The lead adds density to the glass so it is heavier; this weight advantage over ordinary glass also changes the crystal's light diffraction properties and the sound or ring of the crystal when it is struck.

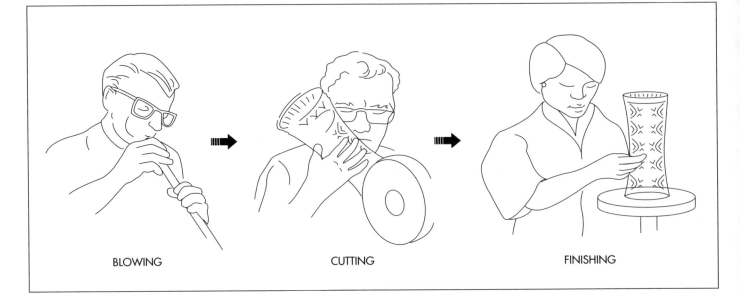

BLOWING　　　　　　CUTTING　　　　　　FINISHING

A team of glass blowers uses physical strength, breath control, and dexterity to create the chamber with a certain thickness. This skill is only developed by experience. The thickness must be suitable to the object itself but also to the depth of the facets that will be cut in the glass.

2 Lead crystal is blown glass, but, because of its thickness, it requires a team of four to seven artists who are as well coordinated as any team of athletes. Wood blocks and molds are used to create the basic shape of the object called a crystal chamber. As soon as the molten glass touches the mold, it begins to cool instantly so glass blowing is a process of seconds. The chamber is also known as hollowware because it has an opening that is the functional part of the beverage glass, vase, or decanter.

3 The team of glass blowers uses physical strength, breath control, and dexterity to create the chamber with a certain thickness. This skill is only developed by experience. The thickness must be suitable to the object itself but also to the depth of the facets that will be cut in the glass. Near the furnace, other parts are added to the chamber. For example, a piece of molten glass is attached to the side of a decanter to form a handle and shaped to the perfect curve in one quick motion. Again, the glass will begin to harden almost immediately in the open air.

4 To slow that hardening process, the blown crystal piece is transferred to an annealing oven. If the crystal cools too quickly, stresses will be induced in the crystal as varying thicknesses cool differently and the crystal will contract too rapidly. The annealing process takes from 2-16 hours depending on the size and configuration of the piece.

5 The cooled crystal is now a "blank canvas" for the glass cutter. To prepare for cutting, the pattern is drawn on the vessel, usually with red-lead and turpentine. The design is then roughed out with a power-driven wheel equipped with different edges depending on the type of cut required. These cut surfaces are coarse and not as long or as deep as the finished cuts. The cutters are true artists who use sight, feel, physical strength, and their extraordinary memory for patterns, details, and cuts. The types of cuts are wedge and flat cuts. Wedge cuts are made with diamond-tipped wheels and produce deep facets. Flat cuts are made with the same equipment but are not as angled; they provide contrast with the deeper wedge cuts.

6 In the smoothing process, a sandstone wheel is used to dress the rough cuts; this part of the process requires the greatest skill because it establishes the finished design.

7 Finally, the finished vessel is polished by dipping it into a mixture of sulphuric and hydrofluoric acids. The acid bath attacks the entire surface of the object and removes a very thin layer while leaving a lustrous and uniform finish.

8 The process of intaglio, which uses tiny copper wheels that revolve slowly to engrave portraits or other illustrations and information into the object, is also used to engrave some pieces. Most engraved pieces,

like large sports trophies, are one-of-a-kind items that are completed on commission. They may require many hours of engraving. The finished piece is carefully inspected and packaged in materials designed to display and protect the object.

Quality Control

Quality is a continuous process at the lead-crystal factory. The artists themselves provide the first level of quality control through their experience and unique skills. No two pieces are identical despite adherence to designs, heritage, and standardization of patterns and cuts. Stringent inspections are performed at each step of the process. This helps avoid wastage in that in imperfectly cooled piece is never sent to the cutter; inspections also ensure conformity with the acceptable range of variations that are inevitable in handwork and among individual cutters and other artists.

Byproducts/Waste

There are no byproducts from production of lead crystal. Waste is avoided in raw materials and by careful control of processes like annealing. Glass factories have the advantage that their products can be melted again in the furnace, so many boast that they produce no seconds because of this luxury of being able to fully recycle an imperfect product.

The Future

Lead crystal has a promising future because it has an enduring association with both handcrafting and elegance. An "ordinary" family or collector without an extraordinary bank account will find it satisfying to build a crystal service or collection of figurines over a lifetime, and this sense of style and worth is inherited by future generations, as the pieces themselves will be. The artistry in each piece of lead crystal is also appreciated in an age of sound bites. Even the most contemporary design represents a long heritage and the skills of the artists who created it. This sense of appreciation shows every sign of flourishing in the next century, as it has in the past.

Where to Learn More

Books

Elville, E. M. *English and Irish Cut Glass.* London: Country Life Limited, 1955.

Elville, E. M. *English Tableglass.* New York: Charles Scribner's Sons, 1951.

Gros-Galliner, Gabriella. *Glass: A Guide for Collectors.* New York: Stein and Day, 1970.

Littleton, Harvey K. *Glassblowing: A Search for Form.* New York: Van Nostrand Reinhold Company, 1971.

Pfander, Heinz G. and Hubert Schroeder. *Schott's Guide to Glass.* New York: Van Nostrand Reinhold, 1983.

Rogers, Frances and Alice Beard. *5000 Years of Glass.* New York: Frederick A. Stokes Company, 1937.

Schrijver, Elka. *Glass and Crystal: Volume I: From earliest times to 1850.* New York: Universe Books Inc., 1964.

Schuler, Frederic. *Flameworking: Glassmaking for the Craftsman.* Philadelphia: Chilton Book Company, 1968.

Webber, Norman W. *Collecting Glass.* New York: Arco Publishing Company Inc., 1973.

Zerwick, Chloe. *A Short History of Glass.* New York: Harry N. Abrams, Inc., Publishers, 1990.

Other

Collection Guide: Swarovski Crystal. http://www.assendorp.nl/.

Crystal World. http://www.crystalworld.com/.

Stegeman Art Gallery. http://www.branson-connection.com/stegeman.htm.

Valaská Belá. http://www.unicorn-connection.co.uk/valaska.

Waterford Crystal. http://www.waterford-usa.com/.

—*Gillian S. Holmes*

Licorice

Licorice represents one of the oldest forms of candy with evidence suggesting it was made as early as the thirteenth century. It continues to be a popular product today making up a significant portion of the over $13 billion, annual non-chocolate candy market.

Background

Licorice is a glossy, gelled candy with a semi-firm consistency that is flavored with licorice root extract. Typically, it is a moderately sweet candy, and is available in a variety of flavors such as black licorice, strawberry, cherry and chocolate. Currently, it is touted as a healthier snack food because it contains almost no fat per serving. It is made in a continuous process, which involves mixing and cooking, forming the candy, cutting it, and putting it in packaging. Licorice represents one of the oldest forms of candy with evidence suggesting it was made as early as the thirteenth century. It continues to be a popular product today making up a significant portion of the over $13 billion, annual non-chocolate candy market. It is anticipated that licorice manufacturers will try to increase yearly sales primarily by increasing the speed at which they can produce the candy.

History

Using sugar-refining techniques, Arabs first produced various types of lozenges for pharmaceutical applications. One type of lozenge was flavored with licorice, which is a native plant of the Mediterranean area. The Arab peoples believed that the licorice root had important medicinal uses. Evidence of this crude predecessor to the contemporary licorice candy suggests that licorice is one of the oldest types of candy known.

During the thirteenth century, licorice root extract was widely used as a medicine for coughs, sore throats, and congestion. It is likely that merchants who sold this medicine combined it with honey to produce the first true licorice candies. Later, when sugar was more readily available, it was used instead of honey. In the late Middle Ages, licorice pastilles, which were cast in rough molds were widely known. Extruded licorice candy is thought to have originated in Holland at the start of the seventeenth century. It became one of the standard confection products for candy producers when the candy industry developed in the mid-1800s.

Raw Materials

Experienced chemists and candy technologists develop licorice candy recipes. By using their knowledge of ingredient characteristics and production processes, they can create a variety of licorice candy types. The ingredients in these recipes are specifically chosen to provide desired characteristics such as texture, taste, and appearance. They are typically mixed together in water to create a homogenous blend, and then much of the water is evaporated off to produce a solid product. The primary ingredients include sweeteners and wheat flour. Other ingredients such as starches, preservatives, colorants, and flavorings are also important.

Sweeteners

Since licorice is a sweet candy, sweetening ingredients make up much of their composition. Sugar and corn syrup are two primary sweeteners. Sugar is sucrose, which is derived from beet and cane sugars. It is supplied as small, white crystals, which readily dissolve in water. Since sugar is not critical to the texture of the licorice, it can be less refined, reducing the cost of the licorice recipe. Corn syrup is typically used in combination with sugar in licorice candy. It is a

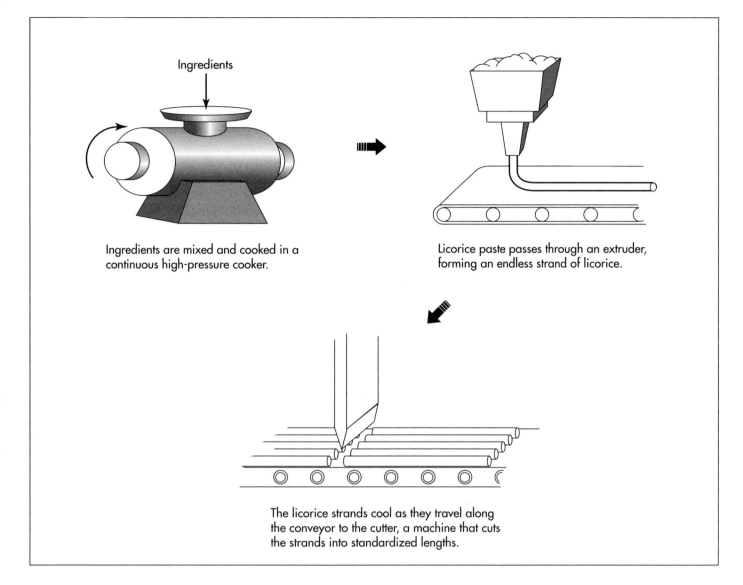

Ingredients

Ingredients are mixed and cooked in a continuous high-pressure cooker.

Licorice paste passes through an extruder, forming an endless strand of licorice.

The licorice strands cool as they travel along the conveyor to the cutter, a machine that cuts the strands into standardized lengths.

modified form of starch, and like sugar, it provides sweetness to the licorice. It also inhibits sugar crystallization, helps control moisture retention and limits microbial spoilage. Beyond sugar and corn syrup, other sweeteners are sometimes incorporated into the licorice recipe. These include molasses, glucose syrups, and other crude sugars. Some low calorie licorice candies incorporate artificial sweeteners like aspartame (Nutrasweet). Sweeteners make up about 60% of the licorice paste.

Flour

All the ingredients in licorice must be bound together to maintain a cohesive product. To accomplish this, candy technologists use wheat flour. Wheat flour is obtained by grinding wheat seeds into a powder. It is primarily composed of starch and protein which when combined with water creates a paste, called gluten, that can be stretched and rolled without breaking. These properties allow the finished licorice paste to be extruded into various sizes and shapes. The flour is also responsible for the licorice's shiny appearance because during licorice manufacture, the starch in the flour is gelatinized. It is typically incorporated into the licorice paste at about 25-40%.

Flavorants, colorants, and other ingredients

Many other ingredients are incorporated into a licorice recipe to produce the familiar candy. To give the candy flavor and color, licorice black juice is included at a level from 1.5-4%. Since this material is quite

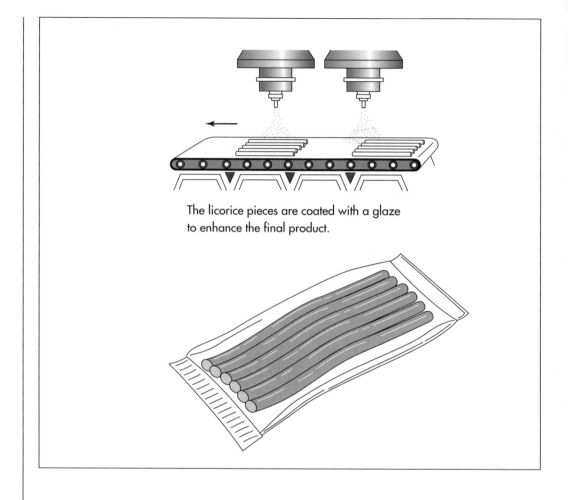

The licorice pieces are coated with a glaze to enhance the final product.

expensive, it is often diluted with aniseed oils. Liquid caramel may also be used for a similar purpose. Other natural flavors obtained from fruits, berries, honey, molasses, and maple sugar have also been used in licorice. The impact of these flavors can be improved by the addition of artificial flavors, which are mixtures of aromatic chemicals. Salt is also included to improve the final flavor of the candy. Additionally, acids such as citric acid, lactic acid, and malic acid can be added to provide flavor. Artificial colors such as certified FD&C colorants are used to modify the color of the final product.

Processing ingredients are important in licorice manufacturing. Cornstarch is a high molecular weight sugar polymer that can have a wide range of functions in a licorice recipe. While it can be a sweetener, texture stabilizer and a gelling agent, it is primarily used as a coating ingredient to prevent pieces of licorice from sticking together. Soybean oil may also be used in the production of licorice. It provides some flavoring

but also acts as a suitable lubricant during processing. Although licorice contains a high concentration of sugar, there is still a potential for microbial contamination. To prevent this type of contamination, potassium sorbate is included as a preservative in some licorice recipes.

The Manufacturing Process

Licorice manufacture begins with making a batch of licorice paste. The batch is then extruded through a nozzle, cooled, cut, and packaged automatically.

Creating the slurry

1 The conventional method of producing licorice begins by making a slurry of the ingredients in the recipe. Factory workers, known as compounders, make slurry batches by pouring, or pumping, the appropriate amount of raw materials into large, stainless steel mixing tanks. These tanks are equipped with mixing, steam heating, and

water cooling capabilities, and can accommodate batch sizes of 1,000-2,000 lb (454-908 kg) or more. The mixers in these tanks must efficiently sweep the sidewalls to prevent burning of the batch. When the slurry passes quality control tests, it can be pumped to the continuous cooker.

Cooking

2 Using traditional techniques, the batches of licorice had to be cooked anywhere from two to four hours. Since the main objective of this cooking process was to reduce the moisture content of the slurry, a faster continuous process was developed. By using a continuous, high-pressure cooking method some manufacturers have cut the cooking time down to a few minutes. In one type of cooker, the slurry of licorice paste is pumped into a vertical evaporator tube, which is surrounded by a steam jacket. Revolving blades inside the tube cause a process of heat exchange (cooking) to occur between droplets of slurry and the cooker. This reduces the moisture level of the slurry, and creates a thin film of licorice candy, which is extruded through an outlet at the bottom of the tube. The candy accumulates in a semi-solid paste, which can then be pumped through an extruder.

Extrusion

3 As the product leaves the cooker, it is pumped through electrically-heated pipes to the extruding machine. The paste goes through the extruder and is expelled on a conveyor as an endless strand of licorice. It can come out in a wide variety of shapes including braids, straws, twists, shoestrings and ribbons to name a few. Currently, extruders can handle over 2,000 lb (908 kg) of licorice per hour.

Cutting

4 The licorice strands then travel along the conveyor to the cutter. As they move toward the cutter, they travel slow enough to cool to an appropriate temperature. This is important because it allows the candy to harden and set. When the licorice strands arrive at the cutter, they are cut "guillotine style" into pieces of a desired length.

Final coating and packaging

5 From the cutter, the licorice pieces are moved along the conveyor to the package equipment. Along the way, they may be further dried and coated with a special glaze, which enhances the product's shine and keeps the pieces from sticking together in the bag. At the packaging stage, the licorice is lined up and stacked. It is placed on a horizontal flow wrapper and when enough pieces are available, the plastic film package is wrapped around the licorice and sealed on both ends. The package is then moved to a stamping machine, which prints a manufacturing tracking code number on it, and then to a boxing machine. Multiple packages are put into individual boxes. The boxes are stacked on pallets and the pallets are shipped on trucks to the food distributor.

Quality Control

The first part of quality control begins with the testing of incoming raw materials. Quality control lab technicians evaluate each ingredient prior to use to ensure that they conform to specifications. Sensory characteristics such as appearance, color, odor, and flavor are all checked. Other characteristics may also be examined such as viscosity of liquids, particle size of solids, and moisture content. Manufacturers depend on these tests to ensure that the ingredients they use will produce a consistent quality batch of licorice.

Beyond testing of the initial ingredients, quality tests are also run on the licorice paste. This includes pH, viscosity, and appearance testing, but it also includes an evaluation of the gelatinization of the batch. It turns out that the quality of the licorice paste is dependent on the extent to which starch gelatinization takes place. If the batch has fully gelatinized, it will have a good gloss and hold the production countlines. A partially gelatinized batch will have a cleaner bite. Since both of these properties are desired, the final batch must be tested to ensure that just the right amount of gelatinization has taken place.

During production, quality control technicians check physical aspects of the extruded candy. The usual method of testing is to compare the newly-made product to an established standard. For example, the color of a randomly sampled licorice stick is compared to a standard licorice that was produced during product development. Other qualities of this sample such as taste, texture, and odor are evaluated by groups of sensory panelists. These are people who are specially trained to notice small differences in tactile properties. Additionally, many instrumental tests developed by the confectionery industry over the years are also performed to complement tests performed by humans.

The Future

The focus of research for licorice producers will be on developing faster, more efficient production methods. Most manufacturers have shifted away from the conventional method to a continuous process because the old batch method is slow and requires too much labor. To reduce manufacturing times, new cookers may be developed in the future. Another way to reduce times is by developing new licorice recipes. These recipes will use substitution ingredients, which can stabilize texture, extend shelf life while still reducing processing times and maintaining a desirable candy taste.

Where to Learn More

Books

Alikonis, Justin. *Candy Technology*. Westport: AVI Publishing Co., 1979.

Booth, R. Gordon. *Snack Food*. New York: Van Nostrand Reinhold, 1990.

Macrae, R. et. al., editors. *Encyclopedia of Food Science, Food Technology and Nutrition*. San Diego: Academic Press, 1993.

Periodicals

Deis, Ronald. "Candy Creations with Starch and Its Derivatives." *Food Product Design* (September 1997): 73 - 88.

—*Perry Romanowski*

Linen

Background

Linen yarn is spun from the long fibers found just behind the bark in the multi-layer stem of the flax plant (*Linum usitatissimum*). In order to retrieve the fibers from the plant, the woody stem and the inner pith (called pectin), which holds the fibers together in a clump, must be rotted away. The cellulose fiber from the stem is spinnable and is used in the production of linen thread, cordage, and twine. From linen thread or yarn, fine toweling and dress fabrics may be woven. Linen fabric is a popular choice for warm-weather clothing. It feels cool in the summer but appears crisp and fresh even in hot weather. Household linens truly made of linen become more supple and soft to the touch with use; thus, linen was once the bedsheet of choice.

While the flax plant is not difficult to grow, it flourishes best in cool, humid climates and within moist, well-plowed soil. The process for separating the flax fibers from the plant's woody stock is laborious and painstaking and must be done in an area where labor is plentiful and relatively inexpensive. It is remarkable that while there is some mechanization to parts of the fiber preparation, some fiber preparation is still done by hand as it has been for centuries. This may be due to the care that must be taken with the fragile flax fibers inside the woody stalk, which might be adversely affected by mechanized processing.

Flax remains under cultivation for linen fiber in a number of countries including Poland, Austria, Belgium, France, Germany, Denmark, the Netherlands, Italy, Spain, Switzerland, and the British Isles. However, the grade of fiber the plants yield in different parts of the world varies. Many believe that Belgium grows the finest-quality flax fibers in the world, with Scottish and Irish linen not far behind. There is no commercial production of linen fabric in any significant quantity in the United States except, perhaps, by individual hand spinners and hand weavers. Thus, the linen fabrics Americans use and wear are nearly all imported into the country from one of these flax-growing and weaving countries.

History

Flax has been cultivated for its remarkable fiber, linen, for at least five millennia. The spinning and weaving of linen is depicted on wall paintings of ancient Egypt. As early as 3,000 B.C., the fiber was processed into fine white fabric (540 threads to the inch—finer than anything woven today) and wrapped around the mummies of the ancient Egyptian pharaohs. Mentioned several times in the Bible, it has been used as a cool, comfortable fiber in the Middle East for centuries as well. Ancient Greeks and Romans greatly valued it as a commodity. Finnish traders are believed to have introduced flax to Northern Europe where it has been under cultivation for centuries.

Both wool and linen were tremendously important fibers in the New World. Relatively easy to grow, American settlers were urged to plant a small plot of flax as early as the seventeenth century. While flax is easy to grow, settlers knew all too well the tedious chore of processing the woody stalks for its supple linen. Before the industrial revolution much sturdy, homemade clothing was woven from linen cultivated, processed, spun, dyed, woven, and sewn by hand. It may be argued

As early as 3,000 B.C., the fiber was processed into fine white fabric (540 threads to the inch—finer than anything woven today) and wrapped around the mummies of the ancient Egyptian pharaohs.

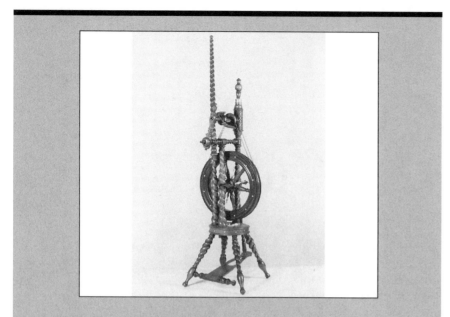

European flax wheel used to spin flax into linen thread. (From the collects of Henry Ford Museum & Greenfield Village, Dearborn, Michigan.)

This is a European "flax wheel" used to spin flax into linen thread within the home. Folklore tells us that it was brought by Henry Ford's Irish grandmother to the New World; it was one of the few family keepsakes Ford had from his Irish ancestors. In fact, it was not unusual for the Scots or Irish to bring such wheels to this country. The British Isles have a long and proud linen tradition, and even decades after others abandoned linen production for cotton in the New World the Irish and Scots here tenaciously clung to their linen-making traditions.

Ford's grandmother placed unspun flax on the tall, vertical, turned distaff and then push the treadle with her foot to power the wheel. The bobbin and flyer mounted horizontally in the center of the wheel would spin the flax and wind it on the bobbin at the same time. The rather small wheel below the bobbin required the spinner to treadle rather fast to keep it moving and because of the small wheel this spinning wheel was not a popular style. It is lovely to look at, though, as this flax wheel is rather fancy, with inlaid bone or ivory set within the wheel. Some refer to this type of European spinning wheel as a "castle" or "parlor" wheel because of its lovely inlays and turnings.

Nancy EV Bryk

that until the eighteenth century, linen was the most important textile in the world.

By the late eighteenth century, cotton became the fiber that was most easily and inexpensively processed and woven in the mechanized British and New England textile mills. By the 1850s, linen production had virtually been abandoned in the United States because it was so much cheaper to buy the factory-made cotton. Some New Englanders of Scot or Irish background continued to cultivate some flax for processing into linen used for fancy domestic linens such as bedsheets, toweling, and decorative tableclothes as their ancestors had for centuries. However, most Americans abandoned the cultivation of the plant in this country and instead chose cheap cotton that was carded, spun, woven, and roller-printed for just pennies a yard. Thereafter and until recently, a different variety of flax plant was raised in this country not for its linen fibers but for its seeds which exude a useful vegetable oil known as linseed oil when pressed.

Raw Materials

All that is needed to turn flax fiber into linen, and then spin and weave the linen fibers into linen fabric is the cellulose flax fiber from the stem of the flax plant. The process for separating the fibers from the woody stalk can use either water or chemicals, but these are ultimately washed away and are not part of the finished material.

Design

The manufacture of linen yarn requires no special design processes. All that has to be determined prior to manufacturing is the thickness of the yarn to be spun. That will depend on the grade of linen in production and the demands of the customer.

The Manufacturing Process

Cultivating

1 It takes about 100 days from seed planting to harvesting of the flax plant. Flax cannot endure very hot weather; thus, in many countries, the planting of seed is figured from the date or time of year in which the flax must be harvested due to heat and the growers count back 100 days to determine a date for planting. In some areas of the world, flax is sown in winter because of heat in early spring. In commercial production, the land is plowed in the spring then worked into a good seedbed by discing, harrowing, and rolling. Flax seeds must be shallowly planted. Seeds may be broadcast by hand, but the

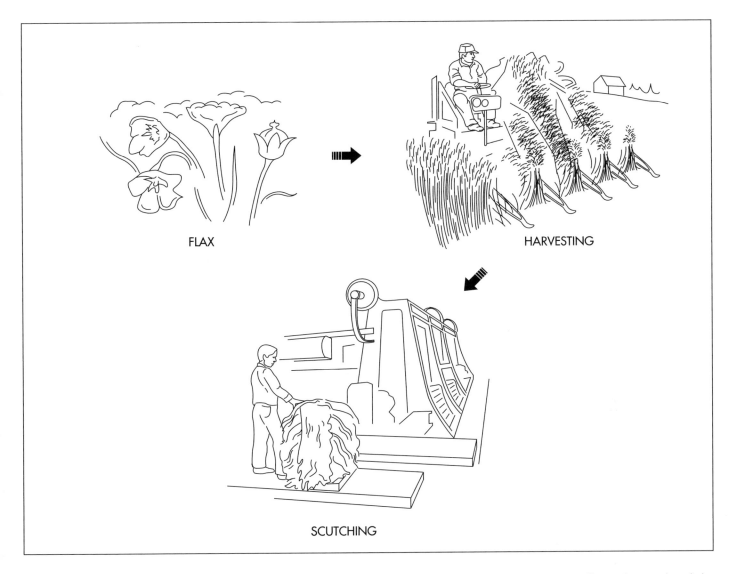

FLAX

HARVESTING

SCUTCHING

seed must be covered over with soil. Machines may also plant the seed in rows.

Flax plants are poor competitors with weeds. Weeds reduce fiber yields and increase the difficulty in harvesting the plant. Tillage of the soil reduces weeds as do herbicides. When the flax plants are just a few inches high, the area must be carefully weeded so as not to disturb the delicate sprouts. In three months, the plants are straight, slender stalks that may be 2-4 ft (61-122 cm) in height with small blue or white fibers. (Flax plants with blue flowers yield the finest linen fibers.)

Harvesting

2 After about 90 days, the leaves wither, the stem turns yellow and the seeds turn brown, indicating it is time to harvest the plant. The plant must be pulled as soon as it appears brown as any delay results in linen without the prized luster. It is imperative that the stalk not be cut in the harvesting process but removed from the ground intact; if the stalk is cut the sap is lost, and this affects the quality of the linen. These plants are often pulled out of the ground by hand, grasped just under the seed heads and gently tugged. The tapered ends of the stalk must be preserved so that a smooth yarn may be spun. These stalks are tied in bundles called beets and are ready for extraction of the flax fiber in the stalk. However, fairly efficient machines can pull the plants from the ground as well.

Releasing the fiber from the stalk

3 The plant is passed through coarse combs, which removes the seeds and leaves from the plant. This process, called

Once flax is harvested and the fiber removed from the stalks, a scutching machine removes the broken outer layer called shives.

The fiber is combed and separated by length. Line fibers (long linen fibers) are spun into linen yarn.

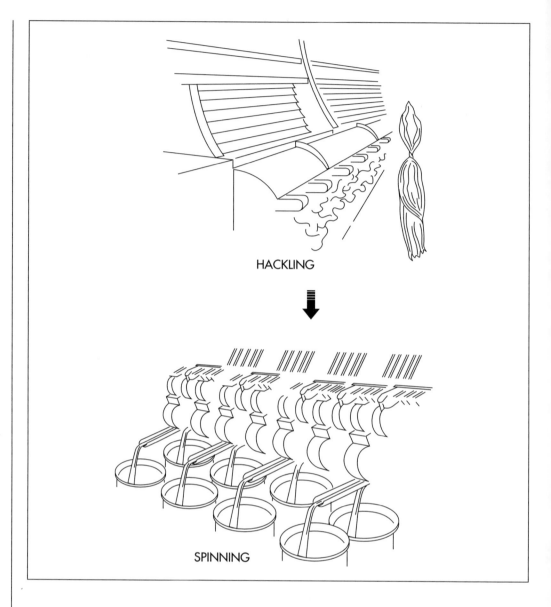

HACKLING

SPINNING

rippling, is mechanized in many of the flax-producing countries.

4 The woody bark surrounding the flax fiber is decomposed by water or chemical retting, which loosens the pectin or gum that attaches the fiber to the stem. If flax is not fully retted, the stalk of the plant cannot be separated from the fiber without injuring the delicate fiber. Thus, retting has to be carefully executed. Too little retting may not permit the fiber to be separated from the stalk with ease. Too much retting or rotting will weaken fibers.

Retting may be accomplished in a variety of ways. In some parts of the world, linen is still retted by hand, using moisture to rot away the bark. The stalks are spread on dewy slopes, submerged in stagnant pools of water, or placed in running streams. Workers must wait for the water to begin rotting or fermenting the stem—sometimes more than a week or two. However, most manufacturers use chemicals for retting. The plants are placed in a solution either of alkali or oxalic acid, then pressurized and boiled. This method is easy to monitor and rather quick, although some believe that chemical retting adversely affects the color and strength of the fiber and hand retting produces the finest linen. Vat or mechanical retting requires that the stalks be submerged in vats of warm water, hastening the decomposition of the stem. The flax is then removed from the vats and passed between rollers to crush the bark as clean water flushes away the pectin and other impurities.

5 After the retting process, the flax plants are squeezed and allowed to dry out before they undergo the process called breaking. In order to crush the decomposed stalks, they are sent through fluted rollers which break up the stem and separate the exterior fibers from the bast that will be used to make linen. This process breaks the stalk into small pieces of bark called shives. Then, the shives are scutched. The scutching machine removes the broken shives with rotating paddles, finally releasing the flax fiber from stalk.

6 The fibers are now combed and straightened in preparation for spinning. This separates the short fibers (called tow and used for making more coarse, sturdy goods) from the longer and more luxurious linen fibers. The very finest flax fibers are called line or dressed flax, and the fibers may be anywhere from 12-20 in (30.5-51 cm) in length.

Spinning

7 Line fibers (long linen fibers) are put through machines called spreaders, which combine fibers of the same length, laying the fibers parallel so that the ends overlap, creating a sliver. The sliver passes through a set of rollers, making a roving which is ready to spin.

8 The linen rovings, resembling tresses of blonde hair, are put on a spinning frame and drawn out into thread and ultimately wound on bobbins or spools. Many such spools are filled on a spinning frame at the same time. The fibers are formed into a continuous ribbon by being pressed between rollers and combed over fine pins. This operation constantly pulls and elongates the ribbon-like linen until it is given its final twist for strength and wound on the bobbin. While linen is a strong fiber, it is rather inelastic. Thus, the atmosphere within the spinning factory must be both humid and warm in order to render the fiber easier to work into yarn. In this hot, humid factory the linen is wet spun in which the roving is run through a hot water bath in order to bind the fibers together thus creating a fine yarn. Dry spinning does not use moisture for spinning. This produces rough, uneven yarns that are used for making inexpensive twines or coarse yarns.

9 These moist yarns are transferred from bobbins on the spinning frame to large take-up reels. These linen reels are taken to dryers, and when the yarn is dry, it is wound onto bobbins for weaving or wound into yarn spools of varying weight. The standard measure of flax yarn is the cut. It is based on the measure of 1 lb (453.59 g) of flax spun to make 300 yd (274.2 m) of yarn being equal to one cut. If 1 lb (453.59 g) of flax is spun into 600 yd (548.4 m), then it is a "no. 2 cut." The higher the cut, the finer the yarn becomes. The yarn now awaits transport to the loom for weaving into fabrics, toweling, or for use as twine or rope.

Byproducts/Waste

Of greatest concern are the chemicals used in retting. These chemicals must be neutralized before being released into water supplies. The stalks, leaves, seed pods, etc. are natural organic materials and are not hazardous unless impregnated with much of the chemicals left behind in the retting process. The only other concern with the processing of linen is the smell—it is said that hand-retted linen produces quite a stench and is most unpleasant to experience.

Where to Learn More

Books

The Irish Linen Guild. *Irish Linen: The Fabric of Elegance.* NY: Elliott & Nelson, 1945.

Jerde, Judith. *Encyclopedia of Textiles.* NY: Facts on File Inc., 1992.

Koob, Katharine. *Linen Making in New England.* North Andover, MA: Merrimack Valley Textile Museum, 1978.

Periodicals

Calhoun, Wheeler and Lee Kirschner. "The Continuous Thread: From Flax Seed to Linen Cloth." *Spin-Off Magazine* (March 1983): 28-35.

—Nancy EV Bryk

Mercury

The EPA has set a goal of reducing the level of mercury found in municipal refuse from 1.4 million lb/yr (0.64 million kg/yr) in 1989 to 0.35 million lb/yr (0.16 million kg/yr) by 2000.

Background

Mercury is one of the basic chemical elements. It is a heavy, silvery metal that is liquid at normal temperatures. Mercury readily forms alloys with other metals, and this makes it useful in processing gold and silver. Much of the impetus to develop mercury ore deposits in the United States came after the discovery of gold and silver in California and other western states in the 1800s. Unfortunately, mercury is also a highly toxic material, and as a result, its use has severely declined over the past 20 years. Its principal applications are in the production of chlorine and caustic soda, and as a component of many electrical devices, including fluorescent and mercury-vapor lamps.

Mercury has been found in Egyptian tombs dating to about 1500 B.C., and it was probably used for cosmetic and medicinal purposes even earlier. In about 350 B.C., the Greek philosopher and scientist Aristotle described how cinnabar ore was heated to extract mercury for religious ceremonies. The Romans used mercury for a variety of purposes and gave it the name hydrargyrum, meaning liquid silver, from which the chemical symbol for mercury, Hg, is derived.

Demand for mercury greatly increased in 1557 with the development of a process that used mercury to extract silver from its ore. The mercury barometer was invented by Torricelli in 1643, followed by the invention of the mercury thermometer by Fahrenheit in 1714. The first use of a mercury alloy, or amalgam, as a tooth filling in dentistry was in 1828, although concerns over the toxic nature of mercury prevented the widespread use of this new technique. It wasn't until 1895 that experimental work by G.V. Black showed that amalgam fillings were safe, although 100 years later scientists were still debating that point.

Mercury found its way into many products and industrial applications after 1900. It was commonly used in batteries, paints, explosives, light bulbs, light switches, pharmaceuticals, fungicides, and pesticides. Mercury was also used as part of the processes to produce paper, felt, glass, and many plastics.

In the 1980s, increasing understanding and awareness of the harmful health and environmental effects of mercury started to greatly outweigh its benefits, and usage began to drop sharply. By 1992, its use in batteries had dropped to less than 5% of its level in 1988, and overall use in electrical devices and light bulbs had dropped 50% in the same period. The use of mercury in paints, fungicides, and pesticides has been banned in the United States, and its use in the paper, felt, and glass-manufacturing processes has been voluntarily discontinued.

Worldwide, production of mercury is limited to only a few countries with relaxed environmental laws. Mercury mining has ceased altogether in Spain, which until 1989 was the world's largest producer. In the United States, mercury mining has also stopped, although small quantities of mercury are recovered as part of the gold refining process to avoid environmental contamination. China, Russia (formerly the USSR), Mexico, and Algeria were the largest producers of mercury in 1992.

Raw Materials

Mercury is rarely found by itself in nature. Most mercury is chemically bound to other materials in the form of ores. The most common ore is red mercury sulfide (HgS), also known as cinnabar. Other mercury ores include corderoite ($Hg_3S_2C_{12}$), livingstonite ($HgSb_4S_8$), montroydite (HgO), and calomel (HgC_1). There are several others. Mercury ores are formed underground when warm mineral solutions rise towards the earth's surface under the influence of volcanic action. They are usually found in faulted and fractured rocks at relatively shallow depths of 3-3000 ft (1-1000 m).

Other sources of mercury include the dumps and tailing piles of earlier, less-efficient mining and processing operations.

The Manufacturing Process

The process for extracting mercury from its ores has not changed much since Aristotle first described it over 2,300 years ago. Cinnabar ore is crushed and heated to release the mercury as a vapor. The mercury vapor is then cooled, condensed, and collected. Almost 95% of the mercury content of cinnabar ore can be recovered using this process.

Here is a typical sequence of operations used for the modern extraction and refining of mercury.

Mining

Cinnabar ore occurs in concentrated deposits located at or near the surface. About 90% of these deposits are deep enough to require underground mining with tunnels. The remaining 10% can be excavated from open pits.

1 Cinnabar is dislodged from the surrounding rocks by drilling and blasting with explosives or by the use of power equipment. The ore is brought out of the mine on conveyor belts or in trucks or trains.

Roasting

Because cinnabar ore is relatively concentrated, it can be processed directly without any intermediate steps to remove waste material.

2 The ore is first crushed in one or more cone crushers. A cone crusher consists of an interior grinding cone that rotates on an eccentric vertical axis inside a fixed outer cone. As the ore is fed into the top of the crusher, it is squeezed between the two cones and broken into smaller pieces.

3 The crushed ore is then ground even smaller by a series of mills. Each mill consists of a large cylindrical container laying on its side and rotating on its horizontal axis. The mill may be filled with short lengths of steel rods or with steel balls to provide the grinding action.

4 The finely powdered ore is fed into a furnace or kiln to be heated. Some operations use a multiple-hearth furnace, in which the ore is mechanically moved down a vertical shaft from one ledge, or hearth, to the next by slowly rotating rakes. Other operations use a rotary kiln, in which the ore is tumbled down the length of a long, rotating cylinder that is inclined a few degrees off horizontal. In either case, heat is provided by combusting natural gas or some other fuel in the lower portion of the furnace or kiln. The heated cinnabar (HgS) reacts with the oxygen (O_2) in the air to produce sulfur dioxide (SO_2), allowing the mercury to rise as a vapor. This process is called roasting.

Condensing

5 The mercury vapor rises up and out of the furnace or kiln along with the sulfur dioxide, water vapor, and other products of combustion. A considerable amount of fine dust from the powdered ore is also carried along and must be separated and captured.

6 The hot furnace exhaust passes through a water-cooled condenser. As the exhaust cools, the mercury, which has a boiling point of 675° F (357° C), is the first to condense into a liquid, leaving the other gases and vapors to be vented or to be processed further to reduce air pollution.

7 The liquid mercury is collected. Because mercury has a very high specific gravity, any impurities tend to rise to the surface and form a dark film or scum. These impurities are removed by filtration, leaving a liquid mercury that is about 99.9% pure. The impurities are treated with lime to

MERCURY PROCESSING

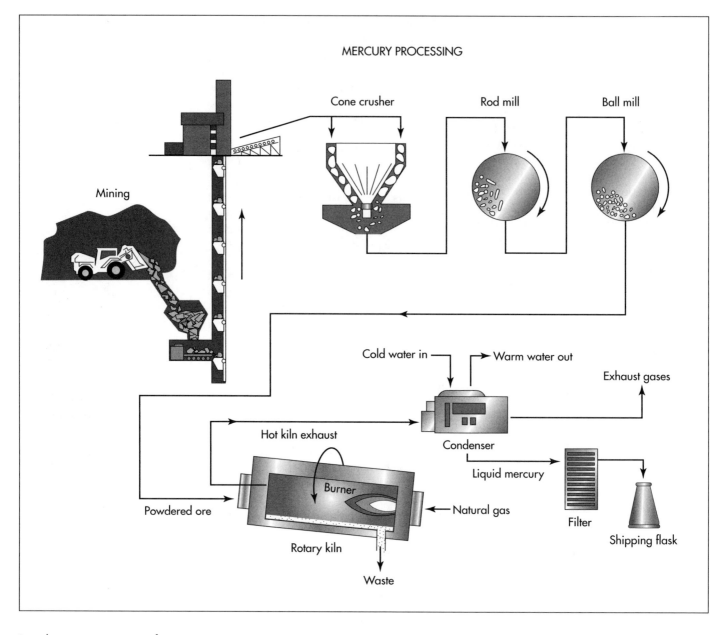

In order to extract mercury from its ores, cinnabar ore is crushed and heated to release the mercury as a vapor. The mercury vapor is then cooled, condensed, and collected.

separate and capture any mercury, which may have formed compounds.

Refining

Most commercial-grade mercury is 99.9% pure and can be used directly from the roasting and condensing process. Higher purity mercury is needed for some limited applications and must be refined further. This ultrapure mercury commands a premium price.

8 Higher purity can be obtained through several refining methods. The mercury may be mechanically filtered again, and certain impurities may be removed through ox-

idation with chemicals or air. In some cases the mercury is refined through an electrolytic process, in which an electric current is passed through a tank of liquid mercury to remove the impurities. The most common refining method is triple distillation, in which the temperature of the liquid mercury is carefully raised until the impurities either evaporate or the mercury itself evaporates, leaving the impurities behind. This distillation process is performed three times, with the purity increasing each time.

Shipping

9 Commercial-grade mercury is poured into wrought-iron or steel flasks and sealed. Each flask contains 76 lb (34.5 kg) of mercury. Higher purity mercury is usually sealed in smaller glass or plastic containers for shipment.

Quality Control

Commercial-grade mercury with 99.9% purity is called prime virgin-grade mercury. Ultrapure mercury is usually produced by the triple-distillation method and is called triple-distilled mercury.

Quality control inspections of the roasting and condensing process consist of spot checking the condensed liquid mercury for the presence of foreign metals, since those are the most common contaminants. The presence of gold, silver, and base metals is detected using various chemical-testing methods.

Triple-distilled mercury is tested by evaporation or spectrographic analysis. In the evaporation method, a sample of mercury is evaporated, and the residue is weighed. In the spectrographic analysis method, a sample of mercury is evaporated, and the residue is mixed with graphite. Light coming from the resulting mixture is viewed with a spectrometer, which separates the light into different color bands depending on the chemical elements present.

Health and Environmental Effects

Mercury is highly toxic to humans. Exposure may come from inhalation, ingestion, or absorption through the skin. Of the three, inhalation of mercury vapor is the most dangerous. Short-term exposure to mercury vapor can produce weakness, chills, nausea, vomiting, diarrhea, and other symptoms within a few hours. Recovery is usually complete once the victim is removed from the source. Long-term exposure to mercury vapor produces shaking, irritability, insomnia, confusion, excessive salivation, and other debilitating effects.

In normal situations, most exposure to mercury comes from the ingestion of certain foods, such as fish, in which the mercury has accumulated at high levels. Although mercury is not absorbed in great quantities when passing through the human digestive system, ingestion over a long period of time has been shown to have cumulative effects.

In industrial situations, mercury exposure is a far more serious hazard. Mining and processing mercury ore can expose workers to mercury vapor as well as to direct contact with the skin. The production of chlorine and caustic soda can also cause significant mercury exposure hazards. Dentists and dental assistants can be exposed to mercury while preparing and placing mercury amalgam fillings.

Because mercury poses a serious health hazard, its use and release to the environment has come under increasingly tight restrictions. In 1988, it was estimated that 24 million lb/yr (11 million kg/yr) of mercury were released into the air, land, and water worldwide as the result of human activities. This included mercury released by mercury mining and refining, various manufacturing operations, the combustion of coal, the discarding of municipal refuse and sewage sludge, and other sources.

In the United States, the Environmental Protection Agency (EPA) has banned the use of mercury for many applications. The EPA has set a goal of reducing the level of mercury found in municipal refuse from 1.4 million lb/yr (0.64 million kg/yr) in 1989 to 0.35 million lb/yr (0.16 million kg/yr) by 2000. This is to be accomplished by decreasing the use of mercury in products and increasing the diversion of mercury from municipal refuse through recycling.

The Future

Mercury is still an important component in many products and processes, although its use is expected to continue to decline. Improved handling and recycling of mercury are expected to significantly reduce its release to the environment and thereby reduce its health hazard.

Where to Learn More

Books

Brady, George S., Henry R. Clauser, and John A. Vaccari. *Materials Handbook,* 14th Edition. McGraw-Hill, 1997.

Heiserman, David L. *Exploring Chemical Elements and Their Compounds.* TAB Books, 1992.

Kroschwitz, Jacqueline I., executive editor, and Mary Howe-Grant, editor. *Encyclopedia of Chemical Technology*, 4th edition. John Wiley and Sons, Inc., 1993.

Stwertka, Albert. *A Guide to the Elements.* Oxford University Press, 1996.

Periodicals

Raloff, J. "Mercurial Airs: Tallying Who's to Blame." *Science News* (February 19, 1994): 119.

Spencer, Peter, and G. Murdoch. "Mercury in Paint." *Consumers' Research Magazine* (January 1991): 2.

Stone, R. "Mercurial Debate." *Science* (March 13, 1992): 1356-1357.

Other

http://www.intercorr.com/periodic/80.htm [This website contains a summary of the history, sources, properties, and uses of mercury.]

—*Chris Cavette*

Milk

Background

Milk is a nutritive beverage obtained from various animals and consumed by humans. Most milk is obtained from dairy cows, although milk from goats, water buffalo, and reindeer is also used in various parts of the world. In the United States, and in many industrialized countries, raw cow's milk is processed before it is consumed. During processing the fat content of the milk is adjusted, various vitamins are added, and potentially harmful bacteria are killed. In addition to being consumed as a beverage, milk is also used to make butter, cream, yogurt, cheese, and a variety of other products.

History

The use of milk as a beverage probably began with the domestication of animals. Goats and sheep were domesticated in the area now known as Iran and Afghanistan in about 9000 B.C., and by about 7000 B.C. cattle were being herded in what is now Turkey and parts of Africa. The method for making cheese from milk was known to the ancient Greeks and Romans, and the use of milk and milk products spread throughout Europe in the following centuries.

Cattle were first brought to the United States in the 1600s by some of the earliest colonists. Prior to the American Revolution most of the dairy products were consumed on the farm where they were produced. By about 1790, population centers such as Boston, New York, and Philadelphia had grown sufficiently to become an attractive market for larger-scale dairy operations. To meet the increased demand, farmers began importing breeds of cattle that were better

suited for milk production. The first Holstein-Friesens were imported in 1795, the first Ayrshires in 1822, and the first Guernseys in 1830.

With the development of the dairy industry in the United States, a variety of machines for processing milk were also developed. In 1856, Gail Borden patented a method for making condensed milk by heating it in a partial vacuum. Not only did his method remove much of the water so the milk could be stored in a smaller volume, but it also protected the milk from germs in the air. Borden opened a condensed milk plant and cannery in Wassaic, New York, in 1861. During the Civil War, his condensed milk was used by Union troops and its popularity spread.

In 1863, Louis Pasteur of France developed a method of heating wine to kill the microorganisms that cause wine to turn into vinegar. Later, this method of killing harmful bacteria was adapted to a number of food products and became known as pasteurization. The first milk processing plant in the United States to install pasteurizing equipment was the Sheffield Farms Dairy in Bloomfield, New Jersey, which imported a German-made pasteurizer in 1891. Many dairy operators opposed pasteurization as an unnecessary expense, and it wasn't until 1908 that Chicago became the first major city to require pasteurized milk. New York and Philadelphia followed in 1914, and by 1917 most major cities had enacted laws requiring that all milk be pasteurized.

One of the first glass milk bottles was patented in 1884 by Dr. Henry Thatcher, after seeing a milkman making deliveries from an open bucket into which a child's

In 1990, the annual production of milk in the United States was about 148 billion lb (67.5 billion kg). This is equivalent to about 17.2 billion U.S. gallons (65.1 billion liters).

filthy rag doll had accidentally fallen. By 1889, his Thatcher's Common Sense Milk Jar had become an industry standard. It was sealed with a waxed paper disc that was pressed into a groove inside the bottle's neck. The milk bottle, and the regular morning arrival of the milkman, remained a part of American life until the 1950s, when waxed paper cartons of milk began appearing in markets.

In 1990, the annual production of milk in the United States was about 148 billion lb (67.5 billion kg). This is equivalent to about 17.2 billion U.S. gallons (65.1 billion liters). About 37% of this was consumed as fluid milk and cream, about 32% was converted into various cheeses, about 17% was made into butter, and about 8% was used to make ice cream and other frozen desserts. The remainder was sold as dry milk, canned milk, and other milk products.

Types of Milk

There are many different types of milk. Some depend on the amount of milk fat present in the finished product. Others depend on the type of processing involved. Still others depend on the type of dairy cow that produced the milk.

The federal Food and Drug Administration (FDA) establishes standards for different types of milk and milk products. Some states use these standards, while others have their own standards. Prior to 1998, the federal standards required that fluid milk sold as whole milk must have no less than 3.25% milk fat, low-fat milk must have 0.5-2.0% milk fat, and skim milk must have less than 0.5% milk fat. Starting in 1998, the FDA required that milk with 2% milk fat must be labeled as "reduced-fat" because it did not meet the new definition of low-fat products as having less than 3 grams of fat per serving. Milk with 1% milk fat could still be labeled as "low-fat" because it did meet the definition. As a comparison, light cream has no less than 18% milk fat, and heavy cream has no less than 36% milk fat.

Other types of milk are based on the type of processing involved. Pasteurized milk has been heated to kill any potentially harmful bacteria. Homogenized milk has had the milk fat particles reduced in size and uni-

formly blended to prevent them from rising to the top in the form of cream. Vitamin-fortified milks have various vitamins added. Most milk sold in markets in the United States is pasteurized, homogenized, and vitamin-fortified.

Grade A milk refers to milk produced under sufficiently sanitary conditions to permit its use as fluid milk. About 90% of the milk produced in the United States is Grade A milk. Grade B milk is produced under conditions that make it acceptable only for manufactured products such as certain cheeses, where it undergoes further processing. Certified milk is produced under exceedingly high sanitary standards and is sold at a higher price than Grade A milk.

Specialty milks include flavored milk, such as chocolate milk, which has had a flavoring syrup added. Other specialty milks include Golden Guernsey milk, which is produced by purebred Guernsey cows, and All-Jersey milk, which is produced by registered Jersey cows. Both command a premium price because of their higher milk fat content and creamier taste.

Concentrated milk products have varying degrees of water removed from fluid milk. They include, in descending order of water content, evaporated milk, condensed milk, and dry milk.

Raw Materials

The average composition of cow's milk is 87.2% water, 3.7% milk fat, 3.5% protein, 4.9% lactose, and 0.7% ash. This composition varies from cow to cow and breed to breed. For example, Jersey cows have an average of 85.6% water and 5.15% milk fat. These figures also vary by the season of the year, the animal feed content, and many other factors.

Vitamin D concentrate may be added to milk in the amount of 400 international units (IU) per quart. Most low fat and skim milk also has 2,000 IU of Vitamin A added.

The Manufacturing Process

Milk is a perishable commodity. For this reason, it is usually processed locally within

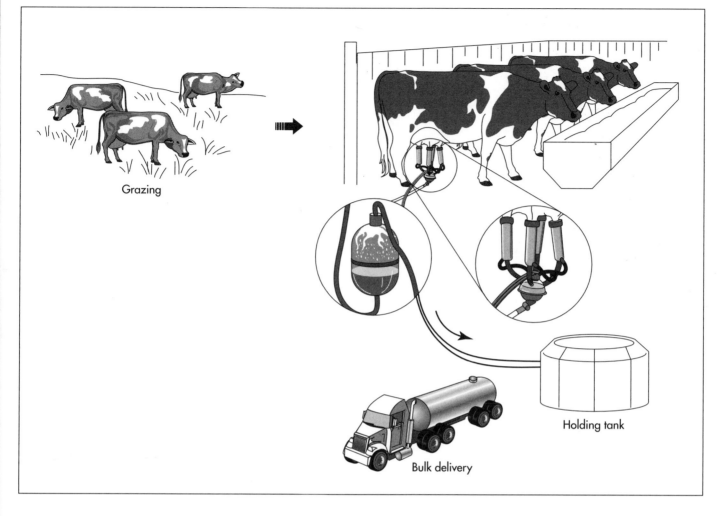

Grazing

Holding tank

Bulk delivery

a few hours of being collected. In the United States, there are several hundred thousand dairy farms and several thousand milk processing plants. Some plants produce only fluid milk, while others also produce butter, cheese, and other milk products.

Collecting

1 Dairy cows are milked twice a day using mechanical vacuum milking machines. The raw milk flows through stainless steel or glass pipes to a refrigerated bulk milk tank where it is cooled to about 40° F (4.4° C).

2 A refrigerated bulk tank truck makes collections from dairy farms in the area within a few hours. Before pumping the milk from each farm's tank, the driver collects a sample and checks the flavor and temperature and records the volume.

3 At the milk processing plant, the milk in the truck is weighed and is pumped into

refrigerated tanks in the plant through flexible stainless steel or plastic hoses.

Separating

4 The cold raw milk passes through either a clarifier or a separator, which spins the milk through a series of conical disks inside an enclosure. A clarifier removes debris, some bacteria, and any sediment that may be present in the raw milk. A separator performs the same task, but also separates the heavier milk fat from the lighter milk to produce both cream and skim milk. Some processing plants use a standardizer-clarifier, which regulates the amount of milk fat content in the milk by removing only the excess fat. The excess milk fat is drawn off and processed into cream or butter.

Fortifying

5 Vitamins A and D may be added to the milk at this time by a peristaltic pump,

Dairy cows are milked twice a day using mechanical vacuum milking machines. The raw milk flows through stainless steel or glass pipes to a refrigerated bulk milk tank.

A clarifier removes debris, some bacteria, and any sediment that may be present in the raw milk. The milk is then fortified and pasteurized.

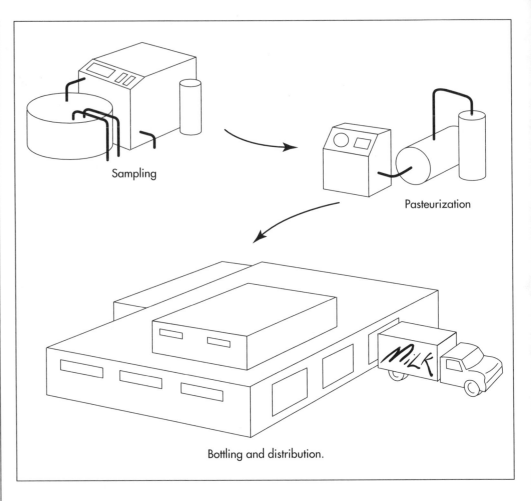

Sampling

Pasteurization

Bottling and distribution.

which automatically dispenses the correct amount of vitamin concentrate into the flow of milk.

Pasteurizing

6 The milk—either whole milk, skim milk, or standardized milk—is piped into a pasteurizer to kill any bacteria. There are several methods used to pasteurize milk. The most common is called the high-temperature, short-time (HTST) process in which the milk is heated as it flows through the pasteurizer continuously. Whole milk, skim milk, and standardized milk must be heated to 161° F (72° C) for 15 seconds. Other milk products have different time and temperature requirements. The hot milk passes through a long pipe whose length and diameter are sized so that it takes the liquid exactly 15 seconds to pass from one end to the other. A temperature sensor at the end of the pipe diverts the milk back to the inlet for reprocessing if the temperature has fallen below the required standard.

Homogenizing

7 Most milk is homogenized to reduce the size of the remaining milk fat particles. This prevents the milk fat from separating and floating to the surface as cream. It also ensures that the milk fat will be evenly distributed through the milk. The hot milk from the pasteurizer is pressurized to 2,500-3,000 psi (17,200-20,700 kPa) by a multiple-cylinder piston pump and is forced through very small passages in an adjustable valve. The shearing effect of being forced through the tiny openings breaks down the fat particles into the proper size.

8 The milk is then quickly cooled to 40° F (4.4° C) to avoid harming its taste.

Packaging

9 The milk is pumped into coated paper cartons or plastic bottles and is sealed. In the United States most milk destined for retail sale in grocery stores is packaged in one-gallon (3.8-liter) plastic bottles. The

bottles or cartons are stamped with a "sell by" date to ensure that the retailers do not allow the milk to stay on their shelves longer than it can be safely stored.

10 The milk cartons or bottles are placed in protective shipping containers and kept refrigerated. They are shipped to distribution warehouses in refrigerated trailers and then on to the individual markets, where they are kept in refrigerated display cases.

Cleaning

11 To ensure sanitary conditions, the inner surfaces of the process equipment and piping system are cleaned once a day. Almost all the equipment and piping used in the processing plant and on the farm are made from stainless steel. Highly automated clean-in-place systems are incorporated into this equipment that allows solvents to be run through the system and then flushed clean. This is done at a time between the normal influx of milk from the farms.

Quality Control

The federal Food and Drug Administration (FDA) publishes the Grade A Milk Ordinance which sets sanitation standards for milk production in most states and for all interstate milk shippers. The composition of milk and milk products is specified in *Agricultural Handbook 52* published by the United States Department of Agriculture. It lists both federal and state standards. Testing of milk products includes tests for fat content, total solids, pasteurization efficiency, presence of antibiotics used to control cow disease, and many others.

The Future

The trend to low-fat dairy products over the last 20 years is expected to continue in the future. Sales of butter are expected to decline, while sales of low-fat yogurt and low- or reduced-fat milk are expected to increase. Overall consumption of liquid milk is expected to increase as the population increases.

Where to Learn More

Books

Giblin, James. *Milk: The Fight for Purity*. Thomas Y. Crowell, 1986.

Hui, Y.H., ed. *Encyclopedia of Food Science and Technology*. John Wiley and Sons Inc., 1992.

Kroschwitz, Jacqueline I. and Mary Howe-Grant, ed. *Encyclopedia of Chemical Technology*, 4th edition. John Wiley and Sons Inc., 1993.

McGraw-Hill Encyclopedia of Science and Technology, McGraw-Hill, 1997.

Other

Dairy Farmers of Ontario. http://www.milk.org.

International Dairy Foods Association. http://www.idfa.org.

National Milk Producers Federation. http://nmpf.org.

—Chris Cavette

Milk Carton

Refillable glass bottles reigned for a long time after milk cartons were introduced, but by 1968, over 70% of milk packaged in the United States went into paper cartons.

Milk cartons are water tight paper containers used for packaging milk for retail distribution. One of the most common supermarket items, and found in nearly every home, the milk carton is nonetheless a precision product, manufactured according to exacting standards.

Background

Up until recent times, milk was not usually available as a retail item. Once milk is removed from the cow, it spoils quickly in heat, and is vulnerable to contamination. Until this century, the most economical and hygienic way to store milk was to leave it in the animal. In Europe, a town cow keeper would bring his or her cow directly to the doorstep of the customer, and milk the animal there into a household container. In some places, milk was sold from a shop next door to the cow stall. In either case, the milk could not be safely stored for anything but a small amount of time. A large metal milk container was developed in Europe between 1860 and 1870. Called a churn, the lidded metal container could hold about 21.12 gal (80 l) of milk. Milk in churns was shipped by rail from farming areas into towns, where the demand for milk was high. Milk in metal churns was also dispensed door to door. Instead of the cow keeper bringing the cow, now the milk was ladled out of the churn into a smaller household bucket or can. The glass milk bottle was invented in 1884. This offered convenience to milk consumers, since the sterilized bottles could be kept sealed until needed. Milk that was pasteurized (quickly heated to above boiling, then cooled) was resistant to bacterial contamination and spoilage for several days. Bottled milk became prevalent across the United States and Europe through World War II, though glass containers are rarely seen now.

The first paper milk carton was introduced in 1933. Wax was applied to the paper, to make it waterproof. In 1940, polyethylene was introduced as the waterproofing material. Refillable glass bottles reigned for a long time after milk cartons were introduced, but by 1968, over 70% of milk packaged in the United States went into paper cartons.

The manufacture of milk cartons is actually a two-step process, at two different locations. The carton manufacturer cuts and prints the carton, which is shipped in a "knocked down" or flattened form to the milk packager. The packager completes the process by forming, filling, and sealing the carton.

Raw Materials

Milk containers are made from paperboard coated with a waterproof plastic, generally polyethylene. The wood pulp that is used to make paperboard for milk cartons is a blend of softwood and hardwood. Softwood is usually a type of pine, though the actual trees used vary depending on the location of the paper mill. Softwood produces long wood fibers that provide strength to the paperboard. Hardwood comes from deciduous trees such as oaks. Hardwood has shorter fibers that make for a better printing surface. Pulp for milk carton board is usually 60% hardwood and 40% soft.

Several other chemicals are used to make milk cartons. One is oxygenated chlorine, which bleaches the wood pulp. Other chemicals specific to each manufacturer are added to the paper to add strength. Chemical pigments in the ink are used for the printing process as well.

The Manufacturing Process

Making the paperboard

1 The heavy paper used for milk cartons is categorized as a type of paperboard. It is typically made on a Fourdrinier machine, one of the oldest and most common types of papermaking equipment. The process begins with wood chips. The chips are heated and bathed in chemicals that soften them and break them into small bits of wood fiber. The pulp is bleached in a bath of oxygenated chlorine. The pulp is then washed and passed through several screens, to remove debris. Next, the pulp is fed through a machine called a refiner, which grinds the wood fibers between rotating disks.

The refined pulp flows into the headbox of the Fourdrinier machine. In the headbox, a mixture of water and pulp is spread across a continually moving screen. The water drains away below through the openings in the screen, leaving a mat of damp wood fiber. The mat is drawn through huge rollers that squeeze out additional water. Next, the paperboard is dried, by passing it over steam-heated cylinders.

Applying waterproof coating

2 The dried paperboard next moves through the rollers of an extruder. As the paperboard is pulled through the rollers, the machine extrudes a small amount of molten polyethylene. The polyethylene clings to both sides of the paperboard in a thin film. Several grades of polyethylene may be combined in the extruder, and the machine actually lays down multiple layers of film in one pass. The different layers accomplish different tasks, such as reducing moisture penetration, reducing oxygen penetration, and aiding in essential oil retention. As the paperboard comes through the extruder, it passes over a chilled roller, which cools both surfaces. The paper now has an extremely glossy, waterproof finish. It is wound into a large roll, to be transported to the printing area. The roll is typically 120 in (3.05 m) wide, too big to fit onto the printing and cutting machine. The large roll is slit into narrower rolls, the width determined by the desired dimensions of the finished carton.

Printing and cutting the blank

3 Printing is usually done by the flexographic method, which uses rubber printing plates attached to steel shells. Workers load the roll of polyethylene-coated paperboard into the press. The press prints the words and images of the milk carton onto the paperboard. A typical milk carton might be printed in anything from one to seven colors. All of the colors are printed at one pass through the machine. Next, the same machine scores the paperboard along what will be the edges of the carton, where the box will fold later. A die lowers, and stamps out the carton. If you cut open an empty milk carton down one side and across the bottom and unfold it, you can see the shape of the cut piece. This flat, scored, and printed piece is called a blank. The high-speed printing and cutting equipment turns out hundreds of blanks per minute.

Sealing the blanks

4 Workers at the carton plant next load the blanks into a sealing machine. The machine takes the flat blank and folds it laterally, creating an overlapping side seam. The seam is then heated and squeezed together. The heated polyethylene bonds and the seam are strong and watertight without any additional glue. Thousands of blanks per minute shoot through the sealing machine. This is the final step at the carton manufacturer. The rest of the process is completed at the dairy. The sealed and folded blanks are loaded into corrugated cartons, and they are shipped.

Forming and bottom-sealing

5 Dairies use specialized machinery to transform the blanks into open containers. Workers first load the blanks into a chute leading into the forming machine. The blanks are pulled by suction down onto mechanical arms called forming mandrels. The forming mandrels snap the carton open along its scored lines, and overlap the two bottom flaps. The mandrels are aligned like spokes on a turning wheel. As the carton on

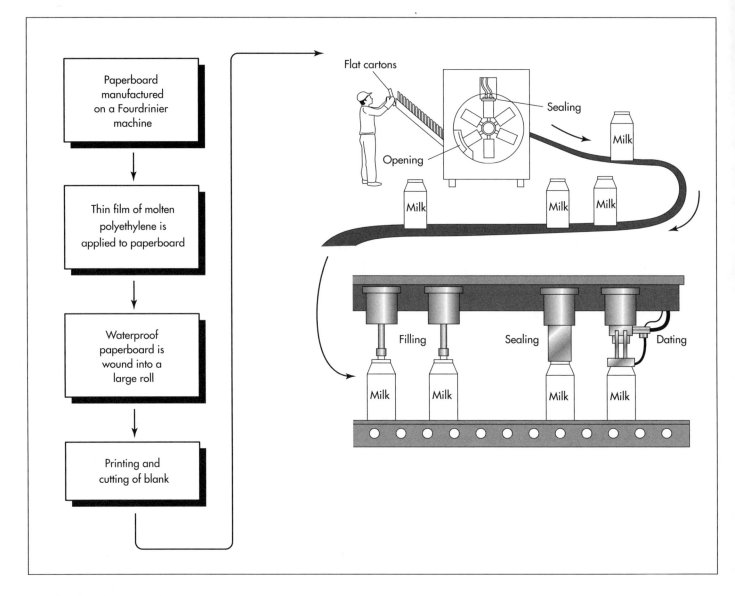

Specialized machinery transforms the milk carton blanks into open containers. First, blanks are loaded into a chute leading into the forming machine. The blanks are pulled by suction down onto mechanical arms called forming mandrels, and the carton is snapped open along its scored lines, and overlap the two bottom flaps. As the carton on the mandrel reaches the top of the wheel, the bottom of the carton is pressed against a hot plate that descends and seals the bottom seam.

the mandrel reaches the top of the wheel, the bottom of the carton is pressed against a hot plate that descends and seals the bottom seam. As the wheel continues to rotate, the bottom-sealed carton moves down, and is pulled by suction off the forming mandrel and set down on a conveyor belt.

Filling and top-sealing

6 The conveyor belt moves the carton to the filling area. Milk from the dairy's storage area descends by pipes to the filling machine. A pre-measured amount of milk fills a chamber above the carton. Then the milk is released through a spout into the carton. The filled carton passes along on the conveyor belt to the top-sealing machine. The top-sealing machine lowers onto the

carton and pinches the top together along pre-scored lines. The shape of the conventional milk carton is called gable-topped. The top-sealer forms the gable, and heats and presses the top seam together. As in all the other seams, the polyethylene bonds to itself, and no additional glue is needed.

Stamping the date

7 All milk cartons must have a date stamped on the top, indicating how long the milk will stay fresh. At the next stop along the conveyor belt, the filled, sealed carton passes under a stamping machine, which impresses the date along the tope edge of the carton. At a big milk-processing plant, the whole operation, from folded blank to date-sealed finished product, takes

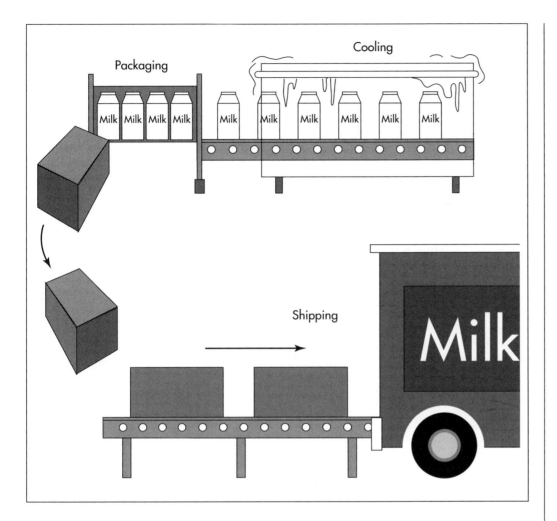

only a few seconds. After the date is stamped, the finished milk carton moves off the conveyor, and is packed either automatically or by hand into a packing case, for shipment to market.

Quality Control

Manufacturers make quality checks at every step along the manufacturing process. The pulp must be inspected to make sure it is the proper color and density, and has the desired fiber characteristics. As the pulp is a blend of long and short fibers, from soft and hardwood trees, batches may differ according to the kind and proportion of trees used. The paperboard must pass numerous quality checks, for different reasons. The Federal Drug Administration (FDA) requires that milk cartons meet strict standards for hygiene and safety. For instance, the FDA must approve any chemicals added to the paperboard and the manufacturer must be

able to prove it is meeting its regulated requirements. The width, thickness, and fiber mix of the paperboard is continually monitored by instruments attached to the papermaking machine, and the board is checked for contaminants as well. At the dairy or milk processing plant, forming and filling of the cartons is done under exacting standards for hygiene and safety.

Byproducts/Waste

The manufacturing process for milk cartons is extremely efficient, and there is very little waste. However, most used cartons are thrown in the trash and end up in landfills. It is possible to recycle them, though, if the appropriate recycling facilities exist. A milk carton recycler collects empty cartons from large users such as schools and hospitals. Then the recycler shreds the cartons, sanitizes them, and ties the shreds into bales. A pulp mill buys the bales from the recycler.

At the mill, the polyethylene coating is separated from the paper, and strained off for re-use by a plastics manufacturer. The shredded cartons are then reprocessed into pulp, and can be used to make high grades of printing and writing paper.

The Future

Milk carton manufacturing has not changed dramatically for many years, because the process is already highly streamlined and efficient. An increasingly popular modification to the tradition gable-topped carton is the addition of a plastic pour spout, but this requires only minor changes in the manufacturing process. As milk consumption falls in the United States, future changes might be in the graphic design of the cartons, as dairies compete harder for customers. Because the gable-topped cartons are very cost-effective to manufacture, packagers are searching for other products that can be sold in them. However, the polyethylene coating for milk cartons is not appropriate for every liquid. For instance, wine and motor oil have different characteristics than milk, and so need different waterproof barriers. Chemists and design engineers are currently researching new plastic coatings, so that other liquids besides milk can use paper cartons.

Where to Learn More

Periodicals

"Milk Carton Recap." *Packaging Digest* (August 1994): 36-37.

"Milk Carton Recycling Does Everybody Good!" *Science Activities* (Winter 1994): 5.

—Angela Woodward

Model Train

Background

"Dear Dad: One thing I want this Christmas more than anything is a Lionel Electric Train. . . . You ought to see the way they run! Like a million dollars. And they whistle too. Real railroad whistle signals by remote control. You can couple and uncouple cars electrically, from a distance, just by touching a button; and reverse the train or speed it up or slow it down. Please get me a Lionel, Dad. We'll have lots of fun together."

This letter was featured in a pull-out section of the 1938 Lionel Trains catalog, leaving space for a boy to note which model number train his father should purchase. Lionel electric model trains were all the rage in 1938, had been for many years prior, and continued to be through the 1950s. Though declining in popularity since the 1960s, the trains are still manufactured and sold throughout the world today. Many adults now collect old Lionel train models, as evidenced by the number of hobby shops and collectors' shows dedicated to the product, and the thousands of members of the Train Collectors Association. Lionel is the largest manufacturer of toy trains in the world.

An electric train runs by transferring a positive current from one track rail through to the motor and then returning the current through the negative track rail. The current is then transferred to a transformer or battery, completing the circuit.

History

Joshua Lionel Cowen claimed to have embarked upon several other inventions prior to his namesake train, including the flashlight, the dry-cell battery and the motorized fan. Whether these claims were true or not is subject to dispute, but there is no argument that Cowen devised one of the first motorized trains as an ad gimmick for a New York City toy manufacturer in the early 1900s. Cowen's idea was not entirely unique; a German toy maker had featured a model electric streetcar at the Columbian World's Fair in Chicago in 1893 and a Cincinnati firm, Carlisle & Finch, came up with a similar invention in 1896. Cowen's original battery-operated invention was not a toy, however. It served solely to draw attention to the other merchandise in the toy shop window and resembled a box on wheels. The words "Electric Express" were embossed on the sides. However, the toy shop customers began requesting the electric car as well as the other merchandise and Cowen began to market his invention.

Cowen soon upgraded his design and began to make a variety of components. In addition to steam locomotives, Pullman sleepers, baggage cars, freight cars and cabooses, he made electric trolleys as well. Trains bearing the logos of various rail lines were available, too, and railroad companies began to submit blueprints of new designs to Lionel in the hopes that he would create a model based on them and give them some free advertising. The transformer was also introduced in the early 1900s.

In 1902, Cowen produced the first of what would become his trademark train catalogs. This 16-page, black-and-white version paled in comparison to the later full-color wish books, but still served as a useful marketing tool. Catalogs and advertising were primary components of the Lionel marketing strategy and for decades the company promoted

In 1920, Lionel train sales topped $2 million, however the Great Depression took its toll on the company. In 1934, the introduction of a handcar operated by Mickey and Minnie Mouse and endorsed by Cowen's friend Walt Disney helped the company bounce back.

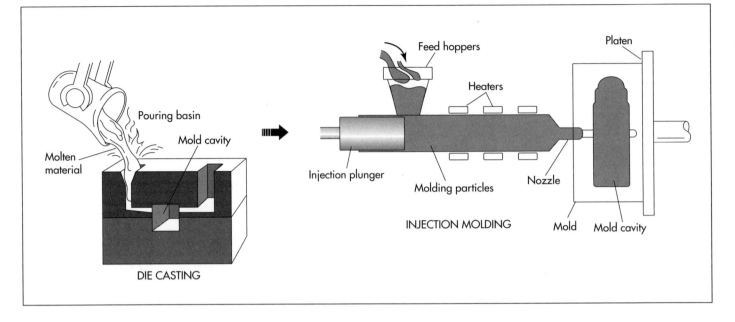

Feed hoppers

Platen

Heaters

Pouring basin

Mold cavity

Molten
material

Injection plunger

Molding particles

Nozzle

INJECTION MOLDING

Mold   Mold cavity

DIE CASTING

The steam engine is made by a process called die-casting, whereby a hot liquid metal is heated and then shot into a highly detailed mold. Plastic cars are made by a process called injection molding.

the sense of importance a boy could feel running his own railroad and the opportunity the product provided for father and son to bond. In 1921, the Lionel Manufacturing Company placed the first-ever advertisement in the color comics section of a newspaper promoting its Lionel Engineers Club for Lionel train owners.

Lionel joined the war effort in 1917, producing compasses, binnacles, and navigating equipment for the U.S. Navy. The company also offered a model war train. By the 1930s, Cowen began to re-think his decision to promote war toys, however. The post-war years were profitable for the Lionel Corporation, as the company was renamed when it was re-organized in 1918. Lionel's sales in 1920 topped $2 million. The Great Depression took its toll on the company, but the 1934 introduction of a handcar operated by Mickey and Minnie Mouse, endorsed by Cowen's friend Walt Disney, helped the company bounce back. Lionel introduced streamlined engines that year to reflect the new Burlington Zephyr and Union Pacific City of Portland in use in the real train world. Lionel introduced remote control operation that year as well. A painstakingly accurate model of the New York Central's Hudson-type steam engine was released in 1937 in an effort to appeal to the burgeoning market among adult model railroad enthusiasts.

World War II halted the production of Lionel trains for a period as all scrap metal was di-

rected toward the war effort, but Lionel remained secure with $5.5 million in government contracts. In order to sustain its popularity in the interim, the company released a paper model train, dubbed the "Wartime Freight Train." Metal trains were back, though, once the war ended and in 1957 Lionel introduced the ill-received pastel pink and blue model train for girls. Cowen retired in 1958, the company's first losing year since the Depression. Nine months later he sold the company to his grand-nephew. With the advent of airplanes, racing cars, and television, model trains dropped in popularity over the next three decades. The company has been purchased several times since Cowen first sold it and is now owned by a group of four investors, one of whom is the rock musician Neil Young, an avid model train collector. Young's interest in Lionel leadership began when he helped the company design a remote control device that could be operated by persons with handicaps affecting their grip. Young hoped to actively share his hobby with his sons, who have cerebral palsy. Since 1970, Lionel trains have been manufactured in Mt. Clemens, Michigan, a suburb of Detroit.

Raw Materials

The primary materials used to manufacture Lionel trains are metals such as steel, aluminum, zinc, and plastic.

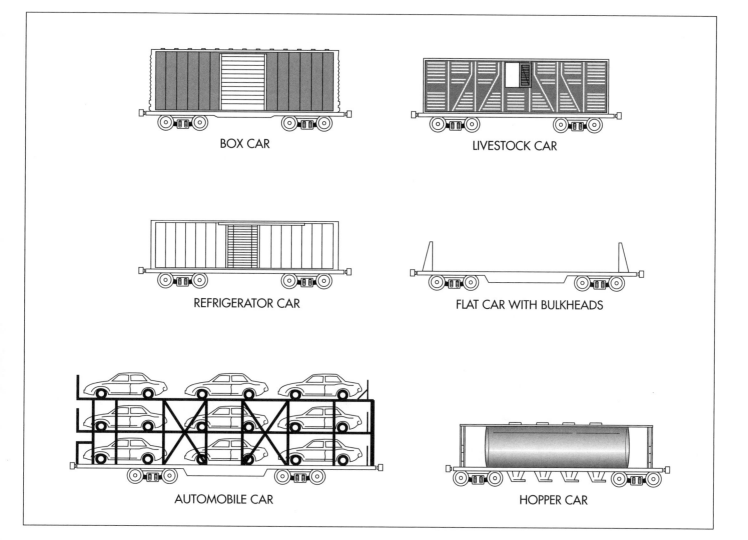

BOX CAR

LIVESTOCK CAR

REFRIGERATOR CAR

FLAT CAR WITH BULKHEADS

AUTOMOBILE CAR

HOPPER CAR

Types of model trains.

The Manufacturing Process

The various components of the Lionel train, such as engines, cabooses, boxcars, and tankers, are designed on a computer.

The engine

1 The Lionel steam locomotive engine is made by a process called die-casting, whereby a hot liquid metal, such as steel or zinc, is heated to 900° F (482.2° C) degrees and then shot into a highly detailed mold.

2 The mold is then placed in a cooling tank where it cools and hardens to form the body of the locomotive.

3 The locomotive body is then trimmed, cut and milled (whereby a high-speed cutter cuts off sections only thousandths of an inch thick to create flat surfaces), both by machine and by hand to form the finished locomotive engine.

4 The locomotive body is placed on the assembly line for painting and final assembly.

The cars

5 Plastic components are formed in a process called injection molding, whereby plastic pellets are melted and shot into a mold. The liquid cools and hardens into the shape of the component. Metal components other than the engine are die-cast.

6 The components are placed on an assembly line and sent to a drilling station, where a combination of manual and automated processes are used to attach hundreds of tiny accessories, created earlier in a

process known as sub-assembly, such as ladders, headlights and handrails.

7 Any die cast parts undergo a process called phosphating, where the component is dipped in a phosphate solution to open up pores in the metal and allow for any paint to soak in.

8 The metal and plastic components are sent to the painting station where they are sprayed with paint by a machine or painted by hand. Components being painted multiple colors undergo a process called masking, where a mask-like guard is placed over areas that a certain color of paint should not reach.

9 At the next station, the components are affixed with lettering and logos through two different processes: hot stamping and pad printing. Hot stamping is used on flat parts and pad printing is used on raised and rounded plastic surfaces and all die-cast surfaces. Hot stamping uses a Mylar-heated rubber dye to emboss print onto the flat surfaces and pad printing uses a dye plate to etch print onto the raised, rounded and die-cast surfaces.

10 The engine is affixed to a chassis, which holds the motor, electrical circuitry and wheels. Each car is affixed to a non-motorized chassis that provides support and holds the wheels.

11 After successful completion of several quality control tests, the finished model train components are sent to a shipping area where they are packaged and prepared for shipping.

The tracks

12 Metal sheets are placed into a forming machine, which cuts the metal into miniature rails and ties.

13 The rails and ties are joined together in a hydraulic press.

14 After quality control inspections, the track segments are sent to a shipping area where they are packaged and prepared for shipping.

Quality Control

At the end of the assembly line, various functioning components of the train are tested, such as whistles and bells. The engine must be able to run off the line under its own power and climb a 30° incline within 5.5 seconds. Paint colors undergo quality control as well. A color spectrometer is used to ensure that all paint is the precisely right color.

The Future

Lionel train components have always reflected the times. Engines have been modeled after real trains in use during various periods in Lionel's history. The cars often reflect the businesses and interests of the times, from the early milk cars to military components during war times to boxcars bearing the names and logos of major businesses of different periods. It is anticipated that Lionel designs will continue to reflect contemporary society and the real-life railroading environment on which the Lionel train is modeled.

Where to Learn More

Books

Hollander, Ron. *All Aboard!: The Story of Joshua Lionel Cowen and His Lionel Train Company.* New York: Workman Publishing, 1981.

—*Kristin Palm*

Motorcycle

Background

The motorcycle is "a form of entertainment that can appeal only, one would think, to the most enthusiastic of mechanical eccentrics," *Engineering* magazine stated in 1901. "We think it doubtful whether the motor cycle will, when the novelty has worn off, take a firm hold of public favour."

Last year, four million motorcycles were in use in the United States alone. Whether relied upon as a primary means of transportation, used to provide weekend recreation, souped up and sped along for racing, or displayed as antique, millions of people across the world have shown that the novelty most definitely has not worn off.

History

As might be imagined, the motorcycle evolved from a vehicle powered by sheer human energy—the bicycle. French bicycle maker Pierre Michaux and his sons Ernest and Henri first fitted a bicycle with cranks and pedals—precursors to the modern-day motor—in 1861. The Michauxes' velocipede was an instant hit and the family became the largest velo producer in Europe with a large factory at Bar-le-Duc in France. Working with Michaux, L.G. Perreaux devised a steam-powered motorcycle engine, called a *velo-a-vapeur*, which was patented in 1868. Sylvester Howard Roper of Roxbury, Massachusetts pioneered a similar invention in the United States around that time as well.

In 1879, Giuseppe Munigotti of Italy patented the first gas-burning internal combustion four-stroke engine for the new mo-

torcycles, although his invention existed only on paper. Meanwhile, two Germans, Dr. Nicolaus Otto and Eugen Langen were developing four-stroke stationary engines, which ran on coal gas supplied from mains. Gottlieb Daimler took the invention further by developing an engine that ran on benzine. Since benzine could carry a vehicle approximately 25 miles on one gallon, only a small tank would need to be attached to the machine. Daimler later abandoned the motorcycle business to concentrate on another invention—the first automobile that became the basis for his company, Daimler Benz, maker of the luxury Mercedes Benz automobile.

Several innovators improved upon these inventions over the next 30 years, and in 1901 the machine that is still regarded as Carl Hedstrom, a Swedish immigrant to the United States, developed the first modern motorcycle. Hedstrom fitted an Indian bicycle with a 1.75-horsepower single-cylinder engine, and the legendary Indian motorcycle was born. Several other U.S. makers came out with similar models, including the company whose name is synonymous with the motorcycle—Harley-Davidson—in 1903. William Harley and Arthur Davidson were students in Milwaukee when they built their first motorcycle on a borrowed lathe from patterns they had made. Davidson's older brothers, both toolmakers, assisted, as did Ole Evinrude, who later became famous as a designer and producer of outboard motor boat engines.

Other makers included Royal, Merkel, Yale, Reading-Standard, Rambler, Tribune and Curtiss. By 1904, motorcycle manufacturers had begun to construct bulkier, sturdier frames, stronger wheels, bigger engines and

In 1997, four million motorcycles were in use in the United States alone.

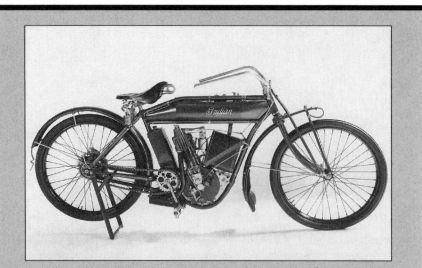

An Indian motor-cycle made by Hendee Manufacturing Co. of Springfield, Massachusetts, circa 1911. (From the collections of Henry Ford Museum & Greenfield Village, Dearborn, Michigan.)

The Europeans took the lead in developing the motorcycle in the early twentieth century. One Englishman proclaimed his countrymen loved the cycle because they enjoyed mechanical things. However, Americans enjoyed their motorcycles as well. An American gentleman could have purchased this home-grown motorcycle manufactured in 1911 by the Hendee Manufacturing Co. of Springfield, Massachusetts. Hendee made early American motorcycles, which featured Native American names indicating a proud ruggedness. Red with gold striping, this one cylinder 3.5 horsepower loop-framed cycle weighs 140 lb (63.6 kg), has a wheel diameter of 23.5 in (59.7 cm) and cost a whopping $225.00 back in 1911. Founded in 1901, the company ceased operation in 1953. However, Indian motorcycles are still beloved—enthusiasts claim that there are still 50,000 Indian motorcycles on the road.

Indian bikes might reach 60 m.p.h., but handbooks cautioned riders to not exceed 10 m.p.h. through town. These early handbooks are full of advice and etiquette for the motorcyclist. Some period gems include: don't ride with the muffler open as "the noise scares restive horses, and worries invalids and nervous people," don't run away in case of accident but "stand by like a man . . . don't get rattled," and don't ride by a motorcyclist who is stalled by the side of the rode as "you may be in the same fix yourself some day."

Nancy EV Bryk

reinforced forks for their bikes and a clear distinction between motorcycles and bicycles emerged. Around this time, the sidecar, affixed to a light, tubular frame extending from the main motorcycle frame, began to be popularized. Based on a similar accessory for the bicycle, his device allowed the driver of a motorcycle to carry a passenger.

By 1905, the focus was on power, and manufacturers begin to beef up their engines. That year, Hedstrom produced a machine boasting a 500 cc twin engine that featured twist-grip control for the throttle and ignition. That same year saw the development of "free engine" devices, which eased the starting and launching of a machine, and variable gears, which eased use on hills and at slow speeds. Chain drive followed, and the stage was set for production of the motorcycle that is currently in use.

Motorcycles continued to grow in popularity for decades, although production for civilians tapered off during World War II. During World War II, however, a need arose for lightweight, collapsible models to be used by parachutists once they had landed. Royal Enfield produced a Flying Flea model for this purpose, while Excelsior came up with its Welbike, which could fit into a small air-drop container. The Welbike was later marketed to civilians as the Corgi, spawning the post-war popularity of the motor scooter, especially in Europe.

The 1950s were regarded as a "golden age" for the motorcycle, with its use being popularized by such prominent figures in popular culture as James Dean in his movie "Rebel Without a Cause." The United States and Europe dominated the motorcycle industry through 1960, at which time Japanese manufacturers, including Honda, Yamaha, Suzuki and Kawasaki, rose to a prominence they maintain to this day.

Raw Materials

The primary raw materials used in the manufacture of the body of motorcycle are metal, plastic and rubber. The motorcycle frame is composed almost completely of metal, as are the wheels. The frame may be overlaid with plastic. The tires are composed of rubber. The seat is made from a synthetic substance, such as polyurethane. The power system consists of a four-stroke engine, a carburetor to transform incoming fuel into vapor, a choke to control the air-fuel ratio, transmission, and drum brakes. The transmission system contains a clutch, consisting of steel ball flyweights and metal plates, a crankshaft, gears, pulleys, rubber belts or metal chains, and a sprocket. The

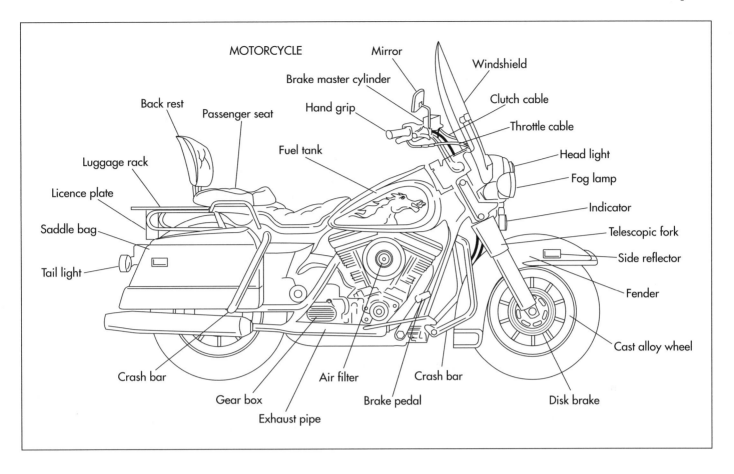

MOTORCYCLE

Mirror
Brake master cylinder
Hand grip
Windshield
Clutch cable
Throttle cable
Head light
Fog lamp
Indicator
Telescopic fork
Side reflector
Fender
Cast alloy wheel
Disk brake
Crash bar
Brake pedal
Air filter
Exhaust pipe
Gear box
Crash bar
Tail light
Saddle bag
Licence plate
Luggage rack
Back rest
Passenger seat
Fuel tank

electrical system contains a battery, ignition wires and coils, diodes, spark plugs, head-lamps and taillights, turn signals and a horn.

A cylindrical piston, made of aluminum alloy (preferred because it is lightweight and conducts heat well), is an essential component of the engine. It is fitted with piston rings made of cast iron. The crankshaft and crankcase are made of aluminum. The engine also contains a cylinder barrel, typically made of cast iron or light alloy.

The Manufacturing Process

1 Raw materials as well as parts and components arrive at the manufacturing plant by truck or rail, typically on a daily basis. As part of the just-in-time delivery system on which many plants are scheduled, the materials and parts are delivered at the place where they are used or installed.

2 Manufacturing begins in the weld department with computer-controlled fabrication of the frame from high strength frame materials. Components are formed out of tubular metal and/or hollow metal shells fashioned from sheet metal. The various sections are welded together. This process involves manual, automatic, and robotic equipment.

3 In the plastics department, small plastic resin pellets are melted and injected into molds under high pressure to form various plastic body trim parts. This process is known as injection molding.

4 Plastic and metal parts and components are painted in booths in the paint department using a process known as powder-coating (this is the same process by which automobiles are painted). A powder-coating apparatus works like a large spray-painter, dispersing paint through a pressurized system evenly across the metal frame.

5 Painted parts are sent via overhead conveyors or tow motor (similar to a ski lift tow rope) to the assembly department where they are installed on the frame of the motorcycle.

A motorcycle engine.

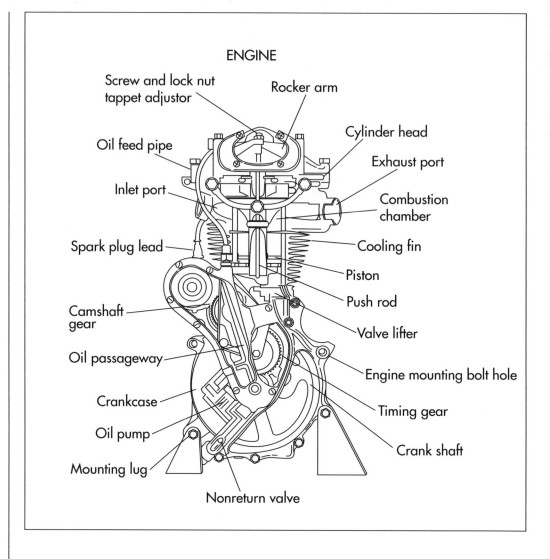

ENGINE

Screw and lock nut tappet adjustor

Rocker arm

Oil feed pipe

Cylinder head

Exhaust port

Inlet port

Combustion chamber

Spark plug lead

Cooling fin

Piston

Push rod

Camshaft gear

Valve lifter

Oil passageway

Engine mounting bolt hole

Crankcase

Timing gear

Oil pump

Crank shaft

Mounting lug

Nonreturn valve

6 The engine is mounted in the painted frame, and various other components are fitted as the motorcycle is sent down the assembly line.

7 Wheels, brakes, wiring cables, foot pegs, exhaust pipes, seats, saddlebags, lights, radios, and hundreds of other parts are installed on the motorcycle frame. A Honda Gold Wing motorcycle, for example, needs almost as many parts to complete it as a Honda Civic automobile.

Quality Control

At the end of the assembly line, quality control inspectors undertake a visual inspection of the motorcycle's painted finish and fit of parts. The quality control inspectors also feel the motorcycles with gloved hands to detect any bumps or defects in the finish.

Each motorcycle is tested on a dynamometer. Inspectors accelerate the motorcycle from 0-60 mph. During the acceleration, the "dyno" tests for acceleration and braking, shifting, wheel alignment, headlight and taillight alignment and function, horn function, and exhaust emissions. The finished product must meet international standards for performance and safety. After the dyno test, a final inspection is made of the completed motorcycle. The motorcycles are boxed in crates and shipped to customers across North America and around the world.

The Future

Motorcycles remain popular and the collecting and riding of antique models is just as popular as riding the new versions. While sleek, new versions will continue to be pro-

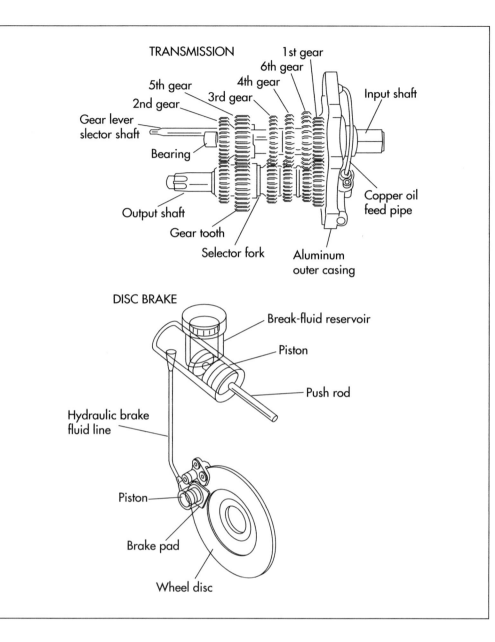

TRANSMISSION

5th gear
2nd gear
Gear lever
slector shaft
Bearing
4th gear
3rd gear
6th gear
1st gear
Input shaft
Copper oil
feed pipe
Output shaft
Gear tooth
Selector fork
Aluminum
outer casing

DISC BRAKE

Break-fluid reservoir
Piston
Push rod
Hydraulic brake
fluid line
Piston
Brake pad
Wheel disc

A motorcycle transmission and disc brake system.

duces, it is anticipated that the value of older models will continue to rise.

Where to Learn More

Books

Ayton, Cyril, Bob Holliday, Cyril Posthumus and Mike Winfield. *The History of Mo-*

torcycling. London: Orbis Publishing, 1979.

Lear, George and Lynn S. Mosher, *Motorcycle Mechanics*. Englewood Cliffs, NJ: Prentice Hall, 1997.

—Kristin Palm

Orange Juice

The current worldwide market for orange juice is more than $2.3 billion with the biggest area being the United States followed by Canada, Western Europe, and Japan.

Background

Orange juice is defined in the United States Code of Federal Regulations as the "unfermented juice obtained from mature oranges of the species *Citrus sinensis* or of the citrus hybrid commonly called Ambersweet." True fresh squeezed juice is difficult to market commercially because it requires special processing to preserve it. Orange juice is commonly marketed in three forms: as a frozen concentrate, which is diluted with water after purchase; as a reconstituted liquid, which has been concentrated and then diluted prior to sale; or as a single strength, unconcentrated beverage called NFC or Not From Concentrate. The latter two types are also known as Ready To Drink (RTD) juices.

Citrus fruits, like oranges, have been cultivated for the last 4,000 years in southern China and Southeast Asia. One variety, the citron, was carried to the Middle East sometime between 400 and 600 B.C. Arab traders transported oranges to eastern Africa and the Middle East sometime between 100 and 700 A.D., and during the Arab occupation of Spain, citrus fruits first arrived in southern Europe. From there, they were carried to the New World by explorers where they spread to Florida and Brazil by the sixteenth century. By the 1800s, citrus fruits achieved worldwide distribution. In the 1890s, the demand for them greatly increased because physicians discovered that drinking the juice of oranges or other citrus fruits could prevent scurvy, a vitamin deficiency disease.

The popularity of orange juice dramatically increased again with the development of the commercial orange juice industry in the late 1920s. In its early days, the juice industry primarily relied on salvaged fruit, which was unsuitable for regular consumption because it was misshapen, badly colored or blemished. In the 1930s, development of porcelain-lined cans and advances in pasteurization techniques led to improved juice quality and the industry expanded significantly. Then, in 1944, scientists found a way to concentrate fruit juice in a vacuum and freeze it without destroying the flavor or vitamin content. Frozen concentrated juices were first sold in the United States during 1945-46, and they became widely available and popular. After World War II, most Americans stopped squeezing their own juice and concentrated juice became the predominant form. With the increase in home refrigerators, frozen concentrate became even more popular. The demand for frozen juices had a profound impact on the citrus industry and spurred the growth of the Florida citrus groves. Frozen concentrates remained the most popular form until 1985 when reconstituted and NFC juices first outsold the frozen type. In 1995, NFC juices were responsible for 37% of the North American market. This is in comparison to reconstituted juice, which held about 39% of the market. Today, commercial aseptic packaging allows RTD juices to be marketed without refrigerated storage. The current worldwide market for orange juice is more than $2.3 billion with the biggest area being the United States followed by Canada, Western Europe, and Japan.

Raw Materials

Fruit

The primary ingredient in orange juice is, of course, oranges. Oranges are members of the rue family (Rutaceae), and citrus trees belong to the genus *Citrus*. Oranges, along with all citrus fruits, are a special type of berry botanists refer to as a hesperidium. Popular types of oranges include navel, Mandarin, and Valencia. A blend of different types of oranges is generally used to provide a specific flavor and to ensure freedom from bitterness. Selection of oranges for juice is made on the basis of a number of factors such as variety and maturity of the fruit. The fruit contains a number of natural materials that contribute to the overall flavor and consistency of the juice including water, sugars (primarily sucrose, fructose, and glucose), organic acids (primarily citric, malic, and tartaric), and flavor compounds (including various esters, alcohols, ketones, lactones, and hydrocarbons.)

Other additives

Preservatives such as sulfur dioxide or sodium benzoate are allowed by federal regulation in orange juice although the amounts are strictly controlled. Similarly, ascorbic acid, alpha tocopherol, EDTA, BHA, or BHT are used as antioxidants. Sweeteners may be added in the form of corn syrup, dextrose, honey, or even artificial sweeteners. More often, though, citric acid is added to provide tartness.

Manufacturers may also fortify juices with extra vitamins or supplemental nutrients such as vitamin C, and less commonly, vitamins A and E, and beta carotene. (Beta carotene is naturally present in oranges, but only to a small degree.) There is some concern about the stability of these added vitamins because they do not survive the heating process very well. Calcium in the form of tricalcium phosphate, is also frequently added to orange juice.

The Manufacturing Process

Harvesting/collection

1 Oranges are harvested from large groves. Some citrus growers are members of cooperative packing and marketing associations, while others are independent growers. When the mature fruit is ready to pick, a crew of pickers is sent in to pull the fruit off the trees. The collected fruit is sent to pack-

ing centers where it is boxed for sale as whole fruit, or sent to plants for juice processing. The oranges are generally shipped via truck to juice extraction facilities, where they are unloaded by a gravity feed onto a conveyor belt that transports the fruit to a storage bin.

Cleaning/Grading

2 The fruit must be inspected and graded before it can be used. An inspector takes a 39.7 lb (18 kg) sample to analyze in order to make sure the fruit meets maturity requirements for processing. The certified fruit is then transported along a conveyor belt where it is washed with a detergent as it passes over roller brushes. This process removes debris and dirt and reduces the number of microbes. The fruit is rinsed and dried. Graders remove bad fruit as it passes over the rollers and the remaining quality pieces are automatically segregated by size prior to extraction. Proper size is critical for the extraction process.

Extraction

3 Proper juice extraction is important to optimize the efficiency of the juice production process as well as the quality of the finished drink. The latter is true because oranges have thick peels, which contain bitter resins that must be carefully separated to avoid tainting the sweeter juice. There are two automated extraction methods commonly used by the industry. The first places the fruit between two metal cups with sharpened metal tubes at their base. The upper cup descends and the fingers on each cup mesh to express the juice as the tubes cut holes in the top and bottom of the fruit. The fruit solids are compressed into the bottom tube between the two plugs of peel while the juice is forced out through perforations in the tube wall. At the same time, a water spray washes away the oil from the peel. This oil is reclaimed for later use.

The second type of extraction has the oranges cut in half before the juice is removed. The fruits are sliced as they pass by a stationary knife and the halves are then picked up by rubber suction cups and moved against plastic serrated reamers. The rotating reamers express the juice as the orange halves travel around the conveyor line.

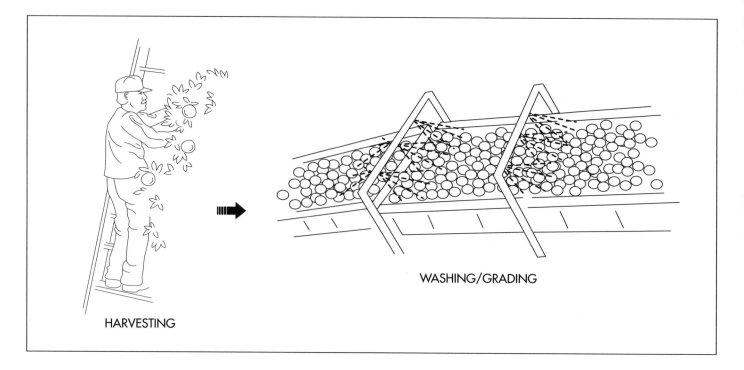

HARVESTING

WASHING/GRADING

When the mature fruit is ready to pick, a crew of pickers pull the fruit off the trees. Once collected, the fruit is sent to plants for juice processing. Before extraction, the fruit is cleaned and graded.

Some of the peel oil may be removed prior to extraction by needles which prick the skin, thereby releasing the oil which is washed away. Modern extraction equipment of this type can slice, ream, and eject a peel in about 3 seconds.

4 The extracted juice is filtered through a stainless steel screen before it is ready for the next stage. At this point, the juice can be chilled or concentrated if it is intended for a reconstituted beverage. If a NFC type, it may be pasteurized.

Concentration

5 Concentrated juice extract is approximately five times more concentrated than squeezed juice. Diluted with water, it is used to make frozen juice and many RTD beverages. Concentration is useful because it extends the shelf life of the juice and makes storage and shipping more economical. Juice is commonly concentrated with a piece of equipment known as a Thermally Accelerated Short-Time Evaporator, or TASTE for short. TASTE uses steam to heat the juice under vacuum and force water to be evaporated. Concentrated juice is discharged to a vacuum flash cooler, which reduces the product temperature to about 55.4° F (13° C). A newer concentration process requires minimal heat treat-

ment and is used commercially in Japan. The pulp is separated from the juice by ultra-filtration and pasteurized. The clarified juice containing the volatile flavorings is concentrated at 50° F (10° C) by reverse osmosis and the concentrate and the pulp are recombined to produce the appropriate juice concentration. The flavor of this concentrate has been judged to be superior to what is commercially available in the United States and is close to fresh juice. Juice concentrate is then stored in refrigerated stainless steel bulk tanks until is ready to be packaged or reconstituted.

Reconstitution

6 When the juice processor is ready to prepare a commercial package for retail sale, concentrate is pulled from several storage batches and blended with water to achieve the desired sugar to acid ratio, color, and flavor. This step must be carefully controlled because during the concentration process much of the juice's flavor may be lost. Proper blending of juice concentrate and other flavor fractions is necessary to ensure the final juice product achieves a high quality flavor.

Pasteurization

7 Thanks to its low pH (about 4), orange juice has some natural protection from

EXTRACTION

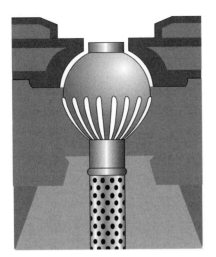

The orange is placed between two metal cups.

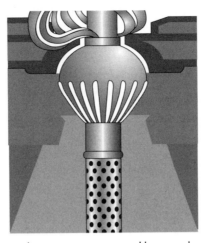

As the orange is compressed between the cups, the peel is removed from the fruit.

Juice is squeezed out of the fruit and forced through perforations in the tube wall.

In an automated process, the juice is extracted from the orange while the peel is removed in one step.

bacteria, yeast, and mold growth. However, pasteurization is still required to further retard spoilage. Pasteurization also inactivates certain enzymes which cause the pulp to separate from the juice, resulting in an aesthetically undesirably beverage. This enzyme related clarification is one of the reasons why fresh squeezed juice has a shelf life of only a few hours. Flash pasteurization minimizes flavor changes from heat treatment and is recommended for premium quality products. Several pasteurization methods are commercially used. One common method passes juice through a tube next to a plate heat exchanger, so the juice is heated without direct contact with the heating surface. Another method uses hot, pasteurized juice to preheat incoming unpasteurized juice. The preheated juice is further heated with steam or hot water to the pasteurization temperature. Typically, reaching a temperature of 185-201.2° F (85-94° C) for about 30 seconds is adequate to reduce the microbe count and prepare the juice for filling.

Packaging/filling

8 To ensure sterility, the pasteurized juice should be filled while still hot. Where possible, metal or glass bottles and cans can be preheated. Packaging which can not withstand high temperatures (e.g., aseptic, multilayer plastic juice boxes which don't require refrigeration) must be filled in a sterile environment. Instead of heat, hydrogen peroxide or another approved sterilizing agent may be used prior to filling. In any case, the empty packages are fed down a conveyor belt to liquid filling machinery, which is fed juice from bulk storage tanks. The filling head meters the precise amount of product into the container, and depending on the design of the package, it may immediately invert to sterilize the lid. After filling, the containers are cooled as fast as possible. Orange juice packaged in this manner has a shelf life of 6-8 months at room temperature.

Byproducts/Waste

Byproducts from orange juice production come from the rind and pulp that is created as waste. Products made with these materials include dehydrated feed for livestock, pectin for use in making jellies, citric acid, essential oils, molasses, and candied peel. Certain fractions of orange oil (known as d-limonene), have excellent solvent properties and are sold for use in industrial cleaners.

Quality Control

Quality is checked throughout the production process. Inspectors grade the fruit before the juice is extracted. After extraction and concentration, the product is checked to ensure it meets a number of USDA quality control standards. The most important measurement in orange juice production is the sugar level, which is measured in degrees Brix (percentages by weight of sugar in a solution). The types of oranges used and the climate in which they were grown effect the sugar level. Manufacturers blend juices with different sugar levels together to achieve a desired sugar balance. The final juice product is evaluated for a number of key parameters include acidity, citrus oil level, pulp level, pulp cell integrity, color, viscosity, microbiological contamination, mouth feel, and taste. A sensory panel is used to evaluate subjective qualities like flavor and texture. Lastly during the filling process, units are inspected to make sure they are filled and sealed appropriately.

The Future

Future processing improvements are likely to come from the use of computer controlled sizing and grading of fruit. Orange juice formulations will see changes as the trend toward adding more nutrition-oriented ingredients, such as antioxidants, continues. In addition, future formulas are likely to be blends of orange juice with other, more exotic, fruit flavors, like kiwi, or even vegetable juices, like carrot.

Where to Learn More

Books

Nelson, P.E. and D.K. Tressler, ed. *Fruit and Vegetable Juice Processing Technology*. Westport, Connecticut: AVI Publishing Co., 1980.

Periodicals

"Juice Up." *Food Product Design* (July 1997).

"Unconcentrated Effort." *Food Processing* (November 1996).

—Randy Schueller

Oxygen

Background

Oxygen is one of the basic chemical elements. In its most common form, oxygen is a colorless gas found in air. It is one of the life-sustaining elements on Earth and is needed by all animals. Oxygen is also used in many industrial, commercial, medical, and scientific applications. It is used in blast furnaces to make steel, and is an important component in the production of many synthetic chemicals, including ammonia, alcohols, and various plastics. Oxygen and acetylene are combusted together to provide the very high temperatures needed for welding and metal cutting. When oxygen is cooled below -297° F (-183° C), it becomes a pale blue liquid that is used as a rocket fuel.

Oxygen is one of the most abundant chemical elements on Earth. About one-half of the earth's crust is made up of chemical compounds containing oxygen, and a fifth of our atmosphere is oxygen gas. The human body is about two-thirds oxygen. Although oxygen has been present since the beginning of scientific investigation, it wasn't discovered and recognized as a separate element until 1774 when Joseph Priestley of England isolated it by heating mercuric oxide in an inverted test tube with the focused rays of the sun. Priestley described his discovery to the French scientist Antoine Lavoisier, who experimented further and determined that it was one of the two main components of air. Lavoisier named the new gas oxygen using the Greek words *oxys*, meaning sour or acid, and *genes*, meaning producing or forming, because he believed it was an essential part of all acids.

In 1895, Karl Paul Gottfried von Linde of Germany and William Hampson of England independently developed a process for lowering the temperature of air until it liquefied. By carefully distillation of the liquid air, the various component gases could be boiled off one at a time and captured. This process quickly became the principal source of high quality oxygen, nitrogen, and argon.

In 1901, compressed oxygen gas was burned with acetylene gas in the first demonstration of oxy-acetylene welding. This technique became a common industrial method of welding and cutting metals.

The first use of liquid rocket propellants came in 1923 when Robert Goddard of the United States developed a rocket engine using gasoline as the fuel and liquid oxygen as the oxidizer. In 1926, he successfully flew a small liquid-fueled rocket a distance of 184 ft (56 m) at a speed of about 60 mph (97 kph).

After World War II, new technologies brought significant improvements to the air separation process used to produce oxygen. Production volumes and purity levels increased while costs decreased. In 1991, over 470 billion cubic feet (13.4 billion cubic meters) of oxygen were produced in the United States, making it the second-largest-volume industrial gas in use.

Worldwide the five largest oxygen-producing areas are Western Europe, Russia (formerly the USSR), the United States, Eastern Europe, and Japan.

Raw Materials

Oxygen can be produced from a number of materials, using several different methods.

When oxygen is cooled below -297° F (-183° C), it becomes a pale blue liquid that is used as a rocket fuel.

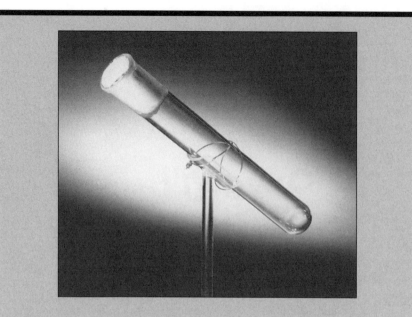

A test tube said to contain the last breath of Thomas Edison and given to Henry Ford, and ardent fan, as a keepsake by Edison's son Charles. (From the collections of Henry Ford Museum & Greenfield Village, Dearborn, Michigan.)

This test tube is one of the most popular artifacts in Henry Ford Museum & Greenfield Village in Dearborn, Michigan. It is said to contain the last breath of Thomas Alva Edison, the great inventor. According to Edison's son Charles, a set of eight empty test tubes sat on the table next to Edison's deathbed in 1931. Immediately after Edison expired, his physician, put several of the tubes up to Edison's lips to catch the carbon dioxide from his deflating lungs. Then, the physician carefully sealed each tube with paraffin and gave the tubes to Charles Edison. Charles Edison knew that Henry Ford's idol was Thomas Edison and presented Ford with one of the tubes as a keepsake. The museum acquired the tube after the death of both Henry and Clara Ford.

There is some discussion among visitors just how much carbon dioxide and how much oxygen currently is contained in the tube. Some ask if anyone evacuated the tube of oxygen before putting the tube to Edison's mouth (very unlikely). If not, how much of Edison's breath could be in the tube? So, they say, it contains both carbon dioxide and oxygen? Nonetheless, it is an unconventional tribute to a great man by those sorry to see his light extinguished.

Nancy EV Bryk

The most common natural method is photosynthesis, in which plants use sunlight to convert carbon dioxide in the air into oxygen. This offsets the respiration process, in which animals convert oxygen in the air back into carbon dioxide.

The most common commercial method for producing oxygen is the separation of air using either a cryogenic distillation process or a vacuum swing adsorption process. Nitrogen and argon are also produced by separating them from air.

Oxygen can also be produced as the result of a chemical reaction in which oxygen is freed from a chemical compound and becomes a gas. This method is used to generate limited quantities of oxygen for life support on submarines, aircraft, and spacecraft.

Hydrogen and oxygen can be generated by passing an electric current through water and collecting the two gases as they bubble off. Hydrogen forms at the negative terminal and oxygen at the positive terminal. This method is called electrolysis and produces very pure hydrogen and oxygen. It uses a large amount of electrical energy, however, and is not economical for large-volume production.

The Manufacturing Process

Most commercial oxygen is produced using a variation of the cryogenic distillation process originally developed in 1895. This process produces oxygen that is 99+% pure. More recently, the more energy-efficient vacuum swing adsorption process has been used for a limited number of applications that do not require oxygen with more than 90-93% purity.

Here are the steps used to produce commercial-grade oxygen from air using the cryogenic distillation process.

Pretreating

Because this process utilizes an extremely cold cryogenic section to separate the air, all impurities that might solidify—such as water vapor, carbon dioxide, and certain heavy hydrocarbons—must first be removed to prevent them from freezing and plugging the cryogenic piping.

1 The air is compressed to about 94 psi (650 kPa or 6.5 atm) in a multi-stage compressor. It then passes through a water-cooled aftercooler to condense any water

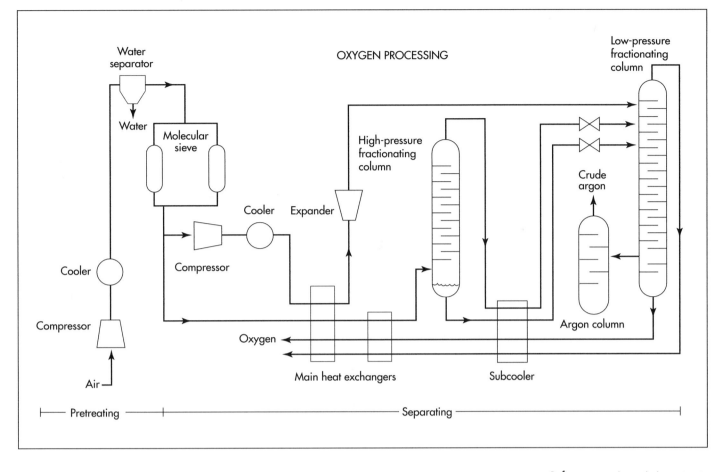

OXYGEN PROCESSING

Pretreating — Separating

vapor, and the condensed water is removed in a water separator.

2 The air passes through a molecular sieve adsorber. The adsorber contains zeolite and silica gel-type adsorbents, which trap the carbon dioxide, heavier hydrocarbons, and any remaining traces of water vapor. Periodically the adsorber is flushed clean to remove the trapped impurities. This usually requires two adsorbers operating in parallel, so that one can continue to process the air-flow while the other one is flushed.

Separating

Air is separated into its major components—nitrogen, oxygen, and argon—through a distillation process known as fractional distillation. Sometimes this name is shortened to fractionation, and the vertical structures used to perform this separation are called fractionating columns. In the fractional distillation process, the components are gradually separated in several stages. At each stage the level of concentra-

tion, or fraction, of each component is increased until the separation is complete.

Because all distillation processes work on the principle of boiling a liquid to separate one or more of the components, a cryogenic section is required to provide the very low temperatures needed to liquefy the gas components.

3 The pretreated air stream is split. A small portion of the air is diverted through a compressor, where its pressure is boosted. It is then cooled and allowed to expand to nearly atmospheric pressure. This expansion rapidly cools the air, which is injected into the cryogenic section to provide the required cold temperatures for operation.

4 The main stream of air passes through one side of a pair of plate fin heat exchangers operating in series, while very cold oxygen and nitrogen from the cryogenic section pass through the other side. The incoming air stream is cooled, while the oxygen and nitrogen are warmed. In some operations, the air may be cooled by passing it

Before processing, air is pretreated to remove impurities that will clog the cryogenic piping. Once pretreated, the air is submitted to fractional distillation. In the fractional distillation process, the components are gradually separated in several stages. Because all distillation processes work on the principle of boiling a liquid to separate one or more of the components, a cryogenic section is required to provide the very low temperatures needed to liquefy the gas components. Once the liquid oxygen is separated, it is purified and stored.

through an expansion valve instead of the second heat exchanger. In either case, the temperature of the air is lowered to the point where the oxygen, which has the highest boiling point, starts to liquefy.

5 The air stream—now part liquid and part gas—enters the base of the high-pressure fractionating column. As the air works its way up the column, it loses additional heat. The oxygen continues to liquefy, forming an oxygen-rich mixture in the bottom of the column, while most of the nitrogen and argon flow to the top as a vapor.

6 The liquid oxygen mixture, called crude liquid oxygen, is drawn out of the bottom of the lower fractionating column and is cooled further in the subcooler. Part of this stream is allowed to expand to nearly atmospheric pressure and is fed into the low-pressure fractionating column. As the crude liquid oxygen works its way down the column, most of the remaining nitrogen and argon separate, leaving 99.5% pure oxygen at the bottom of the column.

7 Meanwhile, the nitrogen/argon vapor from the top of the high-pressure column is cooled further in the subcooler. The mixed vapor is allowed to expand to nearly atmospheric pressure and is fed into the top of the low-pressure fractionating column. The nitrogen, which has the lowest boiling point, turns to gas first and flows out the top of the column as 99.995% pure nitrogen.

8 The argon, which has a boiling point between the oxygen and the nitrogen, remains a vapor and begins to sink as the nitrogen boils off. As the argon vapor reaches a point about two-thirds the way down the column, the argon concentration reaches its maximum of about 7-12% and is drawn off into a third fractionating column, where it is further recirculated and refined. The final product is a stream of crude argon containing 93-96% argon, 2-5% oxygen, and the balance nitrogen with traces of other gases.

Purifying

The oxygen at the bottom of the low-pressure column is about 99.5% pure. Newer cryogenic distillation units are designed to recover more of the argon from the low-pressure column, and this improves the oxygen purity to about 99.8%.

9 If higher purity is needed, one or more additional fractionating columns may be added in conjunction with the low-pressure column to further refine the oxygen product. In some cases, the oxygen may also be passed over a catalyst to oxidize any hydrocarbons. This process produces carbon dioxide and water vapor, which are then captured and removed.

Distributing

About 80-90% of the oxygen produced in the United States is distributed to the end users in gas pipelines from nearby air separation plants. In some parts of the country, an extensive network of pipelines serves many end users over an area of hundred of miles (kilometers). The gas is compressed to about 500 psi (3.4 MPa or 34 atm) and flows through pipes that are 4-12 in (10-30 cm) in diameter. Most of the remaining oxygen is distributed in insulated tank trailers or railroad tank cars as liquid oxygen.

10 If the oxygen is to be liquefied, this process is usually done within the low-pressure fractionating column of the air separation plant. Nitrogen from the top of the low-pressure column is compressed, cooled, and expanded to liquefy the nitrogen. This liquid nitrogen stream is then fed back into the low-pressure column to provide the additional cooling required to liquefy the oxygen as it sinks to the bottom of the column.

11 Because liquid oxygen has a high boiling point, it boils off rapidly and is rarely shipped farther than 500 mi (800 km). It is transported in large, insulated tanks. The tank body is constructed of two shells and the air is evacuated between the inner and outer shell to retard heat loss. The vacuum space is filled with a semisolid insulating material to further halt heat flow from the outside.

Quality Control

The Compressed Gas Association establishes grading standards for both gaseous oxygen and liquid oxygen based on the amount and type of impurities present. Gas grades are called Type I and range from A, which is 99.0% pure, to F, which is 99.995% pure. Liquid grades are called Type II and also range from A to F, although the types and amounts of allowable impurities in liquid grades are different than in gas grades. Type I Grade B and Grade C and Type II Grade C are 99.5% pure and are the most commonly produced grades of oxygen. They are used in steel making and in the manufacture of synthetic chemicals.

The operation of cryogenic distillation air-separation units is monitored by automatic instruments and often uses computer controls. As a result, their output is consistent in quality. Periodic sampling and analysis of the final product ensures that the standards of purity are being met.

The Future

In January 1998, the United States launched the Lunar Prospector satellite into orbit around the moon. Among its many tasks, this satellite will be scanning the surface of the moon for indications of water. Scientists hope that if sufficient quantities of water are found, it could be used to produce hydrogen and oxygen gases through electrolysis, using solar power to generate the electricity. The hydrogen could be used as a fuel, and the oxygen could be used to provide life support for lunar colonies. Another plan involves extracting oxygen from chemical compounds in the lunar soil using a solar-powered furnace for heat.

Where to Learn More

Books

Brady, George S., Henry R. Clauser, and John A. Vaccari. *Materials Handbook*, 14th Edition. McGraw-Hill, 1997.

Handbook of Compressed Gases, 3rd edition. Compressed Gas Association, Inc., Van Nostrand Reinhold Co., Inc., 1990.

Heiserman, David L. *Exploring Chemical Elements and Their Compounds*. TAB Books, 1992.

Kent, James A., editor. *Riegel's Handbook of Industrial Chemistry*, 9th edition. International Thomson Publishing, 1997.

Kroschwitz, Jacqueline I., executive editor, and Mary Howe-Grant, editor. *Encyclopedia of Chemical Technology*, 4th edition. John Wiley and Sons, Inc., 1993.

Stwertka, Albert. *A Guide to the Elements*. Oxford University Press, 1996.

Periodicals

Allen, J.B. "Making Oxygen on the Moon," *Popular Science* (August 1995): 23.

Other

Air Products and Chemicals, Inc. http://www.airproducts.com/gases/oxgen.html.

http://www.intercorr.com/periodic/8.htm (This website contains a summary of the history, sources, properties, and uses of oxygen.)

—Chris Cavette

Parade Float

Today, most major parade floats are designed and constructed by professional builders. Each float costs between $50,000 and $200,000 or more and takes up to a year to create.

Background

A parade float is an elaborately decorated three-dimensional figure or scene, mounted on a wheeled chassis that participates in a procession as part of a specific celebration. Most parade floats are self-propelled, although they may also be towed by another vehicle or pulled by animals. The general shape of the float is such that the underlying structure is not visible, and the figure or scene appears to float on the surface of the street, much as a ship appears to float on the surface of the water. Parade floats are used in a variety of civic and religious celebrations. Two of the best known parades are the Mardi Gras Parade in New Orleans, Louisiana, and the Tournament of Roses Parade in Pasadena, California.

One of the earliest written references to a procession, the predecessor of today's parade, dates to about 1800 B.C. when King Senwosret III of Egypt had his scribes write "I celebrated the procession of the god Up-wawet." Such religious processions may date back to 3200 B.C. or earlier.

The first reference to any vehicle resembling a parade float comes from Greece in about 500 B.C. when a statue of the god Dionysius was carried from his temple in a "festival car" pulled by two men. This procession was part of the opening ceremonies for a stage drama and was designed to gain favor from both the god and the drama critics.

Parades continued to be an important form of celebration and often featured kings, conquerors, and other notables riding in splendidly decorated carriages. The Emperor Maximilian of Germany was one of the first to commission an artist to design "triumphal cars" for his parades in 1515. The cars were decorated with bells, fancy fabrics, and carvings of flowers, fruits, and mythological creatures.

In the United States, parades and parade floats were an important part of American life starting in the early-1800s. Mobile, Alabama, held its first civic parade with floats on New Year's Day in 1831. The first Mardi Gras parade in New Orleans was held in 1857 with two floats. The Ak-Sar-Ben (Nebraska spelled backwards) parade in Omaha, Nebraska, started in 1895 and was the first to use electricity to light and propel the floats. The floats ran on the city streetcar tracks and drew power from the overhead trolley wire.

In Pasadena, California, the first Festival of Roses parade took place in 1890 as a promotion for Southern California's sunny winter weather. Isabella Coleman won second prize in the parade in 1910 and decided to go into the business of building floats in 1913. Her first entry won first prize for her client, and she went on to build Rose Parade floats for the next 59 years. Her success created a small industry of professional parade float builders.

Today, most major floats are designed and constructed by professional builders. Each float costs between $50,000 and $200,000 or more and takes up to a year to create. The floats are built in large warehouses, using a wide variety of materials and construction techniques.

Raw Materials

The main chassis contains the components to power the float, the controls and steering mechanism, and the base for the support structure. Most floats use automotive gasoline engines with automatic transmissions. The engine speed is geared down through one or more auxiliary gearboxes to achieve the desired parade speed of about 2.5 mph (4.0 kph). The engine is cooled by an extra large radiator to ensure that it will not overheat during the long parade. Tires are filled with foam to prevent flats. Two or more drivers sit in hidden positions within the float, where they can control the float's direction of travel.

If the float incorporates parts or figures with extensive or complicated animation, the motion is usually provided by means of hydraulic cylinders and motors powered by hydraulic pumps driven off a second engine. To make the motion appear smooth and realistic, the hydraulic cylinders and motors are actuated by a complex array of valves that are controlled by a computer. Many floats have three or four separate operators surrounded by an array of gauges, manual controls, and computers to monitor the animation effects.

The chassis is constructed of steel plate and tubing. The main supports and framework for the float's characters and backgrounds are made from steel rods and tubing attached to the platform. The various shapes are formed with steel rods that are welded to the main supports and covered with aluminum wire screen. The screen is sprayed with a polyvinyl plastic "cocooning" liquid originally developed to cover and protect ships laid up in inactive reserve. The plastic hardens on the wire screen to form a hard, durable skin.

The decorations themselves may be made from paper, wood, flowers, or a variety of other materials. For the famous Tournament of Roses Parade in Pasadena, the parade rules require that all decorations must be some part of a living plant. The emphasis, of course, is on roses and other flowers, but seeds, petals, bark, leaves, fibers, stems, vegetables, nuts, and almost any other part of a plant are also used. For example, onion seeds are used to give a smooth, black surface. Crushed walnut shells or dried strawberries are blended with cornmeal to create skin tones. Animal fur can be simulated with thistles, palm fibers, or even uncooked oatmeal. Seven different types of glue may be used to hold the flowers in place.

These decorations are enhanced with rigid polyurethane foam pieces that are carved to form detailed objects, and with flexible foam cylinders that are bent to form eyebrows, lips, and decorative molding. Wire is used to make long-stem flowers stand upright, and delicate flowers like roses and orchids are held in narrow plastic vials of water to keep them from wilting.

Design

Each parade float is an original work of art and is designed new from the ground up. Once the theme of the parade has been announced, the builders submit concept drawings for review and approval by the parade committee.

With approved drawings in hand, the builders then start to solicit potential sponsors to fund the construction. Sponsors look for a float that will not only draw favorable attention from the crowd and the judges, but one that will also catch the eye of the many television crews that cover the parade. With an estimated 425 million people in 100 countries watching the Tournament of Roses parade on television, sponsors want the maximum amount of coverage for their money. One way to do this is with floats that incorporate animation. This trend in float design has seen builders turn to movie animation and special effects experts for even more elaborate and dramatic action. As with any complicated system, such animation requires computer controls. Expert computer programmers develop the programs required to properly sequence the motion.

With some float designs, the sheer physical size becomes a problem. In the Tournament of Roses Parade, for example, all floats must pass under a 17 ft (5.18 m) high concrete bridge on one portion of the parade route. Floats that are taller than this must be able to hydraulically collapse in less than 25 seconds to fit under the bridge without delaying the parade. In other cases, weight can be a problem. Designers have to calculate the frame strength for long, cantilevered sections, keeping in mind that the delicate-looking floral decorations can

The steel supports and main framework for each figure and scene are fabricated and welded together.

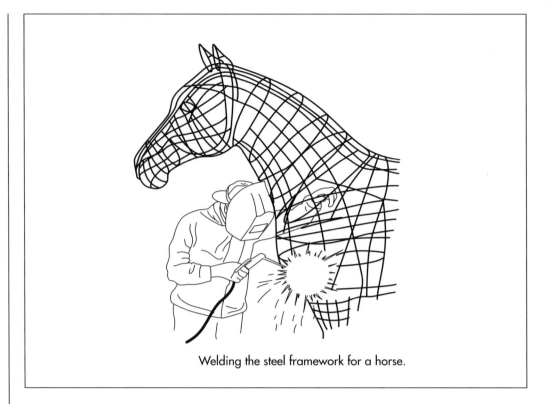

Welding the steel framework for a horse.

triple in weight if nature decides to rain on the parade.

Perhaps the most complex part of float design comes in the selection of materials to achieve the desired colors and textures. This is especially true for floats decorated with flowers and other natural materials. For the Tournament of Roses Parade, most builders employ a floral coordinator to work with the designing artist to help select materials. Designers have learned that some colors do not view well at a distance, and so they balance them with contrasting or outlining colors to bring out their effect. In other cases, the floral coordinator may suggest a more-plentiful, less-costly substitute than what the artist originally planned.

The Manufacturing Process

Float construction starts with the preparation of preliminary design sketches and ends with a frantic flurry of activity as hundreds of people prepare each float for the start of the parade. Here is a typical sequence of operations required to build a float for the Tournament of Roses Parade.

Designing the float

1 In early January, parade officials announce the theme for the following year's parade. Builders immediately begin developing concepts for floats. Each builder is allowed to submit two design concepts for each float proposal.

2 The parade entries committee reviews each design concept to ensure it meets the parade requirements and does not duplicate another entry. From the initial 200-plus submissions, 60 designs are approved for construction. Most professional builders will construct three to 20 floats at the same time.

3 In mid-February, float builders take the approved designs and begin refining the details. The mechanisms for the animated motion are designed. Dimensions and weights are calculated. The best locations for the drivers and animation operators are determined. By March, the builders present their refined designs to potential sponsors for review. After many presentations, each float gets a sponsor who agrees to fund the construction.

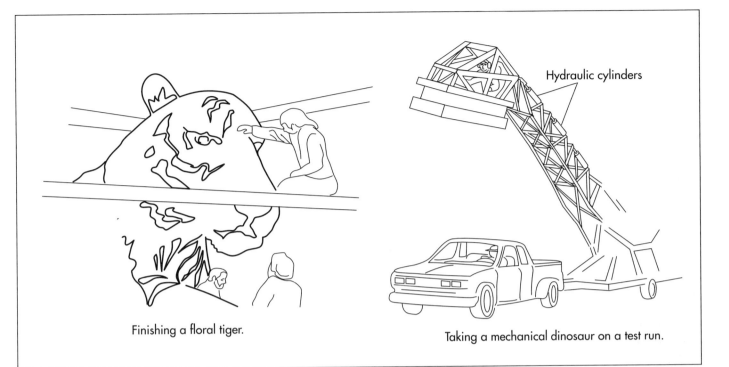

Finishing a floral tiger.

Hydraulic cylinders

Taking a mechanical dinosaur on a test run.

Building the chassis

4 As the final details for each float are worked out with the sponsor, work begins on the float chassis. In some cases, the chassis from one of the previous year's floats can be modified to work. In other cases, and entirely new chassis must be made. The engines, transmission, and axles are installed, followed by the engine controls, steering system, and tires and wheels.

5 By May, the builders are ready to present the final design to each sponsor, including a full-color scale model of the float. Once approved, the float construction proceeds at an accelerated pace.

Forming the figures and scenes

6 Starting in about June, the steel supports and main framework for each figure and scene are fabricated and welded together. Animated portions are fabricated and welded separately, and the hydraulic components are installed. Artists start to bend and weld steel rods to the framework to form the shape of each individual piece. In some cases, they work from dimensioned drawings prepared by the builder, but sometimes they just work from the master artist's rendering of the finished piece.

7 The individual pieces are then welded in place on the chassis, and the hydraulic and electrical systems are connected. The float skeleton may be taken out for a test run at this time to ensure that all the systems are working properly.

8 Aluminum wire screen is cut and molded around the shape of the steel rods to form the outer skin of each part of the float. The screen is glued to the rods and sprayed with a polyvinyl plastic liquid, which hardens to form a solid surface. The finished skin is strong enough to support a person's weight.

Decorating the float

9 As the work gets underway on forming the figures and scenes, the flowers and other decorative materials are ordered in advance. Growers must time their growing cycle exactly so that the flowers are ready to be picked and shipped just a few days before the parade.

10 By September, the main construction is complete, and any small items are fabricated and installed on the float. Deliveries of non-perishable items such as seeds and beans start in October.

Each float requires an average of 10,000 lb (4,545 kg) of flowers and takes 7,000 person-hours or more to decorate.

11 Scaffolding is then erected around each float and the plastic skin and other small items are painted. Each area is painted in a color that closely matches the color of the flowers or other materials that will go there on the finished float. This aids the decorating crews and ensures that if a flower or decoration accidentally falls off, the bare spot will not be noticeable.

12 Flowers start arriving in late-December and are stored in separate tents until they are needed. Approximately 30,000 workers, many of them young people from schools and church groups, report to the various builder's construction sites to begin the round-the-clock job of decorating the floats. Most flowers are prepared by popping the heads off the stems before being glued in place. Delicate flowers are placed in narrow plastic vials filled with water before being pushed in place. Each float requires an average of 10,000 lb (4,545 kg) of flowers and takes 7,000 person-hours or more to decorate.

13 Judging begins on December 30 while the floats are being completed. A second judging takes place on December 31 with all the riders, sound systems, animation, and other portions of the float in parade-ready condition. On New Year's Eve, the floats are slowly towed from the construction sites to the parade staging area. Each float must be in position by 3 a.m. on January 1 or it will be disqualified.

14 At sunrise on January 1, the judges verify the awards for various categories before the parade starts. The parade route covers 5.5 mi (8.8 km), and it takes about two hours for all the floats to pass the starting line. Some one million people line the route, many of them having camped there all night to get good viewing spots.

Salvaging components

15 Although parade floats are never used more than once, many of their inner components are salvaged and reused to make next Year's floats. After the parade, the floats are put on display for several days to allow the public to get a closer look. As the flowers start to wilt, the floats are towed back to the builder's assembly facility and dismantled. The flowers and decorations are discarded, the steel structure is cut up and recycled, and the major components—the engines, hydraulic parts, transmission, tires, wheels, and electronic equipment—are carefully removed and stored for future use.

Quality Control

As with any original work of art, parade floats are constantly inspected by the eye of the master artist. One noted float builder has been known to have an entire float torn down and rebuilt just days before the parade because something did not look right. In addition to the artist's critical eye, each float must meet the requirements of the organization in charge of the parade regarding maximum overall dimensions, travel speed, safety systems, and much more.

The Future

Parades, and parade floats, are expected to remain an important part of celebrations. The floats are expected to become more elaborate and technically sophisticated as builders and sponsors vie for the attention of a worldwide audience. In the Rose Parade, the floral aspect of each float will become more important, as builders search the world for new and unusual floral effects.

Where to Learn More

Books

Hamlin, Rick. *Tournament of Roses.* Mc-Graw-Hill, 1988.

Periodicals

MacCaskey, Michael. "Behind the Scenes at the Rose Parade." *Sunset* (Central West edition), January 1993, pp. 74-79.

Other

Tournament of Roses Association. http://www.tournamentofroses.org.

—*Chris Cavette*

Photograph

Background

A photograph is an image made by a photo-chemical reaction which records the impression of light on a surface coated with silver atoms. The reaction is possible due to the light-sensitive properties of silver halide crystals. In 1556, the alchemist Fabricius was the first to discover that light can photochemically react with these crystals to change the silver ions (Ag+) to elemental silver (Ag0). As the reaction proceeds, the silver atoms grow into clusters, which are large enough to scatter light and produce colors in a pattern identical to that of the original light source. Photography utilizes this chemical principle to record color and black and white images. Silver salt chemistry remains the preferred method of recording high quality images, despite advances in electronic technologies and digital imaging.

One of the first researchers to produce photographic images using silver halide chemistry was Schultze. As early as 1727, he formed metallic silver images by first reacting solutions of silver nitrate and white chalk and then exposing these solutions to light through stencils. Schultze's work was improved upon through the efforts of Louis Jacques Mandé Daguerre who, in 1837, developed a process for printing images on a silver coated copper plate. This type of printed image, called a daguerreotype in honor of its primary inventor, is made by polishing and cleaning a silver-coated copper plate and then reacting the silver coating with iodine vapors to form light-sensitive silver iodide. The silver iodide coated plate is then exposed to light through the optics of a camera that projects and focuses an image

on the plate. In the ensuing reaction, the silver ions are reduced to silver metal. Finally, the plate is treated with mercury to produce an amalgam. In this type of print, the areas of the plate exposed to light appear white and the unexposed areas remain dark. The problem with this method was that it required long exposure times because the intensity of the image depends solely on the strength of the light forming the image.

In 1841, William Henry Fox Talbot overcame this problem by developing a quicker method that did not depend entirely on reflected light to produce the image. He found that silver halide could be exposed in such a way so as to produce a preliminary latent image which required only a small amount of light. This latent image could then be subsequently reacted, without additional light, to produce a final image. Using this technique, known as calotyping, Talbot was one of the first to produce continuous tone images. Unfortunately, these early images were not stable and darkened over time. Fortunately, around the same time Talbot did his work, John Frederick William Herschel discovered a way to stabilize images. His process, known as fixation, chemically converts unexposed silver halide to silver thiosulfate, which can easily be washed off of the image.

The next major advance in photography came with the discovery that certain materials could enhance the sensitivity with which latent images are formed. This enhancement is achieved by coating the silver halide crystals with chemical agents, such as sulfur and gold, which increase the light sensitivity of crystals. Gelatin, which for years had been used as a photographic coating agent,

In 1837, Louis Jacques Mandé Daguerre developed a process for printing images on a silver coated copper plate. This type of printed image, called a daguerreotype in honor of its primary inventor, is made by polishing and cleaning a silver-coated copper plate and then reacting the silver coating with iodine vapors to form light-sensitive silver iodide.

An example of a daguerreotype photograph. (From the collections of Henry Ford Museum & Greenfield Village, Dearborn, Michigan.)

The daguerreotype was the earliest commercial photograph available to Americans. Named for Frenchman Louis Daguerre , who perfected this photographic process in 1837, the daguerreotype was produced directly on coated metal without a negative.

The daguerreotype was easily made in the mid-1800s. Photographic plates were copper faced with silver, polished with flannel and rottenstone, taken to the dark room to be sensitized (coated with thin layers of bromine and iodine). The coated plate was then put in a plateholder and exposed in a camera. The plate was developed in a dark room placed face-down in a vessel filled with mercury at about 120° F (48.9° C). Then, the plate was fixed by washing it a solution of hyposulfite of soda, removing the remaining iodine and bromide. The plate was washed and gilded or toned (some were hand-tinted with color) for that exquisite image.

After 160 years, the daguerreotype remains unsurpassed for its clarity and precision of image. Some claimed that you could count the hairs on the head of the subject, while others complained that the daguerreotype unflatteringly revealed every line and wrinkle. This daguerreotype was likely commissioned by a mother in order to remember her beloved daughter and son just before the Civil War. Others captured houses, farms, siblings, laborers, famous politicians, children alive and deceased, and even scantily-clad prostitutes in these early Victorian photographs.

Nancy EV Bryk

was found to be an effective medium for these light-sensitive materials. In 1888, George Eastman, who pioneered modern film development, coated gelatin-dispersed silver halide crystals onto celluloid sheets. By the next year, Eastman had commercially sold rolls of films prepared by dissolving nitrocellulose with camphor and amyl acetate in a solution of methanol. In the last century, both film processing and camera equipment have improved considerably but these same basic principles are still used to make photographs today.

Raw Materials

Film

Modern film is made by coating light-sensitive ingredients onto a flexible plastic surface. This is a complicated process because a typical roll of film may contain as many as 15 different layers. The first step in the process is to grow microscopic silver halide crystals from silver nitrate and halide ions. After the crystals are grown in solution to a certain minimum size, they are separated and mixed into a gelatin base. This mixture is washed to remove sodium, potassium, and nitrate ions and the resulting silver halide/gelatin emulsion is chilled and allowed to gel. This emulsion is both light and temperature sensitive and must be carefully stored. The emulsion is later melted and the silver grains are coated with chemical agents to enhance sensitivity to certain wavelengths of light. In its molten form, the emulsion is coated onto a support structure, usually a polymeric film. The original film used by Eastman was made from cellulose nitrate and was extremely flammable. Modern film uses solvent-based materials, like cellulose triacetate, and extruded materials, like polyethylene terephthalate. These plastics are safer, stronger, and more chemically stable. As an alternative to plastic film, coated paper is used for some specialty photography.

One common method for coating these plastic films is to dip them into a trough or tray containing the molten emulsion. As the film exits the trough, excess liquid is removed by a knife edge or air jets. Another coating method runs the film below a hopper filled with the emulsion. As the film passes under the hopper, the emulsion is dispensed onto the film. After coating, the emulsion is spread evenly on the film with rollers and is transported to a cooling chamber where the emulsion gels. Finally, the film is sent through a heated chamber which dries and hardens the emulsion. Multiple layers can be coated onto the film in this fashion and specific coatings can be added in order to control how light is reflected/absorbed. Additives used for this purpose include small carbon particles, dyes, or colloidal silver. The last layer is a gelatin overcoat, which seals the film and holds the lower layers in place. In general, the thicker the layers of the emulsion and the larger the silver crystals, the more light sensitive the image. Light sensitivity is gauged by a number known as the ASA (American Standards Association) rating. A low ASA rating means more light is required to record an image; a higher number means less is required. For example, film with an ASA value of 100 (commonly referred to as 100 speed film) is for use in bright sunlight or with a flash. Higher speed film, such as 200 or 400 is more suitable for pictures taken indoors or on overcast days.

After manufacture, film is typically wound onto spools and packaged in light-proof containers. These containers are designed to be opened and loaded into the camera without exposing the film to light.

Developing and printing materials

The chemicals used in developing are designed to grow the microscopic silver atoms into silver centers that are larger enough to be visible to the unaided eye. These developer solutions are composed of reducing agents, restrainers, and preservatives. Hydroquinone is one common reducing agent used for used for black and white film. Bromide ions are commonly used as restrainers, which move the reaction in the opposite direction. Preservatives are added to the mixture to prevent premature oxidation. Sodium sulfite is typically used in this regard.

Printing images requires special paper, which is coated with light-sensitive materials. This paper is available in different grades, which vary smoothness and shine. Printing also requires an enlarger to increase the size of the image and developing and toning solutions, which help control its intensity and color. In addition to the materials described above, developing and printing operations require a variety of equipment

such as trays, measuring glass ware, thermometers, drying screens, timers, mixing pails and stirring paddles, and paper cutters.

The Manufacturing Process

There are three key steps involved in making a photograph: exposing the film to light, developing the image, and printing the photograph. While there are other types of photographic films, such as polaroid and slide films, and other mediums in which to develop photographs, such as film and digital images, the general process of developing 35-mm film into photographic prints is discussed here.

Exposure

1 Once the film is loaded inside the camera it is ready to be exposed. The camera optics focus an image through the lens and onto the emulsion grains. The camera controls the light through a combination of the size of the opening in the lens (the aperture) and the length of time the aperture stays open (the shutter speed). A wide variety of exposure effects can be achieved by varying these two factors. The reaction between the emulsion and the light forms a latent image on the film. The focal length of the camera lens determines the magnification of the latent image, while the penetration of light into the film depends on the combination of lens optics and the chemical properties of the film. The image formed is a negative, meaning it is opposite of how it is seen by the eye. In other words, the areas touched by light are dark and the unexposed areas appear light.

Development

2 After exposure, the film is usually removed from the camera for development, however, there are special Polaroid cameras that use a special self-developing film. This film is unique in that it has the ability to create a photograph in about one minute without any additional development processing. However, the images produced with this film are of poorer quality than those made with standard 35-mm film. Regular film has to go through a complex development process to produce an image. This process involves placing the film in a chemical developer bath to enhance the la-

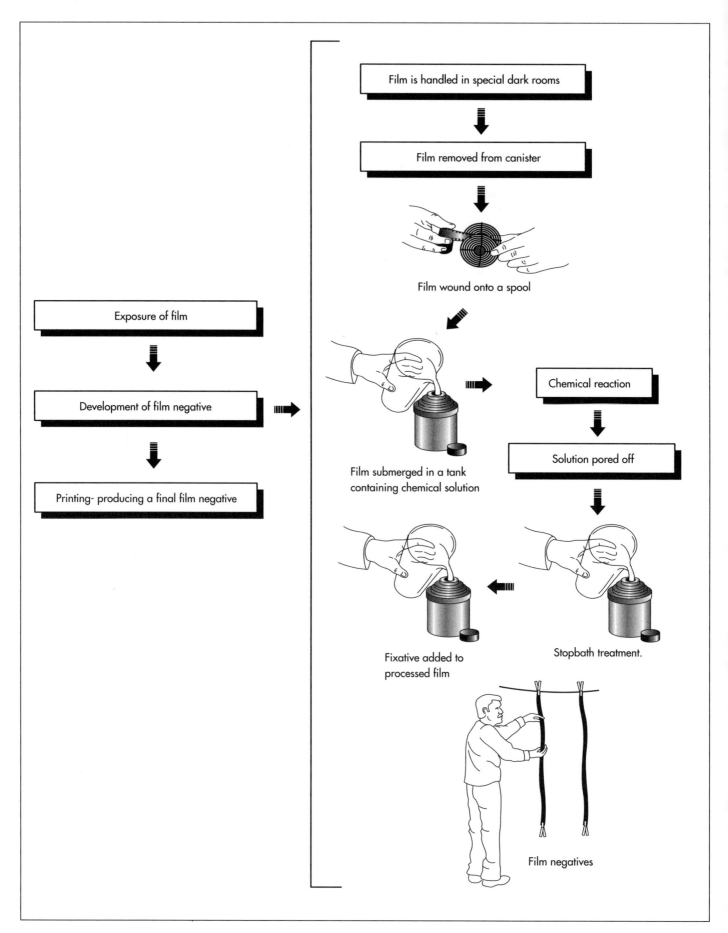

Exposure of film

Development of film negative

Printing- producing a final film negative

Film is handled in special dark rooms

Film removed from canister

Film wound onto a spool

Film submerged in a tank containing chemical solution

Chemical reaction

Solution pored off

Stopbath treatment.

Fixative added to processed film

Film negatives

tent image. This step produces a negative image, which can then be used to print a final picture.

When the film is removed from the camera and taken out of its protective container, caution must be used because the unexposed areas are still light sensitive. Film is handled in special darkrooms, which are illuminated with safe red light that does not effect the film. Once inside the darkroom, the film is removed from its canister, wound onto a spool, and stored in a plastic container to protect it from light and physical damage. The film may then be submerged in a tank containing a solution of the developing chemicals described above. This solution reacts with the exposed areas of the film to amplify the light impressions of the latent image. This process produces variable results depending on the type and temperature of the developer solution used and the level of the original exposure to light. After this stage is complete, the solution is poured off and a stop bath treatment consisting of dilute acetic acid is added to the tank to prevent the film from overdeveloping. After the development is stopped, a fixative can be added to lock in the image. The finished negative then may be washed and rinsed. The reel is then removed from the tank and the fresh negatives are hung up to dry.

Printing

3 Printing is the process of producing a final image from a negative. If photography is the art of taking a picture, printing is the science of making a picture. Printing requires light, a negative, and printing paper. The light source is an enlarger, which uses a lens to focus light through the negative and project it onto light-sensitive paper. The positive image on this paper is then developed in a manner similar to that described above for developing negatives. Finally, the print may be mounted on cardboard or other backing material. Reprints (additional prints of the same image) may be easily produced in a similar fashion from either the original negative or from a previously generated print.

Quality Control

Quality control is a critical element of the photographic process. During film produc-

tion, emulsion coatings must be free from streaks and very uniform in thickness to yield a quality film. The chemistry is exceedingly complicated and is designed to ensure high quality film. Various assays are employed at each step of the film production process to ensure the finished produce is free from defects. Similar care must be used during the development and printing processes to guarantee image quality. Key areas of concern are related to the proper concentrations of chemicals and the time and temperature used in the developing tanks. If the solutions are not the right concentration, the negative or printed photograph may be over or under processed, leading to ghost images or over exposed areas. During processing, the developing solutions must be kept within $5°$ F ($-15°$ C) or the emulsion and film may either expand or contract and produce unwanted patterns on the picture.

The Future

Although photography is a mature technology, advances continue to be made in the way pictures are taken. For example, Kodak has recently introduced a cartridge-based alternative to 35-mm film. This system allows photos of different format to be taken with the same camera, either panoramic or regular prints. Improvements also continue to be made in the automated processes used to develop pictures and have lead to the availability of one-hour photo processing facilities. The real future of photography may lie in the area of digital imagery, a computer-based technology, which produces images electronically. In the future it is likely that methods of capturing and printing digital images may rival the quality of chemical prints. In addition, computer photography offers near-instant results and the ability to manipulate the appearance of images.

Where to Learn More

Books

Birnbaum, Hubert C. *Black and White Dark Room Techniques*. Saunders Photographic, Inc., 1997.

Harris, Ross. *Making Photographs*. New York: Van Nostrand Reinhold Company, 1978.

Opposite page:
Once removed from a camera, exposed film is submerged in a tank containing a solution of developing chemicals. This solution reacts with the exposed areas of the film to amplify the light impressions of the latent image. After this stage is complete, the solution is poured off and a stop bath treatment consisting of dilute acetic acid is added to the tank to prevent the film from overdeveloping. After the development is stopped, a fixative can be added to lock in the image. The finished negative then may be washed and rinsed. The reel is then removed from the tank and the fresh negatives are hung up to dry.

Langford, Michael. *The Story of Photography.* Focal Press, 1997.

Ray, Sidney F. *Photographic Chemistry and Processing.* Focal Press, 1994.

—*Randy Schueller*

Pickle

Background

Pickles are cucumbers preserved in a solution of vinegar, salt, and other flavorings. They are typically fermented with naturally-occurring bacteria prior to vinegar preservation. While pickling technology has been known since ancient times, pickles are still a popular food, with over 5 million lb (2.27 million kg) consumed daily.

History

Pickling of plant and animal foods is a relatively old method of food preservation. It is estimated that the first pickles were produced over 4,000 years ago using cucumbers native to India. The ancient Egyptians and Greeks both have written about the use of pickles for their nutritive value and healing power. Pickles were a common food during the time of the Roman Empire and they soon spread throughout Europe. In America, pickles have always been popular. The first travelers to America kept pickles in large supply because they were nutritious and did not spoil during the long journeys. It is interesting to note that Amerigo Vespucci, America's namesake, was also a pickle salesman. He was the main pickle supplier to many ships. The first large-scale commercial production of pickles did not take place until 1820, when Nicholas Appert began selling pickles in jars. Over the years, the pickle production process has become more automated, however the basic pickling methods have changed very little since the technology was first developed.

While there are many different types of pickles, some characteristics are common to all. In general, pickled cucumbers are crisp vegetables, which can be described as having a strong, biting flavor caused by the vinegar in which they are stored. Different pickle manufacturers normally add spices to give their pickles a unique flavor. Dill-flavored pickles are perhaps the most common of all pickles. There are also sweet pickles, which are packed with added sugar. These are typically used for making relishes. Kosher pickles were pickles that were approved by the Jewish Orthodox Congregations of America, but the word kosher is now often used to describe any garlic flavored pickle.

Raw Materials

There are six basic types of ingredients used for pickle making. The main bulk food is the cucumber. The additional ingredients include acids, flavorings, colorants, preservatives, and stabilizers that make up the liquid, or liquor, in which the pickle is sold. Many of the ingredients are only available at certain times of the year, so steps have to be taken to use fresh materials.

Undoubtedly, the most important ingredient in pickle manufacturing is the cucumber. Special seeds are used to produce cucumbers that are straight, thin skinned, have a predictable number of warts, and are properly sized. These characteristics are important for uniform pickle manufacturing. Technically, pickles can actually be made using all kinds of foods such as onions, peppers, olives, pears, peaches, and even fish and meat. These are usually referred to as pickled foods to indicate the type of processing required to make them.

Acetic acid (vinegar) is the primary ingredient used in pickle manufacturing. After

While pickling technology has been known since ancient times, pickles are still a popular food with over 5 million lb (2.27 million kg) consumed daily.

water, it makes up the bulk of the pickle liquor and contributes significantly to the flavor of the pickle giving it a sour taste. Additionally, it also has a preservative effect and is nontoxic. Vinegar is derived from naturally occurring sugars or starches through a two-step fermentation process. Starch is converted to sugar, which is then yeast fermented to form alcohol. The alcohol is exposed to an acetobacteria, which converts it to vinegar. Vinegar can be obtained from many sources and each one has a slightly different taste. Therefore, depending on its source, the vinegar can have a significant effect on the taste of the final pickle product.

Other ingredients, which impact the final taste of the pickles, are added to the liquor. Sugar is used to provide a sweetness to offset the sour taste of the vinegar. It also helps to make pickles more plump and firm. Artificial sweeteners like aspartame and saccharine can be used for a similar effect without increasing the calorics. Salt is added for flavor and it also has an added preservative effect. Pure granulated salt is typically used since it is devoid of anti-caking ingredients that could make the liquor cloudy.

While vinegar, sugar, and salt make up the bulk of all pickle liquors, it is the various spices and herbs that differentiate between pickle types. Dill weed is the most common type of aromatic spice and is used to make all forms of dill pickles. Other aromatic spices include allspice, cassia, cinnamon, cloves, fennel, fenugreek, and nutmeg. For more potent pickles, hot spices such as capsicum, black pepper, ginger, and mustard are used. Herbs like basil, marjoram, mint tarragon, and thyme are also used to give pickles a unique taste. Flavorful vegetables including onions and garlic are often included in a pickle liquor. Typically, the pickle manufacturer has a standard spice mix made for each type of pickle they manufacture.

Some additional ingredients may be added to ensure the pickles meet standards set by the manufacturer. In general, pickles do not require any colorants because their natural color is acceptable. However, to create a standardized product and overcome the effects of processes such as bleaching, manufacturers often add color. Two common types of colorants are turmeric caramel and cholorphyll. The caramel provides a slightly brown to yellow color and chlorophyll gives a green color. To inhibit color changes in pickles, sulfur dioxide is added. Firming agents such as lime and alum may also be added. These materials help make pickles crispier without significantly impacting the flavor. Surfactants such as polysorbate are also used to couple ingredients in the liquor solution.

The Manufacturing Process

Making cucumber pickles can take up to 42 days depending on the manufacturer's recipe. Production involves four primary steps including harvesting, preservation, pasteurization, and final processing. The process is highly automated once the cucumbers are delivered to the processing plant.

Harvesting

1 Once harvested by field workers, cucumbers are put in large bins and transported to a receiving station. If the cucumbers are transported a long distance, refrigerated trucks are used. This helps to maintain the fresh appearance and flavor of the vegetable. At the receiving station, the cucumbers are poured out onto a conveyor where they are subjected to a cleaning process that removes the excess stems, blossoms, dirt, and other foreign matter. This step is important because trace amounts of bacteria on unwashed cucumbers can ruin the final pickle product. They are then moved to an inspection station where rotten vegetables are removed and the rest are separated by size. From here they are moved to a chiller and stored until they are ready to be used.

Preservation

2 Depending on the manufacturer, conversion of the cucumber into a pickle can be done in one of three ways including fermentation, pasteurization, and refrigeration. The first and oldest method is a process known as fermentation. In this method, the cucumbers are transferred to large, air tight, fiberglass or stainless steel tanks. Some of these containers can hold over 40,000 lb (18,160 kg) of cucumbers. The tanks are filled with a brine solution, which is made up of water and 10% salt. The manufacturer can take

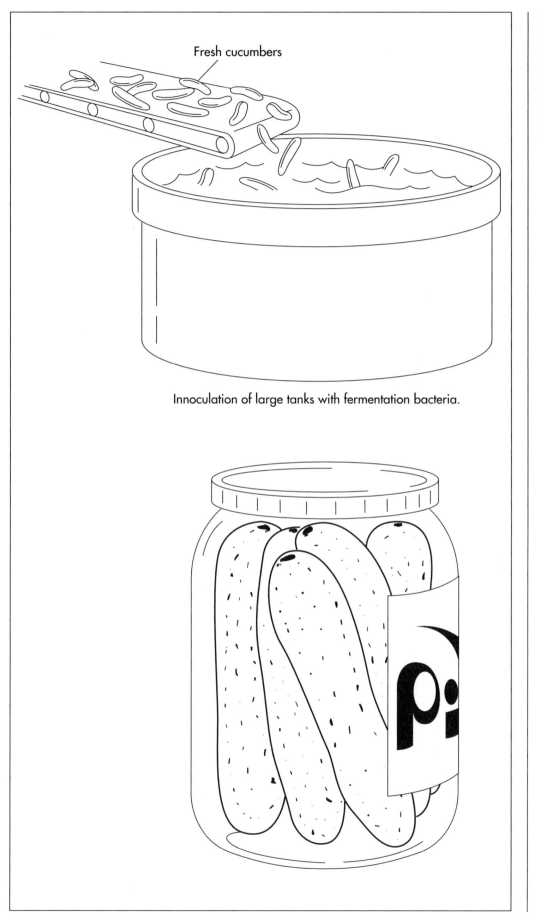

Fresh cucumbers

Innoculation of large tanks with fermentation bacteria.

During a storage period of about five weeks, the fermentation bacteria breakdown the sugars present in the vegetable and produce carbon dioxide. To prevent adverse effects from the carbon dioxide, the tanks are periodically degassed. Pickles made in this way have a shelf life of many months.

advantage of a naturally-occurring bacteria that is present on the cucumbers or innoculate with a specifically desired bacteria. In either case, the bacteria are halophyllic, or salt tolerant. During the storage period of about five weeks, these bacteria breakdown the sugars present in the vegetable and produce carbon dioxide. To prevent adverse effects from the carbon dioxide, the tanks are periodically degassed. Pickles made in this way have a shelf life of many months.

The other two methods of preservation do not require a fermentation step. One method is by direct pasteurization. In this method, the cucumbers are bottled and then exposed to very high temperatures for a set amount of time. This has the effect of killing all of the natural bacteria that may is present. These sterilized cucumbers can then be further processed into pickles. This method of production results in pickles that have a shelf life of only a few months. The third method is by refrigeration and acidification. These pickles depend on the cold temperature and vinegar solution to prevent spoilage. While they are much faster to manufacture, they have a much shorter shelf life.

Processing and packaging

3 After the pickles have adequately fermented, the salt solution is drained. The pickles are then immersed in water to remove all of the salt they may have acquired during the cure. From this point, the pickles are moved along a conveyor to a slicing machine which cuts the pickles to the correct size depending on the type of product desired. They can be cut into slices, chips, or can even be diced. Attempts are made to maintain as clean an environment as possible for the pickles as contamination by microbes could result in an undesirable product.

4 After being cut, the pickles are typically placed in glass jars although cans, plastic bottles, and pouches have also been used. The packing machines are designed to deliver the correct amount of vegetable to each jar. The jars are moved along to a liquid filling machine, which fills them with the liquor. The pickle liquor consists of vinegar, salt, and other materials mentioned previously. This liquor is premixed in a large container prior to filling. To ensure an adequate distribution of spices, these are sometimes filled into the jars before the liquor. From the filling machine, the jars are capped and moved along for pasteurization.

Pasteurization and sealing

5 The problem of spoilage is evident throughout the pickle making process. Cucumbers can spoil during the brining process and even during packing if they are exposed to air for too long. For this reason the pickles are pasteurized. In order to pasteurize the pickles, they are typically exposed to high temperatures for an extended period of time. Depending on how long the pickles are heated, pasteurization can either kill off all of the acetic acid-tolerant organisms or inactivate all of the enzymes in the vegetable. In both cases, pasteurization increases the shelf life of the pickles.

6 Most pickles are vacuum packed which means the air is removed from the jar before it is sealed. This helps maintain the pickle taste and prevents contamination by microorganisms. In order to vacuum pack the pickles, air in the jar is replaced with steam just before the cap is sealed. When the steam cools and condenses, it creates a vacuum, reducing the amount of free oxygen present in the jar. The vacuum seal is responsible for the familiar pop that is heard when a jar of pickles is opened.

7 The jars are next moved along a conveyor to a labeling machine. Labels are automatically affixed and a freshness date is stamped on the jar. From here the jars are moved to automatic packing machines which put them in cardboard boxes. They are transferred to pallets and shipped out to the local retailers.

Quality Control

Quality control is an important part of any food preparation process. It is particularly important in pickle making because poor quality control will result in an unpalatable product. The process begins in the field while the cucumbers are being harvested. Trained workers inspect the cucumbers for any signs of spoilage. If any spoiled cucumbers are found, they are discarded. Most manufacturers set specifications that the cucumbers must meet before use. During production, regular quality control measures include laboratory tests for the level of acid in the pickle liquor. This is done through a titration method using an automatic buret (test tube-like container). Other measurements that are taken on the final pickle liquor are pH, refractory sugar readings, and salt readings. Most of the methods for these tests are described by government regulations in publications by the United States Food and Drug Administration.

The Future

Research focusing on improvements in pickle technology is being done by the various seed companies and universities. One of the primary areas of interest is the development of improved pickling cucumbers. Many university groups are using biotechnology and plant grafting techniques to produce cucumbers that are larger, more plentiful, and resistant to microbial and insect-born diseases. New farming methods concentrating on obtaining a larger harvest with fewer plants are also being tested. In addition, pickle manufacturers are also coming up with new flavors of pickles by varying the composition of the liquor and using different fermentation organisms.

Where to Learn More

Books

Mabey, David. *Perfect Pickle Book.* Parkwest Publications, 1995.

Macrae, R.. editor. *Encyclopedia of Food Science, Food Technology and Nutrition.* San Diego: Academic Press, 1993.

Sudell, Helen. *Country Pickles and Preserves: Gifts from Nature.* Anness Publishing, 1997.

—*Perry Romanowski*

Playing Cards

The first written record documenting the use of playing cards comes from Asia and dates back to the twelfth century.

Background

Playing cards are flat, rectangular pieces of layered pasteboard typically used for playing a variety of games of skill or chance. They are thought to have developed during the twelfth century from divination implements or as a derivative of chess. Cards are produced by the modern printing processes of lithography, photolithography, or gravure. In the future, more computerized methods will likely be adopted promising to generate a substantial increase in the playing card manufacturing industry.

History

The exact story of the emergence of playing cards is debated. Some historians believe that cards were developed in India and derived from the game of chess. Others suggest that they were developed as implements for magic and fortune telling in Egypt. The first written record of the use of playing cards comes from the Orient, dating back to the twelfth century. Playing cards were introduced to Europe during the thirteenth century from the Middle East. Evidence suggests that they first arrived in Italy or Spain and were quickly spread throughout the continent.

Some of these early playing cards were very similar to our modern day cards. They consisted of 52 cards with four suits including swords, cups, coins, and polo-sticks. They also had numerals from one to ten and face cards, which included a king, deputy king and second deputy king. When Europeans began to produce their own cards, they did not produce consistent designs and any number of suits or face cards would be made. In the latter part of the fifteenth cen-

tury, standardized versions of cards began to appear. The modern day system of spades, hearts, diamonds, and clubs first appeared in France around 1480.

The availability of cards became more widespread as production processes improved. The earliest decks of playing cards were hand-colored with stencils. Consequently, they were extremely expensive to produce and were owned almost exclusively by the very wealthy. Cheaper products were also produced, but it is likely that they deteriorated quickly with use. With the advent of new printing processes, production volumes of playing cards were increased. During the fifteenth century, a method of producing cards using wooden blocks as printing templates was introduced in Germany. These decks were quickly exported throughout Europe. The next significant advance in card manufacture was the replacement of wood blocking and hand coloring with copper plate engraving during the sixteenth century. When color lithography was developed in the early 1800s, the production of playing cards was revolutionized. New printing techniques promise to further improve the production of future decks of cards.

Design

A standard deck of playing cards consists of 52 cards which have a rectangular shape, dimensions of about 2.5 x 3.5 in (6.35 x 9 cm), and rounded corners. The cards are made up of layers of paper and are often called pasteboards. The faces of these cards are typically decorated with two colors, red and black, and four suits including clubs, spades, hearts, and diamonds. Each suit has thirteen cards consisting of three face cards (king, queen, jack) and number cards from one (ace) to ten. The face cards are double-ended, which means the same design is on both halves of the card. This eliminates the need to orient these cards in a hand as both ends will automatically be positioned correctly. In the upper left corner of most cards are index numbers and symbols, which make the card value clearly visible when held in a fan position. This is the position most often used during a card game. Two jokers are also typically included with a new pack of cards.

The backs of the cards are decorated with a unique pattern indicative of the card manufacturer. Red and blue are the most commonly used colors, but almost any color or design is possible. Often a company will order a deck of cards as an advertising specialty and have their logo printed on the card back. Some card backs have a white border while the pattern on others extends to the edge of the card. In general, the back patterns are symmetrical so cards have only one real orientation. Notable exceptions are advertising specialty and souvenir cards, which typically have a non-symmetrical picture on the back.

Cards are used for a variety of purposes. The most common use of cards is for playing parlor games. Some of the more popular games include bridge, rummy, and gin. Gambling games such as poker and black-jack also employ standard decks of cards. In addition, specially printed cards are used as game implements for board games. These cards may have trivia questions, words, or symbols on them that are important in game play. Other types of cards are used as teaching aids.

Non-standard decks of cards are also available and used for different reasons. Tarot cards are typically larger and heavy than standard cards. They have 78 cards, 22 of which have symbolic images. They are used for fortune telling and divination purposes. A variety of magic, or trick, cards are produced. One type of trick cards is marked cards. The back design of these cards is subtly changed so that the faces can be determined just by looking at the back. Other trick decks have shortened cards or have tapered ends, which help a conjuror find a selected card. Novelty cards are also

Union playing cards printed at the time of the American Civil War. (From the collections of Henry Ford Museum & Greenfield Village.)

Playing cards may have been used in China as early as the seventh century and perhaps were known in India around this time as well—early European playing cards include Indian motifs associated with Hindu Gods. No one is sure how the playing card moved from Asia to Europe—did Niccolo Polo or his son introduced the playing card and associated games to Italy? Or perhaps the Arabs introduced the Spanish to the colorful hand-painted cards? Nevertheless, we do know that by the thirteenth century the entire continent enjoyed card-playing; the British card-makers petitioned for protection from imported cards, and German printers were block-printing rather than hand-painting cards by the late 1400s.

Dutch, French, and British settlers in the New World brought playing cards with them. Americans and others use the 52-card deck based on the French deck, and include the medieval motifs of the spade, club, diamond, and heart. While the deck motifs have remained relatively unchanged, clothing and appearance of the court cards have been altered according to card designer and intended market. The "Union Playing Cards," shown here, were printed around the time of the American Civil War. They include no depicted European-style royalty, but use politicians and famous Union generals in their place! They were surely printed to bolster pride in the Union cause and thumb their nose at the European royalty at the same time.

Nancy EV Bryk

manufactured. This includes oddly shaped cards or metal cards for outdoor use, which can stick to a magnetized playing surface.

Raw Materials

Playing cards can be made with paper or plastics. To make a card, layered paper is

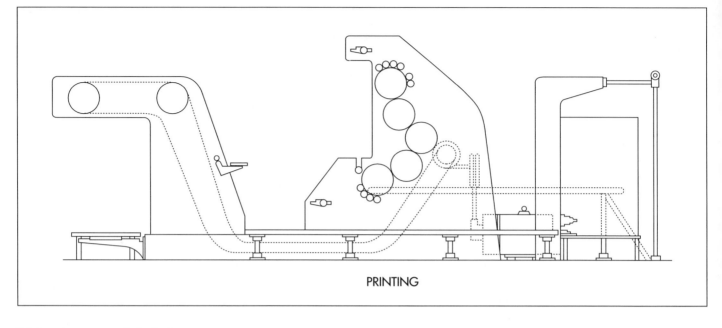

PRINTING

Printing plates are made for both the front and the back of the cards. Once a plate is made for each color represented on the card, they are coated, mounted on the printing press, and a batch is run.

used. Layered paper is produced by putting a number of sheets of paper in a stack and gluing them together. This type of paper is stronger and more durable than standard paper. Higher quality cards may be made from polymeric plastic films and sheets. One material that is often used is a cellulose acetate polymer. This is a semisynthetic polymer that is made into a paper-like sheet by being cast from a solution. This produces a film, which can be stacked and laminated to produce an appropriately thick sheet. This material is much more durable than paper and cards that are made with it last considerably longer. Vinyl plastics are also used in the production of cards. Paper cards are typically of lower quality and wear out more quickly than plastic cards.

The Manufacturing Process

The production of a deck of cards involves the three primary steps including printing the pasteboards, cutting the sheets and assembling the deck. While a variety of printing processes may be employed, lithography continues to be used extensively.

Printing the pasteboards

1 Creating the printing plates is the first step in the production of playing cards. This process begins with camera ready artwork, or electronically created images, which con-

tain pictures of each card that will be included in the deck. A plate is also created for the backs of the cards. Using a photographic process a negative of the image is exposed to a flat plate and coated with a light sensitive material. The plate is developed, and the image area is coated with an oily material that will attract ink but repel water. The non-image area is coated with a mixture, which will attract water and repel ink. One plate must be created for each of the different colors that will be printed on the card.

2 To begin printing, the plates are mounted on rotating cylinders in the printing press. When the press is started, the plate is passed under a roller, which coats it with water. The image area on the plate, previously treated with the oily material, repels the water and remains uncoated. An ink roller is next passed over the plate. Since an oil-based ink is used, it adheres to the plate only on the water-resistant sections.

3 A rubber roller is then passed over the printing plate and the ink from the plate is transferred to it. The card paper is passed under the rubber roller and the ink is transferred to it. The paper is then passed to the next roller assembly where another color may be added. The ink is specially formulated so it dries before it enters the next roller assembly. This process of wetting, inking, and printing is continuous throughout the card manufacturing run.

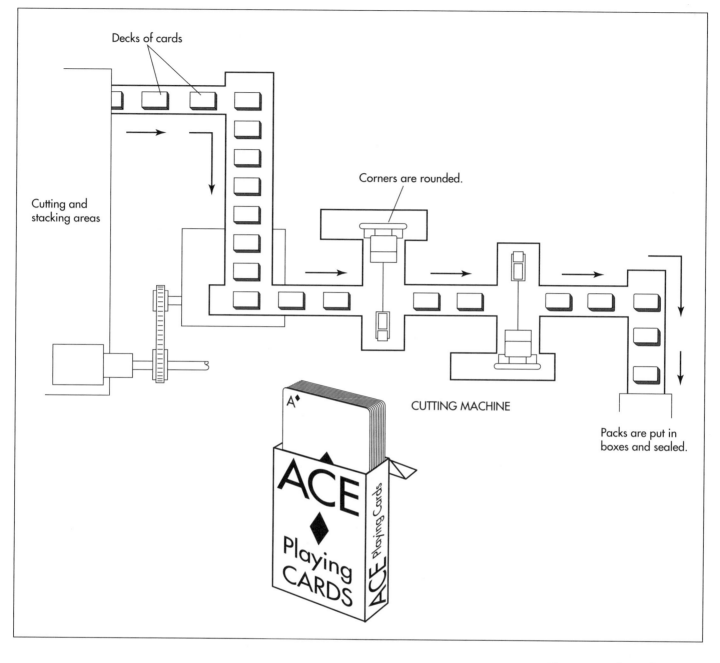

Decks of cards

Cutting and stacking areas

Corners are rounded.

CUTTING MACHINE

A♦

ACE
♦
Playing
CARDS

ACE Playing Cards

Packs are put in boxes and sealed.

---

4 When one sheet of paper exits the printing press, it contains an image on both sides. One side has the image of each card in the deck while the other has the card back image. At this point, the sheet may be coated with a special clear polymer mixture that gives it a slick, glossy look and feel. This coating also helps to protect the cards making them longer lasting.

Cutting and stacking

5 After both sides of the pasteboards are printed, they are transported to a card cutting station. Here precision-cutting ma-

chines cut the cards out from the printed sheets. The cards are cut such that each card is of identical size. They are then assembled into their respective sets and organized into stacks. At this point, the stack contains all of the cards that will end up in the final packaging.

Further cutting and packaging

6 The stack of cards is next transported via conveyor belt to a corner punching station. When it reaches the platform of this station, the stack is pushed up into the punching device, which rounds off the cor-

After printing, the cards are sorted, stacked, cut, and packaged.

ners on one side of each card in the stack. During this phase of production, the stack of cards are held tightly in the punching blades so each card is cut identically. The stack is then removed and transported to another punching station. Here the corners of the other side of the stack of cards are rounded off. After the cards are removed from this station, all four corners are rounded and the decks are ready for final packaging.

7 The stack of cards is returned to the main conveyor and transferred to the packaging station. Here a machine feeds formed boxes onto the assembly line. The cards are then inserted into the box. The boxes are closed and sealed with a sticker at the top. The box is then transported to a shrinkwrap machine where it is wrapped in a clear plastic such as cellophane. The finished deck of cards is then placed in a case with other decks, stacked on pallets and shipped on trucks to distributors.

Quality Control

Quality control begins with the incoming inks and other raw materials used to create the deck of cards. If the manufacturer produces their own stock paper, it is checked to ensure that it measures up to specifications related to physical appearance, dimensions, consistency, and other characteristics. The inks are minimally tested for color, viscosity, and solubility. For materials that are supplied by outside vendors the card manufacturer often relies on the supplier's quality control inspections. Prior to a first printing, the plates are tested to verify they will produce a quality print. During production, the sheets are randomly checked for a variety of printing errors or ink smears. Defective sheets are removed prior to cutting. Line inspectors are also stationed at various points on the production line to make sure that each pack is produced in a flawless manner.

The Future

Future developments in playing card manufacture will focus on new card designs and methods of printing. Since the market for playing cards remains relatively mature, card producers will attempt to increase sales by introducing novel card designs. This might involve using new base materials for the cards, producing three-dimensional designs, or creating novel shapes. With the vast improvements in computer technologies, a variety of new printing methods will be employed. These methods will be used to increase the speed at which cards will be produced. They will also eliminate the need for creating plates as printing can be done directly from computer images. This will make it easier to produce personalized decks quickly and economically.

Where to Learn More

Books

Kirk Othmer Encyclopedia of Chemical Technology. John Wiley & Sons. New York: 1992.

Morley, H.T. *Old and Curious Playing Cards.* Book Sales, 1989.

Seymour, R. and C. Carraher. *Polymer Chemistry.* New York: Marcel Dekker Inc., 1992.

Wowk, K. *Playing Cards of the World.* U.S. Games Systems, 1982.

Other

Pedersen, T. U.S. Patent #4,779,401, 1988.

—*Perry Romanowski*

Plywood

Background

Plywood is made of three or more thin layers of wood bonded together with an adhesive. Each layer of wood, or ply, is usually oriented with its grain running at right angles to the adjacent layer in order to reduce the shrinkage and improve the strength of the finished piece. Most plywood is pressed into large, flat sheets used in building construction. Other plywood pieces may be formed into simple or compound curves for use in furniture, boats, and aircraft.

The use of thin layers of wood as a means of construction dates to approximately 1500 B.C. when Egyptian craftsmen bonded thin pieces of dark ebony wood to the exterior of a cedar casket found in the tomb of King Tut-Ankh-Amon. This technique was later used by the Greeks and Romans to produce fine furniture and other decorative objects. In the 1600s, the art of decorating furniture with thin pieces of wood became known as veneering, and the pieces themselves became known as veneers.

Until the late 1700s, the pieces of veneer were cut entirely by hand. In 1797, Englishman Sir Samuel Bentham applied for patents covering several machines to produce veneers. In his patent applications, he described the concept of laminating several layers of veneer with glue to form a thicker piece—the first description of what we now call plywood.

Despite this development, it took almost another hundred years before laminated veneers found any commercial uses outside of the furniture industry. In about 1890, laminated woods were first used to build doors. As the demand grew, several companies began producing sheets of multiple-ply laminated wood, not only for doors, but also for use in railroad cars, busses, and airplanes. Despite this increased usage, the concept of using "pasted woods," as some craftsmen sarcastically called them, generated a negative image for the product. To counter this image, the laminated wood manufacturers met and finally settled on the term "plywood" to describe the new material.

In 1928, the first standard-sized 4 ft by 8 ft (1.2 m by 2.4 m) plywood sheets were introduced in the United States for use as a general building material. In the following decades, improved adhesives and new methods of production allowed plywood to be used for a wide variety of applications. Today, plywood has replaced cut lumber for many construction purposes, and plywood manufacturing has become a multi-billion dollar, worldwide industry.

Raw Materials

The outer layers of plywood are known respectively as the face and the back. The face is the surface that is to be used or seen, while the back remains unused or hidden. The center layer is known as the core. In plywoods with five or more plies, the intermediate layers are known as the crossbands.

Plywood may be made from hardwoods, softwoods, or a combination of the two. Some common hardwoods include ash, maple, mahogany, oak, and teak. The most common softwood used to make plywood in the United States is Douglas fir, although several varieties of pine, cedar, spruce, and redwood are also used.

The use of thin layers of wood as a means of construction dates to approximately 1500 B.C. when Egyptian craftsmen bonded thin pieces of dark ebony wood to the exterior of a cedar casket found in the tomb of King Tut-Ankh-Amon.

Composite plywood has a core made of particleboard or solid lumber pieces joined edge to edge. It is finished with a plywood veneer face and back. Composite plywood is used where very thick sheets are needed.

The type of adhesive used to bond the layers of wood together depends on the specific application for the finished plywood. Softwood plywood sheets designed for installation on the exterior of a structure usually use a phenol-formaldehyde resin as an adhesive because of its excellent strength and resistance to moisture. Softwood plywood sheets designed for installation on the interior of a structure may use a blood protein or a soybean protein adhesive, although most softwood interior sheets are now made with the same phenol-formaldehyde resin used for exterior sheets. Hardwood plywood used for interior applications and in the construction of furniture usually is made with a urea-formaldehyde resin.

Some applications require plywood sheets that have a thin layer of plastic, metal, or resin-impregnated paper or fabric bonded to either the face or back (or both) to give the outer surface additional resistance to moisture and abrasion or to improve its paint-holding properties. Such plywood is called overlaid plywood and is commonly used in the construction, transportation, and agricultural industries.

Other plywood sheets may be coated with a liquid stain to give the surfaces a finished appearance, or may be treated with various chemicals to improve the plywood's flame resistance or resistance to decay.

Plywood Classification and Grading

There are two broad classes of plywood, each with its own grading system.

One class is known as construction and industrial. Plywoods in this class are used primarily for their strength and are rated by their exposure capability and the grade of veneer used on the face and back. Exposure capability may be interior or exterior, depending on the type of glue. Veneer grades may be N, A, B, C, or D. N grade has very few surface defects, while D grade may have numerous knots and splits. For exam-

ple, plywood used for subflooring in a house is rated "Interior C-D". This means it has a C face with a D back, and the glue is suitable for use in protected locations. The inner plies of all construction and industrial plywood are made from grade C or D veneer, no matter what the rating.

The other class of plywood is known as hardwood and decorative. Plywoods in this class are used primarily for their appearance and are graded in descending order of resistance to moisture as Technical (Exterior), Type I (Exterior), Type II (Interior), and Type III (Interior). Their face veneers are virtually free of defects.

Sizes

Plywood sheets range in thickness from .06 in (1.6 mm) to 3.0 in (76 mm). The most common thicknesses are in the 0.25 in (6.4 mm) to 0.75 in (19.0 mm) range. Although the core, the crossbands, and the face and back of a sheet of plywood may be made of different thickness veneers, the thickness of each must balance around the center. For example, the face and back must be of equal thickness. Likewise the top and bottom crossbands must be equal.

The most common size for plywood sheets used in building construction is 4 ft (1.2 m) wide by 8 ft (2.4 m) long. Other common widths are 3 ft (0.9 m) and 5 ft (1.5 m). Lengths vary from 8 ft (2.4 m) to 12 ft (3.6 m) in 1 ft (0.3 m) increments. Special applications like boat building may require larger sheets.

The Manufacturing Process

The trees used to make plywood are generally smaller in diameter than those used to make lumber. In most cases, they have been planted and grown in areas owned by the plywood company. These areas are carefully managed to maximize tree growth and minimize damage from insects or fire.

Here is a typical sequence of operations for processing trees into standard 4 ft by 8 ft (1.2 m by 2.4 m) plywood sheets:

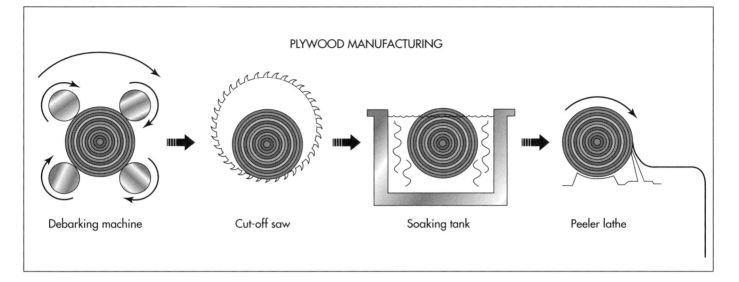

PLYWOOD MANUFACTURING

Debarking machine   Cut-off saw   Soaking tank   Peeler lathe

Felling the trees

1 Selected trees in an area are marked as being ready to be cut down, or felled. The felling may be done with gasoline-powered chain saws or with large hydraulic shears mounted on the front of wheeled vehicles called fellers. The limbs are removed from the fallen trees with chain saws.

2 The trimmed tree trunks, or logs, are dragged to a loading area by wheeled vehicles called skidders. The logs are cut to length and are loaded on trucks for the trip to the plywood mill, where they are stacked in long piles known as log decks.

Preparing the logs

3 As logs are needed, they are picked up from the log decks by rubber-tired loaders and placed on a chain conveyor that brings them to the debarking machine. This machine removes the bark, either with sharp-toothed grinding wheels or with jets of high-pressure water, while the log is slowly rotated about its long axis.

4 The debarked logs are carried into the mill on a chain conveyor where a huge circular saw cuts them into sections about 8 ft-4 in (2.5 m) to 8 ft-6 in (2.6 m) long, suitable for making standard 8 ft (2.4 m) long sheets. These log sections are known as peeler blocks.

Making the veneer

5 Before the veneer can be cut, the peeler blocks must be heated and soaked to soften the wood. The blocks may be steamed or immersed in hot water. This process takes 12-40 hours depending on the type of wood, the diameter of the block, and other factors.

6 The heated peeler blocks are then transported to the peeler lathe, where they are automatically aligned and fed into the lathe one at a time. As the lathe rotates the block rapidly about its long axis, a full-length knife blade peels a continuous sheet of veneer from the surface of the spinning block at a rate of 300-800 ft/min (90-240 m/min). When the diameter of the block is reduced to about 3-4 in (230-305 mm), the remaining piece of wood, known as the peeler core, is ejected from the lathe and a new peeler block is fed into place.

7 The long sheet of veneer emerging from the peeler lathe may be processed immediately, or it may be stored in long, multiple-level trays or wound onto rolls. In any case, the next process involves cutting the veneer into usable widths, usually about 4 ft-6 in (1.4 m), for making standard 4 ft (1.2 m) wide plywood sheets. At the same time, optical scanners look for sections with unacceptable defects, and these are clipped out, leaving less than standard width pieces of veneer.

The logs are first debarked and then cut into peeler blocks. In order to cut the blocks into strips of veneer, they are first soaked and then peeled into strips.

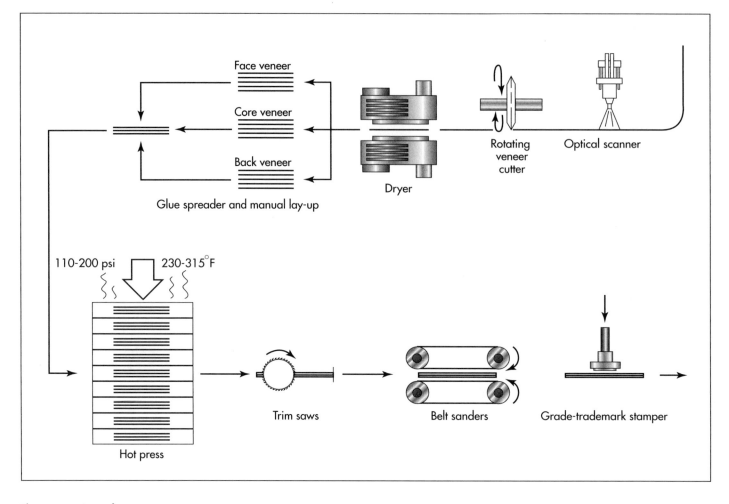

Face veneer

Core veneer

Back veneer

Glue spreader and manual lay-up

Dryer

Rotating veneer cutter

Optical scanner

110-200 psi   230-315°F

Hot press

Trim saws

Belt sanders

Grade-trademark stamper

The wet strips of veneer are wound into a roll, while an optical scanner detects any unacceptable defects in the wood. Once dried the veneer is graded and stacked. Selected sections of veneer are glued together. A hot press is used to seal the veneer into one solid piece of plywood, which will be trimmed and sanded before being stamped with its appropriate grade.

8 The sections of veneer are then sorted and stacked according to grade. This may be done manually, or it may be done automatically using optical scanners.

9 The sorted sections are fed into a dryer to reduce their moisture content and allow them to shrink before they are glued together. Most plywood mills use a mechanical dryer in which the pieces move continuously through a heated chamber. In some dryers, jets of high-velocity, heated air are blown across the surface of the pieces to speed the drying process.

10 As the sections of veneer emerge from the dryer, they are stacked according to grade. Underwidth sections have additional veneer spliced on with tape or glue to make pieces suitable for use in the interior layers where appearance and strength are less important.

11 Those sections of veneer that will be installed crossways—the core in three-

ply sheets, or the crossbands in five-ply sheets—are cut into lengths of about 4 ft-3 in (1.3 m).

Forming the plywood sheets

12 When the appropriate sections of veneer are assembled for a particular run of plywood, the process of laying up and gluing the pieces together begins. This may be done manually or semi-automatically with machines. In the simplest case of three-ply sheets, the back veneer is laid flat and is run through a glue spreader, which applies a layer of glue to the upper surface. The short sections of core veneer are then laid crossways on top of the glued back, and the whole sheet is run through the glue spreader a second time. Finally, the face veneer is laid on top of the glued core, and the sheet is stacked with other sheets waiting to go into the press.

13 The glued sheets are loaded into a multiple-opening hot press. Most presses can handle 20-40 sheets at a time, with each sheet loaded in a separate slot. When all the sheets are loaded, the press squeezes them together under a pressure of about 110-200 psi (7.6-13.8 bar), while at the same time heating them to a temperature of about 230-315° F (109.9-157.2° C). The pressure assures good contact between the layers of veneer, and the heat causes the glue to cure properly for maximum strength. After a period of 2-7 minutes, the press is opened and the sheets are unloaded.

14 The rough sheets then pass through a set of saws, which trim them to their final width and length. Higher grade sheets pass through a set of 4 ft (1.2 m) wide belt sanders, which sand both the face and back. Intermediate grade sheets are manually spot sanded to clean up rough areas. Some sheets are run through a set of circular saw blades, which cut shallow grooves in the face to give the plywood a textured appearance. After a final inspection, any remaining defects are repaired.

15 The finished sheets are stamped with a grade-trademark that gives the buyer information about the exposure rating, grade, mill number, and other factors. Sheets of the same grade-trademark are strapped together in stacks and moved to the warehouse to await shipment.

Quality Control

Just as with lumber, there is no such thing as a perfect piece of plywood. All pieces of plywood have a certain amount of defects. The number and location of these defects determines the plywood grade. Standards for construction and industrial plywoods are defined by Product Standard PS1 prepared by the National Bureau of Standards and the American Plywood Association. Standards for hardwood and decorative plywoods are defined by ANSI/HPMA HP prepared by the American National Standards Institute and the Hardwood Plywood Manufacturers' Association. These standards not only establish the grading systems for plywood, but also specify construction, performance, and application criteria.

The Future

Even though plywood makes fairly efficient use of trees—essentially taking them apart and putting them back together in a stronger, more usable configuration—there is still considerable waste inherent in the manufacturing process. In most cases, only about 50-75% of the usable volume of wood in a tree is converted into plywood. To improve this figure, several new products are under development.

One new product is called oriented strand board, which is made by shredding the entire log into strands, rather than peeling a veneer from the log and discarding the core. The strands are mixed with an adhesive and compressed into layers with the grain running in one direction. These compressed layers are then oriented at right angles to each other, like plywood, and are bonded together. Oriented strand board is as strong as plywood and costs slightly less.

Where to Learn More

Books

Bramwell, Martyn, editor. *The International Book of Wood.* Simon and Schuster, 1976.

Duncan, S. Blackwell. *The Complete Plywood Handbook.* Tab Books, 1981.

Forest Products Laboratory. *Wood Handbook: Wood as an Engineering Material.* United States Department of Agriculture, 1987.

Hornbostel, Caleb. *Construction Materials,* 2nd Edition. John Wiley and Sons, Inc., 1991.

Periodicals

Gould, A.R. "Hardwood Plywood." *Workbench* (October/November 1994): 62-63.

Okrend, L. "Plywood for Construction." *Workbench* (June/July 1994): 44-45.

Russell, J.S. "Picking Structural Panels." *Architectural Record* (October 1992).

—*Chris Cavette*

Polyester Fleece

Background

Polyester fleece is a soft, fuzzy fabric used for sweaters, sweat shirts, jackets, mittens, hats, blankets, and in any other applications where a warm, wool-like material is needed. It is a two-sided pile material, meaning that both the front and back surface of the fabric sprouts a layer of cut fibers, similar to corduroy or velvet. Polyester fleece is an extremely durable fabric that not only holds in warmth but resists moisture and dries quickly. Unlike many other synthetic woolly textiles, polyester fleece does not pill-bunch up into little balls-after extended use. It became popular for outdoor gear in the early 1990s, because backpackers and hikers found it lighter weight and warmer than wool. It is increasingly popular as a fashion fabric, and has found a host of more specialized uses. Polyester fleece has been used to make underwear for astronauts, in deep-sea diving suits, and as ear-warmers for winter-born calves.

Synthetic fibers date back to the nineteenth century, when scientists in England and Germany developed methods of extruding the liquid state of certain chemicals through fine holes, to get thread-like strings. Fiberglass was made this way, and various other chemical fibers that were ultimately not useful as textiles. A Frenchman, Count Hilaire de Chardonnet, invented an artificial silk in the 1880s, using wood cellulose treated with nitric acid and extruded through a nozzle. Chardonnet silk was the first commercially viable synthetic fabric. In the 1920s, chemists at the Du Pont Laboratories in the United States developed nylon, an artificial fiber made of giant string-shaped molecules. British scientists extended the DuPont research in the 1940s, and came up with an-

other polymer made of string-shaped molecules called polyester.

Polyester is made by reacting terephthalic acid, a petroleum derivative, with ethylene glycol, another petroleum derivative (commonly known as antifreeze). When the two chemicals are combined at a very high temperature, they form a new chemical known as a polymer. (Polyester is one of many chemical compounds known as polymers.) As the polymer cools, it becomes thick syrup. This syrup is forced through tiny holes in a metal disk called a spinneret. On contact with air, the streams of liquid polymer dry and harden. The crystalline structure of the polymer is a chain of interlocking molecules forming essentially giant strings. In England, this polymer was called terylene. Du Pont secured exclusive U.S. rights to the polymer in 1946, calling it polyester, with the brand name Dacron.

The chemical name for the polymer, which forms polyester, is polyethylene terephthalate, or PET. If PET is not extruded into fibers, it can be formed into the plastic commonly used for soda bottles. Interest in recycling plastics in the 1980s led to the development of polyester fiber made from used soda bottles. Many polyester fleece garments on the market today are made from a combination of recycled and virgin polyester.

Textile researchers at Malden Mills, a large manufacturer in Lawrence, Massachusetts, developed polyester fleece. Malden Mills had been the leading producer of fake fur fabric in the 1970s, but faced bankruptcy as that market softened by the end of the decade. In the 1980s, Malden's research and development department experimented with a fur-like fabric made from polyester, and this with the advent of polyester fleece. Malden began producing polyester fleece under its trademark names PolarTec and Polar Fleece. Malden's brands comprise most of the polyester fleece on the market today.

Polyester fleece is extremely warm because of its structure. The pile surface provides space for air pockets between the threads, and this goes for both sides of the fabric. Because it is moisture-resistant, it can keep wearers warm even under extreme weather conditions. In the United States, the fabric was first made popular by Patagonia, a leading manufacturer of outdoor clothing and equipment. The firm marketed polyester fleece jackets to mountain climbers, and ardent customers tested the new material up and down many peaks. Other outdoor clothing manufacturers followed with their own polyester fleece garment lines. Gradually the fabric crossed over from its niche as a high-tech, high-performance textile into general use.

Raw Materials

The raw material for polyester fleece is polyester, which is made from two petroleum products: terephthalic acid and ethylene glycol. Some or all of the polyester yarn may be recycled from soda bottles. Various dyes also make up raw materials, as well as finishing substances such as Teflon or other waterproofing chemicals.

The Manufacturing Process

Producing virgin polyester

1 Virgin polyester—fiber that is made from reacting chemicals and not from re-used PET containers—is produced by heating terephthalic acid with ethylene glycol. Workers measure the chemicals into a vat (or in a continuous process, the chemicals may be automatically pumped in). A heating element under the vat raises the temperature of the solution to between 302-410° F (150-210° C). This first reaction creates dihydroxydiethyl terephthalate. This is then pumped into an autoclave, which is a sealed vat much like a pressure cooker. The chemical in the autoclave is heated under pressure to about 536° F (280° C). At this temperature the chemical transforms into PET. As it

An advertisement for Lewis union suits issued by the Lewis Knitting Company during the late 1800s. (From the collections of Henry Ford Museum & Greenfield Village, Dearborn, Michigan.)

Until the late 1800s, women wore chemises, or one-piece shifts, against their skin. Often made of linen, these shifts were not always effective in removing the perspiration that formed against the many layers of clothing worn at the time. By the 1860s, however, there was some concern that women who wore these chemises were continuously damp, thus, in cold weather, these chemise-wearers might catch a chill more easily as they might be soaked with sweat.

Prominent women's rights advocates such as Elizabeth Cady Stanton, urged women to wear "union suits." These suits, essentially a long underwear top and leggings connected at the waist, were worn closest to the skin thus replacing the chemise. They favored the union suit because the knitted suits would absorb moisture away from the skin preventing chills. Particularly favored were wool union suits—even in hot weather—because wool perhaps best draws the moisture from the body. However, short legged and short-sleeved cotton or linen suits were available for summer wear if need be, and those who could afford it might purchase silk union suits.

Nancy EV Bryk

cools, it forms a viscous liquid. This liquid is then extruded through a showerhead-like nozzle, dried, and broken into chips.

Melt spinning

2 The chips of PET are next heated in another vat to 500-518° F (260-270° C). The hot liquid is extruded through very fine

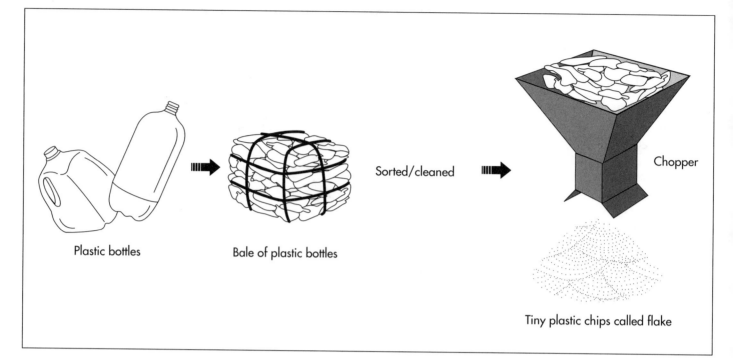

Plastic bottles

Bale of plastic bottles

Sorted/cleaned

Chopper

Tiny plastic chips called flake

Bales of bottles are emptied onto a moving belt. Workers first sort the bottles by color, separating green ones from clear ones. Then workers visually inspect each piece so that the final result is strictly PET bottles. The sorted plastic then moves into a sterilizing bath. The clean containers are dried and crushed into tiny chips.

holes in a metal disk called a spinneret. As the liquid sprays out of the spinneret, it hardens into fiber form. The fibers are wound onto a heated spool. At this point, the fibers form something like a thick rope, which is called tow.

Producing polyester from recycled PET containers

3 When polyester is made from recycled PET, the first step is collecting used PET containers. Yarn makers buy bales of recycled bottles from vendors or from municipal recycling projects.

The bales of bottles are emptied onto a moving belt. Workers first sort the bottles by color, separating green ones from clear ones. Then workers visually inspect each piece, and remove anything, such as non-PET caps or bases, or any foreign objects, so that the final result is strictly PET bottles. The sorted plastic then moves into a sterilizing bath. The clean containers are dried and crushed into tiny chips. The chips are washed again, and the light-colored batch is bleached. Chips from green bottles stay green, and become yarn that will be dyed a dark color.

When the chips are thoroughly dry, they are emptied into a vat and heated, then forced through spinnerets, the same as for virgin

polyester. The finishing steps-drawing, crimping, cutting, baling-are the same as in the process for virgin polyester.

Drawing and crimping

4 The tow from the spool is next pulled through the heated rollers of a drawing machine to three or four times its original length. Drawing increases the strength of the fiber, and helps set the crystalline structure of the PET molecules into smooth strings. The tow then passes through a crimping machine, which compresses the tow and gives it a crinkled, accordion-like texture. This also adds strength. The crimped tow passes to a dryer, and then is cut into lengths of a few inches and baled. At this point, the short, fluffy, hairy fiber looks very much like wool.

Spinning into yarn

5 After the polyester is baled, a sample from each bale is inspected. Fibers are tested for uniformity of strength and thickness. If the bale passes inspection, then the cut tow is sent to a carding machine, which aligns the fiber into thick, rope-like strands. The strands flow out of the machine and are coiled into barrels or open containers. The thick ropes are then fed into a spinning machine. The spinning machine twists the

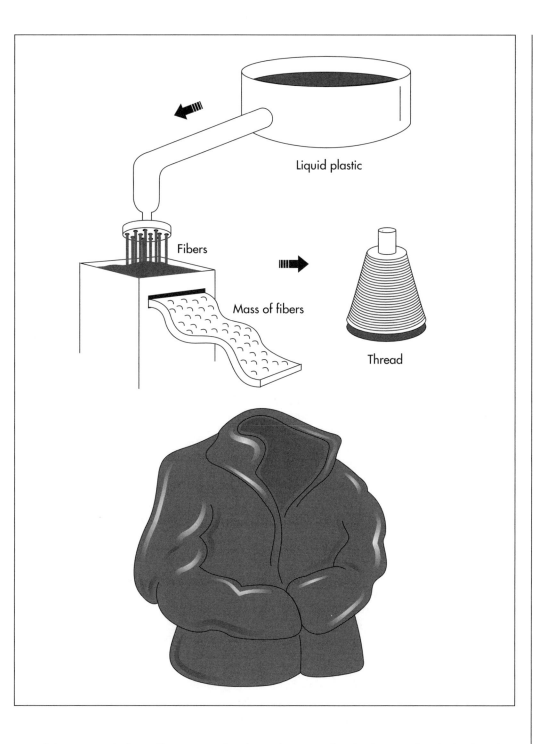

Liquid plastic

Fibers

Mass of fibers

Thread

The chips are emptied into a vat and heated, then forced through spinnerets. The strands flow out of the machine and are coiled into barrels or open containers. The spinning machine twists the strand into a much finer diameter, and collects the finished yarn onto huge spools.

strand into a much finer diameter, and collects the finished yarn onto huge spools.

Dyeing

6 The textile manufacturer buys polyester from the yarn manufacturer on these spools. The yarn is next immersed in heated dye vats in the part of the factory called the dye house. In case of yarn made from green recycled PET bottles, the dye must be a dark hue. Other yarns arrive bleached white, and these can be dyed any color desired. After dyeing, workers feed the yarn through a drying machine.

Knitting

7 The dried yarn is next fed into a particular kind of mechanical knitter called a circular knitting machine. The knitting machine binds the yarn into a continuous tube of cloth. The tube may be approximately

58 in (1.47 m) wide and several hundred yards long.

Napping and shearing

8 To achieve the particular fuzzy texture of fleece, the knitted material is next fed through a napper. The napper runs mechanical bristles along the cloth, raising the surface of the textile. Next, the cloth is sent to a shearing machine, which uses a precision blade to cut the fibers raised by the action of the napper. This same process is used to make velvet, corduroy, and other textured pile fabrics.

Finishing

9 The fabric may next be sprayed with a waterproof material, or with some other chemical finisher that sets the texture of the material. The material is next cut into lengths, according to the customer's needs. The lengths of cloth are wrapped around boards or cardboard planks. These wound lengths are called bolts. At this point, the bolts are ready to be sent to the garment manufacturer. The manufacturer will cut the fabric according to a pattern, and sew the cloth into a garment.

Byproducts/Waste

Making polyester fleece from recycled PET bottles is a significant means to reducing the amount of plastic that is otherwise buried in landfills. One manufacturer estimates that for every meter of polyester fabric made of 80% recycled PET, eight plastic beverage bottles are kept out of landfills. Patagonia, the leading manufacturer of recycled polyester fleece garments, estimates that 25 soda bottles go into each jacket made from the fabric. Recycling PET into polyester is also alleged to be less damaging to the environment even than growing organic cotton, because cotton leaches nutrients from soil and requires so much open space to grow. The energy used to make polyester from recycled PET bottles is also significantly less than that needed to heat the chemicals for virgin polyester.

The Future

Polyester fleece is a remarkably comfortable and adaptable fabric, and will doubtless find many new uses. The future of recycled PET polyester seems to lie in making the recycling process more economically efficient, and in making finer-diameter yarns. Used beverage bottles are very lightweight, and therefore they are expensive to transport, as it takes a large volume of them to make up a ton. Yarn manufacturers must find used bottle sources near the spinning factory in order to make recycling economically viable. The coarser yarns, which are now used primarily for carpets and in tires, are easier to make, but also sell for less than the finer, garment-quality yarns. Manufacturers will continue to refine the recycling process to gain cost advantages. Other developments focus on different recycling processes that do not rely on clean soda bottles. Yarn manufacturers who recycle from PET bottles buy baled bottles from distributors. However, many municipal recycling programs do not separate PET bottles from other recyclables, and this mixed product is more difficult to handle. Several European manufacturers are developing new technology that efficiently removes excess dye, metals, and non-PET plastics from recycled PET. This means that less meticulous hand sorting is needed before the bottles are recycled. As the process is perfected, it will mean that PET and non-PET plastics can be recycled together.

Where to Learn More

Periodicals

Hamilton, Martha M. "Soda-Bottle Chic." *The Washington Post* (April 12,1994): A1.

Lee, Melissa. "Malden Looks Spiffy in New England Textile Gloom." *The Wall Street Journal* (November 10, 1995): B4.

Rotenier, Nancy. "The Golden Fleece." *Forbes* (May 24, 1993): 220.

Sanford, Tobey. "'Cozy, Soft, Warm, Yummy, Fleecy'-Is This Any Way to Describe an Empty Soda Bottle?" *Life* (November 1994): 138-140.

Schut, Jan H. "New Alchemy for PET Arrives." *Plastics World* (August 1995): 27-29.

—*Angela Woodward*

Pottery

Background

Pottery is clay that is modeled, dried, and fired, usually with a glaze or finish, into a vessel or decorative object. Clay is a natural product dug from the earth, which has decomposed from rock within the earth's crust for millions of years. Decomposition occurs when water erodes the rock, breaks it down, and deposits them. It is important to note that a clay body is not the same thing as clay. Clay bodies are clay mixed with additives that give the clay different properties when worked and fired; thus pottery is not made from raw clay but a mixture of clay and other materials.

The potter can form his product in one of many ways. Clay may be modeled by hand or with the assistance of a potter's wheel, may be jiggered using a tool that copies the form of a master model onto a production piece, may be poured into a mold and dried, or cut or stamped into squares or slabs. The methods for forming pottery is as varied as the artisans who create them.

Pottery must be fired to a temperature high enough to mature the clay, meaning that the high temperature hardens the piece to enable it to hold water. An integral part of this firing is the addition of liquid glaze (it may be painted on or dipped in the glaze) to the surface of the unfired pot, which changes chemical composition and fuses to the surface of the fired pot. Then, the pottery is called vitreous, meaning it can hold water.

History

Potters have been forming vessels from clay bodies for millions of years. When no-madic man settled down and discovered fire, the firing of clay pots was not far behind. Pinch pots, made from balls of clay into which fingers or thumbs are inserted to make the opening, may have been the first pottery. Coil pots, formed from long coils of clay that are blended together, were not far behind. These first pots were fired at low temperatures and were thus fragile and porous. Ancient potters partially solved this by burnishing the surfaces with a rock or hard wood before firing. These low-temperature fired pots were blackened by these fires. Decoration was generally the result of incisions or insertions of tools into soft clay. Early potters created objects that could be used for practical purposes, as well as objects that represented their fertility gods.

The civilizations of ancient Egypt and the Middle East utilized clay for building and domestic use as early a 5000 B.C. By 4000 B.C., the ancient Egyptians were involved in pottery on a much larger scale. They utilized finer clays and fired the pieces at much higher temperatures in early kilns that removed the pots from the direct fire so they were not blackened from the fire. Bricks from clay were used as building material as well. The ancient Chinese produced black pottery by 3500 B.C. with round bases and plaited decoration. Closer to 1000 B.C. the Chinese used the potter's wheel and developed more sophisticated glazes. Their pottery was often included in funeral ceremonies. In the first millennium B.C., the Greeks began throwing pots on wheels and creating exquisite forms. Pre-Colombians, ancient Iberians, the ancient Romans (who molded pottery with raised decoration), and the ancient Japanese all created beautiful

The civilizations of ancient Egypt and the Middle East utilized clay for building and domestic use as early a 5000 B.C.

389

A stoneware teapot made by Josiah Wedgwood and Co. of Staffordshire, England. (From the collections of Henry Ford Museum & Greenfield Village, Dearborn, Michigan.)

This lovely, stout stoneware teapot is the work of Josiah Wedgwood and Co., of Staffordshire, England, perhaps the best known of British pottery companies of the nineteenth century. Teapots and associated cups became very popular about the mid-1700s because of the development importance of the "tea" and its ceremony. Thus, a mainstay of potters in the eighteenth century was the teapot and cup sets.

Josiah Wedgwood was not content to simply supply pottery rather haphazardly. He knew there was a large market for high-quality, attractive pottery and he certainly would do his best to regularize the product and develop some new products people just had to have. He was one of the first potters to sell his wares in advance through orders, thus creating a sample or "stock" product. Since his products had to be uniform, he developed glazes that would give consistent results and divided the work process into many different steps so that one worker would not have a tremendous impact on the finished product. Particularly important to Wedgwood was the work of the modeller and the artist, who made the prototype shapes and designs for Wedgwood. Wedgwood discovered that these artists could provide designs for new pottery that looked antique, and these neo-classical pieces were the mainstay of his business for many years.

Nancy EV Bryk

pottery for domestic use as well as for religious purposes.

Until the mid-eighteenth century, European potters generally sold small quantities of completed wares at a market or through merchants. If they wanted to sell more, they took more wares to market. However, British production potters experimented with new body types, perfected glazes, and took orders for products made in factories rather than taking finished goods to the consumer. By the later eighteenth century, many fellow potters followed suit, experimenting with all kinds of new bodies and glazes. Molds were used to make mass quantities of consistent product so that the consumer could be assured of the look of this piece.

Raw Materials

Its primary mineral is kaolinite; clay may be generally described as 40% aluminum oxide, 46% silicon oxide, and 14% water. There are two types of clays, primary and secondary. Primary clay is found in the same place as the rock from which it is derived—it has not been transported by water or glacier and thus has not mixed with other forms of sediment. Primary clay is heavy, dense, and pure. Secondary or sedimentary clay is formed of lighter sediment that is carried farther in water and deposited. This secondary clay, a mixture of sediment, is finer and lighter than primary clay. Varying additives give the clay different characteristics. Clay comes to a production potter in one of two forms—as a powder to which water must be added, or with water already added. Large factories purchase the clays in huge quantities as dry materials, making up the clay batch as needed each day.

Glazes are made up of materials that fuse during the firing process making the pot vitreous or impervious to liquids. (Ceramics engineers define vitreous as a pot that has a water absorption rate of less than 0.5%.) Glazes must have three elements: silica, the vitrifying element (converts the raw pottery into a glasslike form)—is found in ground and calcined flint and quartz; flux, which fuses the glaze to the clay; and refractory material, which hardens and stabilizes the glaze. Color is derived by adding a metallic oxide, including antimony (yellows), copper (green, turquoise, or red), cobalt (black), chrome (greens), iron, nickel, vanadium, etc. Glazes are generally purchased in dry form by production potters. The glazes are weighed and put into a ball mill with water. The glaze is mixed within the ball mill and grinds the glaze to reduce the size of the natural particles within the glaze.

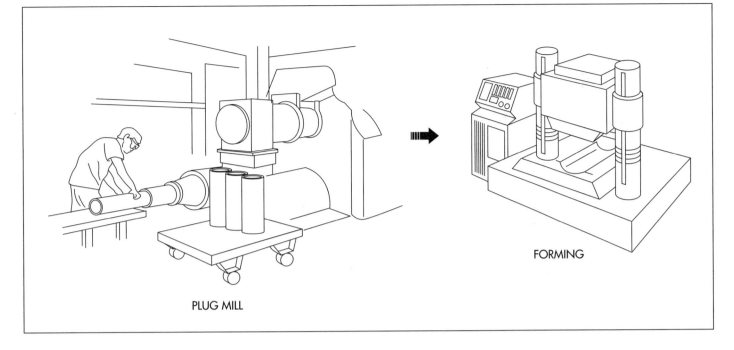

PLUG MILL

FORMING

The cake mixture is formed into plugs and ready for forming.

Design

Pottery factories include art directors whose job it is to conceive marketable goods for the pottery company. Generally the art director, working with marketers, develops or creates an idea of a new creation. (Interestingly, many pottery companies are reproducing old forms popular decades ago such as brightly-colored Fiesta Ware so that new design is not necessary or desirable in all cases.) The art director then works with a clay modeler, who produces an original form of the creation to the art director's specifications. If the form is deemed a viable candidate for production, the mold maker makes a plaster master for the jiggering machine (which essentially traces a master shape onto a production piece) or a hollow into which clay is poured in order to form a production piece.

The Manufacturing Process

Mixing the clay

1 Clay arrives by truck or rail in powder form. The powder is moistened with water and mixed in a huge tank with a paddle called a blunger. Multiple spindles mix and re-mix the clay, in order to evenly distribute water. A typical batch mixed at a large production potter is 100,000 lb (45,400 kg) and they often mix up two batches in a single day. At this point, the slurry is about 30% water.

2 Next, the slurry is filter pressed. A device presses the slurry between bags or filters (like a cider press) to force out excess water. The resulting clay is thick and rather dry and is called cake now and is about 20% water.

3 The cake is then put into a plug mill in which the clay is chopped into fine pieces. This chopping de-airs the clay as pumps suck out air pockets that are exposed by this process. The cake is then formed into cylinders that are now ready to be molded or formed.

Jiggering

4 The fastest way to produce a regular, hollow pot is by using a jiggering machine. Thus, hollowware such as vases is largely made on jiggering machines. The clay cylinders made in the plug mill are sent to the jiggering machine. In order to make a vase, a wet clay cylinder is dropped onto the jiggering machine by a suction arm which positions the clay inside a plaster mold. A metal arm then comes down into the wet clay cylinder forcing it against the interior wall of the plaster mold thus forming the new vessel. The plaster mold, with wet clay

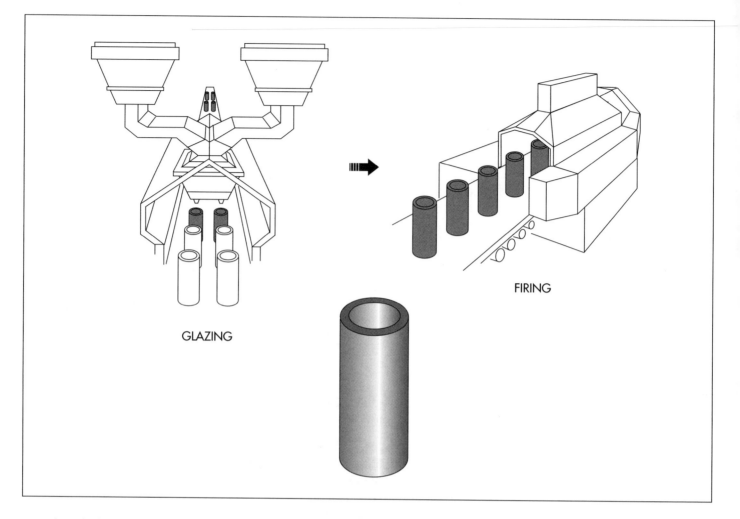

GLAZING

FIRING

Once formed, the greenware is glazed and then fired, creating pottery.

inside, is then lifted off the machine and set in dryer. As the clay heats up and dries slightly the new, wet clay pulls away from the plaster mold and can thus be easily removed. Thus, the factory must have thousands of plaster molds in order to make these vases or other hollowware as a plaster mold is used to make each new vessel. The factory may be able to make as many as 9 pieces of pottery in a single minute.

5 A machine takes the rough edges off the molded piece. The cleaned pieces are placed on a continuously-moving belt which leads to tunnel dryers, which heat the pieces and reduce the water content to under 1% moisture before glazing and firing.

Slip casting

6 Pottery with delicate or intricate silhouette is often formed by slip casting. A pourable slip or slurry is poured into a two-part plaster mold, the excess is poured out,

and the slip is permitted to stiffen and dry. The plaster mold sucks up some of the excess water and helps hasten the drying process. The plaster mold is opened when the greenware (undecorated clay piece still a bit wet) is stiff enough, the piece is cleaned of rough edges and seams from the mold, and the slip-cast greenware is ready for drying in the heated dryers.

Glazing

7 After the pieces have been dried, they are ready for glazing. The pieces may be entirely covered in one color of glaze by being run under a waterfall of glaze that completely coats each piece, or the pieces may be sprayed with glaze. Deep hollowware such as vases have to be flushed with glaze by hand to ensure that they are completely coated on the inside. Glazes are generally applied to a thickness of 0.006-0.007 in (0.015-0.017 cm). Other pieces may be more decoratively glazed. Some

pieces are printed with screen-printing, others have a decorative decal applied by hand, others may have lines or concentric rings applied by machines, and still others may be painted by hand.

Firing

8 Kilns may be heated by gas, coal, or electricity. One large production potter uses tunnel kilns fired with natural gas. Large cars or wagons (about 5 ft or 1.5 m square and nearly 5 ft or 1.5 m tall) are loaded with unfired pottery and sent to the kilns, firing approximately 20,000 dozen pieces

of pottery in a single week. Newer furnaces run at higher temperatures than older kilns and require a shorter firing time—running at about 2,300° F (1,260° C) the pots remain in the kilns about 5 hours—thus allowing the factories to move pieces more quickly through production.

The kiln changes the glaze into a glass-like coating, which helps make the pot virtually impervious to liquid. Single-color production pottery requires only one firing with the new kilns and glazes. (Many glazes require that the greenware be fired once and made into a bisque or dull white, hard body, then glazed and fired again; however, this is not necessary with some new production glazes.)

9 The unglazed foot (or bottom) of the pottery is polished on a machine with a cleaning pad. The piece is then placed in a bin and is sent to packaging, ready to be shipped out for sale.

Quality Control

All raw materials are checked against the company's established standards. Clays must contain the ingredients required by the product and ordered by the company. Glazes must be as pure as possible and are checked for correct shade, viscosity, gravity, etc. Kiln temperature must be carefully monitored with heat cones and thermocouples, etc. And each human involved in production uses their eyes to monitor against inferior products.

Byproducts/Waste

There are no harmful by-products resulting from the production of pottery. Clay scraps and imperfect pieces produced off the jiggering machine or from slip casting may be re-mixed and re-used. Glazes must be lead-free as required by the Food and Drug Administration (FDA), and glazes are tested in-house to assure the FDA that they contain neither cadmium nor lead. All glazes may be touched by the human hand are not harmful in raw state.

Where to Learn More

Books

Barber, Edwin Atlee. *The Pottery and Porcelain of the United States.* New York: G. P. Putnam's Son's, 1893.

Chavarria, Joaquim. *The Big Book of Ceramics.* New York: Watson-Gupthill, 1994.

Forty, Adrian. *Objects of Desire.* New York: Pantheon Books, 1986.

Hiller, Bevis. *Pottery and Porcelain 1700-1914.* New York: Meredith Press, 1968.

—Nancy EV Bryk

Pretzel

A recent market survey found that the pretzel market in the United States is about $560 million a year with over 300 million lb (136.2 million kg) of pretzels and pretzel products being produced.

Background

Pretzels are a snack food, which have unique shapes and a hard, shiny outer surface. They are mass produced using primarily automated machinery. First developed in the seventh century, pretzels have been called one of the world's oldest snack food. A recent market survey found that the pretzel market in the United States is about $560 million a year with over 300 million lb (136.2 million kg) of pretzels and pretzel products being produced. The pretzel market has grown in recent years because pretzels are considered a more healthy, fat-free snack.

The unique, two looped, knot shape of a pretzel is one of its defining characteristics. The typical pretzel has a pleasant cracker-like flavor, a crisp, brittle texture and a brown glossy surface color. Salt crystals are often sprinkled on its surface to make them taste more appealing. Pretzels have a moisture content of anywhere from 2-4% and therefore have a very long shelf life.

While the two-looped knot shaped, hard pretzels may be the most popular kind of pretzel, there are other kinds which are sold. Soft pretzels are also manufactured. These products typically have a much higher moisture content than hard pretzels and are usually larger. They also have a shorter shelf life. Other shapes are also produced such as thick and thin rods, pretzel rings, and loops. Additionally, flavored pretzel such as cheese, rye, caraway, kosher, and butter are also available. Finally, salt-free pretzels called baldies are now made.

History

Some historians have said that the pretzel is the oldest snack food ever developed. It is believed that the pretzel was first developed during 610 A.D. by a monk in southern France or northern Italy. Using the dough left over from bread making, he formed the pretzel shape, which was meant to look like a child's arms folded in prayer. He used these creations as treats for children that learned their prayers. He called the snacks *pretiola,* which means little reward in Latin. The pretiolas eventually found their way into Germany and Austria where they became known as pretzels. The pretzels grew in popularity and are said to have been brought over to America on the Mayflower in 1620. The first pretzels were of the soft variety. Legend has it that one night a baker who was baking a batch of pretzels fell asleep. When he woke up all the moisture was cooked out of them and the hard pretzel was born.

Raw Materials

The primary ingredients in pretzel dough include flour, water, yeast, shortening, and sugar. Each of the ingredients have an important effect on the dough during manufacture and the properties of the final product.

Flour

In pretzel dough making, flour is perhaps the most important ingredient. It is primarily made up of starch and protein. When water is added, the flour protein soaks up the water rapidly and reacts with it to form a mass known as gluten. The gluten can be stretched and formed quite a distance without breaking. This allows pretzels to be formed into the desired shapes. Sometimes the flour is enriched with various nutrients such as thiamin, riboflavin, and iron to increase the nutritional value of the product. Nutrient enrichment is highly regulated by the government so preset limits are required for a flour to be called enriched. The flour used for pretzels is called soft wheat flour and has a protein content of about 9%. It is the largest component in the dough making up about 65-70% of the total recipe.

Yeast and leavening agents

Pretzel dough is unusual in that it contains both yeast and chemical leavening agents. The yeast is put in the dough and it produces carbon dioxide gas as it metabolizes the sugar during fermentation. This gas creates tiny air pockets in the batter, which helps make the pretzels lighter and crispier. Yeast is typically supplied as a dry, granular product. Dry yeast is desirable because it can be stored for a long time at room temperature. Leavening agents have a similar effect as fermenting yeast, however they have less effect on the final taste. Chemical leavening agents include materials such as sodium bicarbonate and ammonium bicarbonate. In the presence of water, they breakdown chemically to produce carbon dioxide gas.

Shortening

Vegetable shortening is a solid form of vegetable fats and oils. Its main purpose in the dough is to inhibit the formation of the gluten. This helps the dough stay softer, increases the volume and gives it a crumbly texture. Fat also allows the dough to remain more palatable for longer. A typical pretzel recipe may call for about 2-3% vegetable shortening.

Other important ingredients

The pretzel dough would not be possible without the use of water. While there is less water in pretzel dough than in bread dough, it still makes up about 30-35% of the recipe. Water is important because it lets the dough flow and allows the yeast to contact the sugars for fermentation. It also causes the chemical leavening agents to activate.

The taste of the pretzel is a result of ingredients such as salt, sugar, and flavorings. Salt is used in the batter at about 1%. It helps to make the dough stronger in addition to improving the taste. Sugars such as sucrose or corn syrup are used for about 2% of the pretzel dough. The sugar has the primary effect of providing food for the yeast cells. Additionally, it will give a slightly sweet taste although this is very minimal in pretzel making. Sugar also contributes to the brown color of the pretzel.

The Manufacturing Process

Today, the manufacture of pretzels is a nearly completely automated process, which converts the raw ingredients into a shaped, finished product. It is estimated that 90% of all pretzels are never touched by human hands during the manufacturing process. The following steps outline the procedure used to make typical hard pretzels. Soft pretzels have a slightly different manufacturing procedure.

Mixing dough

1 The pretzel dough is made by factory compounders in large stainless steel tanks. The flour and warm water are stored in bulk and transferred to the tanks automatically. The yeast is added and the three ingredients are blended with high-speed horizontal mixers. When these are adequately blended, the rest of the ingredients such as sugar, sodium bicarbonate, vegetable shortening, salt and flavorings are added. Compared to many dough products such as bread or crackers, pretzel dough is relatively under mixed. This allows the dough to withstand the punishment of machining without becoming too sticky or misshapen. The dough is then allowed to ferment and rise for about 30 minutes.

Forming pretzels

2 The fermented dough is then transferred to the hopper of the shape-making equipment. Traditionally, pretzels were made by rolling the dough and twisting it into the familiar pretzel shape. However, today most companies have extrusion devices in which the dough is forced through an opening and stamped into shape with a wire cutter. The excess dough is recycled to

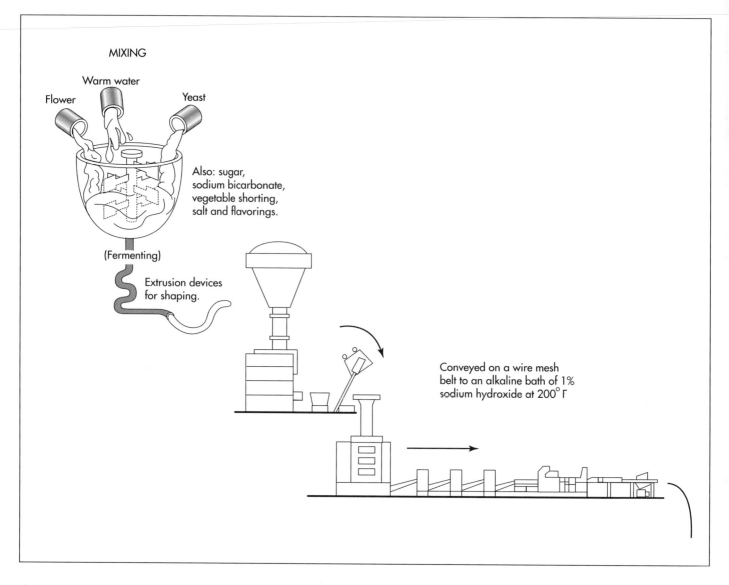

MIXING

Flower

Warm water

Yeast

Also: sugar, sodium bicarbonate, vegetable shorting, salt and flavorings.

(Fermenting)

Extrusion devices for shaping.

Conveyed on a wire mesh belt to an alkaline bath of 1% sodium hydroxide at 200° F

The dough is mixed in large vats by factory compounders. Once fermented, the dough is transferred to a hopper, which feeds the shape-making equipment. The raw pretzels are next conveyed on a wire mesh belt to an alkaline bath.

the hopper while the stamped pretzels are transferred to a conveyor. They are passed under rollers to ensure a flat surface and uniformity of size.

Dipping and salting

3 The raw pretzels are next conveyed on a wire mesh belt to an alkaline bath. It generally takes several minutes for the pretzels to reach the bath. This slow transport is deliberate as it allows the pretzels to undergo another short fermentation or rest period. The alkaline bath is filled with an aqueous solution of either sodium carbonate or lye. The resulting bath has an overall 1% concentration of sodium hydroxide. It also is held at a temperature of about 200° F (93.3° C). The pretzels are dipped in the bath for 10-20 seconds and typically float when they

are finished. This process gelatinizes the starch on the pretzel's surface making it gummy and sticky, allowing the salt to adhere more readily.

After the pretzels leave the hot bath, they are passed under a machine which delivers salt crystals to their surface. Modern pretzel-making lines use a vibrating salter, which consists of a vibrating plate driven by a series of small motors and magnets. The salt is evenly distributed on each pretzel with the excess falling through the wire mesh belt and being recycled. Generally, the aim is to add about 2% salt to each pretzel.

Cooking

4 The in-process pretzels are next transported to long, gas-fired, convection

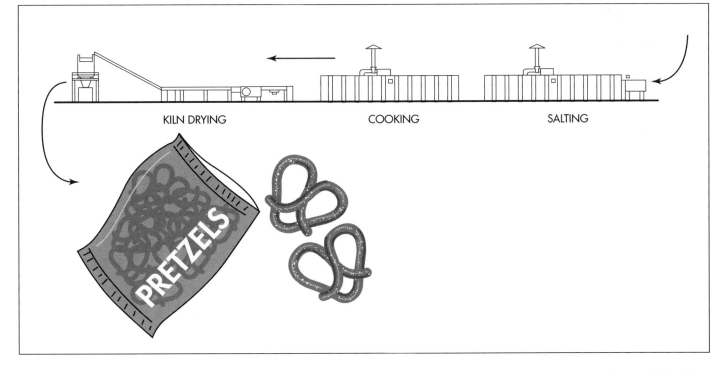

KILN DRYING   COOKING   SALTING

tunnel ovens. The cooking temperature varies from 350-550° F (176.7-286.1° C) and this baking step takes from about 4-8 minutes. In the front of the oven, the temperature is significantly higher than at the end. The initial high heat caramelizes the gelatinized starch, which produces the characteristic dark brown pretzel color. The temperature is gradually raised at the start because if heated too fast, the structure of the pretzels will be weakened which could cause cracking and breaking during shipping. At the end of the oven the temperature is cooler to allow moisture in the pretzel to be released. During this entire baking cycle, the moisture content is reduced to about 15%. In the next baking phase, the pretzels are kiln dried or oven dried at about 250° F (119.4° C) for anywhere from 20-40 minutes. This further reduces the moisture content to below 4%.

Packaging

5 From the ovens, the pretzels are passed along varies conveyors and allowed to cool. They are then moved along to the packaging machines. Here the pretzels are weighed and the correct amount is placed in the packaging. They can be put in many different types of packages including trays, boxes or bags with cellophane or polyethyl-ene protected coatings. It is important that this packaging be air tight to prevent the uptake of moisture by the product. Excessive moisture would cause them to become soft. The package must also have consumer appealing graphics, which help it stand out on a store's shelves. Most major bakeries distribute products to all of the largest cities in the world. Consequently, there are very few people who are unfamiliar with pretzel snacks.

Quality Control

To maintain a high degree of quality, pretzel manufacturers begin by thoroughly testing the raw ingredients. These materials are evaluated by quality control inspectors and subjected to a variety of tests. Various sensory characteristics such as odor, color, and flavor are evaluated. Other factors like the particle size of the flour, thickness of shortening, and pH of liquids are also examined. These tests ensure that the raw materials will produce a consistent batch of pretzels. On the final product, many of the same characteristics that were tested on the raw materials are evaluated. Chemists and technicians check things such as appearance, texture, flavor, and color. This will certify that each batch of pretzels shipped to stores will be of the same quality as those developed in the food laboratory.

Raw pretzels are salted before being transported to long, gas-fired, convection tunnel ovens. The cooking temperature varies from 350-550° F (176.7-286.1° C) and this baking step takes from about 4-8 minutes. Once baked, the pretzels are dried and packaged.

The Future

Future developments in pretzel manufacturing will likely be found in the production of new products and improved machines. Marketers will undoubtedly create new twists on existing products in an effort to make them taste better, appear more healthy, and more unique. Improvements in the automatic pretzel making machines will focus on designs, which increase the baking capacity and speed at which pretzels are made.

Where to Learn More

Books

Booth, Gordon, ed. *Snack Food.* New York: Van Nostrand Reinhold, 1990.

Hoseney, Carl. *Principles of Cereal Science and Technology.* St. Paul: American Association of Cereal Chemists, 1994.

Macrae, R., et al., ed. *Encyclopedia of Food Science, Food Technology and Nutrition.* San Diego: Academic Press, 1993.

Periodicals

Russel, Margie. "Snackmakers Feel the CRUNCH." *Food Engineering* (May 1995): 86-93.

Walter, Andreas. "Food Marketers Display the Future of Eating." *SnackWorld* (June 1997).

—*Perry Romanowski*

Raisins

Background

Raisins are made primarily by sun drying several different types of grapes. They are small and sweetly flavored with a wrinkled texture. The technique for making raisins has been known since ancient times and evidence of their production has been found in the writings of ancient Egyptians. Currently, over 500 million lb (227 million kg) of raisins are sold each year in the United States, and that number is expected to increase because raisins are recognized as a healthy snack.

Most raisins are small, dark, and wrinkled. They have a flavor similar to the grapes from which they are made, but the drying process which creates them concentrates the amount of sugar making them taste much sweeter. They are a naturally stable food and resist spoilage due to their low moisture and low pH.

Raisins are composed of important food elements such as sugars, fruit acids, and mineral salts. The sugars provide a good source for carbohydrates. Fruit acids such as folic acid and pantothenic acid, which have been shown to promote growth, are also significant components. Vitamin B_6 is found in raisins and is an essential part of human nutrition. Important minerals in raisins include calcium, magnesium, and phosphorus. Additionally, iron, copper, zinc, and other nutrients are found in trace amounts in raisins. Considering the composition of raisins and the fact that they have no fat, it is no wonder that this fruit is considered a healthy snack.

The majority of grapes used for making raisins in the United States are grown in California. This area has an ideal climate for grape growing because it has plenty of sun during the summer and very mild winters. Five other countries, which produce a substantial amount of raisins include Greece, Australia, Turkey, Iran, and Afghanistan. Each of these countries have their own variety of raisin that they consistently grow.

History

The technique of drying fruit was likely discovered by accident. It is conceivable that our ancestors came upon fallen fruit, which had dried in the sun, and discovered its sweetness after tasting it. Evidence has shown that raisins were produced by the Egyptians as early as 2000 B.C. Raisins specifically have been mentioned in ancient writings and it suggests that they were used for eating, treating illnesses, and even paying taxes.

Throughout the ages, wine making has been the most important use for grapes, however, a small amount of these grapes have always been made into raisins. During the late 1800s, Spanish missionaries from Mexico introduced grapes into the United States. Many of the vineyards established by these missionaries in California are still producing today. These early vineyards were primarily used to make wines, however in 1873 when the vineyards discovered they could make quicker profits by making raisins, the raisin industry was born.

Raw Materials

The primary raw material for making raisins is grapes. To make 1 lb (453.59 g) of raisins, over 4 lb (1,814.36 g) of fresh grapes are re-

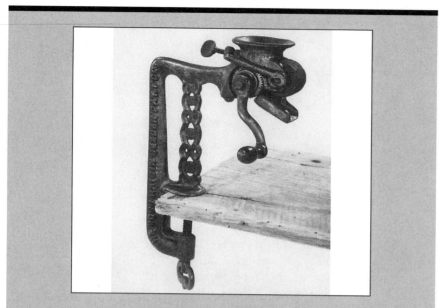

A cast iron raisin seeder made by A.C. Williams of Ravenna, Ohio, circa 1900. (From the collections of Henry Ford Museum & Greenfield Village, Dearborn, Michigan.)

How lucky we are that many of our foodstuffs are already dried, seeded, and otherwise prepared for inclusion in our favorite recipes. We purchase seedless raisins and don't even have the option of purchasing raisins with seeds. However, this was not the case over 100 years ago. Then, seedless raisins (expensive) were sold alongside those with seeds (noted as cheaper and "more commonly used").

One might have saved pennies buying raisins with seeds but invested time in seeding those tiny fruits. How? One cookbook suggests that Valencia raisins be heated slightly with water in order to plump them, and then cut with a knife and de-seeded by hand! However, enterprising manufacturers produced labor-saving devices for women's kitchen chores, including de-seeding raisins. First, the housewife clamped her Boss brand raisin seeder to her kitchen table. Then, she loaded the raisins into the hopper at the top. As the housewife cranked the handle, the raisins were squeezed between two grooved rubber and toothed-metal rollers, which exposed the seeds. The seeds were then forced out a chute at the front (pushed out by the metal-toothed rollers) and the raisins dropped below the rollers into a pile.

Nancy EV Bryk

quired. These grapes must have certain qualities in order to produce quality raisins. For example, they must ripen early and be easy to dry. Additionally, they must have a soft texture, not stick together when stored, have no seeds, and have a pleasing flavor. The most important grapes for raisin production include Thompson Seedless, Black Corinth, Fiesta, Muscats, and Sultans.

By far, the most widely grown raisin grape is the Thompson Seedless variety. They are used in the production of over half the world's raisins. Ninety percent of these come from California. The Thompson was first developed in 1872 by William Thompson, who created it by taking cuttings from an English seedless grape and grafting them with a Muscat grape vine. The resulting plant produced the first Thompson seedless grapes. It is believed that all of the subsequent Thompson seedless vines came from this original grafting.

The Thompson seedless is a white, thin-skinned grape, which produces the best raisins available today. Its small berries are oval and elongated. It does not contain seeds and has a high sugar content. From a raisin production standpoint, Thompson grapes are ideal because they ripen fairly early in the season and do not stick to each other during shipping.

The Black Corinth is a grape that originated in Greece, which has become an important variety of raisin grape. They are about one fourth the size of the Thompson grapes and have a juicy, tangy/tart flavor. These grapes are quite small, spherical in shape, and reddish-black in color. They are thin skinned and nearly seedless. They make good raisins and are excellent for production because they ripen early and dry easily. Because of their flavor, they are more often used for baking cookies, specialty breads, and fruitcakes than for eating.

Next in line of importance to raisin production is the Muscat grapes. These are large, sweet grapes that contain some seeds. Originally grown in Alexandria, Egypt, these grapes were the primary raisin grape before the advent of the Thompson. They were introduced in the United States in 1851. Muscat grapes are juicy, dull green in color, and have a sweet, muscat flavor. They have moderately tough skins and result in excellent tasting, large, soft-textured raisins. When they are used for raisin making, they are subjected to a mechanical process, which removes the seeds after the grapes are dried. These seeds are a significant drawback to using the muscat, and additionally, they do not ship well.

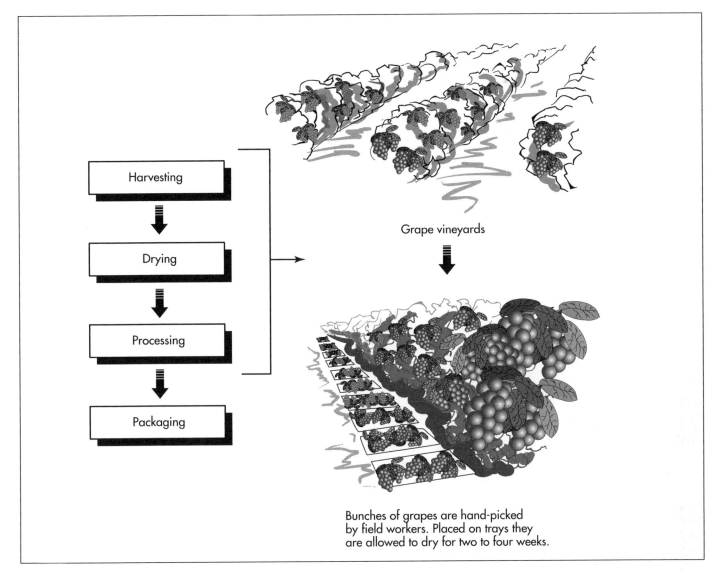

Grape vineyards

Bunches of grapes are hand-picked
by field workers. Placed on trays they
are allowed to dry for two to four weeks.

Harvesting → Drying → Processing → Packaging

Two minor varieties of grape that find some use as raisins include the Fiesta and the Sultana. The Fiesta is a white seedless grape with a good flavor. A major problem with these grapes is that their stems are more difficult to remove. The Sultana grape is nearly seedless, but they make inferior raisins because they are less meaty, have a high acid content, and have some small, very hard seeds. Both Fiesta and Sultana raisins are used more often as baking raisins.

The Manufacturing Process

There are four primary methods for producing raisins including the natural, dehydration, continuous tray, and dried-on-the-vine methods. The most popular of these is the natural method which will be explained in some detail. The basic steps in natural raisin manufacturing include harvesting, processing, and packaging. While a small portion of raisins are made by mechanically dehydrating grapes, the majority of them are produced by sun drying.

Farming

1 The first step to producing good raisins is growing quality grapes in the vineyards. Grape farming is a year-round commitment and includes the practices of pruning, irrigation, fertilization, and pest control. Most of the work done in these vineyards is still done by hand. Pruning involves the removal of parts of the vine to control its growth pattern. This has the benefits of equalizing the quality of grape throughout the vineyard,

Grapes are harvested in August through September. While drying on trays, the grapes' moisture content is reduced from 75% to under 15% and the color of the fruit changes to a brownish purple. After the fruit is dried, the paper trays are rolled up around the raisins to form a package. The rolls are gathered and stored in boxes or bins before being transported by truck to a processing plant, where they are cleaned, inspected, and packaged.

making other farming tasks easier and reducing costs. It is typically done when the vines are dormant between December and March. Irrigation is done during the summer while the vines are growing to keep a continuous supply of water in the vineyard soil. While fertilizers are not needed in all vineyards, some vines respond well to the use of nitrogen and zinc based fertilizers. Fertilization is typically done during the summer growing season.

Vineyards are susceptible to various diseases and insect attacks, so it is important for these factors to be controlled. Chemical and biological agents are used to control mites and other insects. Sulfur dusting is used to prevent the growth of mildew and other fungi. Since these compounds can have an effect on the overall grape quality, attempts are made to minimize the amounts used

Harvesting and drying

2 Starting in late August and continuing through September, the grapes are harvested. At this point in the year they are at their optimum sweetness. Bunches of grapes are handpicked by field workers and placed on paper trays, which are laid out on the ground between the vine rows. To provide a good surface for the trays, the soil between the rows is leveled.

3 Depending on the weather, the grapes are allowed to dry on the trays for two to four weeks. During this time, the moisture content of the grape is reduced from 75% to under 15% and the color of the fruit changes to a brownish purple. At night, the trays are rolled to minimize the accumulation of sand and protect against raisin moth infestation. The paper trays are embedded with a compound, which kills insects that can damage the grapes as they dry. After the fruit is dried, the paper trays are rolled up around the raisins to form a package. The rolls are gathered and stored in boxes or bins before being transported by truck to a processing plant.

Inspection and storage

4 When the rolls of fruit arrive at the manufacturing plant, they are emptied out onto wire screens and shaken to remove dirt and other unwanted debris. They are also inspected to ensure that they meet previously determined specifications. In the United States, dried raisins are inspected by the United States Department of Agriculture to ensure that all state and federal food laws are followed. Factors such as moisture content, color, and taste are all used to evaluate the shipment. Based on their quality, the raisins are graded as either standard or substandard. Only the standard graded raisins can be immediately used.

Whether or not some of the fruit will be stored for later processing or moved to the production lines, is determined by the needs of the manufacturer. If the raisins are moved for storage, they are stacked outside the plant in temporary storage enclosures. These enclosures are constructed with polyethylene sheeting fastened to wooden frames. They are made tight enough to hold the fumigation gasses, which are applied periodically to inhibit insect growth. Methyl bromide and phosphine gases are the primary fumigants used.

Processing

5 The dried grapes are moved from the storage bins to the processing plant. Here they are emptied out onto a conveyor line and mechanically modified. The residual sand and other debris are first removed by running the raisins on a fine mesh screen while air is blown on them. Immature fruit is removed by suction devices. Next, the raisins are separated from the bunch stem by shaking. The cap stems on each raisin are removed by being passed through two rotating conical surfaces. If there are seeds in the raisins, they are mechanically removed. When all these processing steps are completed, the raisins are run through a series of mesh screens to sort them according to size.

6 At this point the raisins can be put into a variety of packaging. These range in size from small half ounce cardboard containers for individual consumption to 1,100 lb (499.4 kg) containers for industrial use. Each package is run through metal detectors, in order to detect any unwanted metal particles, and then checked for the appropriate weight. They are packed onto trucks and shipped to customers. The whole process of receiving the raisins at the factory, processing them and putting them into packaging takes about 10 minutes.

Quality Control

Quality control is an important part of each step in the raisin making process. While the grapes are growing, they are checked for ripeness by squeezing the juice from a grape and using a refractometer. This allows the growers to determine how much sugar is in the grape. They are also tasted and their weight per volume is measured to give a measure of the quality of the fruit. During picking, workers are careful not to place bunches with insects or mold on the trays. They also try not to break berries as the liquid will attract insects. Knives are used to cut down the grape bunches to prevent damage. At the factory, the raisins are thoroughly inspected. They are also subjected to a variety of laboratory analyses to ensure the production of a consistent, high quality product.

The Future

Advancements in raisin production will focus on improvements in raisin yield, variety, and processing. Currently, the amount of grapes that can be produced are limited by the amount of land available. To increase yield, researchers are developing improved farming methods and new, genetically modified vine types. Experimentation is also being done on improving grape variety and characteristics through traditional grafting and biochemical means. It is expected that processing equipment will improve to reduce the amount of time required and improve the quality of the finished product.

Where to Learn More

Books

Densley, Barbara. *Food Preservation Pack : Fun With Fruit Preservation, ABC's of Home Food Dehydration, New Concepts in Dehydrated Food Cookery.* Horizon Publishing Co., 1994.

Macrae, R., et al., ed. *Encyclopedia of Food Science, Food Technology and Nutrition.* San Diego: Academic Press, 1993.

Mullins, Michael, Alain Bouquet, and Larry E. Williams. "Biology of the Grapevine." In *Biology of Horticultural Crops.* Cambridge University Press, 1992.

—Perry Romanowski

Rice Cake

As well as being a source of complex carbohydrates, rice has significant nutritional value. The bran (a layer between the grain and the husk that is removed in processing to make white rice and left in place in brown rice) contains fiber, oils, minerals, protein, and vitamins including E, K, and B-complex.

Background

Rice cakes have literally exploded in popularity as a low calorie, low fat snack. Perhaps this is no coincidence—their production is based on the explosive characteristic of rice (and similar grains like popcorn) when heat and pressure are applied. Although they are considered a "new" and perhaps "high tech" snack, rice cakes have an ancient history and are made by simple processes that are both time- and capital-intensive. So, although producers may relish the pop, snap, and crackle, they fear the breakage that can result from even the slightest imbalance in the careful relationship of moisture, ingredients, time, and temperature that is essential to the production of rice cakes.

History

The history of rice cakes is largely undocumented, but the general principle has existed for perhaps as long as rice has been harvested and relished for its nutritional benefits. Rice itself has been cultivated for more than 7,000 years. The 19 species of this member of the grass family grow best in warm, humid climates; and rice was probably cultivated first in Southern China or Northern Thailand. From Southeast Asia, its cultivation migrated all over the warmer parts of Asia, Southern Europe, and, eventually, the New World. Over half the human race now depends on rice as its main food and grain. As well as being a source of complex carbohydrates, rice has significant nutritional value. The bran (a layer between the grain and the husk that is removed in processing to make white rice and left in place in brown rice) contains fiber, oils, minerals, protein, and vitamins including E, K, and B-complex.

Rice cakes in many variations are known in a number of countries. They are perhaps best known in Japan and the countries of the Pacific Rim where rice production is an economic staple and the grain is the basis for many meals and foods. Soft forms of rice cakes have been popular in Japan for hundreds of years. The rice cake called *mochi* was a sweet confection eaten by the nobility during the Nara Period from 710-794 A.D. when the Chinese greatly influenced Japanese culture. The popularity of rice cakes blossomed during the Kamakura Shogunate from 1192-1333. The variety of cakes included *botamochi*, *yakimochi*, and *chimaki*. During the Edo Period (when Tokyo became the capital of Japan during the Tokugawa Shogunate from 1601-1868), rice cakes became even more popular as treats for festivals and as local specialties. Rice cakes were commonly sold by roadside vendors, a tradition which continues in most of Asia where street vendors sell cakes made of rice and a range of vegetables, seaweed, and seafood. The vendors cook the cakes in hot oil before the buyer's eyes.

Today, the Chinese celebrate New Year with sticky rice balls, rice candy, and the traditional rice cake made of rice steamed for hours and mixed with milk, sugar, pork fat or lard, and flavorful pastes such as date or bean paste. The finished cakes can be fried, baked, or boiled. During the glory days of the British Empire, rice cultivated in India, China, and Southeast Asia was imported to England and Europe and became the basis for many favorite foods including rice pudding. The rice cake tradition was borrowed in the form of rice-bearing pancakes, called rice girdle (British English for griddle) cakes. Puffed rice technology led

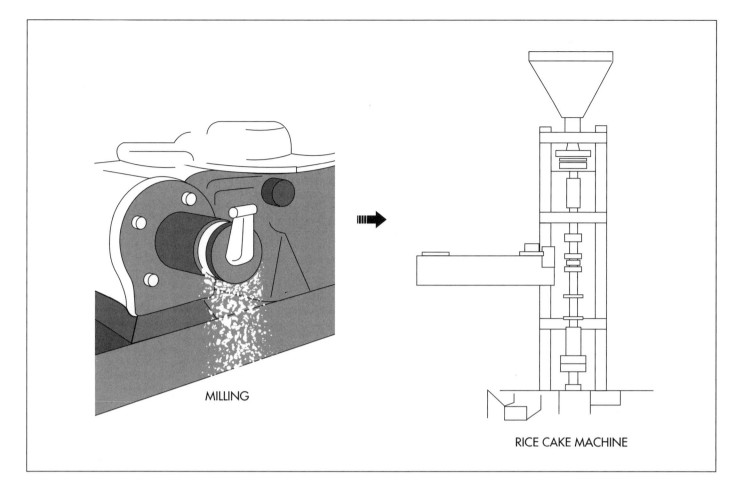

MILLING

RICE CAKE MACHINE

to the packaged rice cakes that are today's darlings of dieters and snackers.

Raw Materials

Rice cakes have only two critical ingredients—rice and water. The rice itself needs certain characteristics to produce the best quality cake and limit breakage. Sticky rice, whether white or brown, tends to work best, while long-grain varieties don't expand during cooking as vigorously. Water is important early in preparation. Other ingredients like salt (added before popping or sprayed on after) and various flavorings are important considerations to taste- and nutrition-conscious consumers but are not significant to the production process.

Product Conceptualization

Ease of production and marketability are major concerns when a new type of rice cake is considered. The popping machines are expensive investments, so the product

must be readily adaptable to the machine. Production trials have shown that additives greatly increase the likelihood of breakage, so spices, herbs, and seeds are not mixed with the rice before the cake is made although they may be added to the surface later. Similarly, salt and flavorings are now sprayed on; earlier methods of adding them to the rice were less than successful in the survival rate of whole rice cakes and in taste. Some manufacturers have also eliminated mini rice cakes from their product line. The novelty of the smaller cakes was more costly to produce than sales warranted. A constant stream of new flavor possibilities and other options are under consideration, but only careful assurance of a contented public and minimal production difficulties justifies a new product line.

The Manufacturing Process

1 The simple process of making rice cakes is based on the fact that rice subjected to the right combination of heat and pressure

After the rice is milled, it is placed in a machine that molds it into cake forms.

The rice cakes are sprayed and packaged.

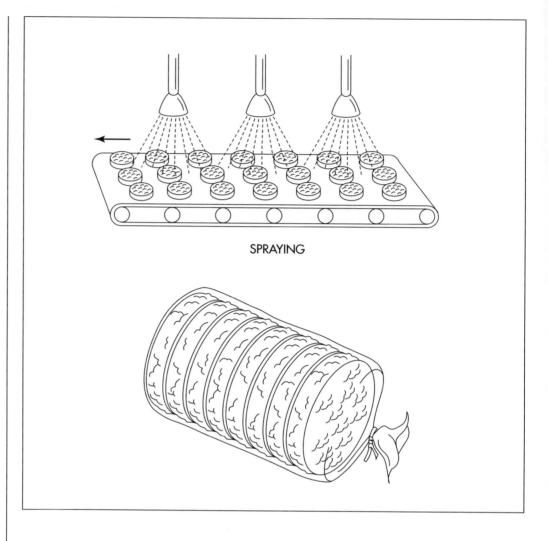

SPRAYING

will expand to fill a given space. The manufacturer's specifically preferred type of raw rice (depending on stickiness, expansion potential, and taste) is soaked in water until the right moisture level is attained.

2 The moist rice is fed into hoppers above popping machines. A major producer of rice cakes may have 80 or 90 machines with one to three cooking heads, each of which produces one cake every 15 seconds.

3 The rice is gravity-fed from the hopper into the cast-iron mold or cooking head in the popping machine. The mold is heated to hundreds of degrees, and a slide plate opens to impose a vacuum on the moist rice mass. After 8 to 10 seconds of exposure to heat at this pressure, the lid of the mold expands, creating an even greater vacuum on the contents. In the last few seconds of heating, the mixture explodes to fill the given space. If the rice forms a large pro-

portion of the exploded mass, it will be more satisfying, have a better texture, and be full of natural flavor. "Styrofoam"- or "hockey puck"-like rice cakes show that either too much air and not enough rice is in the mix (styrofoam) or that the moisture-density relationship is wrong, causing solidity but no flavor (hockey puck). Given the proper chemistry, the bran and other components of the rice bond to each other so the popped mixture sticks together without gumming additives.

4 After the cake has exploded in the popping machine, the cooking head opens and the cake falls gently on a conveyor belt. The belt carries the cake past one or more spraying heads where salt may be added or the cakes are flavor-enhanced. Natural flavors are preferred by consumers and include everything from strawberry, caramel, apple cinnamon, blueberry, and almond to salsa, nacho, taco, salt-only, or Tamari seaweed.

Some rice cake manufacturers will accept orders for private-label flavors.

5 The conveyor, now carrying flavored cakes, passes through a tunnel dryer where the moisture added by the flavor sprayers is driven off.

6 The conveyor moves to the bagging area, where the rice cakes are removed from the conveyor by hand, inspected for any breakage, and stacked, sealed in shrink-wrap, and packaged in an overwrap bag printed with the product identification and sealed. The bags are then packed in cartons for bulk sale.

Quality Control

Quality control at the rice cake plant is a labor-intensive process. Any breakage means lost revenue, and maintenance of moisture levels and popping machines are critical. Moisture throughout the process and the factory is monitored constantly. The ambient (naturally occurring) humidity may alter production; on a dry day, more moisture may have to be added to the rice. The finished cakes will absorb moisture, but this is avoided by completing the process from popping machine to bagging in a few minutes.

The popping machines themselves are cleaned every few hours. If the molds collect moisture or rice, the new cake will stick to the mold and become brittle and break. Because the cakes are individually hand-sorted prior to bagging, damaged rice cakes can be discarded before they reach a bag. Once they are ready for sale, the rice cakes have a remarkably long shelflife of over a year during which they retain taste and texture. If rice cakes have lost their crispness, they can be quickly revived at home by reheating them in a toaster. Even the freshest cakes benefit from a little heat that tends to restore their flavor.

Byproducts/Waste

There is essentially no waste in the process of making rice cakes except for breakage.

Enterprising producers have created markets for the broken cakes by selling them as cold cereal and ingredients in candy bars as well as bags of broken rice cakes for snacking. Flavors that fall out of fashion are removed from the product line and replaced by new flavors that are in development constantly. Chips, crackers, and other snack foods often provide flavor guidelines for rice cake makers.

The Future

Increasing health consciousness bodes well for the future of rice cakes. As the treat has become popular as a snack, buyers have become more discriminating in rejecting Styrofoam/hockey puck products for those with better textures and flavors. The range of designer flavors offers something for every taste, and, of course, the consumer can top the rice cake of choice with fruit, peanut butter, or other enhancements to make an even more varied snack. In an age of energy, health, and time awareness, the 15-second rice cake seems to have harvested its long history.

Where to Learn More

Periodicals

Lundberg Rice Paper. Lundberg Family Farms. Richvale, California.

Other

Foodland Industries MN Inc. http://www. foodlandmn.com.

Kiso Lab, Shinshu University. http:/markus. cs.shinshu-u.ac.jp/.

Lundberg Family Farms. http://www.lundberg.com/.

Sweet Rice Cakes. http://www.ucsc.edu/library/recipe/ricecake.html.

—*Gillian S. Holmes*

Rubber Stamp

Background

The rubber stamp has two faces, one serious and one full of fun. Practical use has been made of the rubber stamp for many years to apply official information to a range of products. On the lighter side, rubber stamps distinguished from their serious brothers as art stamps bear every kind of art work from outlines of Monet's gardens to silly sayings.

Rubber stamps are enjoying a renaissance. They are popular among serious hobbyists, collectors, and want-to-be artists who can't necessarily draw. Rubber stamps are particularly attractive for makers of memory books, scrapbooks, photo albums, and souvenir books of weddings and births because the stamps can be chosen to establish a theme. One or several stamps can be used with a selection of ink colors, special papers, and techniques such as application of embossing powder to slow-drying ink, to communicate the theme with variety and creativity.

History

Primitive stamps that existed long before rubber stamps were made are still produced by hand in other countries. Mud is used in India to make molds that are used directly as stamps. The mud images are painted with colored juice from fruit, flowers, bark, and other plant matter to create a colored image that can be stamped on fabric, paper, and products. Animal hide has also been used by some cultures. Detailed impressions can be cut in thick pieces of hide that are lon-glasting and resilient, like rubber.

There were two essential prerequisites for the origin of the rubber stamp. Rubber was discovered in the Amazon River Basin in 1736 by the French explorer Charles Marie de la Condamine. Cubes were made of the substance and used for rubbing out lead pencil marks, but the material was unstable—when the temperature rose, the cubes turned to jelly. This difficulty was solved in 1839 by Charles Goodyear. Some years before, Goodyear had begun pondering this problem with rubber, and he was determined to solve it. By accident, he spilled a mixture of gum rubber and sulfur on a hot stove. The combination of sulfur as an accelerator and heat cured the rubber. Goodyear named his process vulcanization after the Roman god of fire. Vulcanized rubber was adaptable to thousands of uses, as Goodyear recognized when he patented his process in 1844.

Marking devices similar to rubber stamps but made of other materials were available in the early 1800s. By 1860, mechanical hand stamps made of metal were in common use. The actual inventor of the rubber stamp is a subject of controversy. L.F. Witherell of Knoxville, Illinois, claimed to have invented the rubber stamp in 1866 by having fixed rubber letters on the end of a bedpost for the purpose of marking the wooden pumps he manufactured with identifications. Unfortunately, Witherell never produced the landmark bedpost or other proof. James Orton Woodruff of New York borrowed the vulcanizer used by his uncle, a dentist. Rubber was used in dentistry to mold denture bases, and the small vulcanizers dentists operated were ideal for batch production of rubber stamps. Walnut mounts for Woodruff's stamps and items that have been printed with his stamps remain; the rubber stamps themselves were destroyed by ink that contained solvents.

There are other claimants to the inventor of the rubber stamp, but Woodruff and Witherell have left the best stories, if not convincing evidence. By 1866, rubber stamp businesses were flourishing, and J.F.W. Dorman commercialized the process by manufacturing vulcanizers specifically for stamp makers. By 1892, there were 4,000 rubber stamp manufacturers and dealers in the United States.

Raw Materials

Rubber stamp manufacturers do not produce the elements that make a rubber stamp directly from raw materials. Raw materials include latex rubber, wood for the mounting blocks, adhesive-backed padding that is placed between the rubber and the block, and adhesive-backed labels. All of these items are produced by specialty manufacturers who supply them to rubber stamp makers.

Design

Designs for rubber art stamps originate at the hands of a variety of artists. Large-scale manufacturers of rubber stamps use original art and hire freelance artists to produce unique designs that are copyrighted by the artist and manufacturer. Some makers use existing artwork produced by name artists whose artwork—in rubber stamp form—is then uniquely available from one manufacturer. Some rubber stamp companies will also produce custom designs from artwork provided by the customer, and, of course, manufacturers of rubber stamps for items such as return addresses produce stamps from data supplied by the customer. Usually, this information is provided to fit a template or a specified word or letter limit so that multiple stamps can be produced at once, despite the unique contents.

Almost any kind of artwork can be produced in rubber stamp form. Another rubber stamp revolution has been developed by manufacturing large blocks with families of words (like names of flowers) or greetings (such as "Happy Birthday" repeated in a variety of lettering styles); these blocks are used by stampers to make greeting cards, wrapping paper, and other personal products with a theme or message. By giving classes and showing hobbyists clever ways of using these products, manufacturers fuel demands for their creations.

The Manufacturing Process

1 After a rubber stamp design is selected for manufacture, a drawing of the design is photographed, and multiple copies of the image are transferred to an aluminum plate. The plate is then bathed in acid that eats away the exposed metal so the design is etched or raised above the plate.

2 The engraved plate (a positive image) is placed on a sheet of Bakelite, a registered brand of resins and plastics that can be used to produce finely detailed molds, in the rubber stamp press or vulcanizer. Under heat and pressure in the vulcanizer, the engraved plate leaves a negative mold in the resin sheet. A sheet of rubber is then laid over the Bakelite mold, and both are heated in a vulcanizer to produce the final positive image in rubber.

Vulcanizing is a process in which crude or synthetic rubber is treated with heat and pressure to alter its properties of strength and chemical stability. Vulcanizers were developed as part of the rubber tire industry but are now used in many applications to treat rubber, fibers, polymers, and other materials. Operators of vulcanizers have their own techniques for producing the finest products, and the time and pressure used to vulcanize a sheet of rubber depends on the size of the mold and the operator's experience. Typically, a sheet of rubber stamps is subjected to 900 lb (408.6 kg) per square inch of pressure for about 3 minutes to complete vulcanization. The Bakelite molds are also vulcanized under about the same conditions. The mold or rubber sheet is then removed and set aside to cool.

3 Some manufacturers use sheets of photopolymer resin instead of rubber to make stamps. The resin is applied in liquid form to a negative transparency that is protected by a thin plastic film. A backing sheet is placed over the resin and the whole sandwich-like plate is placed in a photorelief printing plate machine. The machine exposes the negative and the resin to ultraviolet light, which transfers the image. The plate is then placed in a post-exposure unit that dries the resin. The sheet of resin is

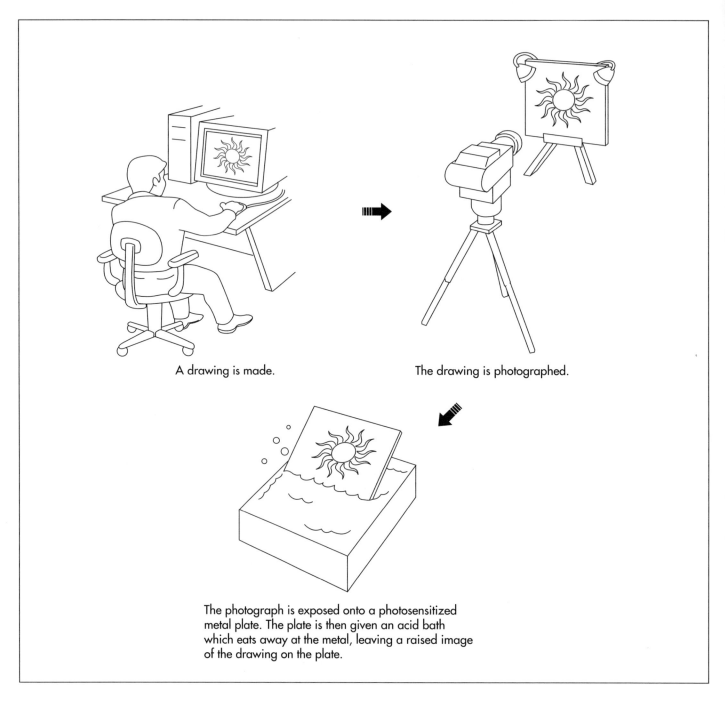

A drawing is made.

The drawing is photographed.

The photograph is exposed onto a photosensitized metal plate. The plate is then given an acid bath which eats away at the metal, leaving a raised image of the drawing on the plate.

then handled just like a sheet of rubber in the following steps to complete a stamp.

4 Most of the other steps involved in producing rubber stamps require skilled hand labor. The sheets of formed rubber are pressed onto sheets of adhering material with adhesive on both sides. The mounted rubber sheets are cut into individual stamps by workers using scissors. They must cut the stamp so that the design is not damaged but also so that enough excess material is trimmed away to limit the likelihood that they will pick up ink and spread unwanted impressions. Some designs are cut on a clicker press, which is a die press that cuts predetermined shapes by applying pressure to dies with razor-sharp edges.

5 The backside of the adhering material supporting the trimmed rubber stamp is mounted on a specially sized block of maple. The wood blocks are made in standard sizes by an offsite manufacturer. Only hard rock maple is used so that the design will not bow or be distorted by natural

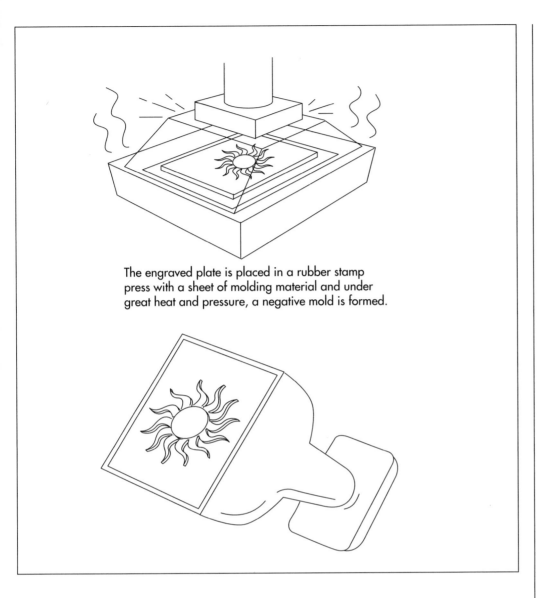

The engraved plate is placed in a rubber stamp press with a sheet of molding material and under great heat and pressure, a negative mold is formed.

aging of the wood or by changes in temperature and humidity. The blocks of wood are hand-rubbed with linseed oil, which helps protect the wood. The designs are mounted by hand, and exceptional care is required to mount the designs squarely. A clear piece of adhesive plastic is placed on the top of the maple block. The stamp design is imprinted on this label, which is called an index. The manufacturer Hero Arts developed a color index to show the design of the rubber stamp and suggest ways in which it can be colored to best convey the design. These labels are also made by outside suppliers.

6 Other steps in the manufacturing process are needed for some specialized commercial stamps. Date stamps and similar stamps with a variety of phrases are made of strips of rubber with the numbers or phrases impressed in them. The rubber-band-like strips are seamless and are placed on metal mounts that can be rotated with a dial or a key on the side of the collection of stamps. Thin sheet metal is used to manufacture the housing for the stamps and the dials or key. A wood handle is attached to the metal mount on the opposite side of the device from the printing image of the stamp. Plastic mounts are used for some commercial stamps with a single stamp for highly repetitive use or with the self-inking kind of commercial stamps that allow room for the stamp to reverse itself in the case and rest against an enclosed ink source.

7 Finished rubber stamps are sorted and stored by type. As orders are received from retailers, the stamps are selected indi-

vidually for packing and shipping. In some cases, sets of stamps are made and prepacked for sale only as sets.

Quality Control

Because so much of the rubber stamp-making process is done by hand, quality control is built into the process. Each laborer sees the work of the previous handler, so substandard stamps can be removed anywhere in the process. Quality is also ensured by using only the finest materials. Manufacturers use 100% Goodyear rubber and maple blocks that are often superior to furniture quality maple. Adhesives used for the adhering material and the indexes are chosen for long life and ability to withstand reasonable ranges of temperature and humidity.

Byproducts/Waste

The process of making rubber stamps produces very little waste. Designs are laid out on the rubber sheets to minimize waste. Although the trimmings from the rubber and adhering material have to be disposed, they amount to a small volume. If a design is discontinued, rubber stamps on hand are not thrown away. Instead, manufacturers donate them to hospitals and charitable organizations where they can be used.

The Future

The future of rubber stamps is both practical and fun. Stamps used by businesses will be required in many kinds of applications until a paperless society is truly achieved. Other practical applications like signature and return address stamps are also not likely to be replaced by computer applications in the near future. As for art stamps and the rubber stamp hobby, this interest appears to be thriving as stampers seek to create individual cards, letters, communications, and keepsakes that convey their personalities and talents.

Where to Learn More

Books

Kaprow, Allan. *Assemblages, Environment & Happenings.* New York: Harry N. Abrams, Inc., 1966.

Miller, Joni K. and Lowry Thompson. *The Rubber Stamp Album.* New York: Workman Publishing, 1978.

Rivard, Karen and Thomas H. Brinkmann. *The Marking Story.* Chicago, IL: Marking Device Association, 1968.

Sloane, T. O'Connor. *Rubber Hand Stamps and the Manipulation of Rubber.* New York: Norman W. Henley & Co., 1891.

Periodicals

Rubberstampmadness. Corvallis, Oregon (a bi-monthly publication for hobbyists).

Other

Hero Arts Rubber Stamps, Inc. http://www.heroarts.com.

Silver Fox. http://www.agate.net/~silvrfox/websites.html.

STAMPCO. http://www.stampco.com.

—*Gillian S. Holmes*

School Bus

Background

A school bus is a motor vehicle, which carries students to and from educational institutions. A vehicle is usually not considered to be a bus unless it can carry at least 10 passengers. About 85% of all school buses in the United States weigh more than 10,000 lb (4,500 kg) and carry more than 16 passengers.

Before the development of motor vehicles, horse-drawn vehicles were used for public transportation. Horse-drawn buses, which could carry 25 to 50 passengers, were used in France as early as 1828. In 1830, the British inventor Sir Goldworthy Gurney designed a bus, which was powered by steam. Despite the early invention of steam-powered vehicles, horse-drawn vehicles continued to be the most important form of public transportation throughout the nineteenth century.

In New York City in 1832, metal rails were first installed to allow horse-drawn vehicles to roll more smoothly over rough city streets. These rails were later used for steam-powered cable cars, steam locomotives, and electric trains. These vehicles, which were limited to fixed routes because they had to follow the rails, were the dominant form of urban transportation until motor vehicles became popular in the early twentieth century.

Long before this happened, however, public schools began to provide transportation systems for their students. The first act of legislation in the United States providing for pubic funds to transport students to and from schools was enacted in Massachusetts in 1869. Usually, local farmers were paid by the state government to carry students in horse-drawn wagons. Vermont passed a similar law in 1876, followed by Maine and New Hampshire. By 1900, eighteen states had such laws, and by 1919 all 48 states had them.

Two factors led to the passage of these laws, which led in turn to the increasing use of school buses. First, compulsory attendance laws required all children to go to school. Second, consolidation laws changed education in rural areas by eliminating small local schools in favor of large central schools, which could provide improved education to more students. The need to transport all children to school, combined with the fact that rural schools now served a much larger area, made school buses a vital part of public education.

During the nineteenth century, the vehicles used to transport students were known as school wagons. The earliest school wagons were simply wooden farm wagons. Later, canvas tarpaulins were used to cover the wagons in order to offer protection from the weather. Stoves were used to heat the wagons in winter.

Meanwhile, motor vehicles began to replace horse-drawn vehicles. In 1860, the Belgian mechanic Étienne Lenoir developed the internal combustion engine. In 1885, the German engineers Karl Benz and Gottlieb Daimler independently used improved internal combustion engines to produce the first automobiles. Automobiles were produced in many European countries and in the United States by the 1890s.

In Germany, an internal combustion engine was used to power a bus, which carried

By the late 1980s, more than 22 million students in the United States were transported on school buses each school day. The United States public school bus system is now the largest public transportation system in the world.

413

eight passengers in 1895. In the same country in 1896, Gottlieb Daimler built the first motor truck. Trucks and buses were built using the same type of chassis (the lower part of a motor vehicle, which is connected to the engine and which propels the vehicle) until the 1920s. In 1921, the Fageol Safety Coach Company of Oakland, California, introduced a special bus chassis, which was wider, longer, and lower than a truck chassis. Although this design soon became the standard for other kinds of buses, school buses are still manufactured using a chassis similar to a truck chassis.

In 1913, American automobile manufacturer Henry Ford revolutionized the industry by introducing the assembly line. Instead of being built one at a time from start to finish, automobiles could now be assembled from standardized parts as the chassis was pulled along by moving assembly belts. This method allowed automobiles to be made in larger numbers, more quickly, and less expensively. Motor vehicles were transformed from luxuries for the rich to affordable transportation for the middle class. By the end of the 1920s, motor vehicles, including school buses, had almost completely replaced horses as a method of transportation.

During the early twentieth century, school buses often consisted of a wooden body attached to a steel chassis. During the late 1920s, steel bodies replaced wooden bodies, resulting in vehicles similar to modern school buses.

School buses became even more important to public education during the second half of the twentieth century, as school consolidation continued. Between the end of World War II and the early 1970s, the number of school districts in the United States decreased from more than 100,000 to about 17,000. By the late 1980s, more than 22 million students in the United States were transported on school buses each school day. The United States public school bus system is now the largest public transportation system in the world.

Raw Materials

The most important raw material used to manufacture school buses is steel, which is an alloy of iron and a small amount of carbon. Steel is used to make the chassis and the body, along with various other components. Steel is made from iron ore, coke (a carbon-rich substance produced by burning coal in the absence of air), and limestone. The coke provides the carbon, which transforms the iron into steel, and the limestone reacts with impurities in the ore to remove them in the form of slag. Oxygen is then blasted into the molten mixture to remove excess carbon and other impurities.

The windows of a school bus are made of laminated glass. Laminated glass consists of two layers of glass surrounding a layer of plastic. The plastic holds the glass in place if the window is broken, adding to its safety.

The tires of a school bus are made from a mixture of natural or synthetic rubber, carbon black, sulfur, and other chemicals, which determine the characteristics of the tires. Natural rubber is obtained from latex, a liquid produced when the bark of a rubber tree is cut. Synthetic rubber is produced from chemicals obtained from petroleum. Carbon black is made by burning petroleum or natural gas in a limited supply of air, resulting in a large amount of fine soot.

Other raw materials used in school bus manufacturing include various metals and plastics. These are used to make the many small parts, which are assembled together with the chassis and the body to make up the completed vehicle.

The Manufacturing Process

Making premanufactured components

1 A school bus is made of hundreds of different components. Many of these components are premanufactured by companies other than the school bus manufacturer. Some components are premanufactured by companies that are owned by the school bus manufacturer.

2 Windows of laminated glass are made by melting together silicon dioxide and various other oxides to produce molten glass. The hot liquid glass is then floated on a pool of molten tin. The flat surface of the

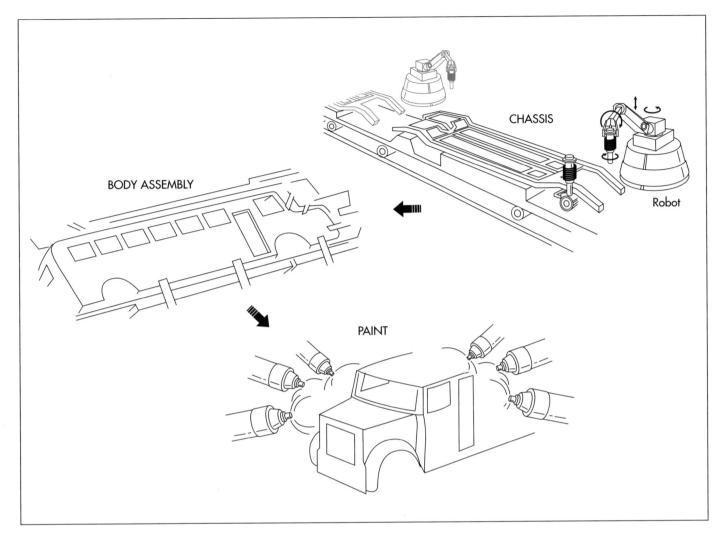

BODY ASSEMBLY

CHASSIS

Robot

PAINT

The chassis is assembled and the body formed. After the body has been assembled, it is painted.

liquid tin causes the molten glass to be transformed into a solid with a flat, smooth surface as it cools. The glass is then quickly heated and cooled to strengthen it, a process known as tempering. Two layers of the tempered glass are bonded to a layer of clear, hard plastic under heat and pressure to form laminated glass.

3 Tires are made by mixing rubber, carbon black, sulfur, and other chemicals together and heating the mixture to form a single compound. Sheets of this rubber compound are wrapped around a rotating drum and glued together to form a tire without treads. This preliminary tire, known as a green tire, is made up of many layers of the rubber compound of many different shapes.

The green tire is then placed in mold, which contains treads on its inner surface. An inflatable bladder is placed inside the tire. The mold is closed and the bladder is filled with steam. The heat and pressure of the steam causes the green tire to take on the shape of the tread pattern inside the mold. The bladder is deflated, the mold is opened, and the treaded tire is allowed to cool.

4 Small metal components are made by using a variety of precision metalworking machines such as drills and lathes. Some metal components, such as those made of aluminum, may be made by melting the metal, pouring it into a mold in the shape of the desired component, and allowing it to cool.

5 Plastic components may be made by injection molding. This process involves melting the plastic into a liquid and forcing it into a mold under pressure, where it cools into the desired component.

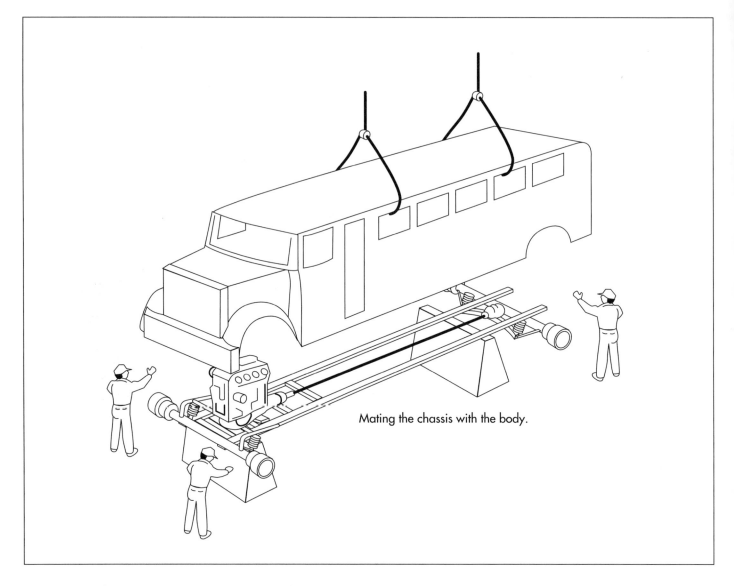

Mating the chassis with the body.

Once assembly of all parts is complete, the chassis and body are put together.

Making the chassis

6 Steel arrives at the school bus factory in the form of sheet metal of the desired thickness. Various cutting and stamping tools are used to produce pieces of steel of the proper shape and size. These various pieces are bolted together as the chassis moves along an assembly line.

7 The frame (the base of the chassis) is bolted together from pieces of steel as assembly begins. As the frame proceeds along the assembly line, the suspension system is attached. Next, the brake and exhaust systems are attached. The engine is then installed, followed by the drive shaft and the wheels, including tires.

The motorized part of the school bus is now completed. A temporary driver's seat can be attached at this point to allow the chassis to undergo a preliminary driving test.

Making the body

8 Like the chassis, the body of a school bus is made of components, which have been molded from sheets of steel of the proper thickness. The various pieces of steel are bolted together or welded together as the body proceeds along an assembly line similar to the chassis assembly line.

9 Steel panels are assembled together to form the bottom, sides, and top of the body. The doors are then joined to the body.

10 The body is cleaned with soap and water, then treated with phosphate to protect it from rust. A coat of primer is sprayed on the body and baked dry in a large oven. Next, a coat of paint is sprayed on and baked dry in a similar manner.

11 Windows are installed in the body. Interior components such as the instrument panel and the seats are then installed. External components such as the door handles and lights are also installed at this point.

Assembling the school bus

12 The body is lifted by a large crane and placed on top of the chassis. The two parts are then bolted together to produce the school bus. Final adjustments such as connecting the electrical wiring are made. The school bus is inspected and shipped to the consumer on special trucks designed to carry large motor vehicles.

Quality Control

The school bus manufacturer inspects all premanufactured parts to ensure that they are free from defects. The steel sheet metal is also inspected and then kept covered during storage to protect it from corrosion. After pieces of steel are cut from the sheet metal, they are inspected to be sure that they are the proper shape and size.

When the chassis is complete it is driven briefly to ensure that the motorized components operate correctly. After the body is attached, the school bus is given a full road test to detect any flaws in operation.

The school bus is sprayed with water to detect any leaks. The entire vehicle is given a detailed final inspection. All the items on a long, written list must be individually inspected and approved before the school bus is ready to be shipped.

Safety is the major quality control concern for school bus manufacturers. The United States government has issued regulations dealing with such items as brakes, emergency exits, floor strength, seating systems, windows, mirrors, fuel systems, and the crashworthiness of the body and chassis. As a result of these regulations, studies have shown that school buses are significantly safer than other forms of transportation used by school-age children.

The Future

The United States government is constantly updating safety standards for school buses. One controversial issue is the possibility of requiring seat belts on school buses. One study done in 1989 predicted that installing seat belts would cost 40 million dollars per year and save one life per year.

Other possible trends include using alternate forms of energy such as natural gas or electricity ot power school buses. School buses are likely to be more comfortable as more of them are equipped with air conditioning. Safety could be improved by replacing the traditional instrument panel with an electronic display panel, which the driver could view without looking down at the dashboard.

Where to Learn More

Periodicals

Mills, Nicolaus. "Busing: Who's Being Taken For a Ride?" *Commonweal* (March 24, 1972): 55-60.

Other

National Highway Traffic Safety Administration. "School Bus Safety Report," May 1993.

—*Rose Secrest*

Shellac

Lac is the name given to the resinous secretion of the tiny lac insect (Laccifer lacca) which is parasitic on certain trees in Asia, particularly India and Thailand. This insect secretion is cultivated and refined because of the commercial value of the finished product known as shellac.

Background

Lac is the name given to the resinous secretion of the tiny lac insect (*Laccifer lacca*) which is parasitic on certain trees in Asia, particularly India and Thailand. This insect secretion is cultivated and refined because of the commercial value of the finished product known as shellac. The term shellac is derived from shell-lac (the word for the refined lac in flake form), but has come to refer to all refined lac whether in dry or suspended in an alcohol-based solvent.

Shellac is primarily used as a wood sealer and finisher today. It has the great advantage of being soluble in ethyl or denatured alcohol, an environmentally-safe solvent. Alcohol solvents also render shellac a quick dry—shellac coatings on wood generally dry in about 45 minutes, as opposed to oil finishes which take many hours to dry. In addition, shellac does not fade in sunlight or oxidize over time. However, shellac has a limited shelf life and may not dry properly if it has exceeded the shelf life recommended by the manufacturer. This shelf life may be as short as six months or as long as three years depending on the manufacturer's additives.

Industrial uses for shellac include floor polishes, inks, grinding wheels, electrical insulations, and leather dressings. This natural, resinous sealer is non-toxic and is Federal Drug Administration (FDA) approved for use to coat candies, pharmaceuticals, fruit, and baby and children's furniture.

Shellac is available at most hardware or paint stores in clear or white shellac or orange shellac, which imparts an orange-red tint to natural wood. Other tints derive their color not from dyes or bleaches, but because of the tree to which the lac bug has attached itself—the sap affects the color of the bug secretions thus altering the color of the refined shellac. Shellac may be applied to wood, over varnish, paint, glass, ceramics, even plastic with remarkable adherence, but it cannot be used under synthetic sealers such as polyurethane.

History

Lac has been cultivated for three centuries. For most of that time, the lac bug secretions were valued for the purple-red dye derived from being soaked in water. This dye was used to color silk, leather, and cosmetics and was cultivated primarily for this purpose until the 1870s. Then aniline or chemical dyes began to supplant these and other natural dyes.

As early as the sixteenth century, references were made to the usefulness of the lac bug secretions as a decorative lacquer for furniture and fine musical instruments. Natives of the Far East had laboriously cultivated and processed the shellac by hand, scraping the branches encrusted with the lac bug secretions, forcing the secretions into muslin, and holding long muslin bags of the secretions over the fire to liquefy and purify it. They pulled it by hand into huge sheets and then broke the sheets into flakes for re-moisturizing later.

Hand processes were partially replaced by the mid-nineteenth century. Just as the lac-derived dye was about to fade in popularity, industrial plants began processing the lac secretions for use as a wood sealer and finish. In 1849, William Zinsser founded Wm. Zinsser & Company in New York. Zinsser's shellacs were soluble in ethyl alcohol and were the first quick-drying, tough, col-

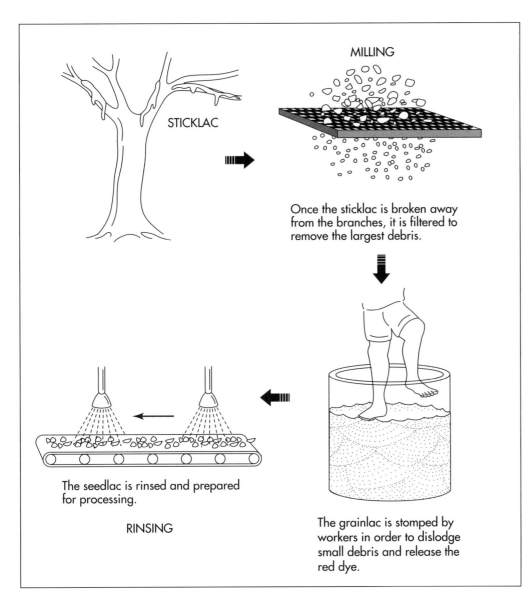

MILLING

Once the sticklac is broken away from the branches, it is filtered to remove the largest debris.

STICKLAC

The grainlac is stomped by workers in order to dislodge small debris and release the red dye.

The seedlac is rinsed and prepared for processing.

RINSING

orless finishes available in the United States. Shellac was particularly popular late in the nineteenth century and in the early twentieth century when houses were being quickly built in early subdivisions at breakneck speed—shellac was an ideal wood finisher because it was so fast to dry and several coats could be applied in a single day. A shellac known as buttonlac, a very dark shellac, imparted a very deep walnut color to inexpensive woodwork that people then found very desirable.

Raw Materials

Shellac is generally made from two ingredients, raw seed lac and ethyl alcohol. In fact, most companies want to purify shellac as completely as possible—impurities from the bug, the cocoon etc. are removed, as are natural waxes. Shellac is generally shipped in dry or flaked form and is re-moisturized with an alcohol solvent, generally denatured alcohol. Some companies add ingredients to lengthen the shelf life of their product but will not reveal these proprietary additives. Shellac that is bleached (or made into clear shellac) are dissolved in sodium carbonate and centrifuge to remove insolubles and then bleached with sodium hypochlorite.

The Manufacturing Process

The role of the lac bug

1 Shellac is produced by a tiny red insect. Swarms of the insects feed on certain

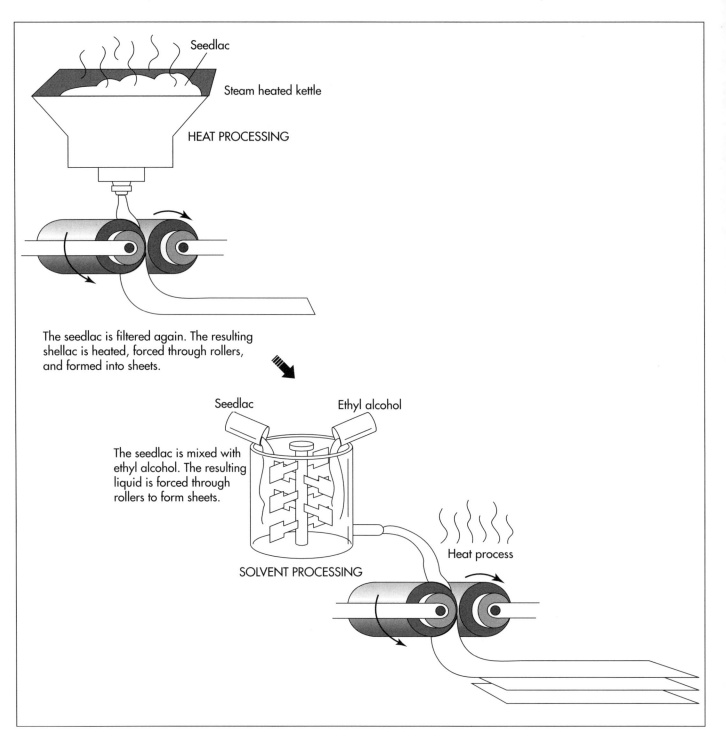

Seedlac

Steam heated kettle

HEAT PROCESSING

The seedlac is filtered again. The resulting shellac is heated, forced through rollers, and formed into sheets.

Seedlac

Ethyl alcohol

The seedlac is mixed with ethyl alcohol. The resulting liquid is forced through rollers to form sheets.

SOLVENT PROCESSING

Heat process

trees, primarily in India and Thailand, known informally as lac trees. The lac bugs' life cycle is only six months, in which time they eat, propagate, and secrete the resin they've taken in from the tree to produce shellac.

In certain seasons of the year, these insects swarm in huge numbers on the trees, settle on branches, and project protrusions into the

tree to penetrate the bark. They suck up the sap and absorb it until they feed themselves to death (called the feast of death amongst the indigenous peoples). At this same time, propagation continues, with each female lac bug laying about 1,000 eggs before dying.

The sap is chemically altered in the lac bug's body and is then exuded onto the tree branch. On contact with the air, the excretion forms a hard shell-like covering over the entire swarm. This covering forms a crust over the twig and insects. As the female lac bug is exuding the ingested sap she is preparing to die and is providing a fluid in which her eggs will mature under protection. The males' role is to fertilize the female, and it is after fertilization that the females' lac output is vastly increased. The adult males and females become inactive, and the young start to break through the crust and swarm out.

Refining the crusty resin

2 Workers cut millions of encrusted branches, called sticklac, for transportation to refineries of some sort (either hand-refined or mechanically refined). Some workers use mallets and break off the crusty coating much as ice is broken from branches in the winter (it is referred to as grainlac).

3 At refining centers, sticklac is scraped to remove the secretions from the twigs. Sticklac and grainlac is ground with rotating millstones. The resulting ground material is quite impure, containing resin, insect remains, twigs, leaves, etc. The mixture is forced through a screen, removing the largest of the impurities.

4 The sifted resin mixture is put into large jars and stomped by a worker to crush granules and force the red dye from the lac seeds and the insect remains will be freed from the resin. Dye water, scum, and other impurities are then washed away in several rinsings. The mixture is spread out on a concrete floor to dry and called seedlac because it resembles seed. Seedlac is the raw material from which both orange shellac and bleached or clear shellac are produced.

Shellac may be made from seedlac by hand or by modern mechanical equipment. Nearly all American-used shellac is refined with the help of machinery, using a heat- or solvent-based process.

Heat process

5 Seedlac is melted onto steam-heated grids. The molten lac is forced by hydraulic pressure through a sieve or screen, either of cloth or fine mesh. The filtered shellac is collected and transferred to a steam-heated kettle, which then drops the molten liquid onto rollers. The liquid is squeezed through the rollers and forced into large, thin sheets of shellac. When dry, this shellac sheet is broken into flakes and transported to another area in which the flakes are combined with denatured alcohol to produce the consumer's shellac.

Solvent process

6 In this process, the seedlac and solvent, usually ethyl alcohol, are mixed in a dissolving tank, refluxed for about an hour and then filtered to remove impurities. The filtered resin is sent through evaporators that remove the alcohol solvent, rendering it a viscous liquid. This liquid is then dropped onto rollers, which force it into sheets. The sheets are then are dried and flaked apart.

Bleached shellac

Despite the removal of much of the red dye from the lac seeds in the refining process, shellac remains an orangish solution after processing is complete. Some consumers prefer a clear shellac finish, so manufacturers have developed a way to bleach the color from the shellac.

7 Bleaching begins with dissolving seedlac, which is alkali-soluble, in an aqueous solution of sodium carbonate. The solution is then passed through a fine screen to remove insoluble lac, dirt, twigs, etc. The resin is then bleached with a dilute solution of sodium hypochlorite to the desired color. The shellac is then precipitated from the solution by the addition of dilute sulfuric acid, filtered, and washed with water. It is dried in vacuum driers and ground into a white powder ready for shipment to a plant that will add liquid to the flakes.

Mixing shellac for the consumer

8 Large shellac manufacturers are shipped the dry shellac flakes. They then remoisturize the flakes by adding denatured ethyl alcohol. Shellac is offered to the consumer in flake form or suspended in denatured alcohol. It is the latter than is most popular with the consumer. Manufacturers of shellac refer to the concentration of shellac flakes to denatured alcohol in terms of pounds of cuts—the number of pounds of

shellac flakes dissolved into a single gallon of denatured alcohol. Thus, a one pound cut of shellac contains one pound of shellac flakes dissolved in a gallon of alcohol—very dilute shellac. The manufacturers' standard cut offered to the consumer pre-mixed is termed a three pound cut. Some consumers then dilute it further with denatured alcohol if they so desire.

The most popular shade of shellac sold pre-mixed is the orange shellac although clear or white shellac is also offered pre-mixed to the consumer. Manufacturers always stamp the date of mixing of the shellac into the can. Each manufacturer has a recommended shelf life for the product and the consumer should heed that the product is not used after the period suggested by the manufacturer. If used after the time span recommended, the shellac may never dry completely.

For woodworkers who prefer the deep rich colors of garnet shellac or buttonlac, the dried flakes of these shellacs may be purchased from the manufacturer and mixed with denatured alcohol by the consumer.

Byproducts/Waste

The denatured ethyl alcohol used in the process of manufacturing shellac is a strictly regulated byproduct and is known as a volatile organic substance (VOC). The most dangerous or hazardous part, perhaps the most polluting, are the insolubles that are refined out of the sticklac and grainlac

such as twigs, cocoons, leaves, bug bodies, etc. saturated with alcohol. The shellac industry is working on building huge evaporators, which will suck all the alcohol out of these insolubles so the volatility will not be an issue. Shellac flakes are all natural and non-toxic. It is the alcohol solvents that are regulated.

Quality Control

Chemical analysis does not assist in determining the quality of shellac. More important are empirical tests such as flow and shelf life that most customers have articulated as of great concern. In addition, carefully examining the purity of the shellac by removing as many of the natural impurities found within the sticklac is of utmost importance (insolubles are defined by the undissolved matter remaining when the resinous compound is mixed with hot alcohol). All refining processes are monitored for their effectiveness in removing these undesirables.

Where to Learn More

Books

Russel, M. *Shellac.* London: Ann Eccles and Son Ltd for Angelo Shellac, 1965.

The Story of Shellac. Somerset, NJ: Wm. Zinsser & Co., 1989.

—*Nancy EV Bryk*

Spandex

Spandex is a lightweight, synthetic fiber that is used to make stretchable clothing such as sportswear. It is made up of a long chain polymer called polyurethane, which is produced by reacting a polyester with a diisocyanate. The polymer is converted into a fiber using a dry spinning technique. First produced in the early 1950s, spandex was initially developed as a replacement for rubber. Although the market for spandex remains relatively small compared to other fibers such as cotton or nylon, new applications for spandex are continually being discovered.

Background

Spandex is a synthetic polymer. Chemically, it is made up of a long-chain polyglycol combined with a short diisocyanate, and contains at least 85% polyurethane. It is an elastomer, which means it can be stretched to a certain degree and it recoils when released. These fibers are superior to rubber because they are stronger, lighter, and more versatile. In fact, spandex fibers can be stretched to almost 500% of their length.

This unique elastic property of the spandex fibers is a direct result of the material's chemical composition. The fibers are made up of numerous polymer strands. These strands are composed of two types of segments: long, amorphous segments and short, rigid segments. In their natural state, the amorphous segments have a random molecular structure. They intermingle and make the fibers soft. Some of the rigid portions of the polymers bond with each other and give the fiber structure. When a force is applied to stretch the fibers, the bonds between the rigid sections are broken, and the amorphous segments straighten out. This makes

the amorphous segments longer, thereby increasing the length of the fiber. When the fiber is stretched to its maximum length, the rigid segments again bond with each other. The amorphous segments remain in an elongated state. This makes the fiber stiffer and stronger. After the force is removed, the amorphous segments recoil and the fiber returns to its relaxed state. By using the elastic properties of spandex fibers, scientists can create fabrics that have desirable stretching and strength characteristics.

The primary use for spandex fibers is in fabric. They are useful for a number of reasons. First, they can be stretched repeatedly, and will return almost exactly back to original size and shape. Second, they are lightweight, soft, and smooth. Additionally, they are easily dyed. They are also resilient since they are resistant to abrasion and the deleterious effects of body oils, perspiration, and detergents. They are compatible with other materials, and can be spun with other types of fibers to produce unique fabrics, which have characteristics of both fibers.

Spandex is used in a variety of different clothing types. Since it is lightweight and does not restrict movement, it is most often used in athletic wear. This includes such garments as swimsuits, bicycle pants, and exercise wear. The form-fitting properties of spandex makes it a good for use in undergarments. Hence, it is used in waist bands, support hose, bras, and briefs.

History

The development of spandex was started during World War II. At this time, chemists took on the challenge of developing synthetic

Spandex fibers can be stretched to almost 500% of their length.

423

Corset designed by Jacob Kindliman of New York City in 1890. (From the collections of Henry Ford Museum & Greenfield Village, Dearborn, Michigan.)

This corset-clad torso was produced by Jacob Kindliman of New York City in 1890. Kindliman, a corsetiere, hardly needed to advertise. At that time, women thought it was necessary to wear a corset and considered themselves indecently dressed without it until early in the twentieth century. Corsets were a combination brassiere-girdle-waist cincher in an all-in-one garment, forming the foundation shape for fashionable dress.

In days before spandex, how did the corset contour the body effectively? In the eighteenth century, thick quilting and stout seams on the corset shaped the body when the garment was tightly laced. In the early nineteenth century, baleen, a bony but bendable substance from the mouth of the baleen whale, was sewn into seams of the corset (hence the term whalebone corsets), however the late 1800s corsets like this were stiffened with small, thin strips of steel covered with fabric. Such steel-clad corsets did not permit movement or comfort. By World War I, American women began separating parts of the corset into two garments—the girdle (waist and hip shaper) and bandeau (softer band used to support and shape the breasts).

Nancy EV Bryk

replacements for rubber. Two primary motivating factors prompted their research. First, the war effort required most of the available rubber for building equipment. Second, the price of rubber was unstable and it fluctuated frequently. Developing an alternative to rubber could solve both of these problems.

At first, their goal was to develop a durable elastic strand based on synthetic polymers. In 1940, the first polyurethane elastomers were produced. These polymers produced millable gums, which were an adequate alternative to rubber. Around the same time, scientists at Du Pont produced the first nylon polymers. These early nylon polymers were stiff and rigid, so efforts were begun to make them more elastic. When scientists found that other polyurethanes could be made into fine threads, they decided that these materials might be useful in making more stretchable nylons or in making lightweight garments.

The first spandex fibers were produced on an experimental level by one of the early pioneers in polymer chemistry, Farbenfabriken Bayer. He earned a German patent for his synthesis in 1952. The final development of the fibers were worked out independently by scientists at Du Pont and the U.S. Rubber Company. Du Pont used the brand name Lycra and began full scale manufacture in 1962. They are currently the world leader in the production of spandex fibers.

Raw Materials

A variety of raw materials are used to produce stretchable spandex fibers. This includes prepolymers which produce the backbone of the fiber, stabilizers which protect the integrity of the polymer, and colorants.

Two types of prepolymers are reacted to produce the spandex fiber polymer backbone. One is a flexible macroglycol while the other is a stiff diisocyanate. The macroglycol can be a polyester, polyether, polycarbonate, polycaprolactone or some combination of these. These are long chain polymers, which have hydroxyl groups (-OH) on both ends. The important feature of these molecules is that they are long and flexible. This part of the spandex fiber is responsible for its stretching characteristic. The other prepolymer used to produce spandex is a polymeric diisocyanate. This is a shorter chain polymer, which has an isocyanate (-NCO) group on both ends. The principal characteristic of this molecule is its rigidity. In the fiber, this molecule provides strength.

When the two types of prepolymers are mixed together, they interact to form the spandex fibers. In this reaction, the hy-

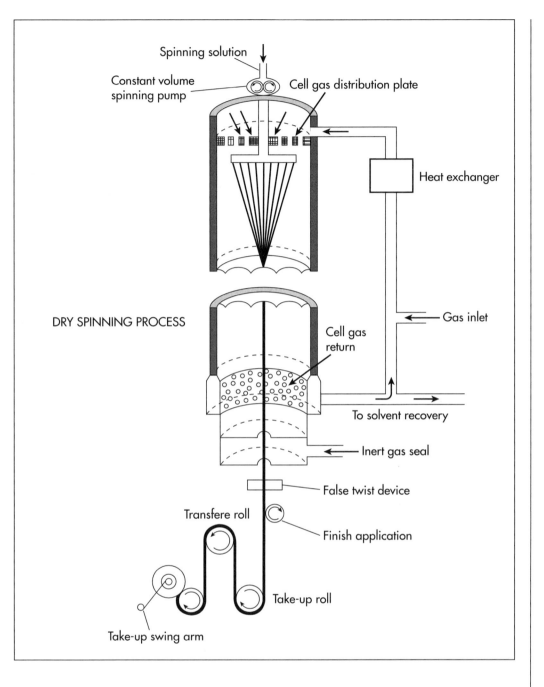

DRY SPINNING PROCESS

droxyl groups (-OH) on the macroglycols react with the isocyanates. Each molecule gets added on to the end of another molecule, and a long chain polymer is formed. This is known as a step-growth or addition polymerization. To initiate this reaction, a catalyst such as diazobicyclo[2.2.2]octane must be used. Other low molecular weight amines are added to control the molecular weight of the fibers.

Spandex fibers are vulnerable to damage from a variety of sources including heat, light atmospheric contaminants, and chlo-rine. For this reason, stabilizers are added to protect the fibers. Antioxidants are one type of stabilizer. Various antioxidants are added to the fibers, including monomeric and polymeric hindered phenols. To protect against light degradation, ultraviolet (UV) screeners such as hydroxybenzotriazoles are added. Compounds which inhibit fiber dis-coloration caused by atmospheric pollutants are another type of stabilizer added. These are typically compounds with tertiary amine functionality, which can interact with the oxides of nitrogen in air pollution. Since spandex is often used for swimwear, antim-

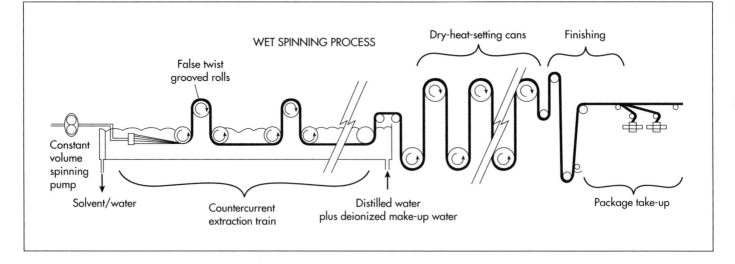

WET SPINNING PROCESS

False twist grooved rolls

Dry-heat-setting cans

Finishing

Constant volume spinning pump

Solvent/water

Countercurrent extraction train

Distilled water plus deionized make-up water

Package take-up

Wet-spinning process.

ildew additives must also be added. All of the stabilizers that are added to the spandex fibers are designed to be resistant to solvent exposure since this could have a damaging effect on the fiber.

When they are first produced, spandex fibers are white. Therefore, colorants are added to improve their aesthetic appearance. Dispersed and acid dyes are typically used. If the spandex fibers are interwoven with other fibers such as nylon or polyester, special dying methods are required.

The Manufacturing Process

Spandex fibers are produced in four different ways including melt extrusion, reaction spinning, solution dry spinning, and solution wet spinning. Each of these methods involve the initial step of reacting monomers to produce a prepolymer. Then the prepolymer is reacted further, in various ways, and drawn out to produce a long fiber. Since solution dry spinning is used to produce over 90% of the world's spandex fibers, it is described.

Polymer reactions

1 The first step in the production of spandex is the production of the prepolymer. This is done by mixing a macroglycol with a diisocyanate monomer. The compounds are mixed in a reaction vessel and under the right conditions they react to form a prepolymer. Since the ratio of the component materials produces fibers with varying char-

acteristics, it is strictly controlled. A typical ratio of glycol to diisocyanate may be 1:2.

2 In dry spinning fiber production, the prepolymer is further reacted with an equal amount of diamine. This is known as a chain extension reaction. The resulting solution is diluted with a solvent to produce the spinning solution. The solvent helps make the solution thinner and more easily handled. It can then be pumped into the fiber production cell.

Producing the fibers

3 The spinning solution is pumped into a cylindrical spinning cell where it is cured and converted into fibers. In this cell, the polymer solution is forced through a metal plate, called a spinneret, which has small holes throughout. This causes the solution to be aligned in strands of liquid polymer. As the strands pass through the cell, they are heated in the presence of a nitrogen and solvent gas. These conditions cause the liquid polymer to chemically react and form solid strands.

4 As the fibers exit the cell, a specific amount of the solid strands are bundled together to produce the desired thickness. This is done with a compressed air device that twists the fibers together. In reality, each fiber of spandex is made up of many smaller individual fibers that adhere to one another due to the natural stickiness of their surface.

Final processing

5 The fibers are then treated with a finishing agent. This may be magnesium stearate or another polymer such as poly(dimethylsiloxane). These finishing materials prevent the fibers from sticking together and aid in textile manufacture. After this treatment, the fibers are transferred through a series of rollers onto a spool. The windup speed of the entire process can be anywhere from 300-500 mi (482.7-804.5 km) per minute depending on the thickness of the fibers.

6 When the spools are filled with fiber, they are put into final packaging and shipped to textile manufacturers and other customers. Here, the fibers may be woven with other fibers such as cotton or nylon to produce the fabric that is used in clothing manufacture. This fabric can also be dyed to produce a desired color.

Quality Control

To ensure the quality of the spandex fibers, manufacturers monitor the product during each phase of production. Inspections begin with the evaluation of the incoming raw materials. Various chemical and physical characteristics are tested. For example, the pH, specific gravity, and viscosity of the diisocyanate may be checked. Additionally, appearance, color, and odor can also be evaluated. Only by having strict quality control checks on the starting materials can the manufacturer be sure that they will produce a consistent end product. After production, the spandex fibers are also tested. These tests may include those that evaluate fiber elasticity, resilience and absorbency.

The Future

The quality of spandex fibers has continually improved since they were first developed. Various areas of research will help continue their improvement. For example, scientists have found that by changing the starting prepolymers they can develop fibers which have even better stretching characteristics. Other characteristics can be improved by using different prepolymer ratios, better catalysts, and various fillers. In addition to spandex fiber improvements, it is likely that advanced fabrics will be produced which incorporate spandex fibers with conventional fibers. Currently, nylon/spandex fiber blends are available. Finally, improvements in manufacturing will also be discovered. These will focus on producing fibers faster and more efficiently.

Where to Learn More

Books

Jerde, Judith. *Encyclopedia of Textiles.* Facts on File, 1992.

Lewin, M. and J. Preston, ed. *High Technology Fibers.* New York: Marcel Dekker, 1985.

Other

Devra, A. U.S. Patent 5,303,882, 1994.

Goodrich, C & W. Evans. U.S. Patent 5,028,642, 1991.

—*Perry Romanowski*

Synthetic Ruby

Mined for 8,000 years or more, natural rubies are found in only a handful of sites around the world, most notably in Myanmar (formerly Burma), Thailand, Sri Lanka, Afghanistan, Tanganyika, and North Carolina.

Background

Diamonds, rubies, sapphires, and emeralds are known as precious gems. Next to the diamond, the ruby is the hardest gemstone; it is also resistant to acids and other harmful substances. Because large, gem-quality rubies are very rare, the value of a fine ruby may be quadruple that of a similar-quality diamond.

Rubies and sapphires are both composed of corundum, which is the crystalline form of aluminum oxide. They differ only in small amounts of color-producing minerals. Chromium gives rubies their characteristic red color, with higher concentrations producing darker shades. Aluminum oxide crystals not containing chromium are called sapphires; they come in many hues including blue, yellow, green, pink, purple, and colorless.

Natural rubies are found in a handful of sites around the world, most notably in Myanmar (formerly Burma), Thailand, Sri Lanka, Afghanistan, Tanganyika, and North Carolina. Beautifully colored, transparent crystals are prized for jewelry use, while translucent or opaque stones are used for ornamental items such as clock bases.

In addition to their decorative functions, rubies serve a broad range of utilitarian purposes. For example, because of their hardness, they make long-lasting thread guides for textile machines. Ruby is even harder than steel, so it is an excellent bearing material for metal shafts in devices like watches, compasses, and electric meters. Rubies have exceptional wave-transmitting properties for the range from short, ultraviolet wavelengths through the visible light spectrum to long, infrared wavelengths. This makes them ideal for use in lasers and masers (laser-like devices operating in non-visible ranges of microwaves and radio waves).

Because many of these industrial uses demand very high-quality crystals of particular sizes and shapes, synthetic rubies are manufactured. With the exception of minor amounts of impurities, synthetic gems have the same chemical, physical, and optical properties as their natural counterparts. Although some are used as gemstones, about 75% of modern synthetic ruby production is used for industrial purposes.

History

Natural rubies have been mined for 8,000 years or more. In many cultures, the gems have been prized not only for their beauty but also for supernatural powers; it was commonly believed that the ruby's red color came from fire trapped inside the stone. Ancient Hindus believed that rubies could make water boil, and early Greeks thought the crystals could melt wax. In other cultures (e.g., Burmese and Native American), the ruby was thought to protect a wearer because of its blood-like color.

Because it was so highly prized, the ruby was the first gemstone to be made artificially. Documented attempts to make rubies date to the experiments of Marc A. Gaudin, a French chemist who produced some synthetic rubies beginning in 1837. They were not of any value as gems, however, because they became opaque as they cooled. After 30 years of experimenting he gave up, admitting defeat in the published notes of his final ruby experiments.

Around 1885, some rubies sold as gemstones were discovered to be manmade (their unusually low price prompted the buyer to have them carefully examined). The method by which these so-called Geneva rubies were made remained a mystery until about 1970, when an analysis of surviving samples showed that they were formed by melting powdered aluminum oxide and a smaller amount of chromium oxide in an array of torches, and letting the molten material solidify.

Actually, the Geneva rubies may have come from an early developmental stage of what is now known as the "flame fusion" method. In 1877, the French chemist Edmond Frémy and a student assistant described how they heated 44.1-66.15 lb (20-30 kg) of a solution of aluminum oxide dissolved in lead oxide in a porcelain vat for 20 days. As the solvent evaporated and chemical reactions took place among the solution, the vessel, and furnace gases, a large number of very small ruby crystals formed on the basin's wall. The rubies were so small and the production costs so high that the crystals could not realistically be used in jewelry.

Later, Auguste Verneuil, another of Frémy's students, developed a somewhat different process that eventually became successful. By 1891 he was producing rubies by flame fusion, although he did not publish a description of his technique until 1902. His assistant exhibited the synthetic rubies in 1900 at the Paris World's Fair, where they were quite popular. His process took only two hours to grow crystals weighing 12-15 carats (2.5-3 g); the stones were roughly spherical, up to 0.25 in (6 mm) in diameter. By the time Verneuil died at the age of 57 in 1913, the process he had invented was being used to manufacture 10 million carats (2,000 kg, or 4,400 lb) of rubies annually.

In 1918, J. Czochralski developed a different method for synthesizing rubies. Known as crystal pulling, this technique is fast, inexpensive, and effective in producing flawless stones. In fact, when cut as gems the stones are so clear that they look like glass imitations. Consequently, this technique is now used primarily for manufacturing industrial-use rubies.

During World War II, it was impossible to get rubies from traditional sources in France and Switzerland. Because these stones were vitally important for use as bearings in military as well as civilian instruments, efforts were made to improve manufacturing techniques. One such improvement, developed by the Linde Division of Union Carbine Corporation, modified Verneuil's flame fusion process to grow thin rods of ruby crystals up to 30 in (750 mm) long. Such rods can easily be sliced into disks to produce large quantities of bearings.

A process developed by Bell Telephone Company in 1958 employed high temperatures and pressures to grow rubies on seeds that had been produced by flame fusion. Refinements of this technique became known as the hydrothermal method. Carroll Chatham, a San Francisco gem manufacturer who developed and used a hydrothermal process, also developed the first commercially successful application of the flux process of ruby manufacture. This technique, first used in 1959, essentially creates roiling magma in a furnace and grows very natural-looking gems in a period of nearly a year.

Methods of Synthesizing

Several methods are currently used to manufacture rubies; each has advantages and limitations. The most popular methods can be categorized into two main types: production from melt, in which powdered material is heated to a molten state and manipulated to solidify in a crystalline form, and production from "solution," in which the required aluminum oxide and chromium are dissolved in another material and manipulated to precipitate into a crystalline form. Verneuil's flame fusion and Czochralski's crystal pulling are the most commonly used melt techniques, while flux growth and hydrothermal growth are the most popular versions of solution processes.

Flame fusion rubies, generally the least expensive, are commonly used for bearings and relatively mundane jewelry like class rings. Pulled rubies, selling for upwards of $5 per carat, are preferred for laser use. Flux rubies, costing $50 or more per carat, are used in finer jewelry. The less-common hydrothermal process is used for industrial applications demanding strain-free crystals or large crystals in something other than a rod shape.

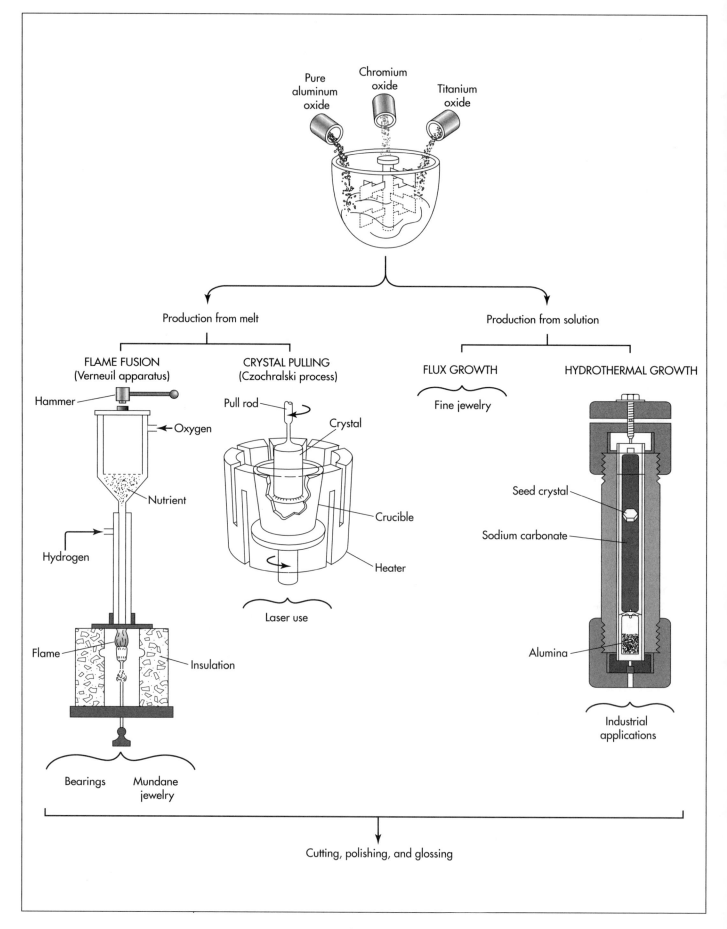

Pure aluminum oxide

Chromium oxide

Titanium oxide

Production from melt

Production from solution

FLAME FUSION
(Verneuil apparatus)

Hammer

←Oxygen

Nutrient

Hydrogen

Flame

Insulation

Bearings   Mundane jewelry

CRYSTAL PULLING
(Czochralski process)

Pull rod

Crystal

Crucible

Heater

Laser use

FLUX GROWTH

Fine jewelry

HYDROTHERMAL GROWTH

Seed crystal

Sodium carbonate

Alumina

Industrial applications

Cutting, polishing, and glossing

Raw Materials

The nutrient (material that will become the ruby crystal) consists primarily of extremely pure aluminum oxide (Al_2O_3); approximately 5-8% of chromium oxide (Cr_2O_3) must be added to produce the essential red color. If an asteriated gem (a star ruby) is being produced, a small amount (0.1-0.5%) of titanium oxide (TiO_2) is also used.

Depending on the method being employed, additional chemicals may be needed. The flame fusion process uses an oxygen-hydrogen torch to melt powdered forms of the two basic components, whereas the Czochralski process uses some form of electrical heating mechanism. The flux method uses a compound such as lithium oxide (LiO), molybdenum oxide (MoO), or lead fluoride (PbF_2) as a solvent for the nutrient. The hydrothermal process uses as a solvent an aqueous (water-based) solution of sodium carbonate (Na_2CO_3). A corrosion-resistant metal such as silver or platinum is used to line the vessel that contains the liquefied ingredients for the Czochralski, flux, and hydrothermal processes.

The Manufacturing Process

Crystal growth

One of the following four methods is typically used to manufacture synthetic rubies.

1 (Flame Fusion) A fine powder of the aluminum and chromium oxides is placed in a hopper at the top of the Verneuil apparatus. A hammer atop the apparatus strikes the hopper repeatedly; each stroke causes a small amount of powder to fall through the fine mesh that forms the hopper's floor. This discharged powder falls into a stream of oxygen that carries it down to a nozzle where it mixes with a stream of hydrogen and is ignited. The intense heat of this flame (around 3,600° F or 2,000° C) melts the nutrient, which falls onto a ceramic pedestal below the flame. Initially, the hammer taps at a rate of 80 beats per minute; after a suitable base for the crystal is formed, the rate is decreased to about 20 beats per minute.

After the base is built up to the desired diameter (about 0.8 in or 20 mm) and formation of the high-quality crystal proceeds, the pedestal is lowered at a rate that just keeps the top of the crystal in contact with the flame. After about five and a half hours, the crystal reaches a length of approximately 2.75 in (70 mm); the gas flow is halted, extinguishing the flame. The crystal, now weighing around 150 carats, is allowed to cool in the enclosed furnace.

2 (Czochralski Process) The nutrient is heated well above its melting point in a crucible that is surrounded by an electric heater. A small ruby crystal is attached to a rod; the desired crystal will grow on this so-called seed crystal. The seed is lowered into the crucible until it is barely immersed in the melt (i.e., the molten nutrient). To maintain a constant contact temperature between the melt and the entire circumference of the seed crystal, the rod is constantly rotated. As nutrient material attaches itself to the seed and crystalizes (a process that is assisted by the seed's attachment to the relatively cooler rod), the rod is slowly raised, pulling the growing crystal out of the melt. The growing tip is kept in contact with the melt until all the nutrient has been used. The rate of growth can be quite rapid, up to a rate of 4 in (100 mm) per hour. Very large crystals can be pulled, with diameters exceeding 2 in (50 mm) and lengths reaching 40 in (1 m) or more.

3 (Flux Growth) Flux is any material that when melted will dissolve another material that has a much higher melting point. Although temperatures in excess of 3,600° F (2,000° C) are needed to melt aluminum oxide, the material will dissolve in certain fluxes at a temperature as low as 1,470° F (800° C). Process temperatures above 2,200° F (1,200° C) are generally used because they produce higher-quality crystals. While dissolved in the flux, ruby molecules can travel freely and attach themselves to a growing crystal. Some manufacturers immerse seed crystals in the solution, and others simply allow the molecules to combine randomly and form an unplanned number of crystals. The temperature is maintained for a period of three to 12 months. Some manufacturers then pour off the still-molten flux to expose the ruby crystals. Other manufacturers cool the material slowly (4° F or 2° C per hour) and then extract the ruby crystals

Opposite page:
There are several processes used to create synthetic rubies. Verneuil's flame fusion and Czochralski's crystal pulling are the most commonly used melt techniques, while flux growth and hydrothermal growth are the most popular versions of solution processes.

by breaking off the solidified flux or dissolving it in acid.

4 (Hydrothermal Process) Powdered or crystalline nutrient is placed at one end of a pressure-resistant tube. A seed crystal is mounted on a wire frame near the other end of the tube. An appropriate water-based solution is placed in the tube, which is sealed shut. The tube is placed vertically in a furnace chamber, with the nutrient-containing end of the tube resting on a heating element. As the floor of the furnace is heated, the bottom end of the tube becomes hotter than the top (about 835° F or 445° C, compared to 770° F or 410° C); dissolved nutrient material migrates toward the seed and crystalizes on its relatively cooler surface. Pressure within the tube can range from 83,000-380,000 kPa (12,000-55,000 lb per sq in), depending on the amount of free space left in the tube when the solvent was inserted.

The tube used for the hydrothermal process can be made in any appropriate size, with a height-to-diameter ratio ranging from 8-16. In an example described in *Synthetic Gem and Allied Crystal Manufacture,* five seed crystals were placed in a 12 in (300 mm) long tube; each crystal grew at a rate of 0.006 in (0.15 mm) per day during the 30-day processing period.

Surface finishing

Whether it will be used as a gem or an industrial device, the ruby must be given a smooth, glossy finish after it has been cut or faceted to the desired shape. The following methods may be used.

5 (Polishing) The surface is rubbed with increasingly fine particles of an abrasive such as diamond powder. This traditional technique leaves only microscopic scratches and pits.

6 (Glossing) After initial polishing, the surface of the stone may be heated rapidly in a gas flame to melt any tiny projections. The surface is then allowed to cool, and the thin layer of molten material solidifies as a smooth surface. Treating ruby rods in this way nearly doubles the rod's tensile strength (resistance to a pulling force).

Comparing Synthetic To Natural

Rubies, grown as rods for industrial use, are readily recognizable as synthetic because of their shape. Manmade stones that are cut as gems are not so easily identified. However, microscopic examination can reveal characteristic patterns of inclusions (foreign particles), bubbles, and striations (growth bands) that can distinguish between natural and synthetic stones, even revealing the location from which a natural stone came or the process by which a synthetic stone was made.

Where to Learn More

Books

Elwell, Dennis. *Man-Made Gemstones.* New York: Halsted Press, 1979.

MacInnis, Daniel. *Synthetic Gem and Allied Crystal Manufacture.* Park Ridge, NJ: Noyes Data Corporation, 1973.

Periodicals

Sunagawa, I. "Gem Materials, Natural and Artificial." *Current Topics in Materials Science* 10 (1982): 353-497.

Tang, Seung Mun. "When Is a Ruby Real?" *Physics World* (October 1992): 21-22.

Ward, Fred. "Rubies and Sapphires." *National Geographic* (October 1991): 100-125.

—*Loretta Hall*

Temporary Tattoo

Background

A temporary tattoo is a decorative image that can be applied to the skin for short periods of time. Most temporary tattoos are novelty items made with a special type of decal. A process known as screen printing is used to create the tattoo image on paper coated with a transfer film. The transfer film allows the image to "slide" off the backing paper and onto the skin when moisture is applied. After drying, the film holds the image on the skin through several washings.

For centuries, men and women have added decorative illustrations to their skin for religious or cultural reasons. One common method of decorating skin is tattooing, a process which involves injecting patterns of dye directly into the skin using a needle. Although this technique was originally practiced in ancient Egypt, the term tattoo is actually derived from a Tahitian word that was most likely spread by sailors in the Pacific. Many other cultures have their own unique tattoo techniques. For example, Eskimos use bone needles to draw soot-covered thread through the skin and the Japanese use fine metal needles to deliver colored pigments. Regardless of which technique is used, all tattoo processes deposit colorants below the surface of the skin to create intense, permanent images. While tattooing remains a popular art form today, it is also expensive, time consuming, and may be somewhat painful. For these reasons, permanent tattoos are not necessarily desirable for every individual.

Temporary tattoos were created as an alternative way for individuals to decorate their skin. Temporary images can be produced by several methods. For example, they can be hand drawn and painted using a brush with water insoluble dyes or pigments. Although this method requires a talented artist to create a high quality image, it does produce a picture which can be removed fairly easily. A better way of achieving a temporary tattoo is by decalcomania, which is the process of applying a decal to the skin. This approach allows the user to apply a preprinted image to the skin at their convenience. Decal-style tattoos are so simple to apply that even a child can use them and the image that is produced can be easily removed with soap and water. Therefore, temporary tattoos can be easily changed to suit the whims of fancy and fashion. Decal-style temporary tattoos are made by printing an image onto special paper coated with a transfer film. To apply the tattoo the user simply moistens the paper and the film slides off the backing layer carrying the image onto the skin.

Raw Materials

Stencil materials

Tattoos can be made by a screen-printing process, which uses stencils to create the image to be printed. These stencils are made from nonporous paper or plastic coated with lacquer, gelatin, or a combination of glue and tusche (a heavy ink-like substance). These materials are used to block portions of the screen during the printing process so the ink only touches the paper in designated spots.

Inks

Because temporary tattoos reside on the skin for relatively long periods of time, all

One common method of decorating skin is tattooing, a process which involves injecting patterns of dye directly into the skin using a needle. Although this technique was originally practiced in ancient Egypt, the term tattoo is actually derived from a Tahitian word that was most likely spread by sailors in the Pacific.

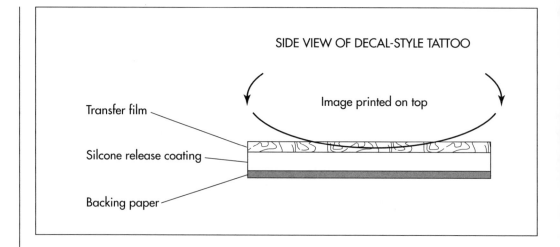

SIDE VIEW OF DECAL-STYLE TATTOO

Image printed on top

Transfer film

Silcone release coating

Backing paper

colorants used in the inks must meet the same requirements as food, drug, and cosmetic colorants. These pigments, which are governed by the Food and Drug Administration (FDA), may be dispersed in water, alcohol, or oil depending on their solubility. Drying agents and extenders are also added to inks to modify their drying behavior.

Backing paper

Temporary tattoos may be printed on paper, plastic films, or combinations of the two. Paper is generally preferred because it is better for printing and processing. This backing paper is coated with a variety of materials using spraying or dipping methods. The coatings can be processed to a uniform thickness by passing the coated paper through a series of rollers outfitted with a knife, which evenly spreads the liquid. The paper can then be passed through a heated tunnel to accelerate drying. The first coating that is applied is typically a sizing agent, which modifies the paper's stiffness and texture. The next layer is a non-stick silicone release coating which helps the image separate from the backing paper. A transfer film is then coated on top of the silicone layer. This film is the layer that the image is printed upon and is composed of gelatin or other polymeric materials such as polyvinyl alcohol or polyvinyl pyrollidone. These materials are designed to be strong enough to adhere to the backing paper during printing yet flexible enough to be easily released during application. Upon drying on the skin, the film should adhere tightly and smoothly to maintain the image quality.

The Manufacturing Process

Stencil preparation

1 An outline of the image to be printed is cut into the gelatin or lacquer layer of the stencil sheet. The lacquer or gelatin is peeled away to expose the areas to be printed. The stencil is then adhered to the screen with a solvent and, after it has dried on the screen; the backing sheet is removed leaving only the film layer. The portions of the stencil that were cut away expose a section of the screen through which ink can be forced.

Screen printing

2 The screen to which the stencil is adhered is typically made of finely woven fabrics (like silk, nylon, and Dacron) or stainless steel mesh. Image transfer is accomplished by forcing the inks through the stencil and onto the printing substrate. A rubber squeegee is used to force ink through openings in the stencil. In this process, only one color can be printed at a time, so the image must pass through the screen press once for every color. The colors are laid down in reverse order, from last to first so the finished tattoo resembles a multilayer sandwich. The bottom layer is the release paper, followed by the transfer film, topped with the detail colors. The background colors are laid down last.

Finishing operations

3 If necessary, the printed tattoo sheets can be coated with another layer of film-forming material to seal in the image. After

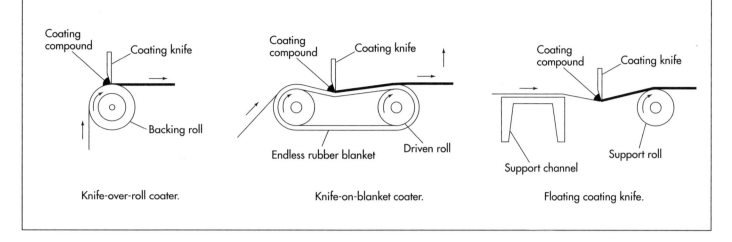

Knife-over-roll coater.

Knife-on-blanket coater.

Floating coating knife.

this final coating step the sheets are cut, or slit, into rolls or individual tattoos. The finished tattoos are then wrapped or boxed for shipping. The packaging materials should be designed to minimize contact with moisture to avoid premature softening of the transfer film.

Transfer process

4 Tattoos are easily transferred from the printed sheet to skin by first lightly dampening the skin. Care must be taken not to saturate the tattoo because the film may begin to dissolve before the image is transferred. The backing paper is then firmly held against the skin either by hand or with a damp cloth or sponge. The paper must be held still to avoid shifting the image during transfer. After one to three minutes the transfer layer will soften and separate from the backing paper. The paper can then be easily peeled away, leaving the transfer film and the printed image intact on the skin. As the film dries, it bonds firmly to the skin.

Quality Control

A number of factors affect the quality of temporary tattoos. First, the stencil must be properly prepared because dull or poor tooling will result in a murky image. Similarly, the printing screen must be carefully maintained to keep ink from clogging the pores in the screen. Inks must be correctly compounded because if they are too thick or too thin they will not pass through the screen properly. Finally, the components of the backing paper must be properly prepared.

The sizing agents, the silicone release layer, and the transfer film must all be coated evenly to minimize problems during printing and to ensure even image transfer. After manufacture is complete, the finished tattoos must be carefully packaged to exclude moisture which could cause ink bleeding or premature softening of the transfer film.

Byproducts/Waste

The decal manufacturing process creates waste in the form of excess lacquer, gelatin, paper, and inks. Some of these waste materials may be flammable or hazardous depending on the solvents used. In many cases, paper may be recycled by repulping, a process which involves shredding the paper and mixing it with water to wash off residual coatings. The repulped paper can be cast into sheet form again and reused to make new tattoos. For all the waste that is generated, manufacturers must comply with all relevant local and federal waste disposal regulations.

The Future

Advances in printing technology are likely to improve tattoo manufacturing processes. One advance that is likely to impact the industry is the use of computer-aided printing technologies, such as ink-jet printing. It is anticipated that ink jet printers could be used to quickly and easily produce decal-style tattoos. However, such printing improvements may not solve the problem of image degradation. Image degradation occurs because of the motion, expansion, and

Once printed, the temporary tattoo can be coated with a release coat and transfer film in three ways.

contraction of the skin, all of which causes tiny fractures in the image film layer. Even though the film remains adhered to the skin, the image quality is severely reduced due to the visible cracks and creases. A recent patent suggests this problem can be solved by using alcohol soluble, water insoluble dyes and a special transfer solution to help maintain the integrity of the transfer film. Reportedly this process results in an improved, longer lasting image. Another patented way to achieve improved temporary tattoos uses an adhesive to hold pictures on the skin. This approach produces an image on a translucent adhesive substrate that can be adhered to the body for long periods of time with a minimal loss of image quality.

Other methods of temporarily tattooing the body are gaining popularity. One such technique is Mehndi, the Indian/Pakistani practice of body painting, which uses copper colored ink made from crushed henna leaves. *mehndi* is becoming widely used in the United States as a way of temporarily adorning the body with elaborate scroll-like decorations. Another emerging trend is the use of stencil-like stickers to prevent sunlight from reaching portions of the body during tanning. When these stickers are worn during sunbathing, a temporary tattoo is created because the covered area does not tan like the surrounding skin. While this method is only able to create monochromatic images, it can be used to produce simple lettering and figures.

Where to Learn More

Books

Sanders, C. R. *Customizing the Body: The Art and Culture of Tattooing,* 1988.

Swerdlow, Robert. *The Step by Step Guide to Screen-Process Printing.* Prentice-Hall, Englewood Cliffs, 1985.

Other

http://www.zibabeauty.com/home/mehndi.html.

US Patent 4169169. "Transfer Process and Transfer Sheet for use Therein," 1979.

US Patent 4594276. "Printed, Removable Body Tattoos on a Translucent Substrate," 1986.

—*Randy Schueller*

Tin

Background

Tin is one of the basic chemical elements. When refined, it is a silvery-white metal known for its resistance to corrosion and its ability to coat other metals. It is most commonly used as a plating on the steel sheets used to form cans for food containers. Tin is also combined with copper to form bronze and with lead to form solder. A tin compound, stannous fluoride, is often added to toothpaste as a source of fluoride to prevent tooth decay.

The earliest use of tin dates to about 3500 B.C. in what is now Turkey, where it was first mined and processed. Ancient metalworkers learned to combine relatively soft copper with tin to form a much harder bronze, which could be made into tools and weapons that were more durable and stayed sharp longer. This discovery started what is known as the Bronze Age, which lasted about 2,000 years. The superiority of bronze tools spurred the search for other sources of tin. When extensive tin deposits were found in England, traders brought the precious metal to countries in the Mediterranean area, but kept the source a secret. It wasn't until 310 B.C. that the Greek explorer Pytheas discovered the location of the mines near what is now Cornwall, England. Much of the impetus for the Roman invasion of Britain in 43 A.D. was to control the tin trade. The chemical symbol for tin, Sn, is derived from the Latin name for the material, *stannum*.

Elsewhere in the world, tin was used in ancient China and among an unknown tribe in what is now South Africa. By about 2500-2000 B.C., metalworkers on the Khorat Plateau of northeast Thailand used local sources of tin and copper to produce bronze, and by about 1600 B.C. bronze plows were being used in what is now Vietnam. Tin was also known and used in Mexico and Peru before the Spanish conquest in the 1500s.

The use of tin as a plating material dates to the time of the Roman Empire, when copper vessels were coated with tin to keep them bright looking. Tinned iron vessels appeared in central Europe, in the 1300s. Thin sheets of iron coated with tin, called tinplate, became available in England during the mid-1600s and were used to make metal containers. In 1810, Pierre Durand of France patented a method of preserving food in sealed tinplate cans. Although it took many years of experimenting to perfect this new technique, tin cans began replacing bottles for food packaging by the mid-1800s.

In 1839, Isaac Babbitt of the United States invented an antifriction alloy, called Babbitt metal, which consisted of tin, antimony, and copper. It was widely used in bearings and greatly assisted the development of high-speed machinery and transportation.

In 1952, the firm of Pilkington in England revolutionized the glassmaking industry with the introduction of the "float glass" method for the continuous production of sheet glass. In this method, the molten glass floats on a bath of liquid, molten tin as it cools. This produces a very flat glass surface without the rolling, grinding, and polishing operations that were required prior to the introduction of this method.

Today, most of the world's tin is produced in Malaysia, Bolivia, Indonesia, Thailand,

Today, most of the world's tin is produced in Malaysia, Bolivia, Indonesia, Thailand, Australia, Nigeria, and England. There are no major tin deposits in the United States.

437

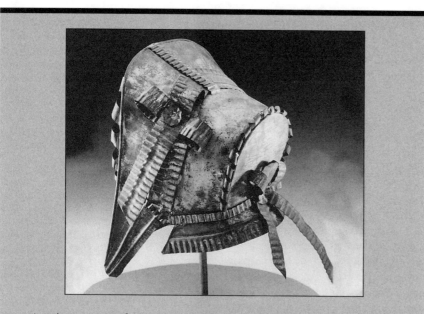

A tin bonnet was often given as a tenth anniversary gift during the 1800s. (From the collections of Henry Ford Museum & Greenfield Village, Dearborn, Michigan.)

In the 1800s, tin was an ordinary household material particularly popular with the working class because of its low cost and bright luster. Made of iron or steel rolled thin and dipped in molten tin, it was easy to manipulate, cut, and solder. Tin was used for nearly everything that copper, pewter, brass, or silver could be used for, but generally did not last as long. Reviewing tin catalogs from about 1870 reveals that tin was used for far more than cookie cutters—it was used to make children's toys, coffeepots, lunch boxes, and even gentlemen's spitoons!

However, it was also popularly used to produce a gift for the tenth anniversary, called the "tin anniversary." While not as well-known as the twenty-fifth, which requires silver gifts, the Victorian housewife knew she might well receive a tenth anniversary gift of tin like the tin bonnet depicted here. Shaped in the form of a "spoon bonnet" popular about 1870, it is likely that this piece dates to that time. Certainly, it can't be worn, but was meant to be displayed on a shelf as a remembrance of that anniversary. Tinsmiths provided whimsical gifts just for this purpose. Museum collections include not only hats but tin shoes and decorative vases that could never be used to hold water.

Nancy EV Bryk

Australia, Nigeria, and England. There are no major tin deposits in the United States.

Raw Materials

There are nine tin-bearing ores found naturally in the earth's crust, but the only one that is mined to any extent is cassiterite. In addition to the ores themselves, several other materials are often used to process and refine tin. These include limestone, silica, and salt. Carbon, in the form of coal or fuel oil, is also used. The presence of high concentrations of certain chemicals in the ore may require the use of other materials.

The Manufacturing Process

The process of extracting tin from tin ore varies according to the source of the ore deposit and the amount of impurities found in the ore. The tin deposits in Bolivia and England are located deep underground and require the use of tunnels to reach the ore. The ore in these deposits may contain about 0.8-1.0% tin by weight. Tin deposits in Malaysia, Indonesia, and Thailand are located in the gravel along streambeds and require the use of dredges or pumps to reach the ore. The ore in these deposits may contain as little as 0.015% tin by weight. Over 80% of the world's tin is found in these low-grade gravel deposits.

Regardless of the source, each process consists of several steps in which the unwanted materials are physically or chemically removed, and the concentration of tin is progressively increased. Some of these steps are conducted at the mine site, while others may be conducted at separate facilities.

Here are the steps used to process the low-grade ore typically found in gravel deposits in Southeast Asia:

Mining

1 When the gravel deposits are located at or below the water level in the stream, they are brought up by a floating dredge, operating in an artificial pond created along the streambed. The dredge excavates the gravel using a long boom fitted either with chain-driven buckets or with a submersed rotating cutter head and suction pipe. The gravel passes through a series of revolving screens and shaker tables onboard the dredge to separate the soil, sand, and stones from the tin ore. The remaining ore is then collected and transferred ashore for further processing.

When the gravel deposits are located in dry areas at or above the water level in the

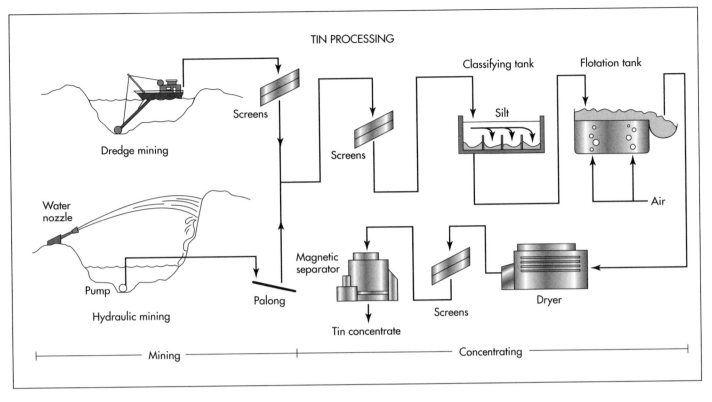

TIN PROCESSING

Mining — Concentrating

stream, they are first broken up with jets of water pumped through large nozzles. The resulting muddy slurry is trapped in an artificial pond. A pump located at the lowest point in the pond pumps the slurry up into a wooden trough, called a palong, which has a gentle downward slope along its length. The tin ore, which is heavier than the sand and soil in the mud, tends to sink and is trapped behind a series of wooden slats, called riffles. Periodically the trapped ore is dumped from the palong and is collected for further processing.

Concentrating

2 The ore enters the cleaning or dressing shed adjacent to the mining operation. First, it passes through several vibrating screens to separate out coarser foreign materials. It may then pass through a classifying tank filled with water, where the ore sinks to the bottom while the very small silt particles are carried away. It may also pass through a floatation tank, where certain chemicals are added to make the tin particles rise to the surface and overflow into troughs.

3 Finally the ore is dried, screened again, and passed through a magnetic separator to remove any iron particles. The resulting tin concentrate is now about 70-77% tin by weight and consists of almost pure cassiterite.

Smelting

4 The tin concentrate is placed in a furnace along with carbon in the form of either coal or fuel oil. If a tin concentrate with excess impurities is used, limestone and sand may also be added to react with the impurities. As the materials are heated to about 2550° F (1400° C), the carbon reacts with the carbon dioxide in the furnace atmosphere to form carbon monoxide. In turn the carbon monoxide reacts with the cassiterite in the tin concentrate to form crude tin and carbon dioxide. If limestone and sand are used, they react with any silica or iron present in the concentrate to form a slag.

5 Because tin readily forms compounds with many materials, it often reacts with the slag. As a result, the slag from the first furnace contains an appreciable amount of tin and must be processed further before it is discarded. The slag is heated in a second furnace along with additional carbon, scrap iron, and limestone. As before, crude tin is formed and recovered along with a certain amount of residual slag.

When gravel deposits are located at or below the water level, they are brought up by a floating dredge, operating in an artificial pond created along the streambed. When the gravel deposits are located in dry areas at or above the water level, they are first broken up with jets of water pumped through large nozzles. Next, the ore enters the cleaning or dressing shed adjacent to the mining operation.

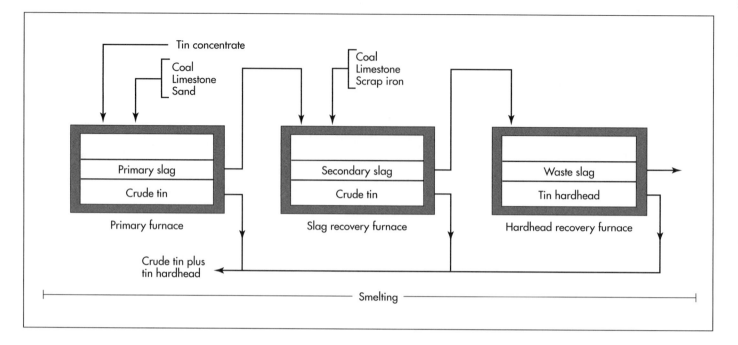

Tin concentrate

Coal
Limestone
Sand

Coal
Limestone
Scrap iron

| Primary slag | Secondary slag | Waste slag |
| Crude tin | Crude tin | Tin hardhead |

Primary furnace   Slag recovery furnace   Hardhead recovery furnace

Crude tin plus
tin hardhead

Smelting

The tin concentrate is placed in a furnace along with carbon in the form of either coal or fuel oil. It is heated and forms a slag along with the crude tin. The slag and crude tin are heated several more times to remove impurities and recover tin hardhead.

6 The residual slag from the second furnace is heated one more time to recover any tin that has formed compounds with iron. This material is known as the hard head. The remaining slag is discarded.

Refining

7 The crude tin from the first furnace is placed in a low-temperature furnace along with the crude tin recovered from the slag plus the hard head. Because tin has a melting temperature much lower than most metals, it is possible to carefully raise the temperature of the furnace so that only the tin melts, leaving any other metals as solids. The melted tin runs down an inclined surface and is collected in a poling kettle, while the other materials remain behind. This process is called liquidation and it effectively removes much of the iron, arsenic, copper, and antimony that may be present.

8 The molten tin in the poling kettle is agitated with steam, compressed air, or poles of green wood. This process is called boiling. The green wood, being moist, produces steam along with the mechanical stirring of the poles. It was from this crude, but effective use of wood poles that the poling kettle got its name. Most of the remaining impurities rise to the surface to form a scum, which is removed. The refined tin is now about 99.8% pure.

9 For applications requiring an even higher purity, the tin may be processed further in an electrolytic refining plant. The tin is poured into molds to form large electrical anodes, which act as the positive terminals for the electrorefining process. Each anode is placed in an individual tank, and a sheet of tin is placed at the opposite end of the tank to act as the cathode, or negative terminal. The tanks are filled with an electrically conducting solution. When an electrical current is passed through each tank, the tin is stripped off the anode and is deposited on the cathode. The remaining impurities, which are generally bismuth and lead, fall out of the solution and form a slime at the bottom of the tank.

10 The cathodes are remelted, and the refined tin is cast in iron molds to form ingots or bars, which are then shipped to the various end users. Lower purity tin is usually cast into ingots weighing 25-100 lb (11-45 kg). Higher purity tin is cast into smaller bars weighing about 2 lb (1 kg).

Quality Control

The processes described have been proven to consistently produce tin at 99% purity and higher. To ensure this purity, samples are analyzed at various steps to determine whether any adjustments to the processes are required.

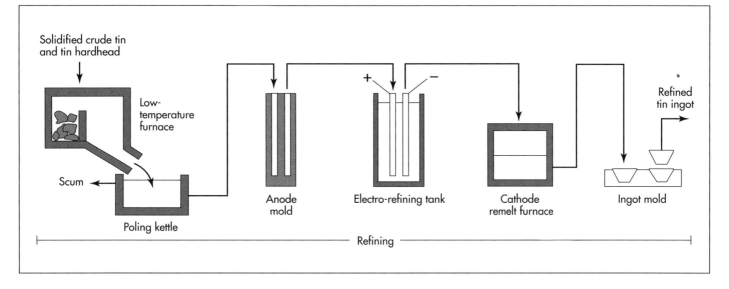

In the United States, the purity levels for commercial grades of tin are defined by the American Society for Testing Materials (ASTM) Standard Classification B339. The highest grade is AAA, which contains 99.98% tin and is used for research. Grade A, which contains 99.80% tin, is used to form tinplate for food containers. Grades B, C, D, and E are lesser grades ranging down to 99% purity. They are used to make general-purpose tin alloys such as bronze and solder.

Byproducts/Waste

There are no useful byproducts produced from tin processing.

Waste products include the soil, sand, and stones that are rejected during the mining and concentrating operations. These constitute a huge amount of material, but their environmental impact depends on the local disposal practices and the concentrations of other minerals that may be present. The slag produced during the smelting and refining operations is also a waste product. It may contain quantities of arsenic, lead, and other materials that are potentially harmful. Tin itself has no known harmful effects on humans or the environment.

The Future

The use of tin is expected to grow as new applications are developed. Because tin has no known detrimental effects, it is expected

to replace other more environmentally harmful metals such as lead, mercury, and cadmium. One new application is the formulation of tin-silver solders to replace tin-lead solders in the electronics industry. Another application is the use of tin shot to replace lead shot in shotgun shells.

Development work is underway to create a tin-based compound for use in refuse disposal landfill sites. This compound will interact with heavy metals, such as lead and cadmium, to prevent rain water from carrying them into the surrounding soil and water table.

Where to Learn More

Books

Brady, George S., Henry R. Clauser, and John A. Vaccari. *Materials Handbook*, 14th Edition. McGraw-Hill, 1997.

Heiserman, David L. *Exploring Chemical Elements and Their Compounds*. TAB Books, 1992.

Hornbostel, Caleb. *Construction Materials*, 2nd Edition. John Wiley and Sons, Inc., 1991.

Kroschwitz, Jacqueline I. and Mary Howe-Grant, ed. *Encyclopedia of Chemical Technology*, 4th edition. John Wiley and Sons, Inc., 1993.

The tin hardhead is further refined, until it is molded into tin ingots.

Stwertka, Albert. *A Guide to the Elements.* Oxford University Press, 1996.

Periodicals

"Bronze Age Mine Found in Turkey," *Science News* (January 15, 1994): 46.

Other

http://www.intercorr.com/periodic/50.htm.

International Tin Research Institute. http://www.itri.co.uk.

—*Chris Cavette*

Topographic Map

Background

A topographic map is a two-dimensional representation of a three-dimensional land surface. Topographic maps are differentiated from other maps in that they show both the horizontal and vertical positions of the terrain. Through a combination of contour lines, colors, symbols, labels, and other graphical representations, topographic maps portray the shapes and locations of mountains, forests, rivers, lakes, cities, roads, bridges, and many other natural and man-made features. They also contain valuable reference information for surveyors and map makers, including bench marks, base lines and meridians, and magnetic declinations. Topographic maps are used by civil engineers, environmental managers, and urban planners, as well as by outdoor enthusiasts, emergency services agencies, and historians.

History

Some of the earliest known maps were made in Mesopotamia, in the area now known as Iraq, where a series of maps showing property boundaries were drawn in about 2400 B.C. for the purpose of land taxation. A Roman map dating from about 335-366 A.D. showed such topographical features as roads, cities, rivers, and mountains. The word topography is derived from the Greek words *topos*, meaning a place, and *graphien*, meaning to write. Thus, topography is the written, or drawn, description of a place.

Although the basics of land surveying were known as early as 1200 B.C., and perhaps even earlier, the use of surveying techniques in preparing maps was limited to cities and other small-scale areas. Larger-scale maps were prepared from sketches or journals kept by explorers and sometimes reflected more imagination than observation. As a result, the exact positions of points on a map were often grossly in error.

In 1539, the Dutch mathematician and geographer Reiner Gemma Frisius described a method for surveying an area by dividing it into triangles. This concept of triangulation became one of the basic techniques of field surveying and is still used today. One of the first large-scale mapping projects using triangulation was started in the 1670s by Giovanni Domenico Cassini, who had been persuaded to make a detailed map of France. After Cassini's death, his children and grandchildren continued to labor on the project. The final result, called the *Carte de Cassini*, was published in 1793 and was the first accurate topographic map of an entire country. Its only shortcoming was the general lack of elevation measurements, other than a few spot elevations determined by measuring the variation in air pressure with altitude using a barometer. The concept of contour lines to show different elevations on a map was developed by the French engineer J.L. Dupain-Triel in 1791. Although this method allowed the accurate depiction of land contours and elevations on a flat, two-dimensional map, it was not widely used until the mid-1800s.

In the United States, the federal government recognized the importance of accurate topographic maps in a rapidly growing country. In 1807, President Thomas Jefferson established the Survey of the Coast to map the Atlantic coastline as an aid to travel and commerce. In 1836, this organization was

Today, the USGS has more than 56,000 topographic maps of the United States in various scales, plus maps of the moon and planets.

443

renamed the U.S. Coast Survey, and in 1878 the name changed to the U.S. Coast and Geodetic Survey. In the meantime, mapping of the interior of the country fell to a variety of individuals and organizations, including the Lewis and Clark expedition in 1804-1806, who mapped their route from St. Louis, Missouri, to the Pacific Northwest. During the period from 1838 until the outbreak of the Civil War in 1861, the Army's Corps of Topographical Engineers made major contributions in mapping the western United States, including a detailed map published in 1848 based on John Fremont's explorations. By the 1870s, so many different groups were conducting surveys that their work began to overlap. To consolidate this effort, the U.S. Geological Survey (USGS) was established in 1879.

Most of the early map making was done by laborious field surveys. Starting in the 1930s, the USGS began using aerial photography techniques to produce and update maps. In the 1980s the use of computers to scan and redraw existing maps significantly reduced the time required to update maps in areas of rapid growth.

Today, the USGS has more than 56,000 topographic maps of the United States in various scales, plus maps of the moon and planets. They also publish specialty maps including geologic, hydrologic, and photoimage maps for a variety of uses.

Map Scales, Symbols, and Colors

In order to be useful, topographic maps must show sufficient information on a map size that is convenient to use. This is accomplished by selecting a map scale that is neither too large nor too small and by enhancing the map details through the use of symbols and colors.

The most common USGS topographic map scale is 1:24,000. In this scale 1 inch on the map represents 24,000 in, or 2,000 ft, (1 cm represents 240 m) on the ground. These maps are called 7.5 minute quadrangle maps because each map covers a four-sided area on the surface of Earth that is 7.5 minutes of longitude wide and 7.5 minutes of latitude high, where 60 minutes equals one degree of angle. Because the distance between longi-

tude lines gets narrower as you move from the equator towards the poles, the widths of the maps also vary. For maps of the United States, the maps measure about 23 in (58.4 cm) wide by 27 in (68.6 cm) high for locations below a latitude of 31 degrees and about 22 in (55.9 cm) wide by 27 in (68.6 cm high) for locations above that latitude. Other common USGS map scales are 1:63,360, 1:100,000, and 1:250,000. These scales cover larger areas than the 1:24,000 maps, but with less detail.

In order to make the topographic maps easier to interpret, symbols and colors are used to represent various natural and man-made features. Some symbols are designed to look like the feature when viewed from overhead. For example, buildings are shown as solid objects in the shape of the building outline. Other symbols are universally recognized representations such as a long line with small cross marks to represent a railroad. Colors play an even more important role. Rivers, lakes, and other bodies of water are shown in blue. Forests and heavily vegetated areas are shown in green. Minor roads and highways are shown in black, while major highways are shown in red. Contour lines, which represent the shape of the ground itself, are shown in brown. Recent revisions to the map are shown in purple.

The Manufacturing Process

The production of an accurate topographic map is a long and complex process that may take as much as five years from start to finish. It takes a skilled team of surveyors, engravers, fact checkers, printers, and others to produce a good map. Here is a typical sequence of operations used by the U.S. Geological Survey to produce a 7.5 minute quadrangle topographic map.

Photographing the area

1 The area to be mapped must first be photographed from the air. Each section of ground is photographed from two different angles to provide a stereoscopic three-dimensional image that can be converted into contour lines. The sky must be clear, and the sun must be at the proper angle for the type of terrain being photographed. Seasonal

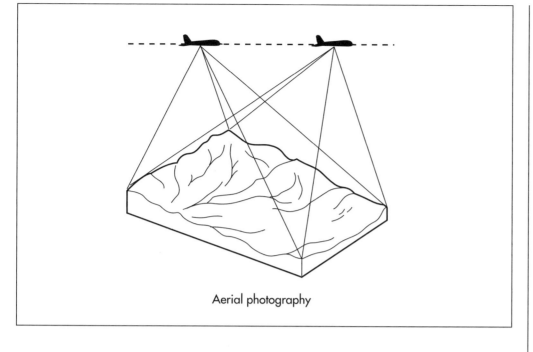

Aerial photography

factors must also be taken into consideration. For example, in areas where there are deciduous trees, the photos are usually taken between late fall and early spring when the trees are bare and the underlying ground features are more visible.

2 The aircraft is flown over the area at a constant altitude in a north-south direction along carefully determined flight paths while special cameras take 10 precisely positioned photographs of each quadrangle. Each camera can cost $250,000 or more.

Surveying the control points

3 To ensure the accuracy of a map, the exact location of various control points must be established by field surveys. Typical control points may be the intersection of two roads or other prominent features within the map area. Horizontal control points are surveyed to determine the longitude and latitude, while vertical control points are surveyed separately to determine elevations. The location and elevation of these control points help the map makers correctly position the aerial photo images and assign values to the contour lines.

4 While the surveyors are in the field, they also look for features, which may require further checking, such as roads or streams hidden beneath overhanging foliage, or buildings that may have been constructed or demolished since the aerial photographs were taken.

Verifying the map features

5 Some map features may require additional verification. For example, some streams may run only intermittently, in which case they would be represented on the map by a dash-dot or lighter-weight instead of a solid line. Certain roads may turn out to be private roads, rather than public roads, and these must be marked. Field checkers go into the area and verify these features by talking with local residents or consulting local property records. Any questionable features noted by the survey crews must also be verified. The correct spelling of place names must be determined.

Compiling the map manuscript

6 After the area has been surveyed and all the features have been checked, the pairs of overlapping aerial photographs are placed in a stereoscopic projector. One image is projected to the operator's left eye and the other image to his right eye. The result is a three-dimensional view of the terrain. Two small beams of light are connected to a pointer and are adjusted to intersect in a tiny white dot corresponding to a given elevation on the three-dimensional terrain image. By

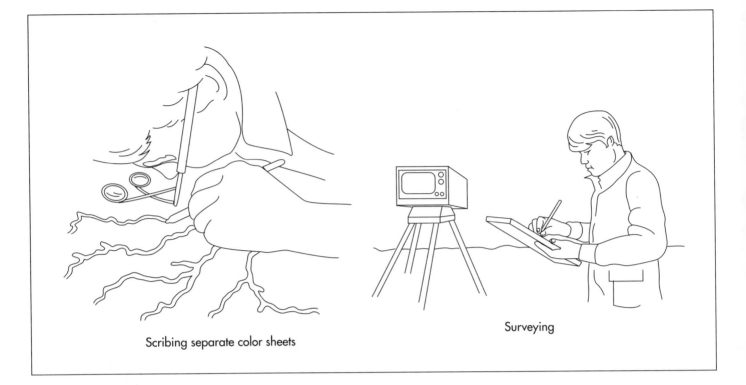

Scribing separate color sheets

Surveying

To ensure the accuracy of a map, the exact location of various control points must be established by field surveys. A separate scribecoat is made for each color used.

moving the pointer while keeping the two beams focused in a dot, the operator traces each contour line of the ground and the location of various features. The pointer is connected to a pen on the tracing table that draws the contour or feature being traced. All contours and features are drawn in black at this point. This process is called compiling the map manuscript.

7 When the tracing is completed, the finished map manuscript is photographed, and a map-sized film negative is made. This negative is photochemically reproduced onto several thin plastic sheets coated with a soft, translucent coating called a scribecoat.

Scribing and editing the map

8 The plastic sheets are taken one at a time and placed on a light table, where a soft light shines up through a white plastic surface. This illumination from below makes the lines of the map manuscript visible through the scribecoat. An engraver carefully cuts away the scribecoat along the lines and areas that are to be a certain color on the finished map. For example, one sheet will have all the lines for rivers, lakes, and other bodies of water that are to be blue. This process is repeated for each color.

9 Separate sheets for the lettering are prepared by placing a clear plastic sheet over each scribed sheet and carefully aligning the lettering with the features to be labeled. Type sizes, styles, and fonts are selected according to standards, which assure consistency and legibility from one map to another. A film negative is then made of each finished type sheet.

10 After the scribed sheets are reviewed and edited several times, a color proof sheet is made by exposing each sheet under different color light to produce a color print that looks very much like the finished map. After further review and editing, the map is ready to be printed.

Printing the map

11 A press plate is prepared for each map color by exposing the scribed sheets and the lettering negatives. Paper is loaded into a lithographic printing press, and the first color is printed. The press plate and ink are changed and the paper is run through the press a second time to print the second color. This process is repeated until all the colors have been printed. Some of the largest presses can print up to five colors in sequence without changing plates or reloading the paper.

Quality Control

The USGS uses the National Map Accuracy Standards set up in 1947. Starting in 1958, the USGS began testing the accuracy of their maps by field checking 20 or more well defined points on about 10% of the maps being produced each year.

For a 7.5 minute map at 1:24,000 scale, the horizontal accuracy standard requires that the locations shown on the map for at least 90% of the points checked must be accurate to within 40 ft (12.2 m) of the actual locations on the ground. The vertical accuracy standard requires that the elevations shown on the map for at least 90% of the points checked must be accurate to within one half of the contour interval on the ground. For a map with 10 ft (3 m) contour intervals, this means the elevations shown on the map must be accurate to within 5 ft (1.5 m) of the actual elevations on the ground. To give you an idea of what these standards mean to map makers, the horizontal accuracy standard requires that the location of at least 90% of the check points on the map must be drawn to within 0.02 in (0.05 cm) of the correct position.

The Future

Most of the topographic maps currently in use were produced manually. For mapmakers, however, the future is here today. A well-established network of navigational satellites form the basis of the Global Positioning System (GPS). This system allows field surveyors to accurately determine horizontal positions within a few feet, even in the most remote terrain where conventional surveying techniques are impossible.

Other satellites carrying a variety of sensors may soon replace the aerial photography method of making maps. The first of a series of Landsat satellites was launched in 1972, and by 1984 they could detect objects on the surface of Earth about 100 ft (30 m) in size. In 1998, an American company was preparing to launch a satellite that could detect objects as small as 3 ft (1 m), which would produce images with as much detail as current USGS 7.5 minute maps. More importantly, these images would be captured and transmitted as digital data, which could then be processed and printed by computers. This would significantly reduce the time required to produce or update maps and would improve the overall accuracy as well.

Where to Learn More

Books

Thompson, Morris M. *Maps for America*, 3rd edition. U.S. Department of the Interior, Geological Survey National Center, 1987.

Periodicals

Pike, Richard J. and Gail P. Thelin. "Building a Better Map." *Earth* (January 1992): 44-51.

Wilford, John Noble. "Revolutions in Mapping." *National Geographic* (February 1998): 6-39.

Other

"Map Accuracy Standards." U.S. Department of the Interior, U.S. Geological Survey, July 1996.

"Map Scales." U.S. Department of the Interior, U.S. Geological Survey, October 1993.

"Topographic Mapping." U.S. Department of the Interior, U.S. Geological Survey.

"Topographic Map Symbols." U.S. Department of the Interior, U.S. Geological Survey.

U.S. Geological Survey. http://www.usgs.gov.

—*Chris Cavette*

Toy Wagon

The most famous toy red wagon is produced by Radio Flyer Incorporated, a company was started by Antonio Pasin just before the 1920s. He was a craftsman who immigrated to the United States from Italy.

Background

A toy wagon is a four-wheeled toy consisting of a main body section and a steering handle. It is produced by a semi-continuous method, which involves making, painting, and assembling the various parts. First introduced as a toy in the 1880s, the basic wagon design has changed little over the years.

History

The best evidence suggests that wagons were first developed during the middle of the fourth century B.C. in Mesopotamia. These wagons were used for transporting various agriculture and building supplies. During the ancient Roman times, a major technical advancement was introduced which made wagons more useful. This was the invention of the pivoting front axle, which provided better steering and turning capabilities.

The toy wagon was first produced in the 1880s. These early toys were handcrafted and made completely of wood. Over the years, steel and plastic replaced wood as the material of choice. The most famous toy red wagon is produced by Radio Flyer Incorporated, a company was started by Antonio Pasin just before the 1920s. He was a craftsman who immigrated to the United States from Italy. The first wagons he produced by hand but soon found that the process was too slow to keep up with demand. When he adapted metal-stamping technology that was developed for the auto industry to his wagon business, he was able to produce large quantities daily. By the 1930s, his company was the world's largest producer of coaster wagons. Wagon production was slowed during World War II because steel used for non-war-related goods was rationed by the government. After the war however, the demand for wagons reached record levels and the company was back on track. Over the years, the manufacturer has continued to make minor improvements in the design of the wagon, which make them safer and more versatile toys.

Design

The typical toy wagon is made up of a main body portion, an undercarriage, wheels, and a steering handle. This design has changed little over the last hundred years. The main body is the riding or hauling part of the wagon. It is usually rectangular in shape. It has a flat bottom with walls on all sides and an open top. The height of the wagon walls is variable depending on the model type. In steel wagons, the top edges of the walls are curled under to prevent injury. Additionally, the corners are rounded for the same reason. The wagon body is rated for a maximum payload. Novel wagon bodies are also produced with such features as removable walls and storage areas.

The undercarriage provides support and an interface between the main body and the wheels. It is made up of support brackets and the wheel axles. The back axle is affixed to the wagon body and lets the wheels move in only one direction. The front axle is attached to a pivoting mechanism, which is fixed to the wagon body. This allows the direction of these wheels to be rotated.

On both axles, the wheels are attached through a ball bearing system, which ensures that they will freely rotate. The ball bearings are made of plastic and require occasional lubrication. The wheels come in various sizes and generally, larger wheels are needed for wagons that have a large load capacity. Some wagons have a braking mechanism, which slows or locks the wheels in place.

The steering handle is a long steel or plastic rod, which extends out from the front of the wagon. At the end of the rod is a handle, which is perpendicular to it. Newer designs have been developed since the traditional handle causes some discomfort. The entire steering device is attached to the front wheel assembly through a ball joint. This enables children, or cargo, to be pulled in the wagon's main body while providing a method for steering. For safety reasons, the turning radius of the wagon is limited to prevent the wagon from tipping over. The steering handle can be lifted up or down depending on the height of the operator. In some wagon models, the steering handle can be pulled or lifted up for operation by the passenger. All of the parts of the wagon are attached via screws, nuts, bolts, and various welds.

Raw Materials

A variety of raw materials are used to produce toy wagons. The main body and other parts of the wagon can be constructed from steel, wood, or plastic. When wagons were first produced as toys, they were constructed almost entirely out of wood. Some wagons are still made this way, however steel has largely replaced wood as the material of choice since it is stronger and more durable. The steel that is now used is a soft, draw quality material which is lightweight and sturdy. Plastic wagons made of high-density polyethylene (HDPE) are a relatively recent development and are a result of improvements in polymer technology.

To put the finishing touches on the wagon, other materials are used. Acrylic-based paints that are durable and non-toxic are used to decorate the main body and other wagon parts. Common colors include red, black, and white. To change the properties of the plastic and make them easier to work with, fillers are often added. Additionally, colorants are added to the plastic for decora-tive purposes. Finally, rubber is used for making the tires.

The Manufacturing Process

The parts of a toy wagon are made using a variety of methods such as metalworking, plastic molding, and die-casting. The pieces can then be assembled by the manufacturer or packaged and sold for home assembly. The exact manufacturing method depends on the material of the final wagon. In this section, the method for producing a steel red wagon will be discussed.

Forming the body

1 The wagon body is produced using a drawn steel stamping process. The sheet metal used for making the wagon body is supplied to the manufacturer rolled up in large coils. The coils are put on a de-spooling device, and the steel is fed through a powered straightener to the production line. The straightener ensures that the steel will have no kinks or twists. On the production line, the steel is moved to the cutting machine, which fashions it into rectangular sheets, which have the proper dimensions. The scrap steel is collected and recycled for later use.

2 The cut steel is conveyored to the molding station. Here, it is forced into a molded cavity, or die, by a high-pressure press. The press is then lifted, and the part is ejected from the die. At this point, it generally conforms to the size and shape of the wagon body. The body is next moved through a trimming device, which cuts off any excess steel. Then it is moved to a crimping machine, which grabs the top edges and rolls them under. This eliminates any sharp edges that might remain. Depending on the wagon type, the body may be welded to make it further reinforced.

3 To finish the wagon body, it must be treated and painted. To accomplish this, it is hung on a moving line. On this line, it is slowly carried through a washing machine to remove dirt and excess steel shavings. It then passes through a drying machine. Next, the bodies are moved through a series of spray painting machines, which cover them with various coatings including the red

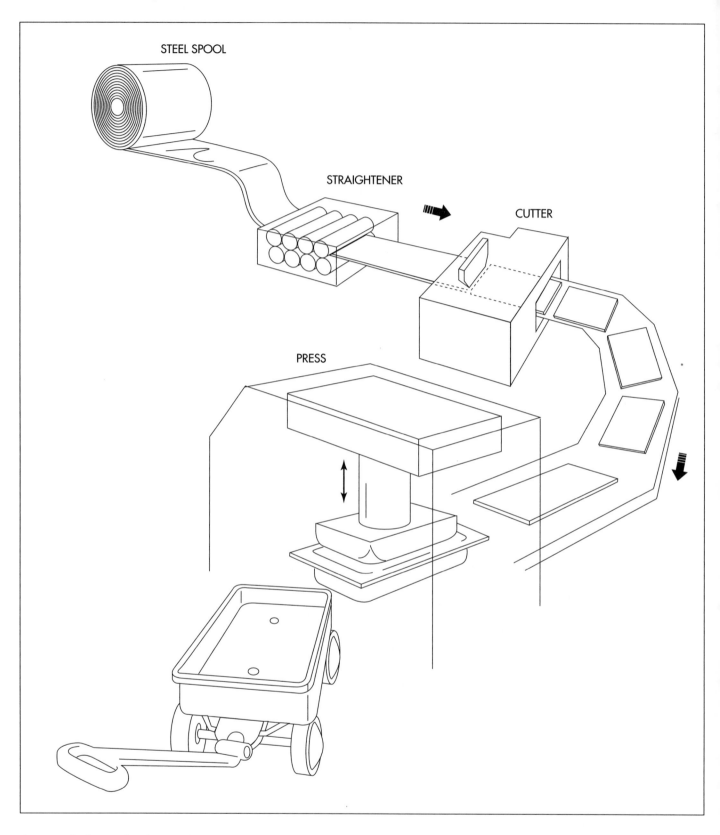

The wagon body is produced using a drawn steel stamping process. The steel is cut into rectangular sheets, which have the proper dimensions, and then molded into the shape of the wagon body.

paint. Finally, the bodies are moved through a large oven, and the paint is baked on. As they exit the oven, a stenciling machine is used to label the wagon. This entire coating process takes about 30 minutes.

Making the wheels and other parts

4 The other wagon parts are made in a variety of ways. The wheels are made with steel and rubber. Just like the wagon body,

the steel for the wheels are cut to the appropriate size and shape. In this case, circular steel shapes are used. After cutting, the steel is pressed to give it a bowl shape. Two of these steel bowls are welded together to form the rim of a single wheel. The wheel rim is painted, and after it dries a rubber tire is put on. Additional steel parts, such as the axles, undercarriage, and steering handle, are also made by using a metal stamping process. Cutting, welding, and painting complete the production of these parts.

5 The plastic parts are made using an injection molding process. This requires a machine, which converts plastic pellets into finished parts. The pellets are first melted and then physically injected into a two-piece mold. Inside the mold, the plastic is held under pressure for the required length of time and then cooled. As it cools, the plastic hardens. The mold is opened and the piece is ejected.

Packaging and assembly

6 When all of the pieces are ready, they are moved to a packing area. For pre-assembled wagons, trained workers put each of the wagon pieces together. The wagon is then covered in plastic and sealed in a box. If the product is sold unassembled, the parts are put in a box. In this process, a cardboard box is put on a continuous belt and workers place the various parts in the box as it passes by. The box is then sealed and then put on a pallet. It is then ready to be shipped from the plant to the local toy store.

Quality Control

The toy industry is directed by specific governmental regulations related to toy safety. Therefore, quality control is an important part of production. It begins with an inspection of the incoming raw materials and any finished parts that are used to produce the wagon. This includes such things as the steel, plastic resin, and paints. During various phases of production, visual inspections of parts, welds, and painting are performed by trained quality control technicians. Parts may be rejected if they are significantly damaged. In such cases, the part is removed from the production line and the steel or plastic is recycled.

The Future

Future developments in wagon production will likely be in the areas of wagon design and increased production speed. Companies such as Radio Flyer continue to receive patents on their new wagon designs. For example, a recent patent describes a toy wagon which has additional storage capacity built in. Another patent describes a wagon, which has a steering handle that is easier to pull. Future designs will have improved safety features or other novel design elements. From a manufacturing standpoint, improvements in steel and plastic technology should be adapted to increase production speed and make the process more automated. Computer design will undoubtedly play a bigger role.

Where to Learn More

Books

Panquin, J.R and R.E. Crowley. *Die Design Fundamentals*. Industrial Press Inc., 1987.

Seymour, R. and C. Carraher. *Polymer Chemistry*. Marcel Dekker Inc., 1992.

Smith, David. *Quick Die Change*. Society of Manufacturing Engineers, 1991.

Periodicals

"Little Red Wagon: Easy to Build & Easy to Love." *The Family Handyman* (November 1992).

Other

Pasin et. al. United States Patent #5,538,267, 1996.

Radio Flyer Inc. Grand Ave, Chicago IL 60639. (800) 621-7613.

von Braucke et. al. United States Patent #5,529,323, 1996.

—*Perry Romanowski*

Vinyl Floorcovering

By the late 1950s, resilient floorcoverings were here at last—these included vinyl flooring which gives slightly as one walks across it. These vinyl floorcoverings were far brighter and more colorful than linoleum because the vinyl floorcoverings were made with a clear vinyl gel which made the printed colors vibrant.

Background

Vinyl floorcovering is defined as either resilient vinyl sheet floorcovering or resilient vinyl tile floorcovering. Vinyl sheet floorcovering is generally available in either 6 ft (1.83 m) or 12 ft (3.66 m) widths and vinyl tiles are generally 12 x 12 in (30.48 x 30.48 cm). Sheet goods are generally retailed with no sticky backing. Thus, adhesives need to be purchased to adhere the flooring to underlayment. Vinyl tiles may be purchased dry or with a pressure-sensitive glue backing protected by a paper covering that must be removed by the installer.

Vinyl sheet flooring, particularly, varies in thickness and in manner decorated. Thinner sheet vinyl is 10-15 mils (mils are a thousandths of an inch) in thickness as opposed to longer wearing sheet vinyl (which is also more expensive) that may be 25-30 mils thick. Patterns may be printed with a rotary press (called rotogravure printing) or with large plates engraved intaglio with the design engraved below the surface of the metal. Both printing methods impress a pattern on top of the gel layer of foam and underneath the wear layer, rendering a relatively durable pattern.

These vinyl floorcoverings are preferred by many homeowners for their ease of installation—many do-it-yourselfers are able to install them with relative ease. However, vinyl sheet floorcoverings that are 25-30 mils may difficult to handle or install for the unskilled homeowner. The vinyl tiles are far easier to install and are the vinyl flooring most often installed by the homeowner. Furthermore unlike other flooring materials, sharp blades will easily cut the vinyl flooring so that it may conform to corners, cabinets, and curves.

In addition, vinyl floorcoverings may be applied over old flooring, and are easily cleaned with a vacuum or a mop with soap and warm water. To the delight of many householders, with proper care, many brands do not require waxing. Polyurethane coatings render a high-gloss finish that emulates a shiny, waxy surface and generally stays rather shiny over the years. If dulled, a special vinyl floor finish may be applied. However, vinyl floorcoverings are not as durable as ceramic tile and will have to be replaced periodically.

As with many household materials, vinyl sheet and tile flooring comes in residential grade and commercial grade. Residential vinyl flooring varies in thickness, method used for imprinting or decorating, and style. Commercial vinyl floorcoverings conform to specifications that require superior durability and stain resistance (particularly useful in hospital settings). Some commercial-grade vinyl floorcovering is non-slip for high public traffic areas.

History

Housewives have long known that a hard, relatively waterproof surface makes a fine floorcovering. Tamped earth mixed with ox blood dried into an easily-swept surface in primitive homes. Painted wooden floors were relatively easily maintained but had to be renewed at some effort. It was particularly troublesome keeping carpeting clean in eating areas such as dining rooms. In the early eighteenth century, floor cloths, which were large squares of fabric, were laid under dining tables to catch wayward crumbs. These floor cloths were simply taken outside and shaken free of crumbs and returned to their spot under the table.

However, later in the 1700s, someone decided that the floor cloth could be improved upon if a relatively sturdy fabric such as canvas, hemp, or linen was coated with and evaporating oil and paint and thus made waterproof. Easy to scrub and sweep, these floor oilcloths were quite an improvement over the fabric floor cloths. Better yet, these oilcloths were inexpensive and could be hand made at home or purchased mass produced later in the nineteenth century. From there it was a quick hop to linoleum which was manufactured from linseed oil, cork gums, and pigment. This leather-like floorcovering was mass produced by 1890 both in the United States and abroad.

Linoleum was tremendously popular from 1900 until after the World War II when floorcovering manufacturers sought to replace linoleum with other long-wearing fashionable, easily maintained floorcoverings. By the late 1950s, resilient floorcoverings were here at last—these included vinyl flooring which gives slightly as one walks across it. These vinyl floorcoverings were far brighter and more colorful than linoleum because the vinyl floorcoverings were made with a clear vinyl gel which made the printed colors vibrant. As a 1960 Sears Catalog proclaimed about vinyl floorcoverings: "All gloss and glow— no hard work!"

Raw Materials

The primary components of vinyl floorcoverings include polyvinyl chloride (or vinyl) resins, plasticizers (high molecular-weight solvents), pigments and trace stabilizers, and a carrier sheet or backing. The backing may be felt or highly filled paper made from wood pulp and calcium carbonate. High-gloss surface vinyl sheets or vinyl tiles have an additional polyurethane coating applied at the end of the process. The glue applied to the back of some vinyl tiles (to make a pressure-sensitive adhesive) is made from organic resins.

Design

The design departments of vinyl floorcovering companies are constantly seeking new inspiration for successful patterns and colors that will work in sheet and tile form. The designers work with marketing groups to determine what colors and styles will capture the public's interest as much as five years from current production. The designs are transferred from drawing to computer, and mock-ups of the different patterns are produced in an array of colors. The designs are then printed off of the computer on full-size paper and in full color.

If a full-scale paper pattern is approved for further development, printing plates either 18 x 24 in (45.72 x 61 cm) or 24 x 36 in (61 x 91.44 cm) in size are created by engravers. These plates are then used to print samples of the pattern on undecorated flooring (called gel stock) as prototypes. The wear layer, or final, often shiny surface of the sheet flooring, is applied over the printed pattern, so the designers have a close approximation of the finished product.

The prototype is either approved as is, retooled, or dropped. The time it takes from design to market varies from as little as three months or as long as six months.

The Manufacturing Process

Making the vinyl sheet floorcovering

1 Vinyl resins and plasticizers are stirred together in a vat to make a plastisol. To this plastisol, AZO compound (which consists of two nitrogen atoms that are united at both ends to separate carbon atoms) is added. When the resins, plasticizers, and AZO compound is heated, the AZO compound decomposes forming nitrogen gas bubbles. From this mixture, a vinyl foam is produced. This vinyl foam has the consistency of pancake batter and can be spread, in a slurry, onto the installation medium or backing.

2 The slurry is laid down on the felt or wood pulp backing via a reverse roll coater—it is poured on and smoothed out. The coated sheet then goes through an oven where the vinyl foam is gelled. The oven is heated just enough for the vinyl resin to absorb the plasticizer and set.

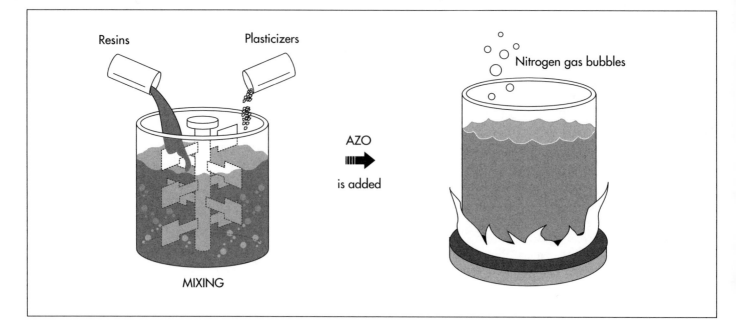

Resins   Plasticizers

Nitrogen gas bubbles

AZO

is added

MIXING

Vinyl resins and plasticizers are stirred together in a vat to make a plastisol, which is then heated to form a batter.

3 At this point, the gel is run through a printing press and is impressed with metal intaglio plates (with pattern carved beneath the surface of the metal). This impresses the pattern into the gel sheet, creating the decorative pattern.

4 A second mixing of plasticizer and vinyl are applied on the printed gel. The gel (with backing) is run through an oven at an even higher temperature. In the oven, the vinyl resin absorbs the plasticizer and melts, creating a clear vinyl. This is known as the wear layer, which takes the brunt of foot traffic. Printed patterns and inlaid patterns are thus protected under this wear layer rendering the pattern durable.

5 If the pattern requires a matte finish, the sheet vinyl is essentially ready to be rolled. However, patterns designed with a high gloss finish receive a layer of polyurethane coating via rollers. The thickness of this coating is controlled with an air knife to insure a consistent thickness. The polyurethane coating is cured photochemically with ultraviolet radiation lamps and is ready to be rolled.

6 The matte or high-gloss vinyl sheet flooring is then cut to rolls that are 12 ft (3.66 m) wide x 1,500 ft (457.2 m) long that can be subdivided based on the needs of retailers.

Creating vinyl tile floorcovering

7 Vinyl tiles are made a bit differently than vinyl sheet flooring. The polyvinyl chloride resins are mixed with calcium carbonate, plasticizers, and pigments in a large industrial mixer.

8 The mixture is heated to melting and consolidated. The friction from the mixing blades produces a compound with the consistency of bread dough. The dough-like substance is put through calendar rollers and the material is squeezed into sheets.

9 The sheets are embossed while still in rolls. Once decorated, the sheet is then cut into individual tiles with a die cutting machine, resembling multiple cookie cutters.

10 The tile squares (12 x 12 in or 30.48 x 30.48 cm) are cooled and put into a box if they are dry sheets (without pressure-sensitive glue on the back). If they are to receive glue for affixing to the floor, a roll-coater carrying organic resins deposits the glue on the tile backing. A paper cover that protects the glue is put over backing. The tiles are boxed (in boxes of 10 or 12) and ready for shipment.

Quality Control

As with most manufactured goods, all raw materials (polyvinyl chloride resins, plasti-

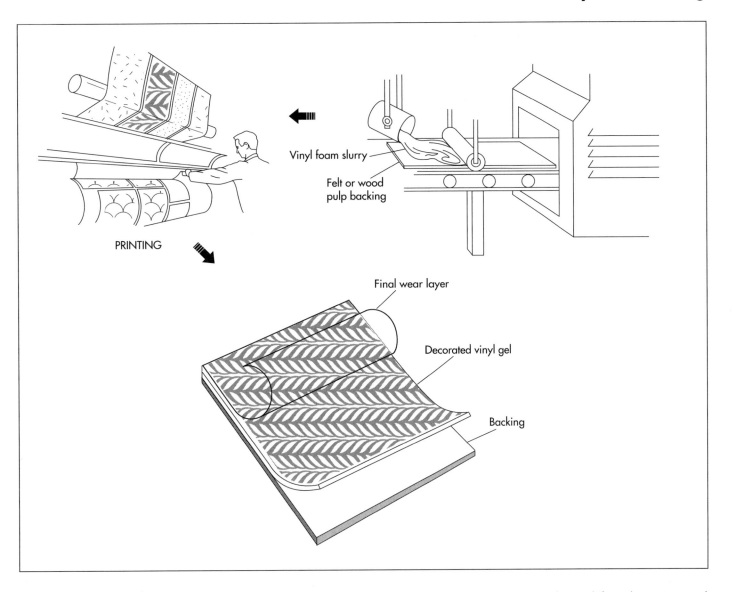

PRINTING

Vinyl foam slurry

Felt or wood pulp backing

Final wear layer

Decorated vinyl gel

Backing

cizers, pigments, stabilizers, and the installation medium) are checked to insure they meet minimum quality standards of production for the company. Felt or paper backings are checked for thickness and tensile strength. Physical tests are performed on coatings—viscosity, lumpiness, etc. are examined. If all ingredients are chemically and physically adequate to render a quality product, the manufacturing can begin.

Throughout all phases of production, intermediary checks are made to insure that standards are met. Members of the production staff perform visual checks on all pieces. If the larger roll does have contain an imperfection, quality control personnel adjust the computerized cut map that informs the cutting machines to cut around imperfections.

Commercial tiles and sheet vinyl used by Housing and Urban Development (HUD) and the Federal Housing Administration (FHA) undergo an array of testing including minimum thickness requirement, durability, flammability, etc.

Byproducts/Waste

No hazardous materials are unleashed into the environment as a byproduct of the manufacture of vinyl floorcovering. Waste products are either recycled at the point of manufacture or sent to a reclaimer for disposal. The heavy metal stabilizers and pigments used by the vinyl floorcovering industry were replaced years ago with those considered more environmentally safe. Presently, the industry is examining ways to utilize

The vinyl foam batter is spread onto the baking and heated, forming a sheet. The pattern is then printed on the flooring and a wear layer is applied.

waste vinyl which results from manufacture as well as recycle or reclaim the scrap vinyl floorcovering discarded by the consumer.

Where to Learn More

Books

Schuler, Stanley. *The Floor and Ceiling Book.* New York: M. Evans and Co., Inc., 1978.

Von Rosenstiel, Helene. *America's Rugs and Carpets.* New York: William Morrow and Company , 1978.

—*Nancy EV Bryk*

Water

Background

Water is a chemical compound needed by most plants and animals on Earth in order to sustain life. Pure water is a tasteless, odorless, transparent liquid. In small amounts it is colorless, but it takes on a bluish tint in larger amounts. Water is an excellent solvent and as a result it usually contains a wide variety of dissolved minerals and other chemicals. It can also carry and support bacteria. Most of the water distributed through municipal water systems is treated to remove harmful substances. Some bottled waters undergo even further treatment to remove almost all impurities. The English word water is derived from the German word *wasser*, which in turn is derived from an ancient Indo-European word meaning to wet or wash.

The controlled use of water dates to at least 8,000 B.C. when farmers in Egypt and parts of Asia trapped floodwaters for crop irrigation. The concept of using irrigation canals to bring water to crops, rather than waiting for a flood, was first developed about 2,000 B.C. in Egypt and Peru. By about 1,000 B.C., the city of Karcho, in what is now Jordan, built two aqueducts to bring an adequate supply of water for the city's population. This is the first recorded instance of a planned municipal water supply.

Early water treatment was surprisingly advanced, although rarely practiced. An ancient Sanskrit manuscript, from what is now India, advises that drinking water should be kept in copper vessels, exposed to sunlight, and filtered through charcoal. Ancient Egyptian inscriptions give similar advice. Many of these methods are still used today.

In about 400 B.C., the Greek medical practitioner Hippocrates suggested that water should be boiled and strained through a piece of cloth. Despite these early references, most people drank untreated water from flowing streams or subterranean wells. As long as there were no sources of contamination nearby, this was a satisfactory solution.

As the population of Europe and other parts of the civilized world grew, their sources of water became increasingly contaminated. In many cities, the rivers that served as the primary sources of drinking water were so badly contaminated with sewage that they resembled open cesspools. Cholera, typhoid, and many other water-borne diseases took their toll. In 1800, William Cruikshank of England demonstrated that small doses of chlorine would kill germs in water. By the 1890s, several municipalities found that slowly filtering water through beds of sand could also significantly reduce the incidence of disease. The public outcry for safe drinking water reached such a crescendo that by the early 1900s most major cities in the United States had installed some sort of water treatment system.

Even with water treatment, water contamination remained a serious concern as an increasing amount of industrial wastes poured into the nation's rivers and lakes. As the adverse health effects of lead, arsenic, pesticides, and other chemicals became known, the United States federal government was obliged to pass the Water Pollution Control Act of 1948. This was the first comprehensive legislation to define and regulate water quality. It was followed by a series of increasingly tougher requirements, culminating in the current Environmental Protection

A typical water district may perform more than 50,000 chemical and bacteriological analyses of the water supply each year to ensure the standards are being met.

Agency (EPA) water quality standards. In addition to the federal standards, most states have their own water quality laws, and some state laws are more stringent than those specified by the EPA.

Types of Water

Pure water is an almost non-existent entity. Most water contains varying amounts of dissolved minerals and salts, plus an abundance of suspended particles such as silt and microscopic organic material. Different types of water are classified by the presence or absence of these impurities.

Tap water, or municipal water, has undergone a series of treatments to kill harmful bacteria, remove sediments, and eliminate objectionable odors. It may also have had one or more chemicals added for a variety of reasons.

Hard water contains high amounts of calcium and magnesium salts. This causes soap to form curds. Hard water is further divided into temporarily hard water and permanently hard water. Temporarily hard water contains bicarbonates of calcium and/or magnesium, which react to form a hard substance called scale when the water is heated. Scale can clog hot water heaters and pipes and leave deposits on cooking utensils. Permanently hard water contains sulphates, chlorides, or nitrates of calcium and/or magnesium, which are not affected by heating. Soft water contains relatively low amounts of calcium and magnesium salts, although the definition of "low" varies. The term "softened water" refers to hard water that has had enough salts chemically removed to avoid forming soap curds. It is high in sodium chloride.

If water contains a large quantity of dissolved minerals, it is called mineral water. Mineral waters can be divided into five main classes: saline, alkaline, ferrunginous, sulphurous, and potable. Saline water has a high level of sodium or magnesium sulphate or sodium chloride. Alkaline water has a high concentration of salts which give it a pH in the range of about 7.2-9.5, where a pH of 7 is neutral and a pH of 14 is highly alkaline. Ferrungious water is rich in iron, which gives it a rusty color. Sulphurous water is rich in sulphur compounds and is distinguished by its rotten egg smell.

Potable water has a mineral content of less than 500 parts per million and is most commonly bottled and sold as a specialty drinking water.

Carbonated water, soda water, and sparkling water all contain dissolved carbon dioxide. This may occur naturally where limestone or other carbonate rocks are present, or the carbon dioxide may be added artificially under pressure.

Spring water and artesian water are distinguished only by the fact that they flow from the ground naturally without the aid of drilling or pumping. Otherwise, there is nothing that makes them different than water from other sources.

Distilled water has been purified by an evaporation-condensation process that removes most, but not all, impurities. Deionized water has been purified by an ion-exchange process, which removes both positive ions, such as calcium and sodium, and negative ions, such as chlorides and bicarbonates. It is sometimes called de-mineralized water. Purified water is municipal water that has undergone carbon filtration, distillation, deionization, reverse osmosis, ultraviolet sterilization, or some combination of these processes to remove almost all minerals and chemical elements, both good and bad.

Raw Materials

A water molecule consists of two atoms of hydrogen bonded to one atom of oxygen. The chemical symbol is H_2O. Water usually also contains a wide range of organic and inorganic materials in solution or suspension.

In the process of treating water for use in a municipal system, several chemicals may be added. These include disinfectants like chlorine, chloramine, or ozone; coagulants like aluminum sulfate, ferric chloride, and various organic polymers; acidity neutralizers like caustic soda or lime; and chemicals to help prevent tooth decay in the form of various fluoride compounds.

The Treatment Process

The specific water treatment process depends on the intended application. Some water, such the water used to irrigate crops,

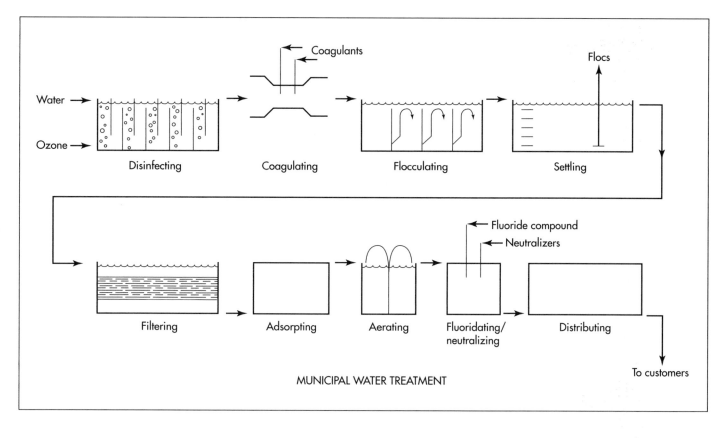

Coagulants

Flocs

Water →

Ozone →

Disinfecting   Coagulating   Flocculating   Settling

Fluoride compound

Neutralizers

Filtering   Adsorpting   Aerating   Fluoridating/   Distributing
neutralizing

To customers

MUNICIPAL WATER TREATMENT

receives no treatment. Other water, such as the water used to make pharmaceuticals, is highly purified.

Here is a typical series of operations used to treat municipal water for distribution to homes and businesses.

Collecting

1 Most municipal water comes from two sources: ground water and surface water. Most ground water is tapped by drilling wells into the underground water-bearing layer called the aquifer. Some ground water rises naturally in the form of springs. Surface water is tapped by impounding rivers behind dams. The surrounding area that drains into the rivers is called the watershed. In many cases, access to and use of the watershed is limited to prevent contamination of the runoff water.

2 From the well or dam, the water is carried to the water treatment plant in open canals or closed pipes. In some cases, the water supply is close to the municipality. In other cases, the water has to be transported many hundreds of miles (km) to reach its destination. Sometimes the water is stored in intermediate reservoirs along the way to ensure that there will always be an adequate supply available to meet a city's fluctuating needs.

Disinfecting

3 In some water treatment plants, the water is initially disinfected by contact with ozone-rich air in a series of chambers. This step is used by most plants in Europe, but only a few plants in the United States. Ozone (O_3) is formed by passing compressed air through a high-voltage electric arc. This causes some of the oxygen (O_2) molecules in the air to split in half and reattach themselves to other oxygen molecules to form ozone. Ozone effectively kills most germs and also destroys compounds, which cause unpleasant tastes and odors. It has a relatively short life, however, and does not remain in the water to protect it during storage and distribution. For this reason, a small dose of chlorine or chloramine is added to the water at the end of the treatment process.

Coagulating/flocculating

4 The water then passes through a flash mixer where chemicals known as coagu-

lants are rapidly mixed with the water. The coagulants alter the electric charge around any suspended particles in the water and make them attract each other and clump together, or coagulate.

5 The water moves slowly through a series of chambers where it is gently mixed by the swirling flow. As the water mixes, the charged particles continue to bump into each other and form even larger particles called flocs.

Settling

6 The water flows into a settling basin or tank where the heavy flocs sink to the bottom. Some settling basins have two levels to double their capacity. The material that settles to the bottom is vacuumed out of the basin with a device like a pool vacuum and is deposited in a solids holding basin. The trapped material from the filter (step 7) is also added to the solids holding basin. These combined materials are sent through a gravity thickener and then a press where most of the water is squeezed out. The remaining solids are loaded into trucks and transported to a landfill for disposal.

Filtering

7 The partially cleaned water passes through several layers of sand and pulverized coal, which trap any very small particles that remain in the water. Some harmful organisms are also trapped this way in those water treatment plants that do not use ozone as an initial disinfectant. The filter layers are back-flushed periodically to remove the trapped material.

Adsorpting

8 In some plants, the water is passed through a bed of activated charcoal granules. Chemical contaminants in the water stick to the surface of the charcoal in a process known as carbon adsorption.

Aerating

9 In some areas where the water contains undesirable amounts of iron and manganese or certain dissolved gases, the water is sprayed into the air from large basins to aerate it. When the water mixes with the air, it picks up oxygen, which causes some of the contaminants to settle out. Other contaminants are removed by evaporation.

Fluoridating

10 In some water treatment plants, a fluoride compound is added to the water to help prevent tooth decay. Fluoride occurs naturally in some water supplies and additional amounts are not required. In the past, fluoridation has been a hotly debated subject, and not every municipality adds fluoride to their water.

Neutralizing

11 Other chemicals may be added to the water to help reduce corrosion in pipes and plumbing fixtures. This is done by adding controlled amounts of certain chemicals to adjust the pH factor to a neutral level.

Distributing

12 As the water leaves the treatment plant, it receives a small dose of chlorine or chloramine to kill any harmful bacteria that may have found their way into the distribution system. If the plant does not use ozone as an initial disinfectant, a larger amount of chlorine or chloramine is added to the water.

13 After the water leaves the plant, it is usually stored in covered tanks or reservoirs to protect it from contamination. In some areas, these storage facilities are located at a higher elevation than the surrounding terrain, and the water is pumped up into the tank or reservoir. This elevated storage position provides the pressure necessary for adequate flow through the water mains and pipes within the city. In other cases, the water is stored in ground-level facilities, and the pressure is supplied by electric pumps that run on demand.

Quality Control

The federal and state water quality standards set maximum contamination levels for more than 90 organic, inorganic, microbiological, and radioactive materials that may be found in water. These standards are further divided into primary standards, which cover materials that may be harmful to humans, and secondary standards, which cover materials and properties that may affect aesthetic qualities such as taste, odor, and appearance. A typical water district may perform more than 50,000 chemical and bacteriological analyses of the water supply each year to ensure the standards are being met.

The Future

The public's concern over safe drinking water is expected to result in even more stringent water quality standards in the future. Ironically, one of the most recent concerns is not about outside contamination, but about the effects of one of the substances commonly used to disinfect water—chlorine. Studies within the last 30 years have shown that chlorine forms certain compounds with the organic materials found in water. The most common compounds are called trihalomethanes, or THMs, which have a 1-in-10,000 risk of causing cancer when ingested or inhaled over a long period. One alternative to using chlorine is chloramine, which is a combination of ammonia and chlorine that does not form THMs as readily. Many water treatment plants have already switched to chloramine. Other alternative disinfectants include ozone, ultraviolet light, chlorine dioxide, and a hybrid of ozone and hydrogen peroxide called peroxone.

Where to Learn More

Books

von Wiesenberger, Arthur. *H₂O: The Guide to Quality Bottled Water*. Woodbridge Press, 1988.

Water Quality Standards Handbook, 2nd edition. United States Environmental Protection Agency, 1994.

Periodicals

Arrandale, T. "A Guide to Clean Water." *Governing* (December 1995): 57-60.

Wasik, J. F. "How Safe is Your Water?" *Consumers Digest* (May/June 1996): 63-69.

Other

"Alameda County Water District Water Treatment Facility." Pamphlet. Alameda County Water District, 1993.

"Layperson's Guide to Drinking Water." Pamphlet. Water Education Foundation, 1995.

Los Angeles Department of Water and Power. http://www.ladwp.com.

—*Chris Cavette*

Wet Suit

An unclothed diver entering water cooled to 50° F (10° C) would only survive in such temperatures for approximately 3.5 hours. A diver wearing a diving suit would survive for approximately 24 hours in water of the same temperature.

Background

Underwater, or deep sea diving is a popular recreational sport, and is also necessary for underwater rescue, salvage, and repair operations. Such activities often require diving to great depths in very cold water. Even in warm climates, the ocean can be very cold at great depths. For protection from such temperatures and the prevention of hypothermia, underwater divers wear diving suits, which keep them warm by preserving their body heat. By way of example, an unclothed diver entering water cooled to 50° F (10° C) would only survive in such temperatures for approximately 3.5 hours. A diver wearing a diving suit would survive for approximately 24 hours in water of the same temperature.

There are two basic classifications of diving suits: the helmet suit, which completely encloses the diver and contains a breathing apparatus that fits over the head, and the scuba suit, also known as the free-diving suit. Scuba is an acronym for Self-Contained Underwater Breathing Apparatus. The Scuba suit is used in tandem with an independent breathing apparatus strapped to the diver's back. There are two types of scuba suits. The dry suit keeps the diver completely dry; a diver can even wear clothes under a dry suit. The wet suit, on the other hand, holds a thin layer of water between the body of the diver and the suit. This water is warmed by the body and serves as insulation, along with the suit, against cold water.

History

The concept of underwater diving evolved with the invention of the diving bell, a large, bell-shaped chamber into which air was pumped from above surface and in which a diver could be transported below water. Early diving bells were made from open-ended metal-rimmed wooden barrels. This invention dates back to antiquity and may have been used by Alexander the Great. Aristotle also tells of the existence of such an invention. In 1665, a diving bell was engaged on a gun salvaging mission to a shipwrecked Armada vessel. English astronomer Edmund Halley is credited with devising the first modern diving bell in the early 1700s. Halley's bell utilizing lead containers filled with fresh air that were lowered to the bell from the surface in order to replenish the air. Halley later developed a helmet that allowed the diver to leave the bell while remaining attached to the air supply system. Toward the end of the century, British engineer John Smeaton incorporated an air pump into the diving bell's design, allowing for a constant supply of fresh air. Later bells were sealed with glass at the bottom. Bells are still used today and can carry up to four divers. They can travel to depths of 1,000 ft (304.8 m).

It also has been recorded that an Egyptian diver named Issa developed a breathing machine for use during the wars between the Crusaders and the Saracen in the twelfth century. Issa's breathing machine included a bellows and allowed him to remain under water for long periods of time. He kept himself just below the water's surface by tying stones to his belt.

Six centuries later, John Lethbridge of Devon created a six-foot-long diving tube. The tube was designed to allow Lethbridge to lay horizontally inside of it with his arms protruding from the apparatus. Air was pumped in from above water using bellows.

Lethbridge stayed underwater inside his apparatus for up to six hours at a time, and was contracted to salvage treasure hulks from underwater areas across the world.

The helmet suit is a variation on this invention and functions as a portable diving bell. Like the diving bell, air is pumped into the helmet from above the water's surface. The suit itself is composed of rubberized fabric. The diver enters the suit through a hold in the neck. The helmet is attached to the suit with a waterproof seal. The air is pumped into the helmet, which has glass ports for vision, at the pressure of the surrounding water. This is known as ambient pressure. Expired air is expelled through an outlet valve. A line is attached to the suit, which allows the diver to be hoisted to the surface. Modern helmet suits are also typically equipped with a telephone line, allowing the diver to maintain voice contact with people above the water.

While the helmet suit allows a diver to stay underwater for long periods, due to the constant supply of air, it does not allow for much mobility. Free-diving, or scuba, suits, on the other hand, are variations on this innovation that allow for increased mobility. They are used in conjunction with fins for the diver's feet, a diving mask, and the independent breathing apparatus, known as an aqualung. The dry suit is loose fitting, allowing for clothing to be worn underneath it, and equipped with waterproof seals at the neck, wrists and, on some, the waist. The dry suit traps air, however, and that air is compressed as the diver swims deeper and the volume of air decreases. This compression makes the suit rigid and inhibits the diver's movement. In addition, the diver's skin can get caught and pinched in the folds of the suit, causing welts. The wet suit, therefore, is preferable in many situations. The dry suit, however, is better suited to extremely cold water temperatures as it allows the diver to wear warm, dry clothing underneath. It is also more protective from elements, which makes it more desirable in polluted water. Disinfectant can also be poured over the dry suit.

The wet suit was adapted from the dry suit and is made from close-fitting, foam rubber-like material. The wet suit is not water tight. Rather, water seeps into and under the suit and is trapped between the suit and the diver's skin. The diver's skin warms the water, and the water acts as a second layer of insulation, with the foam-like substance, which traps air bubbles, providing the first. Warm water may also be poured into the suit before the dive takes place. A disadvantage of the wet suit is that the air bubbles cause buoyancy, requiring the diver to wear a weighted belt. As the diver descends, the ambient pressure shrinks each air bubble, resulting in a loss of both buoyancy and insulation. Thus, the diver becomes much heavier. Products that help compensate for this loss of buoyancy include an adjustable buoyancy life jacket, which is affixed to a cylinder of compressed air. As the diver descends, he or she can let air into the jacket to increase buoyancy, and on ascending the air can be released. A significantly more expensive wet suit utilizes air-filled gas bubbles, rather than the foam bubbles, to help maintain buoyancy.

Recent innovations in diving suit technology include a hot water suit. This sealed suit is provided with hot water from above the surface. The hot water flows through a series of passageways in the suit and exits through valves, to allow for a constant flow of hot water. This suit is commonly used in saturation diving, where the diver breathes a mixture of helium and oxygen. Helium conducts heat faster than air, so a diver breathing this mixture is at greater risk of hypothermia.

In the 1970s, as companies seeking oil below the ocean floor desired to dig their wells even deeper, an old technology was resurrected for divers who repaired the wells. In the 1920s, an engineer named Joseph Peress designed an atmospheric diving suit (ADS), a massive-sized apparatus that allowed a diver to breathe air at normal atmospheric pressure. Peress had used his early ADS to seek shipwrecks in the 1930s and tried to market it to the Royal Navy, which had no use for it. Peress' early ADS was made from cast magnesium and plexi-glass, coated with waterproof sealant. A cushioned ball-and-socket joint system allowed the diver relative freedom of movement. An ADS can carry enough air to last 72 hours. Today, the ADS is made from welded aluminum or glass fiber.

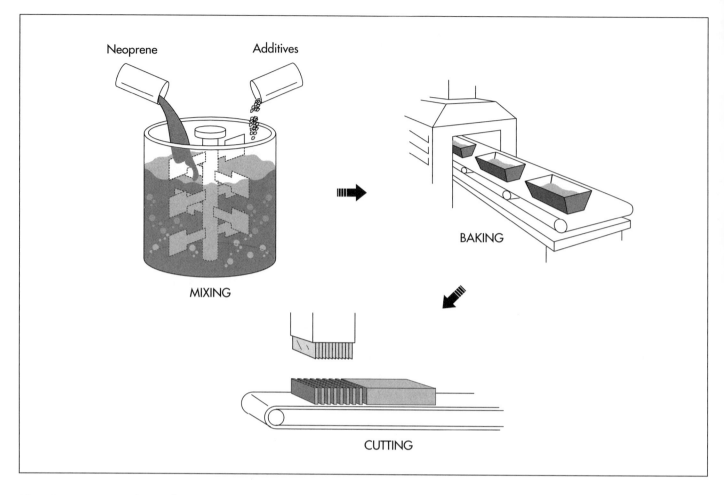

Neoprene   Additives

MIXING

BAKING

CUTTING

After mixing neoprene with several additives, the liquid is baked in loaf-like shapes. Once baked, the material is sliced into preset widths.

Raw Materials

The primary raw material used in the making of a wet suit is a type of sponge rubber known as neoprene. The dry suit utilizes a rubberized fabric. Some metal is also used for zippers.

The Manufacturing Process

The manufacturing process for both types the wet and dry scuba suits is similar. Both are constructed in assembly-line fashion. Here, the process for the manufacture of the wet suit is outlined.

1 The neoprene arrives at the factory in liquid form. The manufacturer adds additives to the liquid and it is mixed with an industrial-sized mixer.

2 Next, the liquid is baked in a large oven. The baked product measures about two feet tall and resembles a huge loaf of rubber bread.

3 The baked rubber is allowed to cool.

4 The cooled rubber is run through a slicing machine, which cuts the large mass lengthwise. The slicing mechanism is set to a specified thickness, typically 0.12, 0.24, or 0.28 in (3, 6, or 7 mm). (Divers wear wet suits of different thicknesses, depending on the type of water they are diving in.)

5 The sheets of rubber, each about the size of a sheet of plywood, are then placed on a conveyor belt where they are lifted and sprayed with glue.

6 The sheets of rubber are laminated with a form of nylon that is squeezed onto the rubber. The laminated nylon is then allowed to dry. Once dry, the nylon-bonded rubber becomes stretchy.

7 The rubber is then inspected and divided up by grade, or thickness.

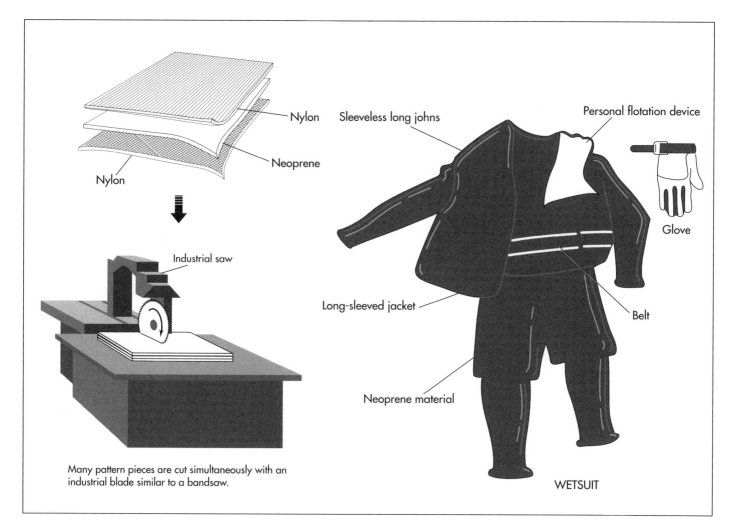

Many pattern pieces are cut simultaneously with an industrial blade similar to a bandsaw.

WETSUIT

8 The sorted rubber is loaded onto a palette and sent to the wet suit makers.

9 The wet suit maker stretches a stack of rubber about 10-15 sheets high and lays a pattern over the top.

10 The pattern is traced with a white crayon.

11 A sawing machine cuts suit panels out of the stacks of rubber, following the crayoned-on pattern. Dry suits are hand-cut.

12 The suit panels are sent to the decal department, where decals are adhered using a heat press.

13 Next, the panels are sent to the zipper department where zippers, pockets, kneepads, and flatwork are sewn or pressed on.

14 The panels are then sent to the gluing department, where each panel is coated with neoprene cement. The open sides of the front and back panels are glued together and arms and legs are attached to the body of the suit.

15 The suits are sent to the final sewing area, where nylon thread is used to stitch the seams of the suit.

16 The suit is cleaned and inspected for quality and tags are adhered using stitching or a hot press.

Quality Control

Most quality control for diving suits is conducted along various stages of the manufacturing process and/or at the end of the line. Finished suits may also be spot-tested for durability and water resiliency.

The nylon-neoprene-nylon layered material is cut into pattern pieces and the wet suit is sewn.

The Future

Diving suit designs have changed fairly little over time. New colors and styles of suits and the patches and logos applied on them occur on the market regularly, and small design alterations may be made to established designs to improve comfort or enhance durability and/or water resiliency. New technology is always being explored, such as that which led to the creation of the hot water suit. The diving suit has also gone high-tech, with the utilization of the ADS, and it is likely further technological advances will lead to updates and alterations of that product.

Where to Learn More

Books

Clinton, Larry. *The Complete Outfitting and Source Book for Sport Diving.* Henry Holt, 1979.

Desiderati, Barbara M. *Pictorial History of Diving.* Best Publishing Co., 1988.

Farley, Michael B. *Scuba Equipment Care and Maintenance.* Marcor Publishing, 1980.

—*Kristin Palm*

Whistle

Background

A whistle is a simple device that produces sound when air is forced through an opening. Their loud, attention-getting blast makes whistles essential for police officers and sports referees. They can save lives when used by lifeguards, lost campers, or crime victims. Innumerable organizers and leaders, ranging from teachers to drum majors, use them to focus attention and demonstrate authority.

In addition to those blown by humans, mechanically operated whistles serve many purposes. For example, a form of whistle can measure the rate of flow of a fluid in a manufacturing process and even control fluid flow. Whistles can signal the escape of some sort of gas, the ordinary tea kettle being the simplest example. Whistles that are activated when a filter becomes clogged have been designed for a variety of devices including home furnaces, automobile catalytic converters, and hair dryers.

Complex aerodynamics are involved in the operation of a whistle. In the familiar design known as the American police whistle, air blown through the mouthpiece travels down a rectangular tube until it encounters a slot at the top of the whistle. The far edge of the slot slices the airflow into two parts. The top portion is deflected upward in curls like rind being peeled off a fruit, forming swirling vortexes of air. This whirling stream causes vibrations in the air above the slot, generating sound waves. The lower portion of the inflowing air is deflected downward into the barrel-shaped chamber of the whistle, where it swirls around the curved wall until it once again reaches the top. It then pushes up through the slot and reinforces the vibrations being caused by the whirling upper layer of the airstream.

Such a whistle does not need a ball in the chamber to produce an effective sound. However, if a ball is present, it too swirls around inside the chamber, alternately blocking and unblocking part of the exit slot. This action produces a warbling alternation of tone that makes the whistle's sound more attention grabbing.

So-called "pea-less" whistles operate without the presence of a ball in the chamber. Often, they use a combination of chamber shapes to produce multiple tones that make the whistle's sound more audible above other noises such as cheering crowds or howling wind. One such whistle is designed so that the different tones are produced out of phase with one another, alternately canceling and reinforcing each other, to produce a trilling effect.

Various manufacturers claim to produce the "world's loudest whistle." Objective loudness is based on a decibel rating, which can be measured for each whistle; the loudest ones produce sound of about 120 decibels. Subjective evaluations of loudness reflect the fact that certain pitches of sound seem to be louder than others even at the same decibel level. Thus, in a noisy indoor arena a higher-pitched whistle might appear to be louder than a lower-pitched whistle that has a slightly higher decibel rating.

History

Whistles made of bone or wood have been used for thousands of years for spiritual,

The modern era of whistle use began in 1878 when a whistle was first blown by a referee during a sporting event. Joseph Hudson, a toolmaker who was fascinated with whistles, fashioned a brass instrument that was used in a match at the Nottingham Forest Soccer Club. This device was found to be superior to the usual referee's signal of waving a handkerchief.

practical, and entertainment purposes. One of the most distinctive whistles is the boatswain's pipe used aboard naval vessels to issue commands and salute dignitaries. It has evolved from pipes used in ancient Greece and Rome to keep the stroke of galley slaves. A medieval version was used during the Crusades to assemble English crossbow men on deck for an attack. The model currently being produced by the Acme Whistle Company of Birmingham, England, was first manufactured in 1868 by the company's founder, Joseph Hudson.

The modern era of whistle use began in 1878 when a whistle was first blown by a referee during a sporting event. Hudson, a toolmaker who was fascinated with whistles, fashioned a brass instrument that was used in a match at the Nottingham Forest Soccer Club. This device was found to be superior to the usual referee's signal of waving a handkerchief.

In 1883, the London police force made it known that it was seeking an alternative noisemaker to replace the heavy, cumbersome hand rattle the officers had been using. Hudson invented a light, compact whistle that produced two discordant tones that could be heard for more than a mile. It was immediately adopted and the same design is still in use today.

The following year, Hudson invented the "pea whistle." Movement of a small ball enclosed in the whistle's air chamber produces the familiar trilling effect now commonly associated with American police and referee whistles. The pea whistle remains the world's largest-selling type.

Raw Materials

Manufactured whistles are made of either metal or plastic. The only metal whistles manufactured in the United States are made of brass (an alloy of copper and zinc); the fact that they are nickel or chrome plated gives rise to a common misconception that they are made of steel. Brass is used because of the same tone and resonance qualities that make it effective in musical instruments. Because brass is a relatively expensive metal, these whistles sell for about $3.

Despite common terminology, the ball contained in pea whistles is actually made of cork (except for inexpensive, plastic versions). American Whistle Corporation has developed a synthetic cork material that behaves like natural cork in every respect except that it does not absorb any moisture. This helps keep the ball from getting stuck inside the whistle and not swirling freely.

Because metal whistles are hard and subject to temperature fluctuations in very hot or cold weather, rubber mouthpieces are manufactured as an optional accessory. Besides providing a cushioned mouth grip and a reliably comfortable temperature, the accessory also provides a mechanism for adding color to the metal whistle.

Plastic whistles were first manufactured in Britain in 1914, after earlier attempts to produce a satisfactory model from vulcanite (hardened rubber) had failed. This allowed design variations and colors that became popular with consumers. The components of modern plastic whistles may be either glued or ultrasonically welded together.

The Manufacturing Process

The following process is used to manufacture the type of metal pea whistle that is commonly associated with American military, police, and sports applications.

1 Sheets of brass are unrolled from supply reels and fed into stamping machines. Blanks are die-cut for the whistle's three primary pieces: the rectangular air-input tube, the barrel-shaped chamber, and a fitting for the back of the whistle that will hold a ring to which the user can attach a lanyard (cord). Different grades (hardnesses) of brass are used for the two main functional pieces of the whistle.

2 The pieces pass through a series of stamping dies and presses that perform a progression of cutting and bending operations to form them into the required shapes. The company name (or a customer's custom logo) is stamped onto what will become the top of the whistle.

3 Custom-designed machines hold the three parts of the whistle in position

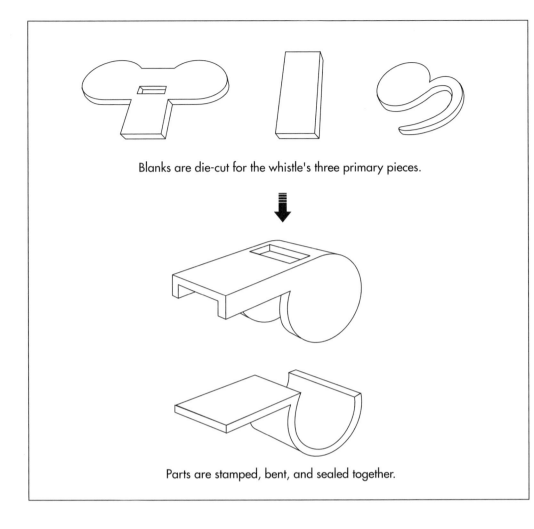

Blanks are die-cut for the whistle's three primary pieces.

Parts are stamped, bent, and sealed together.

Pea whistle blanks are die-cut for the whistle's three primary pieces: the rectangular air-input tube, the barrel-shaped chamber, and a fitting for the back of the whistle that will hold a ring to which the user can attach a lanyard (cord).

while solder is applied to the joints and the assembly is heated, sealing the parts together.

4 Each whistle is mounted in another machine, where a vibratory sander smooths the exterior surface to a pre-plate finish.

5 A nickel or chrome plating is applied to the exterior of the whistle.

6 The cork ball is inserted in the whistle. A machine compresses the cork and shoots it through the slot on top of the whistle. Once inside, the cork returns to its original shape, making it too large to fall out through the slot.

7 A metal lanyard ring is inserted into the holder on the back of the whistle.

8 Standard whistles are packaged in a plastic bubble attached to a cardboard backing. More expensive versions, suitable for use as gifts or awards, are mounted in a plastic box or a wooden case.

Innovations

With the exception of cosmetic touches like a gold-plated, rose-engraved, cubic zirconia-studded safety whistle for a loved one, the metal whistles manufactured by the American Whistle Corporation have not changed substantially in the past 40 years. Similarly, the Acme Whistle Company's European-style tubular, pea-less police whistle is the same model it has been producing for more than 100 years. The designs are classic.

On the other hand, inventors continue to develop new types of whistles. In 1987, Ron Foxcroft, a Canadian basketball referee, invented a plastic, pea-less whistle that produces a warble much like the traditional referee-style whistle. In 1992, the All-Weather Safety Whistle Company was founded to

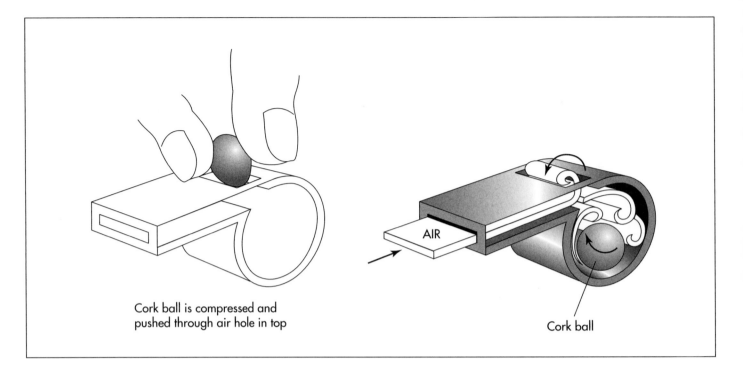

Cork ball is compressed and
pushed through air hole in top

AIR

Cork ball

The cork ball is inserted in the whistle. Once inside, the cork returns to its original shape, making it too large to fall out through the slot. When air and the cork ball collide, it produces the familiar trilling effect.

manufacture and market a whistle invented by Howard Wright, an American dentist; this unusual whistle is designed to work even under water.

U.S. patent applications reveal ingenious adaptations of whistles for a mind-boggling array of applications. Some are developed to meet serious needs, like a device that sounds an alarm tone when there is an insufficient flow of gas in an anaesthetic apparatus. Some serve a practical purpose of less-compelling importance, like a sensor that whistles when a piece of meat has been cooked to the desired doneness. Some are purely entertaining, like a frozen confection with an embedded whistle. And some are multipurpose, like the instant-cooling apparatus for a beverage can that not only chills the contents when activated but also expels a colorful gas while emitting an amusing whistling sound.

Where to Learn More

Periodicals

Chanaud, Robert C. "Aerodynamic Whistles." *Scientific American* (January 1970): 40-46.

Meeks, Fleming. "Whistle Blower." *Forbes* (April 12, 1993): 104-105.

Other

American Whistle Corporation. http://www.americanwhistle.com (March 20, 1998).

"The Fox 40 Pea-less Whistle Story." Fox 40 International Inc. http://www.fox40whistle.com/about1.htm (March 2, 1998).

"Safety Whistles." All Weather Whistle Company. http://www.9am.com/whistles/index.html (March 25, 1998).

"A Whistle Stop Tour Through More Than 125 Years of Sound Development." Acme Whistle Company. http://www.webxpress.co.uk/acme/history.html (March 25, 1998).

—*Loretta Hall*

Wooden Clog

Background

Wooden clogs are heavy work shoes that were typically worn by French and Dutch peasants up through the beginning of the twentieth century. Known in French as *sabots*, and in Dutch as *klompen*, these sturdy shoes protected the feet of agricultural workers from mud and wet and from injury by the sharp tools used in the field. French clogs were often made from a combination of wood and leather. However, the classic Dutch clog is entirely wooden. Wooden clogs are naturally highly water resistant, and therefore they were especially useful in the marshy fields of the Netherlands. Farm workers also wore specially decorated wooden clogs to church and on holidays. In World War I, entrenched soldiers wore wood and leather clogs called sabotines. Up through this time, clogs were typically made by hand. Later, industrialization made leather and rubber shoes more readily available, and wooden clogs became less widespread. However wooden clogs are still worn by Dutch farm workers, and also by Dutch fishermen and steel factory workers. Clogs made a resurgence in the 1960s across Europe and North America, not as a work shoe but as fashion. They are still popular in the 1990s. These modern clogs are usually a leather shoe attached to a wood sole. Clogs made entirely from rubber are also popular as gardening shoes.

Raw Materials

Wooden clogs are usually made from one of three kinds of wood: European willow, yellow poplar, or tulip poplar. These woods are all hard and water resistant. After the lumber is cut, it is not treated in any way, but made into shoes as soon after felling as is practical. No other material is necessary to make wooden clogs, though some shoes are varnished or decorated with paint.

The Manufacturing Process

Wooden clogs were traditionally made entirely by hand, either by their wearers or by specialized artisans. The shoes were roughly carved on the outside, then clamped into a bench that held them vertically, toe down. Then the artisan scooped them out with a long-handled tool. Less than a hundred years ago, a wooden clog factory might consist of dozens of workers making shoes in this same manner, by hand. The introduction of automated machines sped up the process, though machines still required attentive operators.

Making the blanks

1 The willow or poplar trees are felled and sawn into logs. The logs are debarked, then fed into a saw, which cuts them into rough rectangular blocks. Each block, called a blank, will be formed into one shoe. The size of the block varies depending on what size shoe is to be made out of it. For a men's size 8 shoe, the block might be 14.5 x 5.25 x 5.25 in (37 x 13.3 x 13.3 cm).

Shaping

2 Two blanks are placed into a machine called a shaper (also known as a copier or duplicator). This shapes the outside of the shoes. Next to the blanks is a vinyl shoe, which is used as a pattern. Each shoe size has its own vinyl pattern, and the machine operator locks the appropriate pattern into

Known in French as sabots, *and in Dutch as* klompen, *clogs protected the feet of agricultural workers from mud and wet and from injury by the sharp tools used in the field. French clogs were often made from a combination of wood and leather. However, the classic Dutch clog is entirely wooden.*

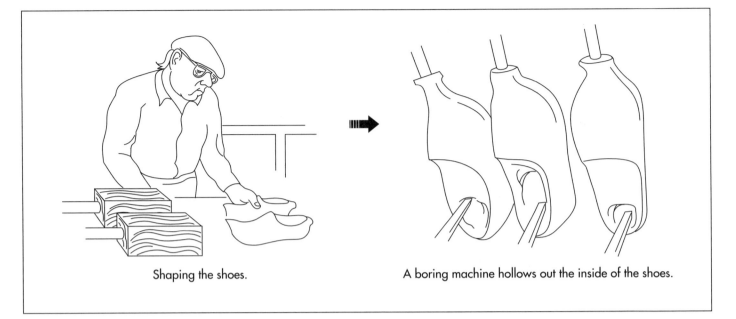

Shaping the shoes.   A boring machine hollows out the inside of the shoes.

the shaper. A pointer is set to ride along the pattern shoe. Attached to the pointer are two electrically powered cutting tools. These are set to the right and left shoe blanks. The machine operator turns the power on, and carefully traces the outline of the pattern shoe with the tracer. The cutting tools follow the motion of the tracer, and carve out the outline of the shoe. The two blanks rotate in opposite directions, allowing a left and a right shoe to be carved simultaneously.

Carving the interior

3 Next, the carved blanks are placed in another machine called a dual action borer. This machine has a three-pronged cutting implement. The center prong is a tracer, and this goes inside another vinyl pattern shoe. The right and left prongs are set to the right and left shoe blanks. Their cutting ends are sharp-edged scoops similar to ice cream scoops or melon ballers. The operator holds a long metal rod attached to the tracer prong, and pushes this along the inside of the pattern shoe. The cutters follow the tracer's movement, and scoop out the wood blocks. This machine carves out the interior of the shoes to its approximate finished dimensions, leaving an extra 0.25 in (0.64 cm) of material all around.

Refining

4 The shoes are placed in a similar machine called a refiner, which is in this

case entirely automatic. Two cutters follow a pointer on a vinyl pattern and scoop out the inside of the shoes, trimming away the excess 0.25 in (0.64 cm) of material left by the previous step. The fine action of this machine leaves the interior of the shoes extremely smooth, and they need very little finishing after this point.

Drying

5 The shoes are left to air-dry for four to six weeks. They may be simply placed in a dry storeroom, or they may set in a low temperature furnace, which circulates warm dry air around them. As they cure, moisture is drawn out of the wood, and the shoes harden.

Finishing

6 After the shoes are completely dry, workers sand them lightly inside and out. At this point the shoes are completely finished and ready to wear. If the shoes are to be decorated, they are painted or varnished after sanding.

Where to Learn More

Books

Rowland, Della. *A World of Shoes.* Chicago: Contemporary Books, 1989.

Once sanded, the wooden clogs are decorated and then varnished.

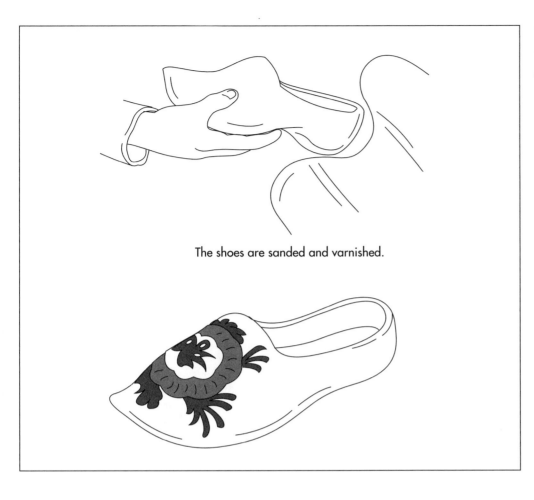

The shoes are sanded and varnished.

Yue, Charlotte. *Shoes: Their History in Words and Pictures*. New York: Houghton-Mifflin, 1997.

Periodicals

Chargot, Patricia. "Clompin' Around." *Detroit Free Press* (March 23, 1998).

Kuniholm, Erin. "Going Dutch: Wearing Clogs Is the Next Best Thing to Going Barefoot." *Women's Sports and Fitness* (October 1997): 82-84.

—*Angela Woodward*

Yogurt

In the body, it is thought that yogurt can encourage the growth of beneficial bacteria in the gut. These organisms help to digest food more efficiently and protect against other, harmful organisms.

Background

Yogurt is a dairy product, which is made by blending fermented milk with various ingredients that provide flavor and color. Although accidentally invented thousands of years ago, yogurt has only recently gained popularity in the United States.

It is believed that yogurt originated in Mesopotamia thousands of years ago. Evidence has shown that these people had domesticated goats and sheep around 5000 B.C. The milk from these animals was stored in gourds, and in the warm climate it naturally formed a curd. This curd was an early form of yogurt. Eventually, a process for purposely producing yogurt was developed.

While yogurt has been around for many years, it is only recently (within the last 30-40 years) that it has become popular. This is due to many factors including the introduction of fruit and other flavorings into yogurt, the convenience of it as a ready-made breakfast food and the image of yogurt as a low fat healthy food.

Manufacturers have responded to the growth in the yogurt market by introducing many different types of yogurt including low fat and no-fat, creamy, drinking, bio-yogurt, organic, baby, and frozen. Traditional yogurt is thick and creamy. It is sold plain and in a wide assortment of flavors. These are typically fruit flavors such as strawberry or blueberry however, newer, more unique flavors such as cream pie and chocolate have also been introduced. Cereals and nuts are sometimes added to yogurts. Yogurt makers also sell products with a varying level of fat. Low fat yogurt, which contains between 0.5% and 4% fat, is currently the best selling. Diet no-fat yogurt contains no fat at all. It also contains artificial sweeteners that provide sweetness while still reducing calories. Creamy yogurt is extra thick, made with whole milk and added cream. Drinking yogurt is a thinner product, which has a lower solids level than typical yogurt. Bio-yogurt is made with a different type of fermentation culture and is said to aid digestion. Yogurt that is made with milk from specially fed cows is called organic yogurt. This type of yogurt is claimed to be more nutritious than other yogurts. Other types of yogurts include pasteurized stirred yogurt that has extended shelf life, baby yogurt made specifically for children, and frozen yogurt.

The yogurt itself has a generally aldehydic flavor, which is a result of the fermentation process. Since it is made from milk, yogurt is rich in nutrients. It contains protein and vitamins and is a rich source of calcium. In fact, a small container of yogurt contains as much calcium as a third of a pint of milk. In addition to these nutritional characteristics, yogurt is also thought to have additional health benefits. One of the suggested benefits of yogurt is that it acts as a digestive aid. In the body, it is thought that yogurt can encourage the growth of beneficial bacteria in the gut. These organisms help to digest food more efficiently and protect against other, harmful organisms. Another health benefit of yogurt is for people that are lactose intolerant. These people have difficulty digesting milk products however, they typically can tolerate yogurt.

Raw Materials

In general, yogurt is made with a variety of ingredients including milk, sugars, stabilizers, fruits and flavors, and a bacterial culture

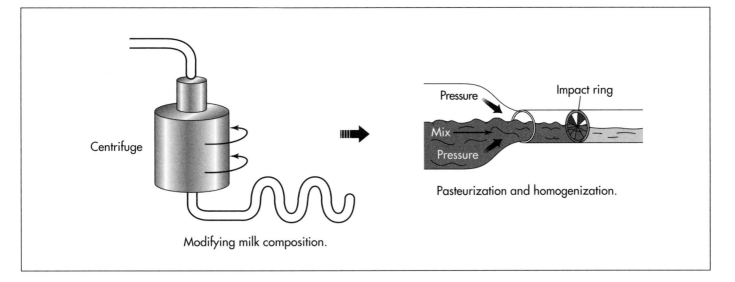

Modifying milk composition.

Pasteurization and homogenization.

(*Lactobacillus bulgaricus*). During fermentation, these organisms interact with the milk and convert it into a curd. They also change the flavor of the milk giving it the characteristic yogurt flavor of which acetaldehyde is one of the important contributors. The primary byproduct of the fermentation process is lactic acid. The acid level is used to determine when the yogurt fermentation is completed which is usually three to four hours. The suppliers of these yogurt cultures offer various combinations of the two bacterial types to produce yogurts with different flavors and textures.

To modify certain properties of the yogurt, various ingredients may be added. To make yogurt sweeter, sucrose (sugar) may be added at approximately 7%. For reduced calorie yogurts, artificial sweeteners such as aspartame or saccharin are used. Cream may be added to provide a smoother texture. The consistency and shelf stability of the yogurt can be improved by the inclusion of stabilizers such as food starch, gelatin, locust-bean gum, guar gum and pectin. These materials are used because they do not have a significant impact on the final flavor. The use of stabilizers is not required however, and some marketers choose not to use them in order to retain a more natural image for their yogurt.

To improve taste and provide a variety of flavors, many kinds of fruits are added to yogurt. Popular fruits include strawberries, blueberries, bananas, and peaches, but almost any fruit can be added. Beyond fruits, other flavorings are also added. These can include such things as vanilla, chocolate, coffee, and even mint. Recently, manufacturers have become quite creative in the types of yogurt they produce using natural and artificial flavorings.

The Manufacturing Process

The general process of making yogurt includes modifying the composition of and pasteurizing the milk; fermenting at warm temperatures; cooling it; and adding fruit, sugar, and other materials.

Modifying milk composition

1 When the milk arrives at the plant, its composition is modified before it is used to make yogurt. This standardization process typically involves reducing the fat content and increasing the total solids. The fat content is reduced by using a standardizing clarifier and a separator (a device that relies upon centrifugation to separate fat from milk). From the clarifier, the milk is placed in a storage tank and tested for fat and solids content. For yogurt manufacture, the solids content of the milk is increased to 16% with 1-5% being fat and 11-14% being solids-not-fat (SNF). This is accomplished either by evaporating off some of the water, or adding concentrated milk or milk powder. Increasing the solids content improves the nutritional value of the yogurt, makes it easier to produce a firmer yogurt and improves the stability of

When the milk arrives at the plant, its composition is modified before it is used to make yogurt. This standardization process typically involves reducing the fat content and increasing the total solids. Once modification occurs, it is pasteurized to kill bacteria and homogenized to consistently disperse fat molecules.

The milk substance is fermented until it becomes yogurt. Fruits and flavorings are added to the yogurt before packaging.

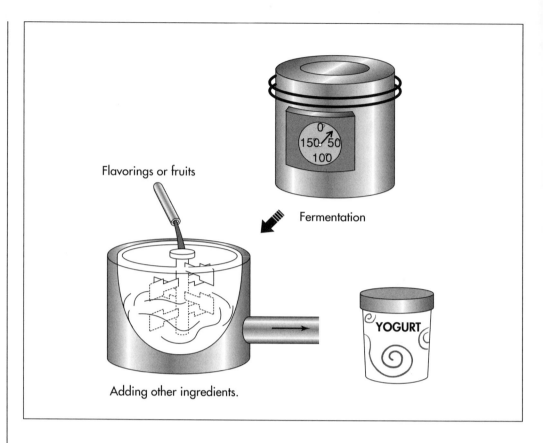

Flavorings or fruits

Fermentation

YOGURT

Adding other ingredients.

the yogurt by reducing the tendency for it to separate on storage.

Pasteurization and homogenization

2 After the solids composition is adjusted, stabilizers are added and the milk is pasteurized. This step has many benefits. First, it will destroy all the microorganisms in the milk that may interfere with the controlled fermentation process. Second, it will denature the whey proteins in the milk which will give the final yogurt product better body and texture. Third, it will not greatly alter the flavor of the milk. Finally, it helps release the compounds in milk that will stimulate the growth of the starter culture. Pasteurization can be a continuous- or batch-process. Both of these processes involve heating the milk to a relatively high temperature and holding it there for a set amount of time. One specific method for batch process pasteurization is to heat a large, stainless steel vat of milk to 185° F (85° C) and hold it there for at least 30 minutes.

3 While the milk is being heat treated, it is also homogenized. Homogenization is a process in which the fat globules in milk are broken up into smaller, more consistently dispersed particles. This produces a much smoother and creamier end product. In commercial yogurt making, homogenization has the benefits of giving a uniform product, which will not separate. Homogenization is accomplished using a homogenizer or viscolizer. In this machine, the milk is forced through small openings at a high pressure and fat globules are broken up due to shearing forces.

Fermentation

4 When pasteurization and homogenization are complete, the milk is cooled to between 109.4-114.8° F (43-46° C) and the fermentation culture is added in a concentration of about 2%. It is held at this temperature for about three to four hours while the incubation process takes place. During this time, the bacteria metabolizes certain compounds in the milk producing the characteristic yogurt flavor. An important byproduct of this process is lactic acid.

5 Depending on the type of yogurt, the incubation process is done either in a large tank of several hundred gallons or in the final individual containers. Stirred yogurt is fermented in bulk and then poured into the final selling containers. Set yogurt, also known as French style, is allowed to ferment right in the container it is sold in. In both instances, the lactic acid level is used to determine when the yogurt is ready. The acid level is found by taking a sample of the product and titrating it with sodium hydroxide. A value of at least 0.9% acidity and a pH of about 4.4 are the current minimum standards for yogurt manufacture in the United States. When the yogurt reaches the desired acid level, it is cooled, modified as necessary and dispensed into containers (if applicable).

Adding other ingredients

6 Fruits, flavors, and other additives can be added to the yogurt at various points in manufacturing process. This is typically dependent on the type of yogurt being produced. Flavor in non-fruit yogurts are added to the process milk before being dispensed into cartons. Fruits and flavors can also be added to the containers first, creating a bottom layer. The inoculated milk is then added on top and the carton is sealed and incubated. If the fruit is pasteurized, it can be added as a puree to the bulk yogurt, which is then dispensed into containers. Finally, the fruit can be put into a special package, which is mixed with plain yogurt upon consumption.

7 The finished yogurt containers are placed in cardboard cases, stacked on pallets, and delivered to stores via refrigerated trucks.

Quality Control

Milk products such as yogurt are subject to a variety of safety testing. Some of these include tests for microbial quality, degree of pasteurization, and various forms of contaminants. The microbial quality of the incoming milk is determined by using a dye reaction test. This method shows the number of organisms present in the incoming milk. If the microbial count is too high at this point, the milk may not be used for manufacture. Since complete pasteurization inactivates most organisms in milk, the degree of pasteurization is determined by measuring the level of an enzyme in the milk

called phosphatase. Governmental regulations require that this test be run to ensure that pasteurization is done properly. Beyond microbial contamination, raw milk is subject to other kinds of contaminants such as antibiotics, pesticides or even radioactivity. These can all be found through safety testing and the milk is treated accordingly.

In addition to safety tests, the final yogurt product is also evaluated to ensure that it meets the specifications set by the manufacturer for characteristics such as pH, rheology, taste, color, and odor. These factors are tested using various laboratory equipment such as pH meters and viscometers and also human panelists.

The Future

The future of yogurt manufacturing will focus on the development of new flavors and longer lasting yogurts. The introduction of new flavors will be driven by consumer desires and new developments by flavor manufacturers. The suppliers of the bacterial cultures are conducting research that hints at the development of uniquely flavored yogurts. By varying the types of organisms in the cultures, yogurt is produced much faster and lasts longer than conventional yogurt.

Additionally, the nutritional aspects of yogurt will be more thoroughly investigated. There is some evidence that has shown consumption of yogurt has a beneficial antibiotic effect. It has also been shown to reduce the incidence of lactose intolerance and other gastro-intestinal illnesses. Other purported benefits of yogurt include the reduction of cholesterol, protection against certain cancers, and even boosting the immune system. The research is still not complete on these benefits however, these factors will likely be important in the continued market growth of yogurt.

Where to Learn More

Books

Helferich, W. and D. Westhoff. *Yogurt: All About It*, 1980.

Hui, Y.H., ed. *Dairy Science and Technology Handbook*. New York: Wiley VCH, 1992.

Robinson, R.K. "Snack Foods of Dairy Origin." In *Snack Food.* Edited by Gordon R. Booth. New York: Van Nostrand Reinhold, 1990, pp. 159 - 182.

Robinson, R.K and A.Y. Tamime. "Recent developments in yoghurt manufacture." In *Modern Dairy Technology.* Edited by B.J.F. Hudson. London: Elsevier Applied Science Publishers, 1986, pp 1-36.

—*Perry Romanowski*

Index

Bold-faced terms indicate main entries in this volume. Entries from past volumes are listed with their volume and page number. For example, an entry on page 46 of the second volume appears as 2:46.